Alternative Liquid Dielectrics for High Voltage Transformer Insulation Systems

IEEE Press
445 Hoes Lane
Piscataway, NJ 08854

Alternative Liquid Dielectrics for High Voltage Transformer Insulation Systems

Performance Analysis and Applications

Edited by

U. Mohan Rao
Université du Québec à Chicoutimi, Canada

I. Fofana
Université du Québec à Chicoutimi, Canada

R. Sarathi
Indian Institute of Technology Madras, Chennai, India

IEEE PRESS SERIES ON POWER ENGINEERING

IEEE PRESS

WILEY

Published by John Wiley & Sons, Inc., Hoboken, New Jersey.

Published simultaneously in Canada.

Library of Congress Cataloging-in-Publication Data

Names: Rao, U. Mohan (Ungarala Mohan), editor. | Fofana, I. (Issouf), editor. | Sarathi, R. (Ramanujam), editor.
Title: Alternative liquid dielectrics for high voltage transformer insulation systems : performance analysis and applications / edited by U. Mohan Rao, I. Fofana, R. Sarathi.
Description: Hoboken, New Jersey : Wiley-IEEE Press, [2022] | Series: IEEE Press series on power and energy systems | Includes bibliographical references and index.
Identifiers: LCCN 2021042127 (print) | LCCN 2021042128 (ebook) | ISBN 9781119800163 (cloth) | ISBN 9781119800170 (adobe pdf) | ISBN 9781119800187 (epub)
Subjects: LCSH: Electric insulators and insulation–Liquids. | Liquid dielectrics.
Classification: LCC TK3441.L5 A58 2022 (print) | LCC TK3441.L5 (ebook) | DDC 621.319/37–dc23/eng/20211007
LC record available at https://lccn.loc.gov/2021042127
LC ebook record available at https://lccn.loc.gov/2021042128

Cover Design: Wiley
Cover Images: (center) © Panther Media GmbH/Alamy Stock Photo, (top) © Sam Robinson/Getty Images

Set in 9.5/12.5pt STIX Two Text by Straive, Pondicherry, India

10 9 8 7 6 5 4 3 2 1

This book is dedicated

to

Professor Issouf FOFANA

Contents

Editor Biographies

Dr. U. Mohan Rao, *SMIEEE, Secretary-IEEE DEIS Technical Committee on Liquid Dielectrics*
Post-Doctoral Fellow and Lecturer, Research Chair on the Aging of Power Network Infrastructure (ViAHT),
Université du Québec à Chicoutimi (UQAC), 555, Boulevard de l'Université, Chicoutimi, QC G7H 2B1, Canada.

U. Mohan Rao (M'15-SM'20) obtained his bachelor's degree in Electrical and Electronics Engineering from Jawaharlal Nehru Technological University, Kakinada, India, in 2010. He obtained his master's and doctoral degrees from the National Institute of Technology (NIT), Hamirpur, India, in 2012 and 2017, respectively. At present, he is a lecturer and postdoctoral fellow at Université du Québec à Chicoutimi (UQAC), Québec, Canada, with the Research Chair on the Aging of Power Network Infrastructure (ViAHT). He is a member of the IEEE DEIS and IEEE SA. Dr. Mohan is also a secretary of the IEEE DEIS Technical Committee on "Liquid Dielectrics." His main research interests include aging phenomena of high-voltage insulation, condition monitoring of electrical equipment, power transformer diagnostics, alternative liquid dielectrics, and AIML Applications. He has authored/co-authored over 60 scientific publications, 2 monographs, and edited 2 books.

Professor I. Fofana, *SMIEEE, Chair-IEEE DEIS Technical Committee on Liquid Dielectrics*
Professor and Director, Research Chair on the Aging of Power Network Infrastructure (ViAHT),
Université du Québec à Chicoutimi (UQAC), 555, Boulevard de l'Université, Chicoutimi, QC G7H 2B1, Canada.

Issouf Fofana (M'05-SM'09) obtained his electromechanical engineering degree in 1991 from the University of Abidjan (Côte d'Ivoire), and his master's and doctoral degrees from École Centrale de Lyon, France, in 1993 and 1996, respectively. He was a postdoctoral researcher in Lyon in 1997 and was at the Schering Institute of High Voltage Engineering Techniques at the University of Hanover, Germany, from 1998 to 2000. He was a Fellow of the Alexander von Humboldt Stiftung from November 1997 to August 1999. He joined Université du Québec à Chicoutimi (UQAC), Quebec, Canada, as an associate researcher in 2000, and he is now a professor there. Dr. Fofana has held the Canada Research Chair, tier 2, of insulating liquids and mixed dielectrics for electrotechnology (ISOLIME) from 2005 to 2015. He is actually holding the research chair on the Aging of Power Network Infrastructure (ViAHT), director of the MODELE laboratory, and director of the International Research Centre on Atmospheric Icing and Power Network Engineering (CenGivre) at UQAC. Professor Fofana is an accredited professional engineer in the province of Quebec and fellow of the IET. He is currently a member of the DEIS AdCom and member of the international scientific committees of few IEEE DEIS-sponsored or technically sponsored conferences (ICDL, CEIDP, ICHVE and CATCON). He is a member of the ASTM D27 committee. He has authored/co-authored over 300 scientific publications, two book chapters, one textbook, edited two books, and holds three patents.

Professor R. Sarathi, *SMIEEE*
Professor and Director, High Voltage Division
Department of Electrical Engineering, Indian Institute of Technology Madras, Chennai - 600036, India

Dr. R. Sarathi is a professor of High voltage engineering with IIT Madras, India. He has rich and varied experience of over 25 years in research, teaching, and consulting. He has been instrumental in conceptualizing and developing the state-of-the-art High Voltage Laboratory at IIT Madras. His research area includes condition monitoring of power apparatus, development of nanocomposite structures for power apparatus and pulsed power technology. He reached HAG scale few years before. He guided 23 PhD scholars and around 15 MS (by research). He published his work in around 175 peer-reviewed international journal publications. He visited Akita University, Japan, as a JSPS Fellow; visited Technical University, Munich, as DAAD Fellow, and served as visiting researcher in leading universities such as Nagaoka University, Japan, Khalifa University, Abu Dhabi, Kyushu University, Japan, University of Strathclyde, Glasgow, and Waseda University, Japan. He received UKIERI project, an international collaboration initiative on Innovation in high voltage insulation for electrical energy networks. He is a member of Indian Standards on High voltage testing techniques, CBIP distribution system, International Reference Group for SweGRIDS, Sweden. He is a senior IEEE member, IEEE Madras Section PES Chair, Fellow, Institution of Engineers, India, FNAE, FIET, and FRSC.

List of Contributors

Carmela Oria Alonso
Electrical and Energy Engineering Department, University of Cantabria, Santander, Cantabria, Spain

Ramazan Altay
BEST transformer, Balikesir, Turkey

A. J. Amalanathan
Department of Electrical Engineering, Indian Institute of Technology Madras, Chennai, India

Niharika Baruah
Indian Institute of Technology, Guwahati, Assam, India

Xin Chen
State Key Laboratory of Power Transmission Equipment & System Security and New Technology, Chongqing University, Shapingba, Chongqing, China

Hans Edin
School of Electrical Engineering and Computer Science, KTH - Royal Institute of Technology, Stockholm, Sweden

Dawei Feng
State Key Laboratory of Power Transmission Equipment & System Security and New Technology, Chongqing University, Shapingba, Chongqing, China

Cristina Fernández Diego
Electrical and Energy Engineering Department, University of Cantabria, Santander, Cantabria, Spain

Inmaculada Fernández Diego
Electrical and Energy Engineering Department, University of Cantabria, Santander, Cantabria, Spain

Alfredo Ortiz Fernández
Electrical and Energy Engineering Department, University of Cantabria, Santander, Cantabria, Spain

I. Fofana
Research Chair on the Aging of Power Network Infrastructure (ViAHT), University of Quebec at Chicoutimi, Chicoutimi, QC, Canada

Chenyu Gao
State Key Laboratory of Power Transmission Equipment & System Security and New Technology, Chongqing University, Shapingba, Chongqing, China

H. Griffiths
HVDM Research Group, APEC Research Centre, Khalifa University, Abu Dhabi, PO, UAE

Jian Hao
State Key Laboratory of Power Transmission Equipment & System Security and New Technology, Chongqing University, Shapingba, Chongqing, China

N. Harid
HVDM Research Group, APEC Research Centre, Khalifa University, Abu Dhabi, PO, UAE

Joyce Jacob
Department of Electrical Engineering, National Institute of Technology Calicut, Kerala, India

T. Jayasree
Research Chair on the Aging of Power Network Infrastructure (ViAHT) University of Quebec at Chicoutimi, Chicoutimi, QC, Canada

Ahmet Kerem Köseoğlu
BEST transformer, Balikesir, Turkey

Sindhu Thiruthi Krishnan
Department of Electrical Engineering, National Institute of Technology Calicut, Kerala, India

Ruijin Liao
State Key Laboratory of Power Transmission Equipment & System Security and New Technology, Chongqing University, Shapingba, Chongqing, China

L. Loiselle
Research Chair on the Aging of Power Network Infrastructure (ViAHT), University of Quebec at Chicoutimi, Chicoutimi, QC, Canada

Mrutyunjay Maharana
Xi'an Jiaotong University, Shaanxi, China

G. D. P. Mahidhar
Department of Electrical Engineering, Indian Institute of Technology Madras, Chennai, India

D. Martin
ETEL Transformers, Auckland, New Zealand

L. McPherson
Essential Energy, Port Macquarie, New South Wales, Australia

Cristina Méndez Gutiérrez
Electrical and Energy Engineering Department, University of Cantabria, Santander, Cantabria, Spain

U. Mohan Rao
Research Chair on the Aging of Power Network Infrastructure (ViAHT), University of Quebec at Chicoutimi, Chicoutimi, QC, Canada

Sisir Kumar Nayak
Indian Institute of Technology, Guwahati, Assam, India

Cristian Olmo Salas
Electrical and Energy Engineering Department, University of Cantabria, Santander, Cantabria, Spain

P. Picher
Institut de recherche d'Hydro-Québec, Varennes, Canada

Preetha Prabhu
Department of Electrical Engineering, National Institute of Technology Calicut, Kerala, India

E. Rodriguez Celis
Institut de Recherche d'Hydro-Québec, Varennes, QC, Canada

Pawel Rozga
Institute of Electrical Power Engineering, Lodz University of Technology, Lodz, Poland

Rohith Sangineni
Indian Institute of Technology, Guwahati, Assam, India

R. Sarathi
Department of Electrical Engineering,
Indian Institute of Technology Madras,
Chennai, India

Nathaniel Taylor
School of Electrical Engineering
and Computer Science, KTH - Royal
Institute of Technology, Stockholm,
Sweden

Lijun Yang
State Key Laboratory of Power Transmission
Equipment & System Security and New
Technology, Chongqing University, Shapingba,
Chongqing, China

Wenyu Ye
State Key Laboratory of Power Transmission
Equipment & System Security and New
Technology, Chongqing University, Shapingba,
Chongqing, China

Acknowledgments

For more than a year my coeditors are working on the progress of this book with certain expert advice. I would like to thank Professor Issouf Fofana from the University of Quebec at Chicoutimi (UQAC), Canada, and Professor Ramanujam Sarathi from the Indian Institute of Technology Madras, India, for being the fellow editors of this book. On behalf of the editors, I would like to take this opportunity to thank all the lead authors and individual contributors of this book for their cooperation, support, and patience at different stages.

A special note of thanks to the PhD graduate students and researchers at the Research Chair on the Aging of Power Network Infrastructure (ViAHT) at UQAC, Canada, and researchers at Institut de Recherche d'Hydro-Québec, Varennes, Canada, for their research collaboration. I would also like to thank Dr. P. Rozga from the Lodz University of Technology, Poland, for his expert collaboration on the topic "Prebreakdown phenomena". The editors of this book would like to thank the anonymous reviewers and project handling staff of the Wiley-IEEE press for their cooperation.

I would like to acknowledge the support and cooperation extended by the beloved family members of my coeditors and all the contributors. In particular, I would like to extend my deepest gratitude to my wife Ms. Jayasree Thota, for her support in the successful accomplishment of this book.

U. Mohan Rao,
Editor

Editorial

Dr. U. Mohan Rao
University of Quebec at
Chicoutimi, Canada

Dreams are dreamt in life, but a few with determination are realized and accomplished by us. This book on the topic of "*Alternative Dielectric Liquids for Transformer Insulation Technology*" is the outcome of that determination I had developed since my acquaintance waith Professor Issouf Fofana at the International Research Center on Atmospheric Icing and Power Network Engineering (CenGivre) at the University of Quebec at Chicoutimi (UQAC). Finally, two and a half years later on the second day of April 2021, I am happy to write this editorial on the realization of this book, which has been enriched with contents derived out of paramount discussions I have had with Professor Fofana at various occasions. In the early 2019, we initiated the idea of delivering a short e-book to the scientific community on the modern knowledge of ester liquids and potential challenges on new-generation insulating fluids. The development of this project kicked off in the mid-2019s after Professor Fofana revitalized the IEEE DEIS Technical committee of Liquid Dielectrics (TCLD) as a technical committee chair. It is also my immense pleasure and pride to have contributed to the TCLD activities as an invited committee member. Further around late 2019, Professor Fofana and I decided to implement the book project and an international study group under the umbrella of the TCLD by inviting potential research groups to contribute. In the second quarter of 2020, I had a pleasure to propose the idea of this book volume to the Wiley-IEEE press, during which I had the privelage to invite Professor Ramanujam Sarathi from the Indian

Institute of Technology, Madras, India, who joined as a fellow coeditor for the present edition. This book includes edited chapters contributed by renowned scientist from around the world from countries, namely, Australia, Canada, China, India, New Zealand, Poland, Spain, Sweden, Turkey, and United Arab Emirates. Initially planned to be published by the end of 2020, this book, like many other projects worldwide, has suffered a slight delay due to the influence of the COVID-19 global pandemic.

Increasing requirements of electricity at an alarmingly rapid rate due to population and industrial growth have caused severe energy crises throughout the world. The shortage of fossil fuels (such as coal, crude oil, and natural gas) amplifies this crises, by progressively diminishing the availability of conventional methods of power generation. In addition, emission of harmful greenhouse gases (by-products of fossil fuels) discourages further consideration of conventional resources as a long-term future solution. The growing concern over issues related to energy security and global warming has resulted in the evolution of renewable energy resources and potential alternatives with environmental and economic benefits. The recent Paris climate accord signed by 195 countries, dealing with the mitigation and adaptation of greenhouse gas emissions and the divestment from fossil fuels, has spurred the exponential development and demand of renewable energy-based technologies. One such green and alternative candidate to mineral insulating oil (extracted from crude petroleum stocks) is ester-based dielectric liquid. Therefore, this book is dedicated to reach the readers across the globe in transferring the contemporary knowledge on these new dielectric fluids. The following paragraphs provide an insight of the scientific contents of the book

The first chapter gives a general introduction to the key topics of the book authored by the editors (U. Mohan Rao and I. Fofana) and co-authored by Dr. Rodriguez Mariela, Institut de recherche d'Hydro-Québec, Canada. This chapter covers the fundamentals of the transformer oil–paper insulation and introduction to various insulating liquids. The second chapter contribution from the Indian Institute of Technology Guwahati, India, and Xi'an Jiaotong University, Shaanxi, China, provides the details on the development and evaluation of natural ester liquids. Given the fact that the insulating liquids should have a high degree of compatibility with solid insulating liquids, the third chapter originating from the University of Cantabria, Spain, and BEST Transformer, Turkey, addresses the compatibility of the ester liquids with cellulose insulating materials. Authors affirm that the esters are not only compatible with cellulosic materials, but they also protect the cellulose, reducing its aging rate in comparison with traditional oil. The fourth chapter, authored by researchers from the high voltage laboratory at Indian Institute of Technology, Madras, India, and HVDM Research Group at Khalifa University, UAE, presents the details on the degradation and characterization of ester liquids.

The fifth chapter is authored by researchers from the Research Chair on the Aging of Power Network Infrastructure from the University of Quebec at Chicoutimi, Canada, which emphasizes the monitoring of the decay products of the new insulating liquids. The authors have demonstrated the monitoring of soluble and colloidal particles in ester liquids and spanned the discussion to the feasibility of using fuller's earth for regeneration of ester liquids. The sixth chapter arises out of a research collaboration between Lodz University of Technology, Poland, the University of Quebec at Chicoutimi, Canada, and the Institut de recherche d'Hydro-Québec, Canada. This chapter is an extended and updated version of the International study group article published in the IEEE Transactions on Dielectrics and Electrical Insulation in Volume: 27, Issue: 5, October 2020, with a primary focus on the pre-breakdown phenomenon of the ester liquids.

The seventh chapter presented by the research group from State Key Laboratory of Power Transmission Equipment & System Security and New Technology, Chongqing University, China, discusses ester liquids' miscibility behavior and mainly focuses on the engineering application of the new mixed liquid. The eighth chapter, focused on the ester-based nanofluid, is a contribution from the National Institute of Technology Calicut, India. The potential areas of future research related to the feasibility of natural ester nanofluids in transformers are discussed. The ninth chapter contributed by researchers from the high voltage laboratory at the Indian Institute of Technology, Madras, India, and KTH - Royal Institute of Technology, Sweden, presents the details of a new stable synthetic ester silica nanofluid with better dielectric strength compared to the synthetic ester fluid.

The tenth chapter, authored by researchers from the Research Chair on the Aging of Power Network Infrastructure from the University of Quebec at Chicoutimi, Canada, details the gassing behavior of ester liquids under corona discharging, arcing, and hotspot conditions. Dissolved gas analysis and diagnostic characterizations have been reported in the sections of this chapter. The eleventh chapter is a contribution from ETEL Transformers, New Zealand, and Essential Energy, Australia, on the in-service experience of natural ester-filled transformers. An overview of different parameters and fluid measurements is presented to benefit transformer owners and utility engineers.

1

Liquid Insulation for Power Transformers

U. Mohan Rao[1], I. Fofana[1], and E. Rodriguez Celis[2]

[1] Research Chair on the Aging of Power Network Infrastructure (ViAHT), University of Quebec at Chicoutimi, Chicoutimi, QC, Canada
[2] Institut de Recherche d'Hydro-Québec, Varennes, QC, Canada

1.1 Background of Liquid-Filled Transformers

Increasing requirements of electricity at an alarmingly rapid rate due to population and industrial growth have caused severe energy crises throughout the world. The shortage of fossil fuels (such as coal, crude oil, and natural gas) amplifies these crises, by progressively diminishing the availability of conventional methods of power generation [1–9]. In addition, emission of harmful greenhouse gases (by-products of fossil fuels) discourages further consideration of conventional power generation as a long-term future solution for increasing electricity demands [8–17]. The growing concern over issues related to energy security and global warming has resulted in the evolution of renewable energy resources (solar, wind, geothermal, and tidal power generation) as potential alternatives because of their environmental and economic benefits [12–15]. Distributed generation (DG) and high-voltage direct current (HVDC) transmission systems are promising solution for renewable power production and usage. The integration of DG and HVDC to the existing grid involves a significant difference in the voltage magnitudes. Power and distribution transformers handle these voltage magnitudes and ensure a reliable operation of the power grid [16, 17]. It is to be mentioned that, a few millions of transformers are connected across the global electric power network. In addition, transformers contribute the major segment of the economy involved in generation, transmission, and distribution of electricity. Therefore, transformer technology is always a high engineering importance to the researchers and utilities.

In a typical liquid-filled transformer, windings are wound on an iron core and the whole assembly is immersed in the insulating oil. Based on the assembly of the core and windings, transformers are classified into Core- and Shell-type transformers. In core-type transformers, windings surround a considerable part of the core whereas in shell-type transformers, core surrounds a considerable portion of the windings [18]. Transformers are also classified on the basis of their purpose of the application as a step-up transformer to increase the voltage level at secondary terminals and step-down transformer to decrease the voltage level at secondary terminals [19]. Insulating papers and pressboards are used for insulating windings within the core assembly. In oil-filled transformers, insulating oil is allowed to circulate for dissipating heat through the cooling tubes mounted on the body of a transformer tank.

Alternative Liquid Dielectrics for High Voltage Transformer Insulation Systems: Performance Analysis and Applications,
First Edition. Edited by U. Mohan Rao, I. Fofana, and R. Sarathi.

Generally, main parts of a power transformer are transported from manufacturers separately and are assembled at the site. Some preinstallation tests like winding resistance tests, sweep frequency response analysis to ensure mechanical integrity of core and windings, and heat oil circulation will be performed to ensure effective operation. Initially, in oil-filled power transformers, main body of the transformer is circulated with hot oil through it in order to remove any ingressed moisture and is filled with insulation oil. After the installation process, the transformer is connected across the supply mains. The successful operation of a transformer is dependent on the proper installation. When the primary winding is connected to ac mains supply, a current flows through it. Since this winding is magnetically linked with the core, current flowing through the primary winding will produce an alternating flux in the core. This alternating flux links with the secondary windings and an EMF is induced in the secondary winding due to mutual inductance. When the load is connected across the secondary terminals forming a closed path, a secondary current is circulated in the secondary windings through the load. In oil-filled transformers, insulation oil and insulation paper together form a composite dielectric medium often called as oil–paper insulation. The aging performance of oil and paper are closely interrelated; deterioration of either of the two leads to premature aging of the other one. Oil–paper insulation in transformers is expected to last for three to four decades in operation. Owing to variable thermal excursions, the degradation rate of oil–paper insulation gets accelerated and thus leading to premature aging. Hence, there is a great need for continuous condition monitoring of in-service transformers to avoid catastrophic failures and prevent subsequent capital as well as human loss in certain situations.

Insulation technology in transformers plays a critical role in judging the performance of the transformer. In oil-filled transformers, insulation oil along with insulation paper is used as an insulating medium. Insulation paper (mostly cellulose fibers) is used to isolate live conductors. Insulation oil (traditionally, mineral insulating oil) is used preliminarily for insulation of the core. The main functions of the insulation oil in an oil-filled transformer are as follows:

- Oil acts as an insulating medium: Provides insulation for the conductors.
- Oil serves as a coolant by dissipating internally generated heat: Heat dissipation is done in conduction, convection, and radiation modes with the help of solid insulation and cooling tubes.
- Oil acts as a diagnostic medium for prognosis of aging characterization of in-service oil-filled transformers: Small volume of the liquid is collected from the transformer to test according to the standards for diagnosis and prognosis in understanding the situations prevailing in the transformer (similar to blood in human body).
- Oil acts as a barrier to the laminated sheet steel core by preventing direct contact with atmospheric air.

Mineral insulating oils extracted from crude petroleum are being used in transformers successively since decades. However, due to various technical demands/benefits, health/safety aspects, and environmental concerns, alternatives for mineral oils are of high demand by the industry. There is a need to shift the transformer insulation technology to a suitable new alternative and biodegradable liquids. Ester-based dielectric fluids, both natural and synthetic fluids, have been subjected to extensive research since decades. The performance of these new insulating liquids is found to be affirmative and comparable to mineral insulation oils. Despite their overwhelming technical benefits, very few utilities started using these alternative insulating fluids. One of the reasons includes the availability of limited diagnostic and prognostic information in the existing knowledge. The reclamation aspects and service experience of these new insulating liquids also remain as a pertinent challenge for the utilities.

Meanwhile, there are some significant aspects of ester fluids that need to be further investigated to improve the existing knowledge. This book aims at discussing the critical aspects that will act as a guide to the industry and researchers. The critical concepts with the significant research findings over the years and the topics that will need to be further emphasized will be digested in the chapters of this book. The knowledge embodied in this book will also act as a quick reference for utility engineers, transformer owners, academia, and researchers those interested in alternative and biodegradable insulating fluids for liquid-filled transformers. The scope of this book will aid to the focus of the IEEE DEIS technical committee on liquid dielectrics and lies within the interests of the IEEE DEIS community and transformer industry across the world. This book will be a unique sample of its kind and will act as a reference pertaining to the biodegradable insulating liquids. The contents of this book are framed in a way to cover the pertinent orbits of ester liquids, key aspects that need to be emphasized to further expand the existing knowledge, and act as a quick guide to the utility engineers. This book also covers the discussions on critical challenges that will be helpful for the engineers in successful operation of the ester-filled transformers. Importantly, reputed and experienced researchers who are working on the individual subtopics for a long time contribute the chapters of this book. This will allow widening the scope and technical orbits of the book in an effective and unique manner. This book will also act in bridging the gap between the researchers and utility engineers concerning the transformer insulation systems.

1.2 Insulation System in Liquid-Filled Transformers

Liquid insulation in transformers plays a critical role in assessing the lifetime of a transformer. Useful life of a transformer relies on the electrical health and effectiveness of insulating oil. Owing to prevalent thermal conditions within an in-service oil-filled transformer, the performance of solid insulation paper, pressboard, etc., and the effectiveness of the insulating oil to serve as an insulator and as a coolant, reduces significantly over a period. High thermal excursions in the transformer tend to accelerate the aging process of the oil–paper insulation. Eventually, insulation paper degrades by releasing certain gases, moisture, furan-based compounds, and suspended particles percolating into the insulation oil; thus, enhances the deterioration of oil. As per ASTM D117-18 [20] standard, commonly adapted parameters of insulation oil to monitor the transformers are classified as electrical, physical, and chemical.

Hence, degradation of insulation paper can be estimated by performing diagnostic and prognostic tests on the insulation oil. Simply, characterization of insulation oil facilitates prediction of performance aspects of insulation and helps estimate the prevailing conditions within the in-service oil-filled transformers. It has been a past practice of the power industry to perform diagnosis tests on the insulation oil in view of condition monitoring of an oil-filled transformer. Insulation system being the heart of the transformer comprises the liquid insulation (typically oils) and solid insulation system (typically cellulose). Traditionally, mineral insulating oils are put up in practice for the transformer insulation technology. Due to the fact that mineral oils are expected to reach depletion, alternative candidates for these oils is a challenge. Further, mineral oils are toxic and nonbiodegradable. Thus, demanding an alternative candidate, which is biodegradable and nontoxic while meeting the technical requirements that an insulating liquid should exhibit. It is desired that a transformer insulating liquid has good dielectric strength, high thermal performance (high fire point), and has good compatibility with the transformer solid insulation. Classification of several insulation parameters that are used by the industry to monitor the performance of transformer insulation is shown in Figure 1.1.

Parameters highlighted in Figure 1.1 are fundamental parameters that majority of the utilities monitor to access the aging level of insulation in the transformers. All the abovementioned parameters of in-service insulation oil are classified based on the nature of liquid insulating medium

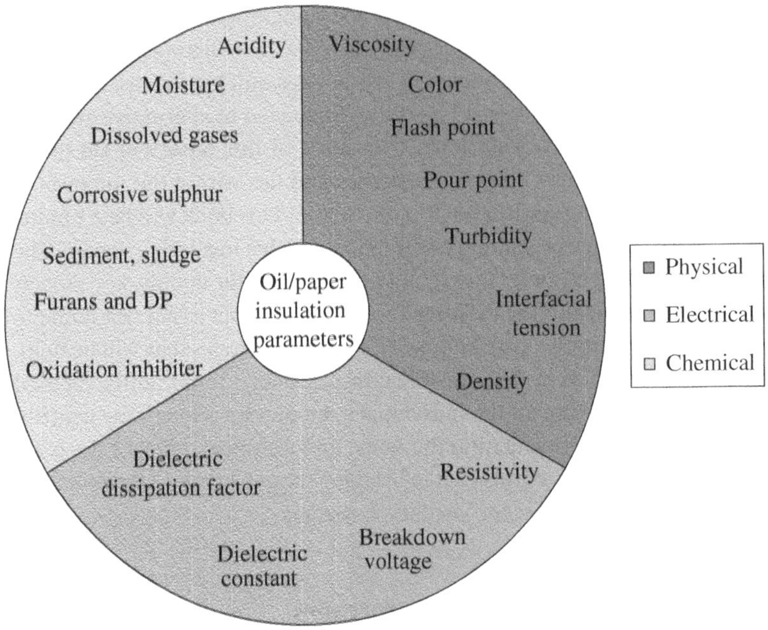

Figure 1.1 Basic insulation parameters for aging assessment of oil–paper insulation.

(property) in oil-filled transformers. However, the aging aspects of transformers are highly interrelated and measuring oil–paper insulation aging has become an established interdisciplinary engineering application field of materials science and chemical engineering through latest instrumentation and aging measurements by various techniques and processes.

1.3 Insulation Aging Phenomena in Transformers

The service life of the oil-filled apparatus is mainly attributable to the rate of degradation of the oil/paper insulation system. The preliminary causes for the degradation of the oil/paper insulation system are its decomposition aspects. The degradation process of the insulation system in oil-filled apparatus is governed by the: (i) heat (dissipated through the core and winding assembly), (ii) moisture (ingressed from the air and liberated from cellulose), and (iii) oxygen (ingressed from air and liberated from oil/paper system). This operating heat will adversely affect the longevity and performance of the insulation system and the life of the apparatus if not dissipated properly in time [21]. As per IEC 60076-2 standard (thermal layout), insulation paper will serve for up to 55 years and more, subject to the absence of dielectric defects in the insulation system. Albeit due care is taken by the manufacturers in designing, the generation of heat from coil–core assembly and subsequent degradation of the insulation system is unavoidable. The degradation of oil/paper insulation generates soluble and colloidal particles in the oil. This increases the oil viscosity, which hampers the flow rate in cooling tubes and affects the heat dissipation efficiency. This excess heat is responsible for the accumulation of hot spots within the transformer, which further intensifies the degradation mechanisms. The degradation mechanisms of an oil/paper insulation system include oxidation, hydrolysis, thermal stressing, and electrical stressing [22, 23]. These mechanisms introduce different by-products that further act in degrading the oil/paper insulation system.

The complete picture associated with the degradation process of oil/paper insulation has been illustrated in Figure 1.2.

Moisture and oxygen in the transformer may be ingressed from the external environment (for breathing transformers) or evolved from cellulose fibers of solid insulation. Moisture, oxygen, metals (components of the transformer), and heat (liberated from coil–core assembly) act as catalysts for deterioration of insulation systems and provide a scope for decomposing by-products and acids. These catalysts and by-products further expedite oxidation, hydrolysis, thermal, and electrical decomposition reactions with respect to operating times. All these degradation reactions are highly interrelated and the by-products of one reaction act in catalyzing the other reactions. For analyzing the degradation perceptions of oil/paper insulation, these reactions may be approached thermally, electrically, chemically, and physically.

Thermally, oil gets oxidized and initiates sludging within the insulation oil while reducing the tensile strength of paper by paralyzing the cellulose fibers. Electrically, oil deteriorates and generates acids and free radicals with decomposition on a large surface of solid insulation. Chemically, the neutralization number of the oil gets adversely affected, and witnessing the evidence of furfural in oil. Physically, the osmotic behavior of moisture migrations through oil and paper with a change in temperature hampers the paper dryness and reduces the degree of polymerization of solid insulation by adding colloidal particles to oil [25, 26]. Ultimately, these degradation mechanisms are all about reducing the integral qualities of the insulation system with operating time. It is to be observed that, these aging mechanisms contribute to sludging, acidity, and decay contents, which reduce dielectric strength and increase the viscosity of the oil. These factors hamper the

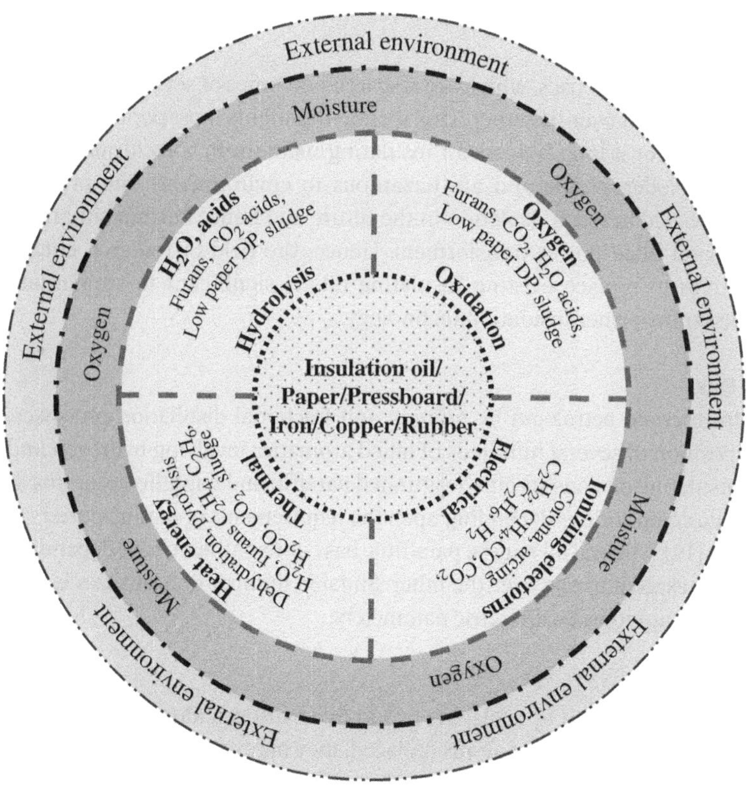

Figure 1.2 Conceptual illustration of degradation in oil filled apparatus. *Source:* Rao et al. [24] / with permission of IEEE.

insulation properties and heat-dissipating nature of oil, which are the primary objectives of the oil. It is to be understood that, these aging mechanisms contribute to the generation of furan contents, reduce the paper degree of polymerization and tensile strength, thus ruining the dielectric and mechanical integrity of solid insulation. Importantly, all the above-discussed processes are also involved in the generation of several gasses that are dissolved in the insulation oil, which further ruins the physicochemical parameters of the insulation system.

Understanding the degradation process and its consequences that influence the performance and life of the oil-filled apparatus indicates the importance of condition monitoring. The current state of knowledge on condition monitoring revealed that degradation/aging is associated with numerous parameters. Henceforth, monitoring and analyzing one single parameter will not reveal the exact situations prevailing with the insulation system. Proper condition monitoring analysis of oil/paper insulation with scheduled maintenance activities enables effective asset management and risk analysis. This also acts in transforming the conditioning monitoring aspects from detective-corrective mode to strategic-preventive mode. ASTM standard test methods are available for monitoring all the oil/paper insulation parameters to assess the quality of oil and the contemporary status of the oil-filled apparatus. Hence, periodical assessment of insulation parameters will lead to early detection of the degradation perspectives of oil/paper insulation. The same analysis will be apparently useful in planning appropriate diagnostics and prognostics.

1.4 Transformer Insulating Liquids

1.4.1 Conventional Liquid Dielectrics

There are many conventional liquid dielectrics, which are research outcomes of several innovative chemical modifications of one product over the other. This section highlights most popular insulation oils that have been survived for a long time as an insulating medium in transformers. The majority of these oils are not biodegradable and are hazardous to environment. The primary sources of these liquids are expected to reach depletion in the future and hence current situations demand perfect replicates for oil-filled power transformers. Hence, the field of research entails vigorous search for alternatives to replace existing insulating oils is significant to strengthen, improve, and sustain oil-filled transformer insulation technology.

1.4.1.1 Mineral Insulating Oils

Mineral oils are extracted from crude petroleum by refining and fractional distillation processes. These oils are complex composition of several hundreds of allied aromatics involving hydrogen and carbon molecules. Mineral insulating oil consists either saturated paraffin and naphthenes or unsaturated aromatics in appropriate compositions depending upon the requirements of manufacturers to satisfy clients and applications [19]. Mineral oil may be paraffinic base or naphthenic base depending on the ratio of their proportion exceeding one over the other. Suitable number of aromatics is also added to these oils for developing appropriate dielectric parameters.

1.4.1.2 Polychlorinated Biphenyl

Polychlorinated biphenyl (PCB) is produced by replacing two to five hydrogen atoms of a benzene ring with chlorine atoms. Based on the number of atoms replaced, they may be called as di, tri, tetra, and penta chlorinated biphenyl [27]. The generic name of these liquids is askarels, which means fire resistant. PCBs were known by their different commercial names in several countries across the world, e.g. Pyranol in the United States, Sowol in Russia, and Aroclors in many parts of the world.

1.4.1.3 High-Temperature Hydrocarbons

High-temperature hydrocarbons (HTHs) are also known as high molecular weight hydrocarbons (HMWH). HTHs are derived naturally as well as through synthesis. The former are paraffinic base hydrocarbons that are obtained from petroleum crude similar to mineral oils, except fractioned from higher levels of distillation. The latter are developed through polymerization of olefins and hence are called as polyalpha olefins (PAO) [27]. These liquids are also known as high fire point liquids.

1.4.2 Alternative Liquid Dielectrics

In this section, liquid dielectrics that are having high thermal performance (having high flash points and fire point) and high environmental performance (nontoxic and biodegradable) have been discussed. The dielectric liquids discussed in this section may be naturally produced, or they may be synthesized. The key advantage of using these insulating oils is their high fire points, readily biodegradable (ecofriendly), and are mostly extracted from sustainable and renewable sources [28]. These insulating oils are also known for their better emission profile and fire characteristics. Some important insulating oils of this category that can be further promoted for use in oil-filled transformers as a replicate to mineral oils are discussed in the following sections.

1.4.2.1 Natural Ester Liquids

Basically, these are ester group compounds that are produced from glycerine and sebacic acids [29]. They are usually fatty acids having high fire point, high breakdown voltage, and good biodegradability. The natural esters are usually triglycerides and are found to have low oxidation stability. The general molecular structure of natural esters is similar to the molecular structure of vegetable oils [30]. These oils were restricted to distribution class transformers or breather-free transformers with the concern toward their low oxidation stability. Nevertheless, attempts are being in pipeline for further improving the oxidation stability of natural esters.

1.4.2.2 Vegetable Oils

Vegetable oils are extracted from agro seeds that are accumulated as fatty acid within the seeds. These are also called as natural esters. They are available from renewable sources and are treated as eco-friendly and biodegradable liquids [29]. Vegetable oils have higher viscosity and high pour point, which degrades the effectiveness to serve as a coolant [31]. At present, the research in the field of natural esters is on to emphasize reduction of viscosity and pour points with improved oxidation stability.

1.4.2.3 Synthetic Ester Liquids

Synthetic esters are such organic compounds that are being developed by synthesizing acids and alcohols. Synthetic esters are available in various chemical compositions depending on the acids and alcohols used for synthesis. Synthetic esters are biodegradable and are developed for high thermal performances with low viscosity and better oxidation stability. Moreover, ester group insulation oils participate actively in hydrolysis, which resists the rate of degradation of the oil.

The complete journey of the liquid dielectrics concerning the transformer insulation technology is outlined and discussed by I. Fofana [30] and U. Mohan Rao et al. [31, 32]. Given the fact that the high demand for alternative biodegradable insulating liquids allowed researchers focus on the performance analysis and behavior of ester liquids. Researchers across the globe have reported numerous studies; this book has been framed to give a complete overview of ester liquids' performance, behavior, and compatibility for current and foreseen transformer insulation technology. To accomplish the same, various topics have been organized in the subsequent chapters of this book. Authors of different chapters aimed to discuss particular topics in a detailed manner. Some fundamental concepts concerning the ester liquids have been spanned to multiple chapters. This is due to the fact that different chapters have been prepared by different research groups. However, due care is taken not to duplicate the sections and technical viewpoints. Therefore, this book editors and authors expect that this book would be useful for transformer owners, utility engineers, and researchers concerned with transformer insulation technology.

References

1 Tang, Y., Zhong, J., and Liu, J. (2016). A Generation adjustment methodology considering fluctuations of loads and renewable energy sources. *IEEE Transactions on Power Systems* 31 (1): 125–132.

2 Malinowski, M., Milczarek, A., Kot, R. et al. (2015). Optimized energy-conversion systems for small wind turbines: renewable energy sources in modern distributed power generation systems. *IEEE Power Electronics Magazine* 2 (3): 16–30.

3 Berseneff, B., Perrin, M., Quoc, T.T. et al. (2014). The significance of energy storage for renewable energy generation and the role of instrumentation and measurement. *IEEE Instrumentation & Measurement Magazine* 17 (2): 34–40.

4 Díaz, N.L., Luna, A.C., Vasquez, J.C., and Guerrero, J.M. (2017). Centralized control architecture for coordination of distributed renewable generation and energy storage in islanded AC microgrids. *IEEE Transactions on Power Electronics* 32 (7): 5202–5213.

5 Wei, W., Liu, F., Mei, S., and Hou, Y. (2015). Robust energy and reserve dispatch under variable renewable generation. *IEEE Transactions on Smart Grid* 6 (1): 369–380.

6 Wang, Z., Chen, Y., Mei, S. et al. (2017). Optimal expansion planning of isolated microgrid with renewable energy resources and controllable loads. *IET Renewable Power Generation* 11 (7): 931–940.

7 Mohan, V., Suresh, R., Singh, J.G. et al. (2017). Microgrid energy management combining sensitivities, interval and probabilistic uncertainties of renewable generation and loads. *IEEE Journal on Emerging and Selected Topics in Circuits and Systems* 7 (2): 262–270.

8 Yang, P. and Nehorai, A. (2014). Joint optimization of hybrid energy storage and generation capacity with renewable energy. *IEEE Transactions on Smart Grid* 5 (4): 1566–1574.

9 Li, N. and Hedman, K.W. (2015). Economic assessment of energy storage in systems with high levels of renewable resources. *IEEE Transactions on Sustainable Energy* 6 (3): 1103–1111.

10 Wong, S. and Pinard, J.P. (2017). Opportunities for smart electric thermal storage on electric grids with renewable energy. *IEEE Transactions on Smart Grid* 8 (2): 1014–1022.

11 Baker, K., Guo, J., Hug, G., and Li, X. (2016). Distributed MPC for efficient coordination of storage and renewable energy sources across control areas. *IEEE Transactions on Smart Grid* 7 (2): 992–1001.

12 Hill, C.A., Such, M.C., Chen, D. et al. (2012). Battery energy storage for enabling integration of distributed solar power generation. *IEEE Transactions on Smart Grid* 3 (2): 850–857.

13 Djairam, D., Morshuis, P.H.F., and Smit, J.J. (2014). A novel method of wind energy generation-the electrostatic wind energy converter. *IEEE Electrical Insulation Magazine* 30 (4): 8–20.

14 Abdelsamad, S.F., Morsi, W.G., and Sidhu, T.S. (2015). Impact of wind-based distributed generation on electric energy in distribution systems embedded with electric vehicles. *IEEE Transactions on Sustainable Energy* 6 (1): 79–87.

15 Wang, H. and Huang, J. (2016). Cooperative planning of renewable generations for interconnected microgrids. *IEEE Transactions on Smart Grid* 7 (5): 2486–2496.

16 Betie, A., Rao, U.M., Fofana, I. et al. (Dec. 2019). Influence of cellulose paper on gassing tendency of transformer oil under electrical discharge. *IEEE Transactions on Dielectrics and Electrical Insulation* 26 (6): 1729–1737.

17 Leila, S., Zafour, H., Rao, U.M., and Fofana, I. (2019). Regeneration of transformer insulating fluids using membrane separation technology. *Energies* 12 (3): 368.

18 Theraja, B.L. and Theraja, A.K. (2014). *Electrical Technology*, vol. II. New Delhi, India: S. Chand Publishers.

19 Kulkarni, S.V. and Khaparde, S.A. (2004). *Transformer Engineering: Design and Practice.* CRC Press.

20 ASTM (2018). *D117-18, Standard Guide for Sampling, Test Methods, and Specifications for Electrical Insulating Liquids.* West Conshohocken, PA: ASTM International.

21 Chakravorti, S., Dey, D., and Chatterjee, B. (2013). *Recent Trends in the Condition Monitoring of Transformers*, 1ste. London: Springer-Verlag.

22 Kassi, K.S., Fofana, I., Meghnefi, F., and Yeo, Z. (2015). Impact of local overheating on conventional and hybrid insulations for power transformers. *IEEE Transactions on Dielectrics and Electrical Insulation* 22 (5): 2543–2553.

23 Rao, U.M., Sood, Y.R., and Jarial, R.K. (2016). Physiometric and Fourier transform infrared spectroscopy analysis of cellulose insulation in blend of mineral and synthetic ester oils for transformers. *IET Science, Measurement & Technology* 11 (3): 297–304.

24 Rao, U.M., Fofana, I., Betie, A. et al. (Nov.-Dec. 2019). Condition monitoring of in-service oil-filled transformers: case studies and experience. *IEEE Electrical Insulation Magazine* 35 (6): 33–42.

25 Martin, D., Saha, T., Gray, T., and Wyper, K. (2015). Determining water in transformer paper insulation: effect of measuring oil water activity at two different locations. *IEEE Electrical Insulation Magazine* 31 (3): 18–25.

26 Martin, D., Saha, T., Dee, R. et al. (2015). Determining water in transformer paper insulation: analyzing ageing transformers. *IEEE Electrical Insulation Magazine* 31 (5): 23–32.

27 Claiborne, C.C. and Pearce, H.A. (1989). Transformer fluids. *IEEE Electrical Insulation Magazine* 5 (4): 16–19.

28 Plessner, K.W. and Reynolds, E.H. (1976). New pure or blended liquids for high voltage application. CIGRE Report No. 15-07.

29 Bartnikas, R. (1994). Electrical insulating liquids. In: *Engineering Dielectrics*, vol. 3. American Society for Testing and Materials.

30 Fofana, I. (2013). 50 Years in the development of insulating liquids. *IEEE Electrical Insulation Magazine* 29 (5): 13–25.

31 Rao, U.M., Fofana, I., Jaya, T. et al. (2019). Alternative dielectric fluids for transformer insulation system: research progress, challenges, and future prospects. *IEEE Access* 7 (1): 1–20.

32 Rao, U.M., Sood, Y.R., and Jarial, R.K. (2015). Review on ester based dielectric liquids for transformer insulation technology. In *2015 International Conference on Condition Assessment Techniques in Electrical Systems (CATCON).*

2

Processing and Evaluation of Natural Esters

Niharika Baruah[1], Rohith Sangineni[1], Mrutyunjay Maharana[2], and Sisir Kumar Nayak[1]

[1] *Indian Institute of Technology, Guwahati, Assam, India*
[2] *Xi'an Jiaotong University, Shaanxi, China*

2.1 Introduction

Transformers are the critical elements of the electrical grid and their effective operation is vital to the health of the whole network. It is of utmost importance to sustain the continuity of supply and prevent financial losses. The mineral oil (MO) is most generally used in the transformer as the dielectric fluid alongside the solid insulation like the pressboards and kraft papers. The development of environment friendly vegetable oils (VOs) or natural ester oils (NEOs) for transformers in place of MO gives us an alternative insulation system. MO is procured from the petroleum-based fossil fuels and will deplete in the long run. Also, the biodegradability of MO is poor and hence causes environmental concern if spillage occurs. Alternatively, NEOs are achieving prominence because of their environment friendly nature, better biodegradability, higher fire and flash points. NEOs also have thermal conductivity better than MO; however, the viscosity is higher than MO, which is a disadvantage as it prevents flowability of the oil.

NEOs are derived from plant sources and there is a variety of them available in the market. Some of them are the linseed oil, castor oil, corn oil, Pongamia pinnata oil (karanji oil), soybean oil, punna oil, neem oil, Jatropha Curcas oil, palm oil, sunflower oil, olive oil, peanut oil, cottonseed oil, rapeseed oil (canola oil), coconut oil, hazelnut oil, and mustard oil [1–8]. The experimentation with NEOs for insulation purposes dates back to the twentieth century as these oils are readily biodegradable, have higher fire points, and are sustainable. The NEO also has higher water absorption capacity than the MO. A little concern over the use of NEOs made from edible oils like peanut and sunflower is that abundant use of NEOs can lead to a food crisis; therefore, these oils should preferably be made from nonedible seeds [9].

The crude oil, when extracted from the seeds, is highly acidic and viscous in nature. The acidic nature and viscosity can be reduced by following the transesterification process of the NEOs [1, 2]. However, the transesterification process reduces the flash point of the NEO. To overcome the reduction in flash points, the NEO is blended with its transesterification-processed oil. The blended oil has low viscosity, higher flash point, and higher impulse breakdown voltage (BDV) [2]. The proper blending of NEOs with MO can also be done for lowering the pour point at the same biodegradability [10].

Alternative Liquid Dielectrics for High Voltage Transformer Insulation Systems: Performance Analysis and Applications, First Edition. Edited by U. Mohan Rao, I. Fofana, and R. Sarathi.

A typical fatty acid composition of few NEOs is tabulated in Table 2.1. The unsaturated fatty acids are at risk of oxidation. Thus, the oxidation stability is lower for NEOs as compared with MO. Though these unsaturated fatty acids help in absorbing the gases released, the oxidation stability needs to be addressed. Another major drawback of the NEOs is the high values of pour point. The high pour points and high viscosity of VO are due to the presence of the fatty acids [14]. The viscosity, however, decreases to a certain extent with increase in temperature in a running transformer. The problem of high pour point and low oxidation stability can be dealt by using additives. Some of the additives used as antioxidants are tert-butyl hydroxy quinone (TBHQ) and butylated hydroxytoulene (BHT). One of the additives used to lower the pour point is polyalkyl methacrylate [8]. The other disadvantage of the NEO is poor dissipation factor. NEOs are proven to be useful only in high-voltage transformers and capacitors [1, 15]. The NEOs may be best utilized in power network with transformers in the coastal areas, where the oil spillage may result in serious threat to coral life and the aquatic life.

The different kinds of studies like electrical, chemical, and physical analyses of the oil are performed to estimate the life of the insulation. The aging study of oil can give information on the cellulose aging, low-level partial discharge (PD), contamination, static electrification, circulating currents, and faults in the transformer. Aging and degradation studies are conducted to assess the gases, furans, phenols, cresols, and metal particles that are formed in the oil during the life of the

Table 2.1 Typical fatty acid composition of some vegetable oils.

		Unsaturated fatty acids (%)		
Vegetable oil	**Saturated fatty acids (%)**	**Mono-**	**Di-**	**Tri-**
Rapeseed/Canola oil[a]	7.9	55.9	22.1	11.1
Corn oil	12.7	24.2	58	0.7
Cottonseed oil	25.8	17.8	51.8	0.2
Peanut oil	13.6	17.8	51.8	0.2
Olive oil	13.2	73.3	7.9	0.6
Safflower oil	8.5	12.1	74.1	0.4
Safflower oil, high oleic	6.1	75.3	14.2	—
Soybean oil	14.2	22.5	51	6.8
Sunflower oil	10.5	19.6	65.7	—
Sunflower oil, high oleic	9.2	80.8	8.4	0.2
Jatropha Curcas oil	19.1	46	34.9	
Rice oil	12.4	15.9	71.7	
Pongamia oil	13.76	60	23	
Neem oil	40	20.9	39.1	
Mustard oil	12.9	64.5	22.6	
Punna oil	39	58.9	2.1	
Castor oil	1.1	97.8	1.1	

[a] Low erucic acid variety of rapeseed oil; more recently canola oil containing over 75% monounsaturate content has been developed.
Source: Based on Beltrán et al. [8]; Oommen [11]; Dung and Huong [12]; Kumar et al. [13].

insulation [11]. The acidity in MO increases with oxidation, whereas the acid generation in NEOs is dominated by the hydrolysis phenomenon. The NEOs are preferred to be used in sealed transformers due to the drawbacks of oxidation and hydrolysis [16]. It is also observed that the paper insulation immersed in NEOs have longer life than the paper immersed in MO [15]. The degree of polymerization (DP) of paper is lower for MO immersed paper insulation with aging compared with NEO immersed paper insulation. This is due to the higher moisture retention capability of the NEO [17]. The BDV is also observed to be higher in NEOs as compared with the MO.

The bubble inception temperature in the insulating oil is dependent on the moisture content of the paper insulation. As the moisture content of paper increases, the bubble inception temperature decreases. The bubble inception temperature also depends on the type of paper (kraft paper/thermally upgraded paper), gas content in oil, age of oil, rate of temperature rise, apart from the moisture content [18]. Any air bubble entrapped due to higher viscosity of the NEOs may reduce the impulse BDV of the NEO [19].

The NEOs are thermally more stable than the MO and the dissolved gas content is higher in the MO. The heat transfer characteristics of the oils can be improved by adding the nanoparticles (NPs). The NPs also help in increasing the BDV of the oils because of the electron-scavenging nature. As the NEOs have already enhanced BDV as compared with MO, their BDV increases further with the stable dispersion of the NPs [4, 14, 20]. The moisture equilibrium is also an important factor when considering the use of NEOs for power applications. It is the balance between the amount of moisture quantity in paper and the moisture quantity in the oil and is temperature dependent. The shape variation of moisture equilibrium curve for paper–NEO is similar to the paper–MO while the moisture content in NEO is considerably higher than that of MO for same moisture content of paper [21]. The NEOs mostly comprise of triglycerides, and these triglycerides convert to long chain fatty acids on water ingression from paper to oil. These long chain fatty acids in the presence of cellulose, main composition of solid insulation, get transesterified. This leads to the lower moisture levels in the transformer, and thereby the degradation of insulation is slower. Another explanation for the lower degradation is the formation of the gel around the solid insulation, which avoids further degradation of the solid insulation [3, 5]. But if such gel formation happens, it reduces the thermal conductivity in the transformer and the hotspot temperature increases leading to winding damage.

The NEOs are reported to have higher breakdown strengths compared with the conventional MO. The stopping length of the streamers propagating in NEOs is known to be longer than that of streamers in MO when compared at the same value of testing voltage. The pre-breakdown phenomenon in NEO needs to be studied extensively for condition monitoring and maintenance of the transformer liquid insulation. Pongamia oil methyl ester (POME) derived from karanji seeds is produced by transesterification of crude karanji seed extract and the POME shows enhanced properties when compared with MO on oxidative aging [16]. However, interfacial tension (IFT) and pour point are not superior for POME when compared with MO.

Dissolved gas analysis (DGA) is an important tool in diagnosis of oil-filled transformers. The main objectives of DGA is to monitor the condition of equipment, operating conditions, and possible localization of faults [22, 23]. For low energy faults, same ratio of hydrogen and acetylene are released for NEO and MO, whereas for the low thermal faults, ethane is the main constituent [22]. The DGA response is the same in NEOs and MOs, but the percentage or gas ratios needs to be redefined because the rate of gas generation is lower for NEOs, which means the NEO is thermally more stable than MO [23, 24]. Stray gassing is observed in FR3 fluid at temperatures between 40 and 50 °C [25]. The use of antioxidants decelerates the aging of NEOs and kraft paper [12].

2.2 Significant Natural Ester Liquids

The liquid insulation in a power/distribution transformer is used for various purposes:

i) A cooling agent – During the operational time, the core and windings get heated, so the liquid insulation helps to transfer the heat outside.
ii) An insulating agent – The liquid insulation helps to insulate the different conducting parts of the transformer.

Owing to the concerns regarding environment and sustainability, researchers started looking for other alternate options to the conventional MO [27]. NEOs are a perfect substitute to the conventional MO and other synthetic liquids. These oils are completely biodegradable and nontoxic in nature. However, the downside of using these oils is that they have high pour point, higher viscosity, and oxidize easily. Some of these oils have greater content of unsaturated fatty acids which lowers the viscosity and aids in faster oxidation. Whereas, some others consist of high percentage of saturated acids, which improves oxidation stability. Thus, the oil chosen should be such that it provides a finest balance for smooth operation of an apparatus. Pongamia and jatropha oils are chosen for discussion because of their nonedible nature and they serve as potential substitutes for the conventional MO. Along with them, some other NEOs are also briefly described below.

2.2.1 Soybean Oil

Soybean oil is one of the most prominent NEO, which is being used as an insulating agent in transformers [28, 29]. It is extracted from soybean seeds and mainly consists of unsaturated fatty acids like the polyunsaturated α-linolenic acid, linoleic acid, and monounsaturated oleic acid [30]. This oil has lower oxidation stability and its viscosity is also on the higher side. This oil polymerizes easily contributing to even higher levels of viscosity. With the addition of antioxidants, it shows relatively better oxidation stability and lower viscosity [31]. Efforts are carried out to obtain a better soybean oil with enhanced properties. This oil shows better oxidation stability when used with antioxidants like citric acid and the tert-butyl hydroquinone [31]. The oil is also winterized to lower the pour point to be suitable for use in colder climates. Studies have shown that some of the properties of soybean oil are superior to that of MO [32].

2.2.2 Pongamia Pinnata Oil

Pongamia pinnata oil or commonly known as the karanji oil belongs to the species of family Leguminosae and genus Pongamia. This plant is native to the tropical and temperate regions including parts of India, China, Japan, Malaysia, and Australia. This plant is unaffected by drought and grows up to 20 m in height with soft, shiny leaves. This plant is also semi-deciduous and leguminous tree, and has the property of nitrogen fixing. This plant starts to flower and seeds are obtained, in around a time frame of five years. The inspection article of Germplasm Resources Information Network United States Department of Agriculture shows that 9–90 kg Pongamia pinnata seeds can be produced from a single tree per year [33]. The prospective yield per hectare is around 900–9000 kg/ha. As per available data, Pongamia oil has got a potential of 135 000 million tons per year and only 6% is being exploited. Traditionally, this plant has been used in Indian and south-east Asian regions as a source of Ayurvedic medicines for skin diseases, rheumatism, animal fodder, green manure, and timber. In recent times, Pongamia pinnata has been accepted as a practicable source of nonedible oil for the biofuel industry [34]. A sample of Pongamia pinnata seeds is shown in Figure 2.1a. The pongamia oil has higher flash point and breakdown voltages than the MO, but a comparable viscosity.

2.2.3 Jatropha Curcas Oil

Jatropha Curcas oil is another type of NEO, which is extracted from the seeds of the jatropha plant. This oil is majorly used in the fields where biodiesel is required. It is a product from a renewable resource and it is environment-friendly, aids in decreasing greenhouse gas emissions, and is able to combat the ever-growing energy emergency [35]. The nonedible nature of this oil is the most important advantage as it can be used for the profit of humankind without distressing the food consuming population. This plant, therefore, does not need very much fertile land to be grown as it is not for human consumption purpose. Also, the rainfall requirement of the jatropha plants is much lesser than the other seed crops like sunflower, canola, and soybeans [36]. The jatropha seeds are shown in Figure 2.1b. The basic component of crude Jatropha Curcas oil is triglyceride or triacylglycerol. The jatropha oil also has higher BDV and flash point than the conventional MO. However, the viscosity of jatropha is higher due to its higher molecular mass.

2.2.4 Palm Oil

Palm oil is derived from the palm tree fruit. This oil has a higher percentage of saturated fat content, thus forms a jelly like substance at room temperature with high viscosity [37]. The palm oil is mostly available from plantations in Indonesia and Malaysia. The basic component present in this oil is triacylglycerol and rest of the 95% is composed of glycerol with three fatty acids. Researches have shown that some of the electrical and thermophysical attributes of this insulating oil are better than the conventional MO [38]. Palm oil has an optimal balance of unsaturated and saturated fatty acids. It also has other benefits like a marginally higher flash point, higher BDV, and better PD properties. It is also biodegradable and nontoxic in nature as the oil is derived from palm fruit. Studies have shown that this oil can be used as an insulating agent in electrical apparatus [39–42]. The viscosity of palm oil is lower than most of the NEOs and somewhat lower than even MO, which gives it an added advantage [43]. A study conducted by some researchers showed the effect of moisture content on BDV and characteristics of palm oil-based insulating oils by FTIR spectroscopy and compared with MO [44]. They showed that the moisture saturation limits of palm oil are higher than that of MO.

(a) (b)

Figure 2.1 (a) Pongamia seeds, (b) Jatropha seeds.

2.2.5 Rapeseed Oil (Canola Oil)

The rapeseed oil or canola oil is extracted from the crude rapeseed oil after three refinement procedures, which are alkaline refinement, vacuum distillation, and bleaching [45, 46]. The basic attributes of the refined oil are shown in Table 2.2. The refined form of rapeseed oil exhibits lighter color, lower acid value, and enhanced electrical properties when compared with the crude form of the oil. The flash point of the rapeseed oil is above 300 °C, which is nearly twofold the flash point value of the conventional MO used in transformers. The dipolar characteristic of ester molecules produces higher values of dielectric permittivity (ε_r), and dielectric dissipation factor ($\tan\delta$) of the insulating rapeseed oil than transformer MO.

2.3 Processing and Pretreatment

The potential methyl ester-rich biological seeds such as Pongamia pinnata, Jatropha Curcas, etc., are the primary feedstock for the preparation of natural ester. Therefore, before processing the seeds for the extraction of the crude oil, pretreatment is required. This process involves a series of steps mentioned in Figure 2.2.

2.3.1 Extraction of Oil

From Figure 2.2, it is seen that the biological seeds are collected when they mature. Thereafter, the thick cover on the oil nut is removed by peeling. Once the seeds are peeled off, they are allowed to

Table 2.2 Basic properties of NEO [2, 6–8, 13, 21, 26].

Properties Oils	Kinematic viscosity at 40 °C (cSt)	Flash point (°C)	Acid value (mg KOH/g)	AC BDV (kV)	Pour point (°C)
Jatropha oil	33.8	260	0.06	73	−3
Crude Pongamia oil	32	250	1.3	87.5	−25
Pongamia oil Methyl Ester	12	284	0.9	82	−1
Rapeseed oil	43	325	0.03	73	−18
Biovolt A (mineraltec)	36.1	312	0.05	55	−21
Soya oil	33.1	318	0.08	42	−12
Rice oil	32.4	318	0.04	37	−15
Sunflower oil	38	312	0.06	37	−6
Coconut oil	25.7	298	0.005 – 0.05	35	23
Neem oil	108.79	288	—	43	—
Mustard oil	95.74	289	—	41.6	—
Punna oil	129.66	280	—	42.3	—
Castor oil	155.71	280	—	31	—
Palm oil	5.06	186	0.005	81	−32.5
Mineral oil	8.13	146	0.01	35	−40

Source: Based on Rakesh and Thomas [2]; Wilhelm et al. [6]; Kumar et al. [13]; Maharana et al. [26].

Figure 2.2 Processing of natural esters from seeds.

undergo heat treatment for drying. The drying process is of two types; one is the natural method of drying by exposing the seeds to sun light, and another is the laboratory method using the heating ovens. After that, the treated seeds are dried completely, and removal of dead seeds and solid impurities are carried out through separation technique. Once the separation is over, the seeds are now ready for extraction of crude oil.

2.3.1.1 Mechanical Extraction

Mechanical extraction is a conventional yet widely used crude oil extraction process followed for biological oil production. This method of extraction is suitable for industrial and large-scale crude oil production. The treated seeds are permitted to flow into the mechanical extruder and allowed to crush the seeds through a twin screw extruder as seen in Figure 2.3a. In this process, the crude oil is collected at the bottom of the extruder whereas the by-products are separated out.

2.3.1.2 Chemical Extraction

Initially, the treated biological seeds are collected and crushed using a grinder and converted into fine tiny particles. Next, the fine amorphous particles of seeds are mixed with the petroleum benzene. The solvent is allowed to stabilize for 24 hours and the yield of oil is collected, which is around 25–30% [47]. The collected oil from the solvent is treated using rotavapor for the removal of moisture as seen in Figure 2.3b. Thereafter, the crude oil is collected through filtration. The extracted crude oil has the rich content of triglycerides, which makes the crude ester highly viscous and of weak flowability. The triglyceride configuration is formed by condensation of three hydroxyl groups with carboxyl groups of three fatty acids, as shown in Figure 2.4a. Since the extracted crude oil from the treated seeds through mechanical and chemical treatment has a high viscosity and acidity content, it is not suitable as liquid dielectric. Therefore, the transesterification process is followed for the crude oil for suitable liquid dielectrics.

2.3.2 Transesterification

Crude vegetable oil does not meet the dielectric liquid standard, as per IEEE C.57.147 and ASTM D6871 [48] for industrial application. Hence, the extracted crude biological oil has to undergo the

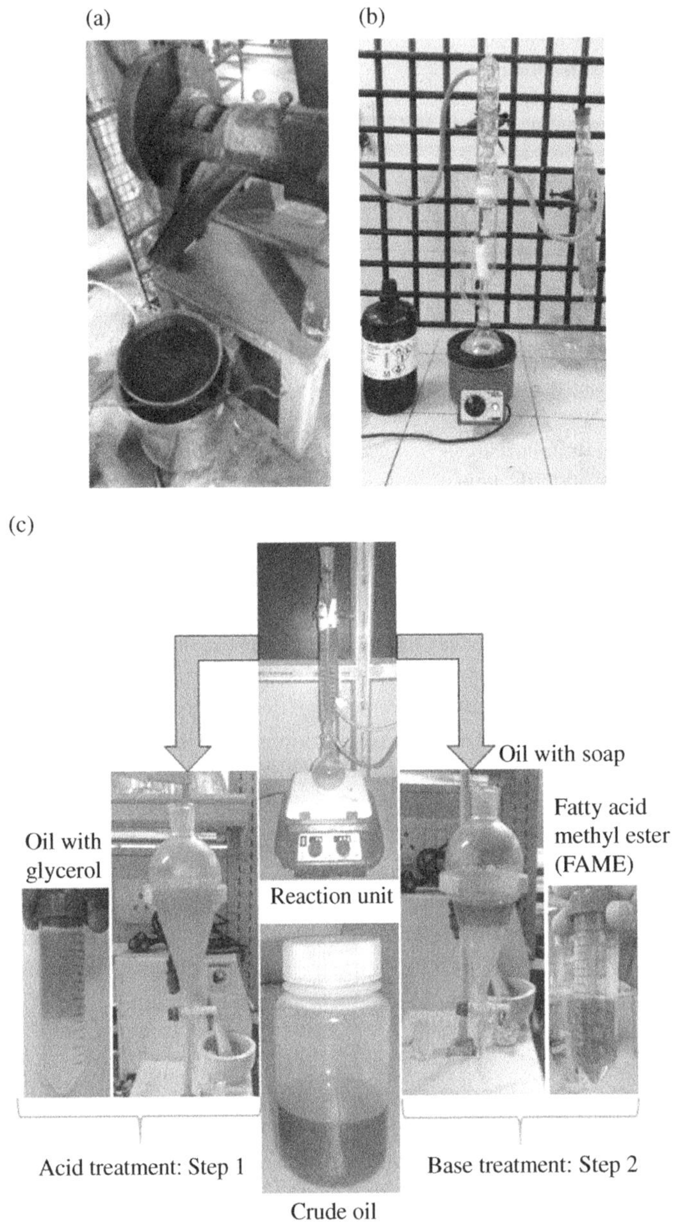

Figure 2.3 (a) Mechanical extraction. (b) Chemical extraction. (c) Transesterification setup.

transesterification process to meet the standard. In this process, moisture removal is the primary step where the crude oil sample is allowed to heat up to 100–130 °C in an oven for one hour. In order to ascertain the number of steps required for the transesterification process, free fatty acid (FFA) measurement is carried out for the dried oil. FFA is measured as per ASTM D5555 [49] using titration method considering 0.1N potassium hydroxide (KOH) solution. It is observed that if the measured FFA is more than 1%, then two-step transesterification is required to obtain the fatty acid methyl ester (FAME), which is the final refined form of the oil. Transesterification is the procedure

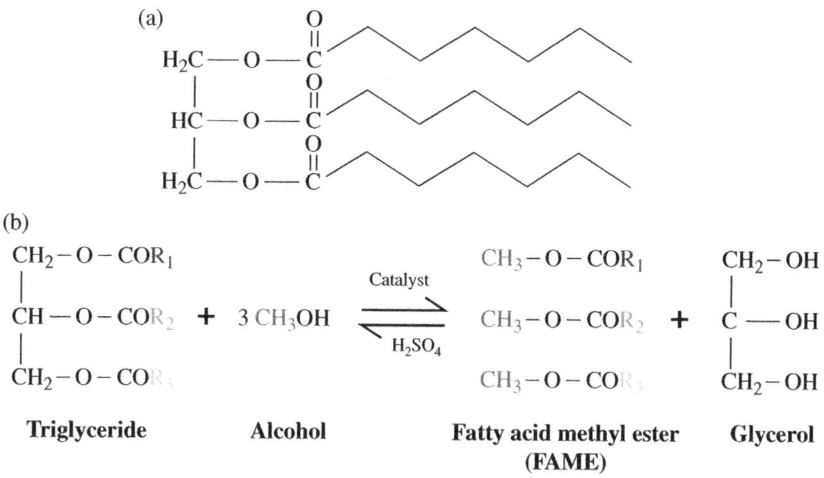

Figure 2.4 (a) Triglyceride structure. (b) Formation of fatty acid methyl ester.

that comprises the making of FAME from the crude oil to lessen the acid and viscosity values and make those similar to that of MO for transformer application. The transesterification reaction process is given in Figure 2.4b. The alcohol reacts with the fatty acids to form three molecules of mono-alkyl ester and crude glycerol.

A two-step transesterification reaction unit, comprising acid and base esterification, is presented in Figure 2.3c. In this process, the FFA is reduced using acid treatment or acid esterification technique. During the acid treatment, sulphuric acid (H_2SO_4) is used as a catalyst. The experiment is carried out using a round bottom flask with reflux condenser. The temperature of the reaction is maintained and controlled using a magnetic stirring hotplate. The separating funnel is used to separate the glycerol out from the crude oil. During the transesterification, a standard molar ratio mixture of 6:1 is chosen for methanol to crude biological oil, respectively. When the temperature of the mixture reaches 60 °C, 1 vol.% of H_2SO_4 is poured into the round bottom flask as a catalyst [50, 51]. The solution is poured into the flask and constantly stirred at 600 rpm for a session of two hours with the help of a magnetic stirrer. Then, the prepared mixture is poured into the separating funnel and allowed to stabilize for 24 hours for the acid separation. Due to the variation of density from oil to water, the water precipitates at the lower part, the lower fatty acid-based oil forms at the middle part, and the mixture of sulphuric acid and methanol lies in the top part. The settled water is removed out using a drain tap attached to the separating funnel and finally the lower FFA oil is collected. The calculated FFA value is the indicator of the decrement of acidity percentage of the crude oil.

If the FFA value lies above 1%, it has to be further treated using base transesterification, in which the mixture obtained from the earlier step is shifted to a round bottom flask and heated up to a temperature of 60 °C. The same molar ratio from the acid transesterification is considered and an adequate amount of base catalyst, that is KOH (1% by wt.), is poured into the mixture. Thereafter, the same solution is heated and stirred simultaneously for three hours. Then, oil separation technique is followed by keeping the solution in the separating funnel for 24 hours for the settlement of high-density glycerol. Top layers are formed in the separating funnel, the upper is FAME and the bottom layer is the glycerol. The solution is continuously washed by using hot distilled water that removes the unwanted impurities. It is then centrifuged at 6000 rpm for 10–15 minutes to obtain the FAME. Due to the difference in density, the low-density oil floats at the top of the centrifuge

Table 2.3 Composition of Pongamia.

FAME	Molecular formula	Retention time (min)	Molar mass (g/mol)
Hexadecanoic acid	$C_{17}H_{34}O_2$	22.48	270
11-Octadecenoic acid	$C_{19}H_{36}O_2$	24.33	296
Octadecenoic acid	$C_{18}H_{34}O_2$	24.46	282.47
CIS-11-Eicosanoic acid	$C_{20}H_{38}O_2$	26.06	310.51

Table 2.4 Composition of Jatropha.

FAME	Molecular formula	Retention time (min)	Molar mass (g/mol)
9-Hexadecanoic acid	$C_{17}H_{34}O_2$	36.31	268
Hexadecanoic acid	$C_{17}H_{34}O_2$	36.525	270
Pentadecanoic acid	$C_{16}H_{32}O_2$	36.55	256
9-Octadecenoic acid	$C_{19}H_{36}O_2$	38.52	296
Heptadecanoic acid	$C_{19}H_{38}O_2$	38.741	298

tube while the unwanted glycerol settles at the bottom. Thus, the FAME can be separated easily from the top.

The composition of Pongamia and Jatropha oils is presented in Tables 2.3 and 2.4, respectively.

2.4 Properties and Evaluation of Natural Esters

There are many properties of natural esters that are superior than the conventional MOs used in transformers. However, in some areas, the natural esters do show some drawbacks which need to be improved to utilize it to its full potential as an insulating liquid. A few of the properties are listed below.

2.4.1 Electrical Properties

2.4.1.1 AC Breakdown Voltage (ACBDV)

The quality of the dielectric insulating oil is studied and assessed by performing the AC breakdown voltage (ACBDV) test. It is conducted using BAUR DTA 100C as per the standard IEC 60156. ACBDV of insulating liquid is the prime parameter in defining its electric characteristics. The ACBDV is affected adversely by conducting particles, moisture content, and air/gas bubbles formed in the transformer insulating liquid. The measured ACBDV indicates the integrity of the oil and gives a superior value when the oil is impurity free. The oil acts as an insulating agent and a coolant in a power transformer. Apart from the solid insulation like kraft paper and pressboard, the composite oil–paper system gives an effective insulation system to the transformer. Some studies reported enhanced values of NEOs to conventional MO in comparison [47]. Some natural esters have an ACBDV of more than twice the values of the conventional oil, which indicates that natural esters can provide better resistance to high electrical stress in comparison to MO [52]. There is a huge variation in the level of water solubility of natural ester and MO, and ACBDV greatly depends on these moisture levels [53]. The tests conducted on samples of NEOs have a higher value of water content in comparison to conventional oil. This happens because of the chemical structure of the natural esters, which has higher saturation limits in water content than that of MO. Water content level of 200 ppm in NEO has a relative humidity value of 26.7%, which is equivalent to 10 ppm of water content in MO [54]. It is observed from Figure 2.5 that the ACBDV of NEOs like pongamia, jatropha, and palm oil is better than MO. Hence, with all these evidences, it is clear that NEO has comparatively higher moisture saturation levels than MO, and performance of NEO is significantly better than MO.

2.4.1.2 Dielectric Dissipation Factor (DDF)

The DDF is a measure of power loss when an external field is applied to an insulating medium. It gives tangent of the dielectric loss angle which is also known as loss factor or tan-δ. The insulation (of any HV equipment) can be represented as a series or parallel combination of a capacitance and a resistance.

2.4.1.2.1 Parallel Representation

The complex permittivity of any dielectric medium can be represented as ε_c, is denoted in (2.1),

$$\varepsilon_c = \varepsilon - j\frac{\sigma}{\omega} = \varepsilon_r' - j\varepsilon_r'' \tag{2.1}$$

where σ is the conductivity and ω is the frequency of the source. The real part ε_r' is the measure of how much energy is stored by the material and the ε_r'' represents how dissipative a material can be in reaction to an external electric field. On the application of an AC sinusoidal voltage source to the lossy dielectric in parallel combination with resistance R_p and capacitance C_p, the resultant current I is composed of two components: the conduction current I_c and the displacement current I_d, as seen in Figure 2.6a and b, which is related to the DDF as given in (2.2),

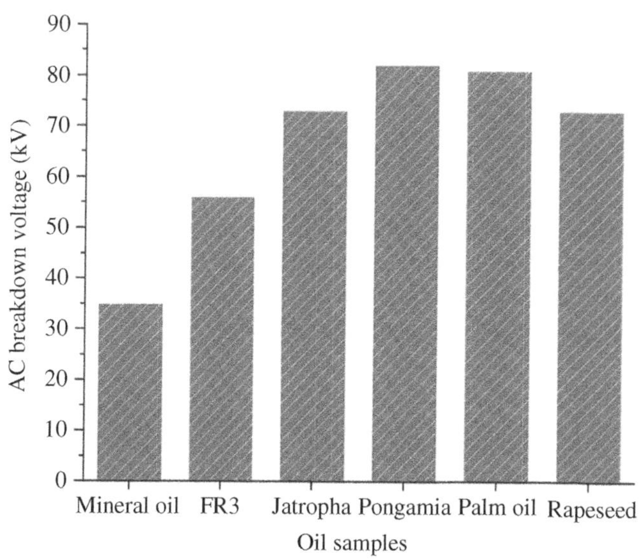

Figure 2.5 Comparison of ACBDV values of different oil samples.

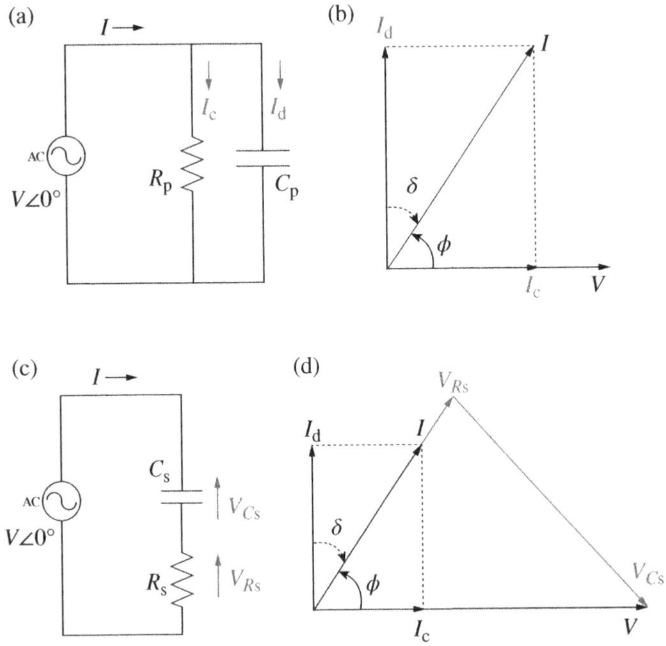

Figure 2.6 (a) Parallel representation of the insulation. (b) Phasor diagram for parallel circuit. (c) Series representation of the insulation. (d) Phasor diagram for series circuit.

$$\tan\delta = \frac{\varepsilon_r''}{\varepsilon_r'} = \frac{I_c}{I_d} = \frac{V/R_p}{V\omega C_p} = \frac{1}{\omega R_p C_p} \tag{2.2}$$

The power loss P can be represented in Eq. (2.3),

$$P = \frac{V^2}{R_p} = \frac{V^2}{\dfrac{1}{\omega C_p \tan\delta}} = V^2\omega C_p \tan\delta \tag{2.3}$$

2.4.1.2.2 Series Representation

The application of an AC sinusoidal voltage source to the lossy dielectric in series combination with resistance R_s and capacitance C_s gives two components of voltage V_{R_s} and V_{C_s} across the resistor and capacitor, respectively, as seen in Figure 2.6c. The power factor angle is represented by ϕ in Figure 2.6d. The cotangent of the power factor angle is given by the equation:

$$\cot\varphi = \frac{\cos\varphi}{\sin\varphi} = \frac{\cos(\pi/2-\delta)}{\sin\varphi(\pi/2-\delta)} = \cot(\pi/2-\delta) = \tan\delta \tag{2.4}$$

So, the loss tangent may be written as:

$$\tan\delta = \frac{V_{R_s}}{V_{C_s}} = \frac{IR_s}{I/\omega C_s} = \omega R_s C_s \tag{2.5}$$

The power loss is given by:

$$P = I^2 R_s = \left(\omega C_s V_{C_s}\right)^2 R_s \tag{2.6}$$

$$= \left(\omega C_s V_{C_s}\right)^2 \frac{\tan\delta}{\omega C_s} \tag{2.7}$$

$$= V_{C_s}{}^2 \omega C_s \tan\delta \tag{2.8}$$

From both representations, the loss is expressed in terms of the capacitance (C) and the loss angle (tanδ). Therefore, the factor $C\tan\delta$ is an important parameter to ascertain the quality of an insulating medium.

For a nonideal dielectric, the δ angle is greater than 0°. This happens because of the existence of some contaminants and some conducting channel is formed in the dielectric. An increase in the value of tanδ indicates that the oil is contaminated because of oxidation and aging and ingress of excess moisture. Generally, it is not possible to have an ideal dielectric in nature. Due to some kind of impurity, a dielectric always has a loss angle less than 90°. The temperature significantly increases when the power dissipation is higher. The lower the DDF value, the better the integrity of the oil. Higher DDF values indicate the presence of contaminations in the oil. Formation of sludge and water leads to decay in dielectric properties in due course of time and increases the DDF value. Better dielectric properties can be achieved when minimum moisture is contained in the insulating oil. Many studies have shown that with aging, the dielectric properties degrade and this results in the increase of the DDF [55]. Oxidation reaction in the oil produces a variety of by-products, in which some of them further form acids, sludge, and suspended particles [56]. DDF is a measure of the polarity of the oil molecules, and so NEOs cannot be directly equated with MOs, because the nature of the polarity in the triglyceride molecules of NEOs is different than MOs.

2.4.1.3 Dielectric Constant

Dielectric constant of insulating oil is a significant parameter of the dielectric spectroscopy analysis. The degradation of these properties makes the oil inappropriate to be used as an insulating medium. It is observed from many studies that the dielectric constant of NEOs is greater than that of the MO. This is because of the triglycerides present in the NEOs, which makes the ester oils strong polar dielectric.

2.4.2 Chemical Properties

2.4.2.1 Water Content

The water content in any dielectric liquid affects the performance of the apparatus containing it. The water holding limit of VO is much higher than that of MO, which is an added advantage in case of NEOs. It also absorbs the moisture present in solid insulation as compared with MO, thus keeping the paper dry and slows down its degradation process in the long run. Natural ester fluids undergo hydrolysis at higher temperatures, consuming the available water from kraft paper and improving its aging characteristics. The water saturation limit of MO is approximately 60 mg/kg [57] whereas vegetable ester oil has 1000 mg/kg [58]. When the moisture content is increased, the dielectric strength of the liquid insulation decreases. When the MO ages, the moisture transfers to the solid insulation like kraft paper and pressboard and degrades its dielectric properties, and thereby decreasing the performance and efficiency of the power transformer. During the operating conditions of a transformer, temperature of the oil increases and the paper then releases water to the transformer oil by undergoing oxidation reaction. This further augments the neutralization value and degrades the oil quality.

2.4.2.2 Sulphur Content

The production of corrosive sulphur is a major problem in transformers filled with MO. However, with natural esters, this is not a problem as they are bio-based and extracted from plant seeds. Hence, using natural esters for insulation will not cause corrosion problems in any equipment. The corrosive sulphur is not present in NEOs, hence performing corrosive tests can confirm if there are any kind of additives added to the NEOs on purchase. During the operational lifetime, corrosive sulphur may be present in the transformer oil, even after using NEOs. This may be due to improper handling of oil during filling, reaction of the conducting parts, and solid insulation with the oil [59]. Most of the sulphur compounds are in stable form, but under certain circumstances, these stable compounds can be transformed into reactive compounds. In a study, comparison is carried out among the properties of transformer filled with MO and some NEOs [60].

2.4.2.3 Total Acid Number (TAN)

Total acid number (TAN) or neutralization value gives the measure of acidity in the sample. It indicates the number of milligrams of KOH required to neutralize the H+ ions present in 1 g of oil. The total acid number provides the purity of the dielectric oil. Stability toward oxidation of a liquid insulation can be analyzed by determining the neutralization value, DDF, and specific resistance. Natural esters undergo oxidation and decompose by the process of hydrolysis and generate a variety of by-products like acids and alcohols. Some reports showed that the acid number of natural esters increased correspondingly with passing time [61]. The acidity for some of the oils ranged from 1 to 3 mg of KOH/g when stored for a period of three months [62]. According to ASTM D4625 (30 °C/50 weeks), the total acidity of Pongamia biodiesel rose up to 6 mg KOH/g. Studies have shown that Low Molecular Weight Acids (LMA) is higher than High Molecular Weight Acids (HMA) in MO, while HMA is higher than LMA in NEOs. Also, the total acid is much greater in NEO and hence the absolute quantity of LMA in a natural ester is higher than MO. This might be because that natural ester is much more polar and hence the interface between solid insulation and natural ester favors the stay of LMA in natural ester.

2.4.2.4 Oxidation Stability

The liquid insulation that is used in power transformer must have higher stability toward oxidation. In terms of oxidation stability, natural esters are not known to perform very well as they tend to oxidize very easily and thus they are not suitable for free-breathing transformers. When these oils come in contact with oxygen, they produce sludge and also the viscosity changes. The addition of antioxidants, however, can slow down the oxidation process. The degradation reaction occurring in the MO leads to the breakage of bonds between two carbon atoms and the formation of alkenes. This happens mainly because of oxidation, dehydrogenation, and cracking. In natural esters, oxidation and hydrolysis occur forming CO, CO_2, and other by-products. The method of oxidation varies between natural esters and MO as the by-products formed in NEOs do not solidify and settle down, but it increases the viscosity.

Though antioxidants can be added to NEOs, they still tend to show lower oxidation stability due to higher biodegradability than MO, and so it is essential to use natural esters in sealed transformers [63]. It is always essential to choose a NEO with the optimum combination of saturated and unsaturated triglycerides. It is to be noted that ester oils with higher content of saturated triglycerides exhibit higher pour points and viscosity, whereas the ones with higher unsaturated triglycerides are more prone to oxidation. Stability toward oxidation of natural esters depends on the distribution of fatty acids, refining process, and the presence of natural antioxidants. The refining process of natural esters also affects the oxidation stability as there are many intermediate processes that lead to decrease in amounts of natural antioxidants, if present.

2.4.3 Physical Properties

2.4.3.1 Pour Point

Pour point is defined as the temperature at which a liquid just starts to flow under certain conditions. Its value should be minimum in order to use the fluid at a very low temperature as per ASTM D97. Natural esters have pour point in the range of −6 to −25 °C which is higher than that of MO. This is because of the presence of triglyceride structure in the natural esters. When crude oil is trans-esterified, these triglycerides form mono-glycerides and support the flowability of the liquid and make it comparable to that of MO [54, 64]. Some comparison of the pour point values of different oils can be seen in Figure 2.7a.

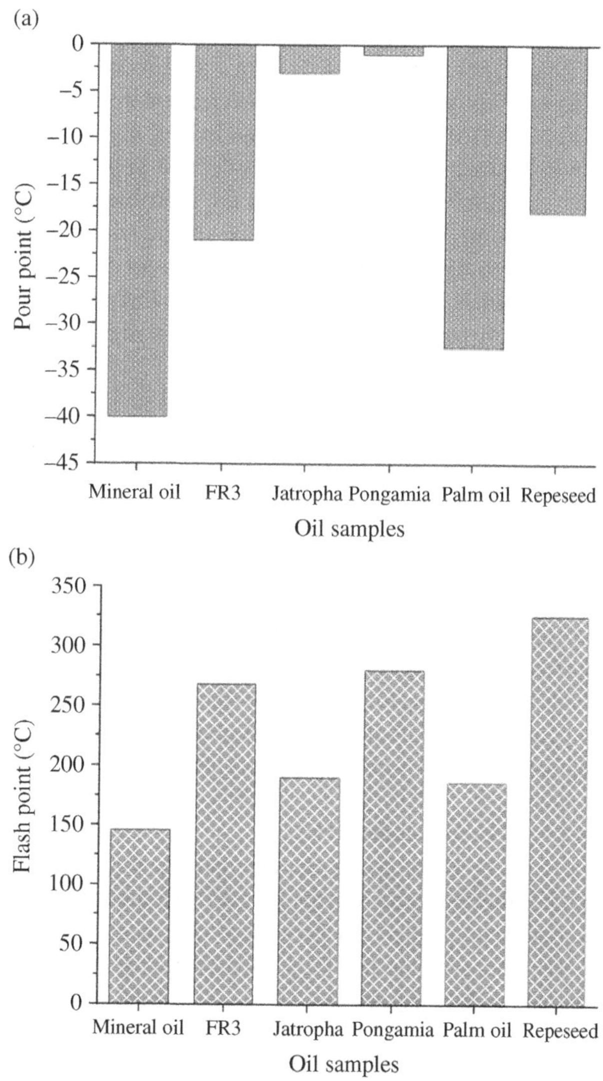

Figure 2.7 Comparison of (a) pour point values of different oil samples and (b) flash point values of different oil samples.

The problem of higher pour point can also be improved with the addition of pour point suppressants, winterization, or blending with other fluids having lower pour points. In case of natural esters, operating at low temperatures becomes difficult as they have high pour points and tend to solidify faster than MO. So, the aspect of pour point must be taken into consideration when designing a transformer or other apparatus for operation in the colder regions. Blending of natural ester with some other compatible fluids may show lower pour points. Many transformer specifications require lower ambient temperatures of less than −20 to −25 °C. In general, the temperature of a running transformer is sufficient to keep the insulating liquid flowing. However, for outdoor installations, especially in colder regions, maintaining a free flow of the liquid becomes a concern as the temperature of the atmosphere drops below 0 °C. There are certain properties of the NEO, which determine the pour point like the acid chain length, level of unsaturation, and type of branching in the oil. Unsaturated fatty acids help in reducing the pour point and also the existence of aromatic groups in natural esters aids in maintaining a lower pour point value.

2.4.3.2 Flash and Fire Point

The growing demand of increased fire safety requires a fluid with high flash and fire points. Flash point can be defined as the temperature at which a combustible liquid can be heated to give off sufficient vapor to momentarily form a flammable mixture with air when a small flame is applied under specified conditions. This is determined by employing a Pensky–Martens closed cup apparatus as per ASTM D93. The NEOs tend to have higher flash points when compared with the MO as seen in Figure 2.7b. The fire point is defined as the temperature at which the liquid itself catches fire. The primary benefits of NEOs are their higher flash and fire points than the conventional MO. Fire point plays a key role during shipment of oils and installation of transformers in indoor as well as outdoors, with lesser concern about the fire safety protocols. Natural esters have fire points of above 300 °C and meet the requirements of "K" class insulators. These oils are widely used in many practical installations and have regulatory advantages in many sites [65].

2.4.3.3 Interfacial Tension (IFT)

The IFT value of insulating liquids provides important information for detection of impurities or polar contaminations in the oil. The poor value of the IFT leads to drop in the quality of the oil, which is caused by the generation of oxides and peroxides in the insulating oil during its service. The value of IFT is measured in accordance with the standard ASTM D971. IFT is determined by the differences of the interactions between the molecules of one fluid to another fluid [66]. It is also observed that the IFT of NEO is lower than that of MO because of the molecular structure of NEO, which contains unsaturated fatty acid chains and the moisture content present in the oil. The NEOs have variances in the fatty acid structure which differ in their carbon chain lengths and in the number of double bonds or unsaturation. IFT is a physical attribute that is closely associated with the molecular configuration. The number of unsaturated fatty acids and the length of the fatty acid hydrocarbon chain affect the IFT value and the tension also increases when the chain length increases.

2.4.3.4 Thermal Conductivity

One of the most significant characteristics to understand the heat transfer property of any insulating oil is thermal conductivity. In order to understand this property of cooling in the transformer, thermal conductivity is measured by using the device KD2 pro at room temperature. Single probe transient hot-wire method is used to observe the thermal conductivity. The governing equation for thermal conductivity is

$$k = \frac{q}{4\pi\left(T_1 - T_2\right)} \ln \frac{t_1}{t_2} \tag{2.9}$$

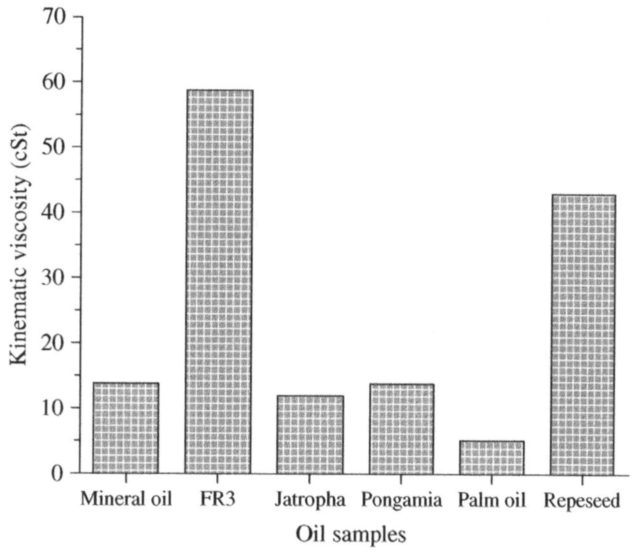

Figure 2.8 Comparison of viscosity values of different oil samples.

where "k" is the thermal conductivity, "q" is the heat flow per unit length of the source, and "T_1" and "T_2" are the temperatures of the heat source at times "t_1" and "t_2", respectively [67]. The thermal conductivity of natural ester is higher than MO because it contains the triglyceride molecular structure.

2.4.3.5 Viscosity

Viscosity of an oil is defined as the measure of the internal friction of the flowing liquid. The viscosity of an insulating fluid influences its ability to transfer heat from the interior of the transformer to the environment. With a higher viscosity, it becomes difficult for the liquid to flow and transfer heat quickly, which results in rise of the hot-spot temperatures within the transformer. Generally, at operating temperature of a power transformer, NEOs show lesser viscosity value than silicone oils but higher than MOs. The natural esters having higher viscosity may indicate that their cooling capability is lesser than MOs. But this is not the case, as it is observed from many studies, that a high specific heat capacity and high thermal conductivity upgrade natural esters to exhibit better cooling performance [68]. As seen from Figure 2.8, the viscosity values of pongamia, jatropha, and palm oil are comparable to MO. However, more research is required when using high-viscosity fluids in power transformers designed for MO insulation.

2.5 Degradation of Different Vegetable Oils

$$RH \rightarrow R\bullet + H \quad \text{Initiation} \tag{2.10}$$

$$\left.\begin{array}{l} R\bullet + O_2 \rightarrow RO_2\bullet \\ RO_2\bullet + RH \rightarrow ROOH + R\bullet \end{array}\right\} \text{Propagation} \tag{2.11}$$

$$\left.\begin{array}{l} ROOH \rightarrow RO\bullet + OH\bullet \\ RO\bullet + RH + O_2 \rightarrow ROH + RO_2\bullet \\ OH\bullet + RH + O_2 \rightarrow H_2O + RO_2\bullet \end{array}\right\} \text{Chain Reaction} \tag{2.12}$$

$$\left.\begin{array}{l} RO_{2\bullet} + RO_{2\bullet} \rightarrow ROOR + O_2 \\ R\bullet + R\bullet \rightarrow R - R \end{array}\right\} \text{Termination} \tag{2.13}$$

$R\bullet$: radical

$RO_{2\bullet}$: peroxy – radical

$$\underbrace{C_3H_8(OOCR)_3}_{\text{Trigylceride}} + H_2O \leftrightarrow \underbrace{C_3H_8(OH)(OOCR)_2}_{\text{Digylceride}} + \underbrace{R - COOH}_{\text{Free fatty acids}} \tag{2.14}$$

$$C_3H_8(OH)(OOCR)_2 + H_2O \leftrightarrow \underbrace{C_3H_8(OH)_2(OOCR)}_{\text{Monogylceride}} + R - COOH \tag{2.15}$$

$$C_3H_8(OH)_2(OOCR) + H_2O \leftrightarrow \underbrace{C_3H_8(OH)_3}_{\text{Glycerol}} + R - COOH \tag{2.16}$$

The degradation in natural esters occurs mainly because of oxidation and hydrolysis reaction. The thermal aging test setup is shown in Figure 2.9a and the change in color of the oil samples after laboratory aging is shown in Figure 2.9b. The oxidation reaction is a radical chain reaction involves three main phases: initiation, propagation, and termination. For oxidation, any conducting material like copper acts as catalyst. The strength of hydrogen–carbon bonds next to the carbon–carbon double bond sites in fatty acid groups is weaker, so a free radical is easily formed in natural esters by eliminating a hydrogen atom from the methylene group next to a double bond [69, 70]. This may be the result of high temperatures or an application of high voltage. Some have reported that natural esters having polyunsaturated fatty acid oxidize even at room temperature whereas the natural esters with mono-unsaturated fatty acids oxidize only at higher temperatures. The degree of unsaturation and the temperature rise are the main reasons that determine the oxidation stability of natural esters.

As shown in Eqs. (2.10)–(2.13), triglyceride hydroperoxide is formed by the reaction of free radical, oxygen, and oil molecules [70, 71]. In the chain reaction, hydroperoxides break down to more free radicals. A variety of compounds like smaller oxygen-containing by-products such as alcohols, aldehydes, ketones, and high molecular acids are produced by decomposition of triglyceride hydroperoxides. In the termination reaction, stable high molecular compound ROOR is formed by combination of the peroxy radicals RO_2. At the last stage of oxidation, secondary nonvolatile substances of oxidation are subjected to polymerization which leads to the formation of high

(a)

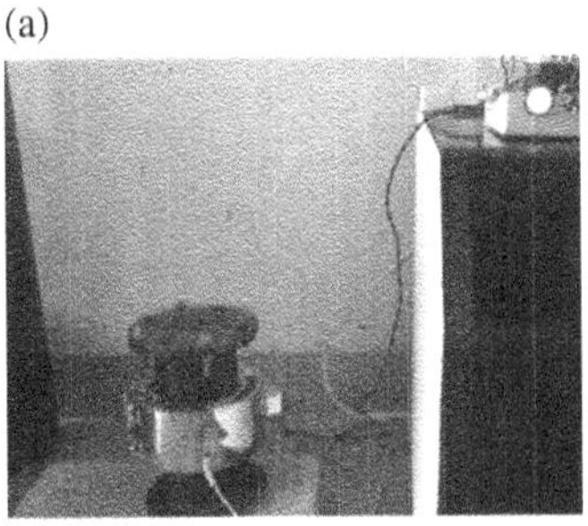

(b)

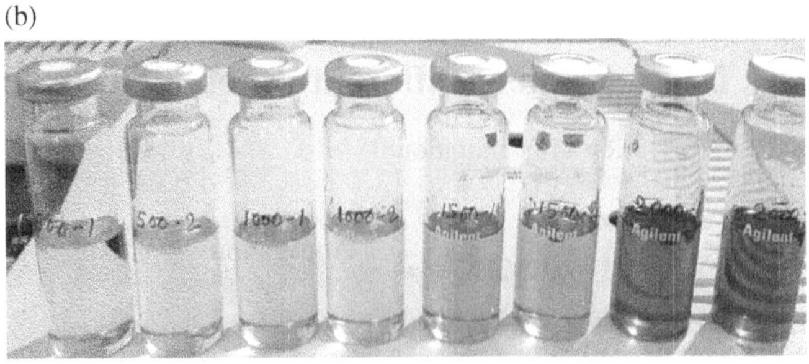

Figure 2.9 (a) Aging test setup. (b) Oil samples depicting change of color due to aging.

molecular weight gelatinous compounds, which tend to stick to the walls of the transformer or bottom of the aging vessels in laboratory experiments. Hydrolysis is another phenomenon which occurs simultaneously with the oxidation process in the natural esters. This is an autocatalytic reaction, because FFA molecules themselves accelerate the hydrolysis reaction. The steps involved in hydrolysis reaction, which are reversible in nature, are shown in Eqs. (2.14)–(2.16) [72]. The hydrolytic degradation mainly increases the acidic level of NEOs in due course of time. Temperature contributes to a large increase in the acidity of NEOs under high heat due to pronounced hydrolytic degradation.

For carrying out the aging experiment to understand the degradation of the oils, the single temperature technique in sealed beaker or an open beaker technique may be used simulating the transformer interior as per the relevant standards. During this process, thermal stress is applied on the composite system of the oil and the corresponding quantity of solid insulation, at a fixed temperature. The duration of thermal stress is changed and samples are collected after the set time intervals for performing the experiments. Complete drying of the solid insulation is carried out to make them moisture-free. The maximum temperature to carry out the aging process of oils should be as per IEEE C57.147 and preferably kept below 180 °C to avoid fluid scorching. The oil dilatation must be taken into consideration while filling the vessels for aging of oil samples as they expand on application of heat. The coefficient of thermal expansion of natural esters is approximately 0.0007/°C, so there is a 1% increase in volume for every 14 °C rise in temperature, by using the expression,

$$\Delta V = \beta V \Delta t \tag{2.17}$$

where ΔV is the change in volume, β is the thermal expansion coefficient, V is the initial volume, Δt is the temperature difference.

A few studies to ascertain as to whether the structure of the functional groups has changed in the oil samples are given below.

2.5.1 Fourier Transform Infrared Spectroscopy (FTIR)

The Fourier Transform Infrared Spectroscopy (FTIR) analysis is performed to identify the existence of the diverse functional groups in the fresh and aged oil samples. This technique aids in analyzing the uniformity of the oil after degradation in other properties because of temperature and duration. In this process, the infrared (IR) radiation is passed through the oil, out of which some percentage is absorbed and some percentage is transmitted. There are different values of frequencies which indicate different functional groups and thus the resulting spectrum represents the molecular absorption and transmission, creating a molecular fingerprint of the sample. When the IR radiation is absorbed, the molecules of the oil vibrate and it leads to bending or stretching of the bonds. It is observed from studies that the transmittance of the a few natural esters like FR3 and Jatropha (JAT) remain unaffected after aging for 2000 hours at 150 °C [73]. As seen in Figure 2.10a and b, the existence of alkanes is denoted by peaks near to 3000 cm^{-1} wavenumbers for C–H stretching. For both oils, the existence of esters is indicated by peaks near to 1740 cm^{-1} for C=O stretching. The CH_2 bending is witnessed at wavenumbers around 1460 cm^{-1} for both the oil samples. The peaks at 1000–1260 cm^{-1} for both oil samples for C–O stretching indicate the presence of alcohol. Aging in oil leads to the formation of carbonaceous particles. Nonetheless, the integrity of the oil is intact as all the functional groups are present even after long hours of aging. There is no significant difference observed for a particular oil in various aging times considered. Thus, the FTIR study might provide evidence if the aging is performed for a longer duration. The

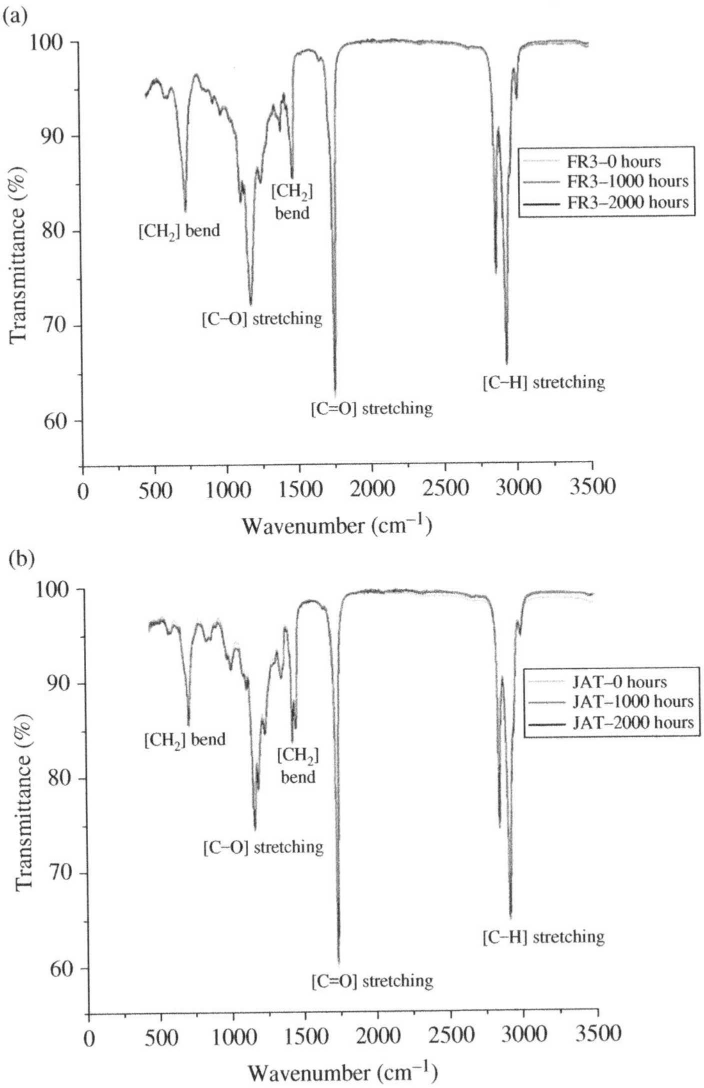

Figure 2.10 FTIR spectrum of (a) FR3 aged at 1000 and 2000 hours along with fresh FR3 and (b) Jatropha aged at 1000 and 2000 hours along with fresh JAT. *Source:* Baruah et al. [73] / with permission of IEEE.

chemical structure of the oils is not affected but the performance of the oil degrades after aging, as the other thermophysical and electrical properties deteriorate.

2.5.2 Nuclear Magnetic Resonance (NMR) Study

The alteration of the chemical assembly of the oil can be studied using the nuclear magnetic resonance (NMR) study. In the NMR spectrum, multiplets are observed for CH_3, CH_2, and allylic protons of the fatty acid fragments in the section of 0.87–2.78 ppm, for both the oil samples, as seen in Figure 2.11. In the range of 4.12–5.35 ppm, two multiplets are identified meant for the CH_2 and CH protons of the glycerol moiety. The existence of methyl ester group is established at 2.3 ppm. The peak that represents the chloroform solvent for both the oils is observed at 7.28 ppm, and this is

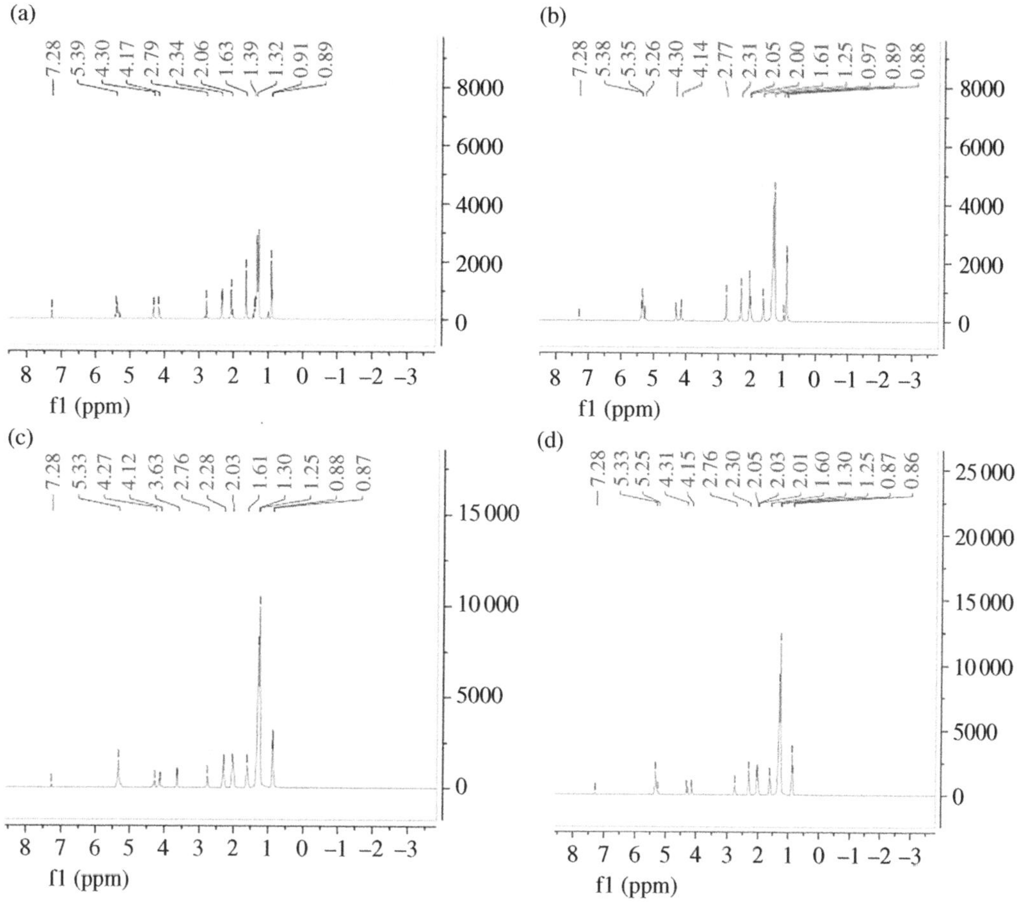

Figure 2.11 NMR analysis of (a) new FR3, (b) aged FR3 for 2000 hours, (c) new JAT, and (d) aged JAT for 2000 hours. *Source:* Baruah et al. [73] / with permission of IEEE.

considered the reference peak [74]. The NMR spectroscopy shows if there are any changes in the structure of the samples over a long duration. However, no significant difference is observed for the oil spectra of NEOs after accelerated aging hours.

2.6 Dissolved Gas Analysis in Natural Esters

Occurrence of faults in the electrical network is very much likely, but proper monitoring can help avert any catastrophic incidence. The two main categories of faults arising in the transformers are thermal and electrical. These faults lead to breakdown of insulation, and release gases in the interior of the transformer, which are detrimental to the overall functioning of the apparatus. Thus, it is very much important to measure the health of the transformer at fixed durations. DGA is an indispensable method to assess the condition of a transformer to gauge the severity of the incipient faults. The key gases developing in the transformer as a result of fault and aging are: (i) hydrogen and hydrocarbons – hydrogen (H_2), ethane (C_2H_6), methane (CH_4), ethylene (C_2H_4), acetylene (C_2H_2), (ii) Carbon oxides – carbon monoxide (CO) and carbon dioxide (CO_2), and (iii) propylene

(C_3H_6) and propane (C_3H_8). The increase in temperature leads to the evolution of these gases and with progressing time, the concentration of these gases change. Detecting the individual concentration and applying the various standard methods can help identify the incipient faults that might occur inside the transformer.

The standard IEEE C57-104 is dedicated to MO whereas the standard IEEE C57-155 is about Interpretation of Gases Generated in Natural Ester and Synthetic Ester-Immersed Transformers. As all the gas ratio techniques are elaborated in IEEE C57-104, it is used for analysis. The Duval Triangle method is mentioned as per IEEE C57-155 and it is used with regards to the stray gassing phenomenon in natural esters under the effect of the thermal stress.

2.6.1 Standard Gas Ratios

There are several methods to infer the DGA results of a transformer, with some of them listed below. The most important hydrocarbon gases are methane (CH_4), ethane (C_2H_6), hydrogen (H_2), ethylene (C_2H_4), and acetylene (C_2H_2). These gases are taken into consideration when analyzing the gas ratios in IEC, Rogers, Doernenburg, and Duval's triangle methods. All these gas ratio methods indicate the types of faults likely to occur in a transformer after the oils are subjected to thermal or electrical stress. The five gas ratios according to standard are: Ratio 1 (R1) = CH_4/H_2, Ratio 2 (R2) = C_2H_2/C_2H_4, Ratio 3 (R3) = C_2H_2/CH_4, Ratio 4 (R4) = C_2H_6/C_2H_2, and Ratio 5 (R5) = C_2H_4/C_2H_6.

2.6.1.1 IEC Gas Ratios

The IEC 60599 standard is one of the prevalent approaches for an elucidation of the faults occurring in a transformer, which is based on ratios of five key gases: CH_4, H_2, C_2H_4, C_2H_6, and C_2H_2. In this method, the ratios R1, R2, and R5 are measured to make the interpretation of the faults as per Tables 2.5 and 2.6. A grouping of the individual codes of R1, R2, and R5 indicates the type of incipient fault. However, this method does not give very accurate results for all fault types.

2.6.1.2 Doernenburg Ratio Method

This technique uses the four ratios R1, R2, R3, and R4 for diagnostics of faults in the transformer. In this method, it is primarily determined whether a fault exists in the transformer by examining the quantity of each gas associated to a minimum concentration limit L1 as given in Table 2.7. There is a fault in the transformer if any of the gases from H_2, CH_4, C_2H_2, and C_2H_4 exceeds double

Table 2.5 IEC gas ratios.

Gas ratio	Value	Code
	$R2 < 0.1$	0
$R2 = C_2H_2/C_2H_4$	$0.1 \leq R2 \leq 3$	1
	$R2 > 3$	2
	$R1 < 0.1$	1
$R1 = CH_4/H_2$	$0.1 \leq R1 \leq 1$	0
	$R1 > 1$	2
	$R5 < 1$	0
$R5 = C_2H_4/C_2H_6$	$1 \leq R5 \leq 3$	1
	$R5 > 3$	2

Table 2.6 Types of faults.

		Code		
No.	**Type of fault**	**R2**	**R1**	**R5**
1	No fault	0	0	0
2	Partial Discharge with low energy density	0	1	0
3	Partial Discharge with high energy density	1	1	0
4	Discharge (arc) with low energy	1→2	0	1→2
5	Discharge (arc) with high energy	1	0	2
6	Thermal faults of temperatures < 150 °C	0	0	1
7	Thermal faults of temperatures between 150 and 300 °C	0	2	0
8	Thermal faults of temperatures between 300 and 700 °C	0	2	1
9	Thermal faults of temperatures >700 °C	0	2	2

Table 2.7 Doernenburg gas ratio method.

Key gases	**Minimum concentration L1 (ppm)**
Hydrogen (H_2)	100
Methane (CH_4)	120
Carbon monoxide (CO)	350
Acetylene (C_2H_2)	1
Ethylene (C_2H_4)	50
Ethane (C_2H_6)	65

Table 2.8 Types of faults by Doernenburg ratio method.

No.	**Type of fault**	**R1**	**R2**	**R3**	**R4**
1	No fault	Conc. (H_2 or CH_4 or C_2H_2 or C_2H_4)>2L1 and Conc. (C_2H_6 and CO) <L1] or [Conc. (H_2 or CH_4 or C_2H_2 or C_2H_4)< 2L1]			
2	Thermal decomposition	R1 > 1	R2 < 0.75	R3 < 0.3	R4 > 0.4
3	Low-intensity partial discharge	R1 < 0.1	R2 = ND	R3 < 0.3	R4 > 0.4
4	High-intensity arcing	0.1 < R1 < 1	R2 > 0.75	R3 > 0.3	R4 < 0.4

the recommended limit L1 and the quantity (ppm) of any one of the other two gases (C_2H_6 and CO) exceeds this limit L1. This procedure is reliable only if the quantity of at least one of the gases in each ratio exceeds the limit value, otherwise oil samples may be again collected for repeated analysis. If the ratio analysis is valid, then each successive ratio is compared in the order of R1, R2, R3, R4 and the fault type is ascertained as given in Table 2.8.

2.6.1.3 Rogers Ratio Method

In this technique, three gas ratios R1, R2, and R5 are used for the explanation of the incipient faults in a transformer. The boundaries of the gas ratios suggesting a specific fault type is given in Table 2.9. Rogers method is not dependent on specific gas concentrations for the analysis to be valid. This technique is not always suitable as the accuracy is not very good in identifying faults.

2.6.1.4 Duval's Triangle

The Duval Triangle uses three hydrocarbon gases only, namely methane (CH_4), ethylene (C_2H_4), and acetylene (C_2H_2), to detect the fault types [75, 76]. The concentration of the gases is plotted in a triangular coordinate system. The types of faults detected by the Duval's triangle method are given in Table 2.10. The coordinates of the fault are denoted by the points A1=%p1, A2=%p2, A3=%p3 for the percentage concentration of CH_4, C_2H_4, and C_2H_2 gases, respectively. The concentration of gases in ppm are considered for $CH_4 = p_1$, $C_2H_4 = p_2$, and $C_2H_2 = p_3$ and are converted to triangular coordinates. The relative proportion of the three gases is calculated as below:

$$\%p_k = \frac{p_k}{\sum_{i=1}^{3} p_i} \times 100 \tag{2.18}$$

Table 2.9 Types of faults by Rogers ratio method.

No.	Type of fault	R1	R5	R2
1	No fault	$0.1 < R1 < 1$	$R5 < 1$	$R2 < 0.1$
2	Partial Discharge with low energy density	$R1 < 0.1$	$R5 < 1$	$R2 < 0.1$
3	Arcing with high energy discharge	$0.1 \leq R1 \leq 1$	$R5 > 3$	$0.1 \leq R2 \leq 3$
4	Low temperature thermal faults	$0.1 < R1 < 1$	$1 \leq R5 \leq 3$	$R2 < 0.1$
5	Thermal faults of temperatures < 700 °C	$R1 > 1$	$1 \leq R5 \leq 3$	$R2 < 0.1$
6	Thermal faults of temperatures >700 °C	$R1 > 1$	$R5 > 3$	$R2 < 0.1$

Table 2.10 Types of faults.

No.	Type of fault	Symbol
1	Partial discharge with low energy density	PD
2	Discharge (arc) with low energy	D1
3	Discharge (arc) with high energy	D2
4	Mixture of electrical and thermal faults	DT
5	Thermal faults of temperatures <300 °C	T1
6	Thermal faults of temperatures between 300 and 700 °C	T2
7	Thermal faults of temperatures >700 °C	T3
8	Overheating	O
9	Stray gassing	S
10	Hot spots with carbonization of paper	C

where $k = 1, 2, 3$ as per the three gases mentioned above.

The quantity of cases with NEO-filled transformers is less, so to assess the state of transformer with evolution of the various gases, creating the conditions replicating the inside of the transformer offers fruitful results. The different combinations of gases evolve depending on the temperatures created by the faults. PD is indicated by the formation of H_2 and CH_4 at low temperatures, whereas arcing faults are indicated by the formation of C_2H_2 at high temperatures. When the solid insulation deteriorates and reacts with the insulating oil, the gases like CO and CO_2 are formed. A gas chromatograph (GC) is used to measure the extracted gases in the oil. Some studies show the results of DGA of two types of NEOs, namely FR3 and JAT, in Figure 2.12a and b [73]. The total gas content (TGC) is calculated using the formula:

$$\text{Total Gas Content}(\text{TGC}) = \frac{\begin{array}{c}\text{Vol. of gas extracted}(\text{ml}) \times \\ \text{atm.pressure}(\text{cm}) \times 293 \times 10^6\end{array}}{\begin{array}{c}76\text{cm} \times (\text{temp.of gas extracted} + 273) \\ \times \text{quantity of oil}(\text{ml})\end{array}} \mu l/l \tag{2.19}$$

The H_2 production is not very significant in both oils, but it is higher in the case of FR3. The insignificant concentration of H_2 gas even after 2000 hours of aging makes the JAT less flammable compared with FR3. A few works show that the higher concentration of C_2H_6 may be due to faults at low temperatures below 150 °C. C_2H_4, which is the chief indicator for high-energy thermal faults, is comparable for both oil types. The presence of solid insulation on reaction with the oil produces CO and CO_2 in significant amounts, which indicates cellulose degradation. This happens because paper starts degrading at temperatures above 105 °C and the aging atmosphere influences the deterioration of oil [73]. The gas levels are quite substantial in the case of jatropha oil which might indicate that solid insulation is better safeguarded by FR3. Also, in general, the concentrations of all the gases in this oil increased with the aging duration. The small variations in gas concentrations can be ascribed to the different origination of these oils. The Duval Triangle 3 (recommended for non-mineral oils) is used for both NEOs, as formulated by Michel Duval. The two NEOs are based on the different compositions of saturated and unsaturated oleic and fatty acids, which result in different patterns of gas generation. In the Duval Triangle 3, CH_4, C_2H_4, and C_2H_2 are selected as the sides 3. The fault gas data are replaced into it, and it is observed from Figure 2.12c, that all of the data points for both oil types fall in corresponding thermal faults region. In some circumstances, the data points lie in the boundary regions and it becomes challenging to identify the actual faults. Also, many times stray gases are produced at temperatures below 200 °C in the PD, T1, or T2 zones, and consequently may affect the correct detection of these faults [22]. For analysis purpose of low-temperature faults, Duval Triangle 6 uses the low-energy gases (H_2, CH_4, and C_2H_6) as represented in Figure 2.12d. It is seen that stray gassing occurs in FR3 for all aging durations, however, for JAT, overheating is observed.

2.7 Challenges in Using Natural Esters as Insulating Liquid

The challenges faced by natural esters to be used as transformer insulating liquids are mainly in the area of oxidation stability, viscosity, acidity, and pour points. They have higher pour points, lower oxidation stability, higher acidic nature, and high viscosity. Generally, higher viscosity results in lower heat transfer capability. The lower the heat dissipation, the higher is the hotspot temperature. However, some literatures have suggested that having a high specific heat capacity and high

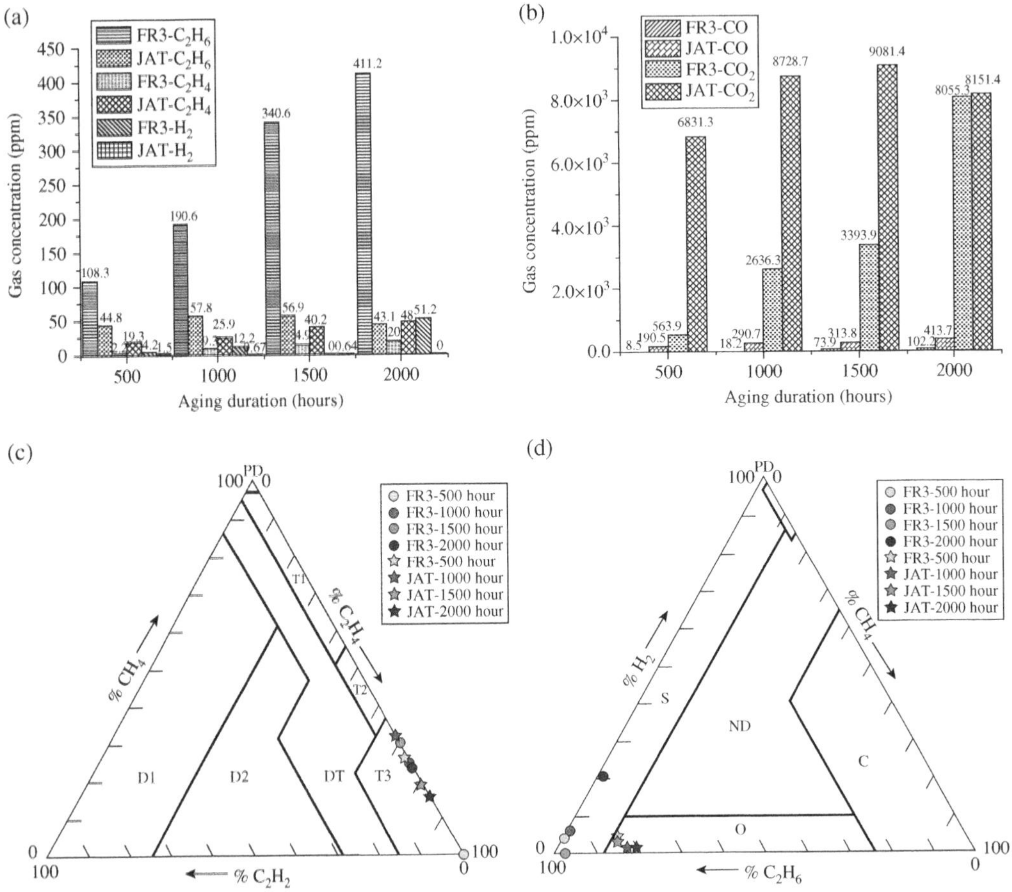

Figure 2.12 (a) DGA analysis of aged FR3 and JAT – ethane, ethylene, and hydrogen, (b) carbon monoxide and carbon dioxide, (c) Duval Triangle 3 for the aged FR3 and JAT oil samples, and (d) Duval Triangle 6 for the aged FR3 and JAT oil samples. *Source:* Baruah et al. [73] / with permission of IEEE.

thermal conductivity upgrades natural esters to exhibit better cooling performance [68]. The higher hotspot temperature leads to faster degradation of paper and oil. The cost involved in processing the NEO is higher than that of MO; therefore, the final price of NEO is higher than the MO. However, the growing needs for an efficient and eco-friendly insulating liquid for transformers have been on the rise, so the requirements for the natural esters are increasing. Additives are used for lowering pour points and increasing the oxidation stability; however, these additives lead to the nonuniform electric field in the liquid insulation. The adding of additives also leads to increase in viscosity. The production of natural esters from edible sources will also lead to concern regarding the food economy. The NEO must therefore be processed from the nonedible seeds to avoid the food crisis. The time required for growing nonedible vegetable seeds from the plants is a long process, whereas MO can be quickly extracted from the crude oil.

The chemical structure of natural esters and MOs are not the same, but they comprise analogous bonds, like C–H bonds, so the fault gasses are similar for both the fluids. DGA helps in analyzing the incipient faults in MO, which too can be used for fault detection in natural esters, under thermal and electrical fault conditions. The formation of fault gases takes place when the chemical bonds are broken. The formation of gases in natural esters is greater during PD and lower in arcing, when compared with MO. As the standard for gas formation is mainly concentrated toward MO, it becomes essential to critically analyze the gases generated in natural esters. The key fault gases are significant for interpretation of faults in the transformer. Thus, it is critical to recognize the key gases of natural esters under diverse types of faults so that accurate identification is done. As different types of ester oils come from a variety of seeds and fruits, the content of triglycerides vary in each one of them, and thus the evolution of dissolved gases is different. Therefore, it is imperative to develop and establish a standard method by which dissolved gases can be estimated for all the natural esters effectively.

The solid insulation needs to be impregnated properly with the liquid insulation for enhanced performance by the composite dielectric system. Natural ester-based fluids have greater viscosities than the conventional MO, which leads to more impregnation time by the ester oil. When the thickness of pressboard is higher, and the insulating oil is a natural ester, it creates a challenge as the viscosity of the liquid is higher and the time taken for impregnation increases accordingly. Laboratory experiments were carried out to study the impregnation process of thicker solid insulation by both NEOs and MO and it was observed that with the rise temperature of the oil, viscosity is reduced and thus decreases the impregnation time. Some studies have ascertained that adequate impregnation of solid insulation with natural esters can be attained at 60 °C within identical time as compared with MO at 20 °C [77].

The design of the transformer or other electrical apparatus, employing MO, also needs to be redesigned to use NEOs as an insulating medium. The design needs to be reviewed and approved and should meet all the requirements to come into the market. The function of the oil is to provide insulation and act as a coolant in the apparatus, and this is regarded as one of the most important areas to be looked into while altering the conventional MO and replacing it with a new fluid. The reaction with the solid insulation, conducting parts, and the thermal losses are to be taken into account while designing the new apparatus. It is desirable to take all these precautions in mind while evaluating the design procedure as the whole insulation system will have an impact on the electrical network, its operation and maintenance, as well as the economic system.

The natural esters have an inherently higher viscosity than the MO, which decreases the flowability of the liquid. This hampers the heat transfer across sections inside a transformer and to the exterior. Apparently, this indicates that the heat transfer by the natural esters is lower than the MO. However, this is not always true as for a higher heat transfer coefficient, the thermal conductivity, specific heat, and density are expected to be of higher value, whereas viscosity should be lower. It is certain that the viscosity of the NEOs is higher than the MO. But the higher viscosity of natural esters may be compensated by a much higher thermal conductivity value. Therefore, an optimum point has to be attained where the heat transfer is higher in case of natural esters.

2.8 Conclusions and Future Scope

Insulating liquids form the major part of any insulation system in many types of power apparatus, including transformers, cables, and capacitors. The characteristics of the insulating liquids are different for different applications. MOs are the conventional insulating liquids used in transformers and they have established itself well in this area. They are low-cost oils with good performance, so for a long time there was no competition to the MO. However, in the recent times, fire safety and environmental concerns have driven the researchers to look for other alternatives to the MO. From the aforementioned observation about the processing and evaluation of NEOs, the following conclusions are summarized:

- The petroleum-based MO has low environmental safety specifications which motivates the scientists for the development of an alternative liquid dielectric for insulation and cooling. Over the

years, scientists have found out that the natural ester-based liquid dielectric is an alternative option to overcome the demerits of petroleum-based MO.
- Natural esters are mostly extracted from the biological seeds such as soybean, pongamia, jatropha, palm fruit, etc., which make them a biodegradable material. Pretreatment of the biological seeds is a mandatory process before extraction of crude natural ester and for better yield.
- The extraction of crude oil is divided into two categories, namely mechanical and chemical extraction. Mechanical extraction is used for bulk production of crude oil, whereas chemical extraction is mostly applicable for the laboratory scale. The crude ester oil is considered to be unfit for usage in industrial dielectric liquid applications because of its high viscosity, high acid number, and poor pour point.
- Processing of the crude ester oil is required for industrial application. Therefore, the transesterification process is followed for the crude oil to prepare the natural ester for dielectric applications.
- In order to ascertain the quality of the natural ester, physicochemical, thermal, and electrical characterization is carried out.
- Since natural ester is biodegradable in nature, the chemical and thermal degradation analysis is carried out to understand the degradation behavior of the natural ester.
- Postoperative condition of the natural esters in power apparatus are analyzed using the DGA by which incipient faults are detected in the aged oil. These methods include the gas ratio techniques and the Duval's triangle technique.
- So far the dielectric liquid is concerned, natural ester has no hazardous effect to the environment. Therefore, in comparison to the MO, natural ester is considered as a prospective replacement for dielectric applications.

References

1. Fofana, I. (2013). 50 years in the development of insulating liquids. *IEEE Electr. Insul. Mag.* 29 (5): 13–25.
2. Rakesh, C. and Thomas, M.J. (2016). Pongamia oil, an eco-friendly alternative for mineral oil used in high voltage transformers. In *Proc. 2016 IEEE Int. Conf. Dielectr. ICD 2016*, vol. 2, pp. 959–962.
3. Martins, M. (2010). Vegetable oils, an alternative to mineral oil for power transformers-experimental study of paper aging in vegetable oil versus mineral oil. *IEEE Electr. Insul. Mag.* 26 (6): 7–13.
4. Farade, R.A., Wahab, N.I.B.A., Mansour, D.E.A. et al. (2020). Investigation of the dielectric and thermal properties of non-edible cottonseed oil by infusing h-BN nanoparticles. *IEEE Access* 8: 76204–76217.
5. Martins, M.A.G. (2010). Correction to vegetable oils, an alternative to mineral oil for power transformers–experimental study of paper aging in vegetable oil versus mineral oil. *IEEE Electr. Insul. Mag.* 26 (6): 7–13.
6. Wilhelm, H.M., Stocco, M.B.C., Tulio, L. et al. (2013). Edible natural ester oils as potential insulating fluids. *IEEE Trans. Dielectr. Electr. Insul.* 20 (4): 1395–1401.
7. Das, A.K., Shill, D.C., and Chatterjee, S. (2020). Potential of coconut oil as a dielectric liquid in distribution transformers. *IEEE Electr. Insul. Mag.* 36 (6): 36–46.
8. Beltrán, N., Palacios, E., and Blass, G. (2017). Potential of Jatropha curcas oil as a dielectric fluid for power transformers. *IEEE Electr. Insul. Mag.* 33 (2): 8–15.
9. Sitorus, H.B.H., Beroual, A., Setiabudy, R., and Bismo, S. (2015). Pre-breakdown phenomena in new vegetable oil – based jatropha curcas seeds as substitute of mineral oil in high voltage equipment. *IEEE Trans. Dielectr. Electr. Insul.* 22 (5): 2442–2448.

10 Trnka, P., Mentlik, V., and Svoboda, M. (2014). Ecologically acceptable insulating liquids for electrical appliances. In *Proc. 2014 IEEE 18th Int. Conf. Dielectr. Liq. ICDL 2014*, pp. 3–6.

11 Oommen, T.V. (2002). Vegetable oils for liquid-filled transformers. *IEEE Electr. Insul. Mag.* 18 (1): 6–11.

12 Dung, N.V. and Huong, H.L. (2020). The effect of antioxidants on the physical and chemical properties of rice oil, corn oil, peanut oil and Kraft paper. *IEEE Trans. Dielectr. Electr. Insul.* 27 (5): 1698–1706.

13 Kumar, S.S., Iruthayarajan, M.W., Bakrutheen, M., and Kannan, S.G. (2016). Effect of antioxidants on critical properties of natural esters for liquid insulations. *IEEE Trans. Dielectr. Electr. Insul.* 23 (4): 2068–2078.

14 Du, B.X. and Li, X.L. (2017). Dielectric and thermal characteristics of vegetable oil filled with BN nanoparticles. *IEEE Trans. Dielectr. Electr. Insul.* 24 (2): 956–963.

15 Carcedo, J., Fernandez, I., Ortiz, A. et al. (2016). Quantitative study on the aging of kraft paper in vegetable oils. *IEEE Electr. Insul. Mag.* 32 (6): 29–35.

16 Xu, Y., Qian, S., Liu, Q., and Wang, Z. (2014). Oxidation stability assessment of a vegetable transformer oil under thermal aging. *IEEE Trans. Dielectr. Electr. Insul.* 21 (2): 683–692.

17 Carcedo, J., Fernández, I., Ortiz, A. et al. (2015). Aging assessment of dielectric vegetable oils. *IEEE Electr. Insul. Mag.* 31 (6): 13–21.

18 Perkasa, C.Y., Lelekakis, N., Wijaya, J., and Martin, D. (2012). Investigating bubble formation in vegetable and mineral oil impregnated transformer paper insulation systems. In *Universities Power Engineering Conference (AUPEC)*, pp. 1–5.

19 Rapp, K.J., Corkran, J., McShane, C.P., and Prevost, T.A. (2009). Lightning impulse testing of natural ester fluid gaps and insulation interfaces. *IEEE Trans. Dielectr. Electr. Insul.* 16 (6): 1595–1603.

20 Nor, S.F.M., Azis, N., Jasni, J. et al. (2017). Investigation on the electrical properties of palm oil and coconut oil based TiO2 nanofluids. *IEEE Trans. Dielectr. Electr. Insul.* 24 (6): 3432–3442.

21 Li, J., Zhang, Z., Grzybowski, S., and Zahn, M. (2012). A new mathematical model of moisture equilibrium in mineral and vegetable oil-paper insulation. *IEEE Trans. Dielectr. Electr. Insul.* 19 (5): 1615–1622.

22 Perrier, C., Marugan, M., and Beroual, A. (2012). DGA comparison between ester and mineral oils. *IEEE Trans. Dielectr. Electr. Insul.* 19 (5): 1609–1614.

23 Jovalekic, M., Vukovic, D., and Tenbohlen, S. (2011). Dissolved gas analysis of alternative dielectric fluids. In *IEEE International Conference on Dielectric Liquids (ICDL)*, pp. 2–5.

24 Wang, Z., Yi, X., Huang, J. et al. (2012). Fault gas generation in natural-ester fluid under localized thermal faults. *IEEE Electr. Insul. Mag.* 28 (6): 45–56.

25 Martin, D., Lelekakis, N., and Davydov, V. (2010). Preliminary results for dissolved gas transformer. *IEEE Electr. Insul. Mag.* 26 (5): 41–48.

26 Maharana, M., Nayak, S.K., and Sahoo, N. (2018). Karanji oil as a potential dielectrics liquid for transformer. *IEEE Trans. Dielectr. Electr. Insul.* 25 (5): 1871–1879.

27 McShane, C.P. (2002). Vegetable-oil-based dielectric coolant. *IEEE Ind. Appl. Mag.* 8 (3): 34–41.

28 Bremmer, B.J. and Larry, P. (2008). *Biobased lubricants market study*, United Soybean Board.

29 Azmi, K., Ahmad, A., and Kamarol, M. (2015). Study of dielectric properties of a potential RBD palm oil and RBD soybean oil mixture as insulating liquid in transformer. *J. Electr. Eng. Technol.* 10 (5): 2105–2119.

30 Dušica, I.S., Jovanka, L.D., and Slavica, S.A. (2010). Fatty acid composition of various soybean products. *Food Feed Res.* 2: 65–70.

31 Cannon, G.S. and Honary, L.A.T. (2000). Soybean based transformer oil and transmission line fluid. US 6159913A.

32 Egbuna, S.O., Ude, O.C., and Ude, C.N. (2016). Suitability of soybean seed oil as transformer oil. *Int. J. Eng. Sci. Res. Technol.* 5 (10): 105–112.

33 Masarakall, V.H., Sikdar, D.C., and Madalageri, S.B. (2015). Development of new dielectric liquid from Pongamia oil as alternative for transformer oil. *Int. J. Tech. Res. Appl.* 3: 304–309.

34 Bobade, S.N. and Khyade, V.B. (2012). Detail study on the properties of Pongamia pinnata (Karanja) for the production of biofuel. *Res. J. Chem. Sci.* 2 (7): 16–20.

35 Berchmans, H.J. and Hirata, S. (2008). Biodiesel production from crude Jatropha curcas seed oil with a high content of free fatty acids. *Bioresour. Technol.* 99: 1716–1721.

36 Sitorus, H.B.H., Setiabudy, R., Bismo, S., and Beroual, A. (2018). Jatropha curcas methyl ester oil obtaining as vegetable insulating oil. *IEEE Trans. Dielectr. Electr. Insul.* 23 (4): 2021–2028.

37 Zamiri, R., Zakaria, A., Ahangar, H.A. et al. (2010). Fabrication of silver nanoparticles dispersed in palm oil using a laser. *Int. J. Mol. Sci.* 11 (11): 4764–4770.

38 Rajab, A., Sulaeman, A., Sudirham, S., and Suwarno (2011). A comparison of dielectric properties of palm oil with mineral and synthetic types insulating liquid under temperature variation. *ITB J. Eng. Sci.* 43 (3): 191–208.

39 Abdelmalik, A.A. (2014). Chemically modified palm kernel oil ester: a possible sustainable alternative insulating fluid. *Sustain. Mater. Technol.* 1–2: 42–51.

40 Suwarno, F., Stitinjak, S.I., and Imsak, L. (2003). Study on characteristic of palm oil and it's derivative as liquid insulating materials. In *7th International Conference on Properties and Applications of Dielectric Materials*, June 1–5, Nagoya, pp. 495–498.

41 Abdullah, U.U., Bashi, S.M., Yunus, R., and Mohibullah, N.A. (2004). The potentials of palm oil as a dielectric fluid. In *IEEE National Power & Energy Conference*.

42 Aditama, S. (2005). Dielectric properties of palm oils as liquid insulating materials: effects of fat content. *IEEE Electr. Insul. Mater.* 1: 91–94.

43 Azis, N., Jasni, J., Kadir, M.Z.A.A., and Mohtar, M.N. (2014). Suitability of palm based oil as dielectric insulating fluid in Transformers. *J. Electr. Eng. Technol.* 9 (2): 662–669.

44 Ghani, S.A., Muhamad, N.A., Chairul, I.S., and Jamri, N. (2016). A study of moisture effects on the breakdown voltage and spectral characteristics of mineral and palm oil-based insulation oils. *ARPN J. Eng. Appl. Sci.* 11 (8): 5012–5020.

45 Xiaohu Li, Jian Li, Caixin Sun (2006). Properties of transgenic rapeseed oil based dielectric liquid. In *2006 IEEE Southeast Conference*, Memphis, USA.

46 Sun, C., Li, J., Li, X., and Grzybowski, S. (2006). Electric properties of vegetable oil-based dielectric liquid and lifetime estimation of the oil paper insulation. In *2006 Annual Report Conference on Electrical Insulation and Dielectric Phenomena*, pp. 680–683.

47 Baruah, N., Maharana, M., and Nayak, S.K. (2019). Performance analysis of vegetable oil based nanofluids used in transformers. *IET Sci. Meas. Technol.* 13 (7): 995–1002.

48 Standard Specification for Natural (vegetable oil) Ester Fluids Used in Electrical Apparatus, ASTM D 6871, 2017.

49 Standard Test Method for Determination of Free Fatty Acids Contained in Animal, Marine, and Vegetable Fats and Oils Used in Fat Liquors and Stuffing Compounds, ASTM D 5555, 2017.

50 Nabi, M.N., Hoque, S.M.N., and Akhter, M.S. (2009). Karanja (Pongamia pinnata) biodiesel production in Bangladesh, characterization of karanja biodiesel and its effect on diesel emissions. *Fuel Process. Technol.* 90: 1080–1086.

51 Musa, I.A. (2016). The effects of alcohol to oil molar ratios and the type of alcohol on biodiesel production using transesterification process. *Egypt. J. Pet.* 25: 21–31.

52 Muhamad, N.A., Phung, B.T., and Blackburn, T.R. (2011). Dissolved gas analysis for common transformer faults in soy seed-based oil. *IET Electr. Power Appl.* 5 (1): 133–142.

53 Toudja, T., Moulai, H., Nacer, A. et al. (2014). Moisture and electrical discharges effect on naphthenic mineral oil properties. *IET Sci. Meas. Technol.* 8 (6): 588–594.

54 Bertrand, Y. and Hoang, L.C. (2004). Vegetable oils as substitute for mineral insulating oils in medium-voltage equipments. *CIGRE Session*: 1–6.

55 Cygan, S. and Laghari, J.R. (1987). Dependence of the electric strength on thickness, area and the volume of polypropylene. *IEEE Trans. Electr. Insul.* 22 (6): 835–837.

56 Lelekakis, N., Martin, D., and Wijaya, J. (2012). Ageing rate of paper insulation used in power transformers part 2: oil/paper system with medium and high oxygen concentration. *IEEE Trans. Dielec. Electr. Insul.* 19: 2009–2018.

57 IEEE guide for acceptance and maintenance of insulating oil in equipment, C57.106-(2002) Institute of Electrical and Electronics Engineers, Piscataway, 2002.

58 Lewand, L. (2001). Laboratory evaluation of several synthetic and agricultural-based dielectric liquids. In *Proceedings of the 86th Annual International Conference of Doble Clients*, Doble Engineering Company, Watertown, MA, USA.

59 Lewand, L.R. (2002). Report on the role of corrosive sulfur in transformers and transformer oil. Doble Company.

60 Rapp, K., Lemm, A., Orozco, L., and C. McShane (2008). Corrosive sulfur phenomena mitigation by using natural ester dielectric fluids-field Experience in Latin America. In *Transmission and Distribution Conference and Exposition: Latin America*, pp. 1–6.

61 Ashraful, A.M., Masjuki, H.H., Kalam, M.A. et al. (2014). Study of the effect of storage time on the oxidation and thermal stability of various biodiesels and their blends. *Energy Fuel* 28 (2): 1081–1089.

62 Obadiah, A., Kannan, R., Ramasubbu, A., and Kumar, S.V. (2012). Studies on the effect of antioxidants on the long-term storage and oxidation stability of Pongamia pinnata (L.) Pierre biodiesel. *Fuel Process. Technol.* 99: 56–63.

63 Martin, D., Wang, Z.D., Darwin, A.W., and James, I. (2006). A comparative study of the chemical stability of esters for use in large power transformers. In *Conference on Electrical Insulation and Dielectric Phenomena (CEIDP)*, pp. 493–496.

64 Li, N., Mao, G., and Shi, X. (2017). Advances in the research of polymeric pour point depressant for waxy crude oil. *J. Disp. Sci. Tech.* 38: 1–7.

65 Classification of insulating liquids according to fire point and net calorific value, EN 61100:1992, 1992-07-15.

66 Dong, L. and Johnson, D. (2003). Surface tension of charge-stabilized colloidal suspensions at the water-air interface. *Langmuir* 19: 10205–10209.

67 Maharana, M., Baruah, N., and Nayak, S.K. (2018). Effect of oxidative ageing on the thermophysical and electrical properties of the nanofluid with statistical analysis of AC breakdown voltage. *IET Sci. Meas. Technol.* 12 (8): 1074–1081.

68 Mazzaro, M., De Bartolomeo, D., Calcara, L. et al. (2017). Power transformer fire and environmental risk reduction by using natural esters. In *IEEE Int. Conf. Dielectr. Liquids (ICDL)*, pp. 1–4.

69 Tenbohlen, S. and Koch, M. (2010). Aging performance and moisture solubility of vegetable oils for power transformers. *IEEE Trans. Power Deliv.* 25: 825–830.

70 Fox, N.J. and Stachowiak, G.W. (2007). Vegetable oil-based lubricants – a review of oxidation. *Tribol. Int.* 40 (7): 1035–1046.

71 Ciuriuc, A., Vihacencu, M.S., Dumitran, L.M., and Notingher, P.V. (2012). Comparative study on power transformers vegetable and mineral oil ageing. In *International Conference on Applied and Theoretical Electricity*, pp. 1–6.

72 Rooney, D. and Weatherley, L.R. (2001). The effect of reaction conditions upon lipase catalysed hydrolysis of high oleate sunflower oil in a stirred liquid–liquid reactor. *J. Process Biochem.* 36: 947–953.

73 Baruah, N., Dey, S.S., and Nayak, S.K. (2020). Evaluation of dissolved gas analysis and long-term performance of non-edible natural ester. *IEEE Trans. Dielectr. Electr. Insul.* 27 (5): 1561–1569.

74 Gomez, N.A., Abonia, R., Cadavid, H., and Vargas, I.H. (2011). Chemical and spectroscopic characterization of a vegetable oil used as dielectric coolant in distribution transformers. *J. Braz. Chem. Soc.* 22 (12): 2292–2303.

75 Duval, M. (Dec. 2008). The Duval Triangle for load tap changers, non-mineral oils and low temperature faults in transformers. *IEEE Electr. Insul. Mag.* 24 (6): 22–29.

76 Duval, M. (Jun. 2002). A review of faults detectable by gas-in-oil analysis in transformer. *IEEE Electr. Insul. Mag.* 18 (3): 8–17.

77 Dai, J. and Wang, Z.D. (2008). A comparison of the impregnation of cellulose insulation by ester and mineral oil. *IEEE Trans. Dielectr. Electr. Insul.* 15 (2): 374–381.

3

Compatibility of Esters with Cellulosic Insulation Materials

Cristina Méndez Gutiérrez[1], Carmela Oria Alonso[1], Cristina Fernández Diego[1], Inmaculada Fernández Diego[1], Cristian Olmo Salas[1], Ahmet Kerem Köseoğlu[2], Ramazan Altay[2], and Alfredo Ortiz Fernández[1]

[1] *R&D Division, Electrical and Energy Engineering Department, University of Cantabria, Santander, Cantabria, Spain*
[2] *R&D Division, BEST transformer, Balikesir, Turkey*

3.1 Introduction

3.1.1 Types of Solid Insulation

Insulation is a vital element in any power transformer that consists of all the materials and components used to insulate parts at different potentials from one another. Each transformer winding needs to be insulated turn from turn and from coil to coil [1]. Paper and board are the most common materials used as solid insulation, due to their high dielectric strength, good cooling properties, and cost-efficiency. Insulation made from pure Kraft pulp provides outstanding oil impregnation characteristics and, thus, excellent performance in electrical fields [1]. That is why Kraft paper, which can be normal or thermally upgraded, is, by far, the most widely used insulation material.

The different types of solid insulation comply with different purposes inside a transformer and can be classified according to their manufacturing processes, see Section 3.1.1.1, and there are also special insulation materials, see Section 3.1.1.2, such as, for instance, synthetic insulating materials that can be used to insulate parts of the transformer or transformers specially designed when the operating temperatures are high, see [2].

3.1.1.1 Classification According to Manufacturing Processes

Unbleached softwood Kraft pulp is the main product used for the manufacturing of paper and pressboard for electrical insulation. The cellulose is processed by the so-called sulphate or Kraft process, and the composition of the insulation obtained is about 78–80% cellulose, 10–20% hemicellulose, and 2–6% lignin [1].

The initial manufacturing process, from unbleached sulphate wood pulp of high-purity electrical quality, is the same for paper and pressboard insulations. First, the cellulose is dissolved in filtered water and stored in vats. Second, the fibers are refined with brushes, and their surface is increased, creating a fine ramification which contributes to achieving a high mechanical strength. The fibers are then cleaned in water, to remove metallic and mineral particles and any insufficiently opened-up fiber bundles [1]. The mixture of water and cellulose is introduced into the sieve

Alternative Liquid Dielectrics for High Voltage Transformer Insulation Systems: Performance Analysis and Applications, First Edition. Edited by U. Mohan Rao, I. Fofana, and R. Sarathi.

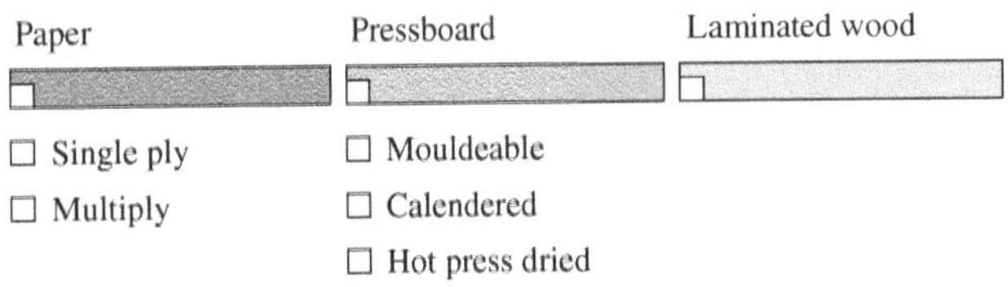

Figure 3.1 Different types of insulation according to manufacturing processes. *Source:* Based on Lundgaard et al. [1].

of the paper machine, forming a paper ply in which more fibers are aligned parallel with (in MD) than perpendicular (in CD) to the flow of the material through the paper-machine, and there are nearly no fibers aligned with the through-thickness direction (ZD) [3].

Figure 3.1 shows a possible classification of solid insulation products according to the different manufacturing processes. *Single ply paper* is pressed, dried, and rolled; later on, it can be calendered, creped, and/or slit. *Multiply papers* can be produced in paper machines with more than one former, by placing the wet plies together to one layer, which is then dewatered in the paper machine [1].

In the case of *pressboard*, the paper web constituted by individual plies is transported from the sieve and introduced on the forming roll, to obtain the required wet thickness, which is couched without a binder. The wet sheet is separated from the forming roll by cutting [1]. With *mouldable pressboard*, the wet sheet is fed into the press and compressed and then dried without applied pressure in a continuous oven. *Calendered pressboard* is produced by feeding the wet sheet into the press and cold pressing, which is then dried without pressure in a continuous oven and finally compressed on a heated calender. The *hot press dried pressboard* is manufactured by drying and compressing the wet sheet at the same time in a heated press under high pressure and at high temperature, and the result is a very hard and rigid board [1].

Laminated pressboard is produced by laminating individual sheets of high-density pressboard together with either a synthetic adhesive (polyester resin) or casein glue [4]. It is common in modern power transformers to use relevant amounts of *laminated wood* (or "plywood"), produced by joining pieces of wood veneers (typically beech) together with a thermosetting synthetic adhesive (typically phenolic) under pressure and heat [1], instead of laminated pressboard, mainly for economic reasons. There are different grades of laminated wood products for transformer use, depending on the density of the wood and the directionality of the beech veneers [5].

3.1.1.2 Special Types of Paper Insulation

It is possible to reduce the aging rate of the paper by a factor of between 1.5 and 3 over the lifetime of the transformer by using *thermally upgraded* (or uprated) paper [1], which better retains the mechanical strength during operation. The water-forming agents (melamine, dicyandiamide, and polyacrylamide) are partially neutralized and the water formation is inhibited by the use of stabilizing agents (cyanoethylation), but the influence of oxygen in the aging is less influenced in upgraded papers [1]. There is a wide range of processes for obtaining upgraded paper, which employ different additives and all of them result in an increased content of nitrogen from 0.5 to 4%. Ordinary Kraft and thermally stabilized Kraft papers have thermal classes of 105 and 120 °C, respectively, which are the temperatures for 50% reduction of tensile strength after 20 000 hours of aging, see [6, 7].

Insulating crepe papers, consisting of 100% pure sulphate wood pulp, are made from electrical-grade Kraft papers and are available in standard or thermally upgraded quality. The greater extensibility of crepe paper makes it possible to insulate rings, bended tubes, and conductor leads.

Besides, there are crepe papers with high tensile energy absorption that can be used as the external wrapping around continuously transposed conductors (CTC) [1], with strains at breakage higher than 20% when the material is not aged [8].

In *diamond dot printed papers*, a thin layer of resin is applied in dots forming squares, arranged in a diamond pattern and printed to either both sides or only one side of the paper surface. The paper is heated in an oven, melting and curing the epoxy resin, which forms a bond resistant to the solvent effect of transformer oil providing a very high mechanical stability. This paper is also available in standard or thermally upgraded quality. The epoxy resin adheres to copper and aluminum conductors, generating solid winding blocks frequently used for distribution transformers. Occasionally, it is used as external wrap on CTCs [1].

3.1.2 Mechanisms of Paper Degradation

Many factors can affect the correct performance of power transformers reducing their lifespan, the deterioration of the insulation being one of the most significant ones. This is due to the high temperatures and mechanical stresses (produced, for example, by short circuits) to which the transformer is subjected [3]. Besides, chemical reactions take place in the dielectric oil in which the paper is impregnated, producing an irreversible aging process in the cellulosic insulation, which cannot be renewed [1]. As the insulation ages, both its electrical properties (such as resistivity or dielectric breakdown strength) and its mechanical properties (such as tensile strength or elongation to breakage) gradually deteriorate [3]. Aging of power transformers is a relevant concern of utilities around the world, as power systems get older and demands on power quality become ever more important [1].

The typical insulating liquids used in power transformers over many years have been mineral oils, with different composition and refining, containing different oil molecules (aromates, naphtenes, and paraphines), sulphur compounds, inhibitors, passivators, etc. In recent years, new insulating liquids such as silicone oils, synthetic esters, and vegetable oils have become more common [1]. The aging processes over the paper insulation can change considerably depending on the different chemical properties of those insulating liquids.

3.1.2.1 Processes That Cause Degradation of the Cellulosic Insulation

Following the study of degradation processes in paper insulation of transformers over many years, it is commonly accepted that the main factors playing a role are temperature, water, oxygen, acids, and contaminants [1]. The following processes cause the cellulose chain scission reaction in insulation paper:

- **Hydrolysis:** The rate of chain scissions in the cellulose depends on the carboxylic acids dissociated in water. As both water and carboxylic acids are produced during aging of cellulose, this process auto accelerates during the aging duration. At the beginning, the water content in the paper is less than 0.5%, and that value increases up to 5% during the lifetime, which can increase the rate of degradation about 20 times [9]. A relevant advantage of natural esters used as insulating liquids is that the water reacts with the triglycerides, producing long-chain fatty acids and consuming dissolved water, drying the paper and reducing aging due to thermo-hydrolytic degradation [10].
- **Oxidation:** The rate of oxidation increases with the oxygen content. The O_2 concentration in transformer oil can reach about 30 000 ppm when fully saturated, but most free-breathing transformers in service only contain 20 000 ppm as a result of the dissolution of the gas from the air [1]. If the oxygen level in the oil is held below 2000 ppm, the rate of degradation of the full insulation system is reported to be five times lower than that of a free-breathing transformer [11].

Other researches claim that aging of paper in oil with access to oxygen is only about 2–3 times higher than aging under vacuum [12, 13]. The reactions are catalyzed by hydroxyl radicals (HO–), produced by decomposition of hydrogen peroxide (H_2O_2) and of organic hydroperoxides (ROOH). Oxidation promotes the accumulation of acids, which also are aging accelerators.

- **Pyrolysis:** This decomposition process happens without the access of water and/or oxygen or other agents. At normal operating or overload temperatures, below 140 °C, its relevance is not marked [1]. At high temperatures, produced by any electrical fault, pyrolysis may occur, followed by CO and CO_2 generation. It has been reported that the degradation of oil-impregnated insulation above about 140–150 °C is different from that observed below 150 °C [14]. That could be due to different factors: maybe the water produced as a by-product of the degradation begins to play a more important role in the degradation of the cellulose at higher temperatures, increasing the rate of bond scission [14]. Another factor could be the formation of large defects in the cellulose due to the production of furans.

In an operating transformer, the above processes act simultaneously, and the dominant mechanism will depend on temperature, oxygen, water, and acid content. It is also probable that there is a synergetic relationship among them [15], which could be represented by a nonlinear Arrhenius plot, involving the complexity of all the degradation processes [1]. Many studies have tried to quantify the degradation rates with respect to water, oxygen, and temperature. According to [9], oxygen and water are both synergistic with temperature, the synergism being stronger with water. In [16–18], realistic transformer operating conditions were considered. Heat and water were found to have a similar contribution into the acceleration of the aging of oil-impregnated insulation, and they are three times more effective than oxygen. Other factors such as electric fields and oil degradation products are considered to be of secondary importance.

According to [19], hydrolysis, oxidation, and pyrolysis contribute to a reduction in the average molecular weight of the cellulose and, particularly, to a decrease in the concentration of the high molecular weight chains relative to those of lower weight. That loss of the high molecular weight chains significantly changes the molecular weight distribution of the polymer, influencing the macroscopical mechanical response which can be examined through tensile strength testing [14].

3.1.2.2 Degradation Products from Cellulosic Insulation

Several substances appear in a power transformer as a result of the aging process. These products can subsequently act as catalysts for the degradation process, and their relative content can be measured for the troubleshooting of a power transformer.

Water is the main product of the degradation of the solid insulation. The hydrolysis of cellulose will consume water, while the thermal and oxidative degradation of the paper and the oil will produce it [1]. Ambient humidity can also ingress the transformer from the outside. Most of the water generated will be retained in the solid insulation, and less than 1% will be dissolved in the oil. The change in the water content differs with mineral oil and vegetable esters. Many studies report a higher initial water content in vegetable oil, and an increase in the content due to aging. However, as the aging proceeds at high temperatures, the ester liquids consume some of the water by hydrolysis, thus preventing the increase of water content in the solid insulation and slowing down its deterioration [8, 20].

Both free acids and the acidic groups attached to the cellulose molecular chain are produced by hydrolysis, which causes scission of the cellulose chain, and is followed by dehydration, where 5-hydroxymethyl-2-furfuraldehyde is a major product [1]. Oxidation also produces acids. If the oil becomes acidic, water solubility in it increases. The concentration of acid in oil can be determined by an amount of potassium hydroxide (KOH) needed to neutralize the acid in 10 g of oil [19].

Furanic compounds refer to a whole family of furane derivatives, whose determination in insulating oil is described in [21]. The most abundant is 2-furfural (2FAL), but 2-acetylfuran (2ACF), 5-methyl-2-furfural (5MEF), 5-hydroxymethyl-2-furfural (5HMF), and 2-furfurylalcohol (2FOL) have also been found in oil and paper insulation. These compounds are produced by the degradation of paper insulation aged at a temperature of 130 °C or more, and they subsequently migrate to the oil. The content of furanic compounds increases with aging temperature, and concentrations of 2FAL and 5HMF have been reported approximately 3–4 times higher in mineral oil than in vegetable ester [22].

CO and CO_2 are ultimate degradation products of all the constituents of paper and board, and to a lesser extent, oxidation products of the oil, which can be measured by dissolved gas analysis and whose content increases with aging temperature. The production rates and relative amounts of CO_2 and CO permit the interpretation of the results [1]. The increase in CO concentration has been reported to be more pronounced in mineral oil than in vegetable oil [22].

3.1.3 Effect of Paper Deterioration on Transformer Performance

The degradation of the solid insulation in a power transformer due to aging modifies its overall performance, changing both its electrical and mechanical properties and can cause particular defects which could even result in the end-of-life of the power transformer. The probability of suffering an electrical fault varies in different transformers. Some transformers are unlikely to be affected by short-circuit currents higher than those for which they are dimensioned, but others are exposed to frequent and high short-circuit stresses (for instance, due to line-to-ground failures on power lines due to lightning, or when moved from one place to another) [1]. In those transformers, the conditions around the hot-spot region within the top end of the winding are the most critical, where both aging and electromechanical forces will be higher, this being the most likely place for a failure to initiate [1].

The paper or board insulation must combine mechanical and dielectric performances, which are interlinked and synergetic. The mechanical properties of the insulation are strongly influenced by aging [1]. According to [23], the main cause for insulation-related faults in power transformers is mechanical failure of the paper, which leads to electrical breakdown, as the mechanical properties of solid insulation change more than the dielectric ones when the material is deeply aged. The deformations caused by electromagnetic forces acting over power transformers can result in breakage of the paper insulation, which is a mechanical phenomenon, and that can cause a subsequent electrical failure, if the insulation stops fulfilling its dielectric purpose [3]. According to [1], aging of the cellulosic insulation reduces the mechanical rigidity of a winding, and the major failure scenario would be winding movement due to high mechanical stresses during a short circuit or multiple inrush currents. The winding clamping forces are expected to change during service, and it is very difficult to predict the remaining clamping force of a given transformer because many parameters play a role (design, load, environment, oil preservation system, maintenance, time in service, etc.) [24]. The embrittled paper insulation wrapped around the conductors can break due to mechanical vibrations and stresses, particularly along the edge of the conductor [14].

In addition to the mechanical performance, several aging effects are essential in a risk assessment. Moist cellulose fibers in the solid insulation due to the increase in water content due to aging may initiate breakdown streamers [25], reduce breakdown voltages [26], and reduce the partial discharge inception voltage (PDIV) [1]. A local increase in water content can produce dielectric losses and formation of blisters in pressboard. The products of cellulosic degradation, together with sludge, can obstruct oil channels and impede the correct cooling of the winding. The formation and growth of copper sulphide, which is a conductive substance, within and on the paper, can also result in failure of the transformer [1].

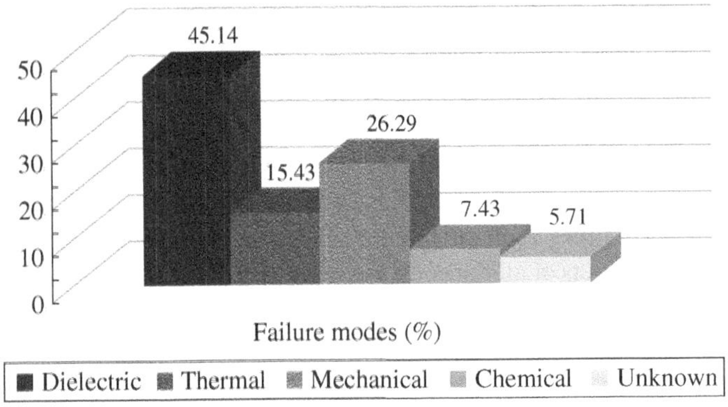

Figure 3.2 Failures classified according to failure modes. *Source:* Based on Tenbohlen et al. [27].

The failure rates that occurred between 2004 and 2009 of 15 509 transformers were analyzed and classified in [27], and are shown in Figure 3.2. Failure classification showed that the most common failure locations were the windings, and mechanical and dielectric types of failure are the most typical failure modes. This is due to the impact of short-circuit forces on the windings and their insulation, which produce winding movement or deformation, and premature aging of the winding insulation [27]. Most transformers used in generation and transmission failed in the 36–40 years age band, with inter-turn failures due to aged insulation [27]. It is probable that unusual system events, such as short circuits, lightning strikes, or switching transients have a relevant contribution to the failure of old transformers, particularly when there are design or manufacturing weaknesses [27].

3.2 Procedure of Accelerated Thermal Aging

Despite the high performance of transformers, heat is produced during their operation due to energy losses. Continuous exposure to operating temperature, along with hot spots and moments of high temperature, produces thermal degradation of the cellulosic insulation materials. Considering the effect of paper deterioration on transformer loss of life, as explained in 3.1.3, it is necessary to investigate the evolution of solid insulation when using esters as liquid dielectrics. However, it takes years to deteriorate the paper under normal operating conditions. To reduce times, the aging process is accelerated in the laboratory to simulate transformer operation. There are two main standards that set procedures for thermal evaluation of oil-filled transformers.

3.2.1 IEEE Std. C57.100

The IEEE Std. C57.100 [28] establishes a method to carry out the thermal aging in real transformers or in prototypes at rated power. The moisture content of solid insulation at the beginning of the aging must be between 0.25 and 0.5% and the transformer must be sealed without air and moisture, in a nitrogen atmosphere (or another gas that preserves the liquid). Heat must be produced by the electric currents of the windings and the temperature has to be controlled in the hot spot and maintained ±1.5% of the test temperature. At least, three temperatures must be considered in the aging. When there is insufficient prior information, tests should be made at minimum three temperatures, 20 °C intervals. When industry information is available, three temperatures should be analyzed, considering the thermal rating respect to the industry tests (150, 165, and 180 °C). The test periods must be selected as 10% of anticipated life expectancy and the power source must be disconnected at least 4 times during each test period, allowing the top oil temperature to reduce up to 25–40 °C. The end-period tests that have to be carried out depending on the type of transformer, but they are based on dielectric measurements.

Considering the investment costs of a transformer or a prototype, this standard also describes a method of thermal aging in test vessels. The test tubes are typically of stainless steel, but alternatively, may be glass. The tubes are typically about 28 cm long with an inside diameter of 4.0 cm, and a wall thickness of 5.0 mm. The moisture content of solid insulation must be between 0.25 and 0.5%, measured by Karl Fischer. An appropriate air, or gas space, should be left in each tube. The gas used to seal the system should be representative of the insulation system under evaluation. For the industry-proven system, this should be sealed with nitrogen. All the materials used in the transformer type being simulated should be represented in each tube – including paper of the type used in the transformer as well as pressboards and other components including core steel and conductor. The proportions of the materials are collected in Table 3.1. This option is useful when new or improved materials are investigated. If changes over materials are minor, single-point test at 160 °C for 706 hours is enough. If changes are significant or the material is new, three-point test as specified in the general requirements is necessary.

3.2.2 IEC 60216

IEC 60216-1 [6] is a complete standard that describes in detail the aging process requirements and the results management and interpretation. Three temperatures must be considered in the experiment, in a way that, at the lowest temperature, the end point is reached in more than 5000 hours and, at the highest temperature, the aging finishes in more than 100 hours but less than 500 hours. The initial values are measured when samples have been exposed to the lowest temperature for 48 hours. According to IEC 60216-2 [7], the end-point criterion for paper, paper-based, and woven materials is the most restrictive of these: reduction of 50% of tensile strength, degree of polymerization (DP), or breakdown voltage (BDV), all of them measured following the corresponding IEC standard. This standard is applicable to samples of 0.1–0.05 mm thickness (IEC 60216-3 [31]). The ovens used during the thermal aging must have air circulation, to promote the exchange and to prevent the lack of oxygen and the accumulation of aging products (IEC 60216-4 [32]).

Nevertheless, despite the effort to standardize the aging process, different methods and procedures can be found in the bibliography, as will be explained below. Depending on the insulating

Table 3.1 Proportions of materials for thermal aging according to IEC 62332-1 [29] and NEMA MW-1000 [30].

Material	Power transf. [29]	Distribution transf. [30]
Insulating liquid	200 cm^3	200 cm^3
0.05–0.10 mm conductor insulation	6.4 cm^3	
0.13–0.38 mm layer insulation		11.2 cm^3
1.00–3.00 mm low-density pressboard		1.2 cm^3
2.00–8.00 mm high-density pressboard	16.4 cm^3	
Ratio liquid to solid	8.8 to 1	16.3 to 1

Sources: Based on IEC [29] and NEMA MW-1000 [30].

liquid, cellulosic material, or laboratory equipment, different conditions of thermal aging are applied.

3.2.3 Accelerated Thermal Aging Conditions

3.2.3.1 Temperature

Aging temperature is the parameter that mostly affects the paper degradation, since life expectancy is directly related to hot-spot temperature, as shown in the Eq. (3.1), known as Arrhenius relationship [28]:

$$\text{LIFE} = \text{EXP}^{\left[\frac{B}{T+273}-A\right]} \tag{3.1}$$

Despite the requirements established in the standards, a wide range of aging temperatures is found in the bibliography, as shown in Table 3.2. Aging temperature is a determining factor in the aging time and in the degradation processes that appear (e.g. pyrolysis is not significant below 140 °C).

3.2.3.2 Atmosphere

Two kinds of transformers should be considered. In free-breathing transformers, air and oil are in contact. Air is usually filtered, degassed, and dried before entering the transformer, so oxygen and moisture contents are controlled. Some authors carry out the thermal aging by sealing the test vessels with air. In order to simulate a failure in the air entrance system, moisture and oxygen contents in the air can be increased. When aging the samples in air atmosphere, oxidation and moisture have a relevant effect on the insulation degradation. The higher the moisture and oxygen levels, the higher the aging rate of cellulosic materials ([9, 15]).

On the other hand, accelerated thermal aging can be carried out in an inert atmosphere, usually nitrogen. In this situation, oxidation and changes in moisture content of the insulating materials depend only on the interaction between the components of the transformer considered in the test and their aging products.

Contrary to moisture or temperature, both air and nitrogen atmospheres can be used during the thermal aging. Nevertheless, it is important to take it into account when comparing the results obtained in different works, since the atmosphere affects the aging rate and the moisture and oxidation levels. Nitrogen has a protective effect of both liquid and solid insulation ([9, 39]).

3.2.3.3 Moisture

Previous treatment of both liquid and solid insulation affects the initial conditions of thermal aging. When constructing the power transformers, manufacturers dry the components of the machine. Drying and impregnating processes are important since the initial moisture content of the

Table 3.2 Aging temperatures found in the bibliography.

	Aging temperature	References
Low temperature	<110 °C	[33–38]
Medium temperature (standard)	110–180 °C	[20, 33–35, 38–53]
High temperature	>180 °C	[41, 44]

Table 3.3 Initial moisture content of solid insulation.

Moisture content of cellulosic material	References
<0.35%	[54]
0.35% ≤ moisture ≤ 0.5%	[15, 33, 34, 39, 49, 55–57]
0.5% < moisture ≤ 1%	[41, 51, 58, 59]
>1%	[42, 55, 60]

insulation is essential to ensure its correct performance. According to the IEEE Std C57.100 [28], the moisture content of solid insulation must be higher than 0.35%, but smaller than 0.5% at the beginning of thermal aging, measured by the Karl Fischer method. The required moisture level is difficult to achieve so it is common to find that the research is carried out with higher moisture contents, as shown in Table 3.3. The differences in the initial moisture have a relevant effect on the results obtained in the aging process, since the higher the initial moisture content, the higher the aging rate [9, 15, 43]. Moreover, it affects the dielectric properties, increasing the dielectric losses [55].

3.2.3.4 Other Materials

IEEE Std C57.100 [28] recommends considering the influence of copper in the aging studies, in order to realistically evaluate the performance of transformer solid insulation. Most researchers carry out the aging process with only liquid and solid insulation, both in glass bottles and stainless-steel vessels. This approximation can be useful to compare the performance of different insulating materials since the aging is only caused by the thermal degradation and interaction between the insulation. Nevertheless, it should be considered that the aging will be slower than when adding other materials [45, 61] since these components are catalysts of the paper and oil degradation. Moreover, the presence of more solid insulators than the one that is going to be analyzed (for example, adding wood) produces an increase in moisture content in comparison with aging without this additional material [45].

As can be seen in Table 3.4, most authors only use oil and paper during the aging process and the ratio of oil:paper varies from 10 : 1 to 30 : 1, whereas other researchers do not give any information about the proportion of materials. When considering copper, the paper can be wrapped around it ([42]).

Some authors have also studied the compatibility of esters with other materials of the transformer during the thermal aging. It is important to study not only the effect of these components on the

Table 3.4 Material ratio in the aging process.

Materials	Proportions	References
Oil:paper	≈10 : 1	[33, 38, 42, 51, 52, 58, 61]
	≈20 : 1	[46]
	≈30 : 1	[41]
Oil + paper + copper	10 : 1	[33, 42]
	20 : 1 : 0.875	[46]
	75 g copper/915 g oil	[40]
Oil + paper + more materials (steel, iron, zinc, +cellulose)	3 g materials/l oil	[47, 58]
		[44, 45, 61]

properties of solid and liquid insulating but also the effect the esters have on them [44]. In [62], the compatibility of FR3 and corn oil with the Kraft paper, varnish, steel, and rubber of the transformer is studied and compared with that of mineral oil. Results show that the dielectric properties of the esters have been affected by the rubber and varnish of the transformer, since the results during the thermal aging have variations greater than the standard limits. These variations do not appear when aging with steel or Kraft paper, so the results are provoked by the rubber–esters and varnish–esters interaction. The effect of the esters on the transformer materials is studied by analyzing their mechanical properties. Results showed that they are not affected by the esters but by the mineral oil. In view of these results, it is necessary to consider all the materials of the transformer during the thermal aging in order to ensure the suitability of using esters with certain components.

3.2.3.5 Electrical Stress

Apart from the temperature effect on paper degradation, electrical stress can also reduce the life expectancy of solid insulation. Some authors, such as L. Cui et al. [57], have analyzed the changes in the dielectric properties of cellulosic materials under electrical stress, simulated through exposure of the insulation to partial discharges (PD) over a long period of time. Results show that after some time of electrical stress, the PD pulses have a dramatic increase, which evidences the cellulosic degradation. However, electrical stress is not commonly found in cellulosic materials and the wide variety of electrical failures (range of voltages, currents or time) makes it difficult to predict the exposure of the cellulosic to this stress during the life of the transformer. Because of that, considering the electrical failure as something rare and less important than thermal degradation, it is not common to analyze the effect of electrical stress on paper insulation.

3.3 Assessment of Liquid Degradation

3.3.1 Physicochemical Properties

There are some properties of the insulating liquid that are essential to ensuring the safe operation of the power transformer.

One of the disadvantages of the esters is their high viscosity (>28 mm^2/s) [50, 63, 64], except for palm oil (5.06 mm^2/s) [45], which implies a worse cooling of the transformer when using esters in comparison with mineral oils (≈8 mm^2/s). Nevertheless, this problem is solved with an appropriate design of the transformer. Because of that, it is important to study the evolution of this property with the aging of the fluid in order to ensure that it remains within the limit values. There are few studies that measure the viscosity of oils during the aging, but those available show that viscosity slightly increases with the aging of esters [49, 51, 61]. In [49], the authors have also studied the differences of the aging of the ester with and without paper. In both cases, the viscosity of the ester is increased with aging, but this increase is higher with the presence of paper, due to the appearance of more aging subproducts. Also, the higher viscosity of esters affects the impregnation process of the cellulose, so the temperature, ambient pressure, and impregnation time must be adapted to the requirements of these new liquids. According to [65], the same impregnation level is reached at 25 °C with mineral oil, 70 °C with synthetic ester, and 84 °C with natural ester, under the same ambient pressure and time.

Other crucial properties are the flash and fire point, which are the main advantages of esters (>300 °C) over mineral oils (<150 °C) [63]. With the aging of the fluid, both flash and fire point are reduced [63], but they always remain higher than in mineral oil.

FTIR is based on the specific vibration frequencies, which have the chemical bonds of the substances. Analyzing the changes in the FTIR response of the esters with the aging, specifically in the functional groups (O–H, CH, C=O, C–O), it can be deduced that there is an alteration in the ester condition, due to the increase or reduction of the absorption intensity [40, 51].

During thermal aging, acids are produced due to the degradation of liquid dielectrics. Acidity of oils is an indicator of the aging level, but the admissible limits of esters and mineral oil are different. In [61], three natural esters (sunflower HOSO, FR3, and Midel eN), one synthetic ester (Midel 7131), and one mineral oil were analyzed. The results showed that the esters always have a higher acidity than mineral oil, both when new and aged [39, 61], due to their chemical structure. The aging conditions affect the acids generated during the aging, since tests in air and sealed systems and with and without additional materials were carried out. When adding additional materials to the test vessel, the acids produced were increased. Moreover, hydrolysis of esters generates acids during the aging, so acidity is more increased during the aging process when using esters than when using mineral oil [20, 48, 51, 61]. Nevertheless, not only is the acid content important, but also the kind of acids. Acids interact with cellulose through acid catalysis hydrolysis (H_3O^+ + cellulose). The acids produced by the esters are high molecular acids, whereas mineral oil generates low molecular acids [51]. Low molecular acids generate H^+, which produces hydrolysis of the cellulose and, consequently, its degradation. Besides these acids are hydrophilic, so they can be dissolved in the paper [33]. High molecular acids do not generate H^+, so they do not degrade the cellulose. Moreover, these acids react with the cellulose through transesterification [35], [53]. O–H groups of cellulose are esterified, and a water barrier is produced [38], so the cellulose is protected from degradation [33].

Water content of ester is important since it affects the dielectric properties and the aging rate of both cellulose and ester itself. Natural esters are hydrophilic, since they have triglycerides that attract water [51]. Their saturation moisture is higher than that of mineral oil [61] and the lowest moisture content that can be reached after the drying process is higher than the minimum moisture content reached in mineral oil [33, 51]. During the aging process, the esters absorb water from the cellulose due to the hydrolysis process of the ester [66]. Acids are produced in this process. Because of that, the degradation process is slower when aging with esters than with mineral oil, as will be explained in 3.4. This reaction is shown in Eq. **(3.2)** [20].

$$R_1COOR_2 + H_2O \rightarrow R_1COOH + R_2OH \tag{3.2}$$

The water content of both esters and mineral oil increases during the thermal aging. This increase is higher at the beginning of the aging in the natural esters [33, 48]. As aging continues, most of the available water has reacted with the ester, so the water content in ester starts to decrease [20, 56], even to lower values than at the beginning of the aging. Nevertheless, the relative water content of the ester is lower than that of mineral oil [20, 33]. Due to the higher capacity of esters to absorb and retain water, reflected in their saturation moisture, dielectric properties are less affected during the aging in comparison with the mineral oil, as it is explained in Section 3.3.2. Moreover, the degradation of esters is smaller and slower than in mineral oil [39].

3.3.2 Dielectric Properties

Dielectric properties of esters determine their suitability to be used as insulating liquids. The most studied properties are BDV, PD, dielectric loss factor (tan δ), dielectric permittivity (ε), and resistivity (R). BDV and PD can be studied both in AC and DC. The others can be studied at power frequency (50 or 60 Hz) or in frequency domain. It is important to remember that the dielectric properties are highly dependent on moisture content of the liquid itself and, during the aging process with cellulosic materials, atmosphere humidity and moisture of the paper will have a great influence on the dielectric response of the ester [61].

BDV is the most important dielectric property. It is used as an end-point criterion for the aging process. As all the dielectric properties, its value is highly influenced by the moisture content of the fluid [20]. But it also depends on the configuration of electrode system (gap length, type of electrode), the rise of the voltage applied (kV/s) or measurement conditions (temperature, rest time) [67]. Despite the differences between the results obtained by different authors, most of them agree in the higher BDV of esters compared with mineral oil [33, 61]. Respecting its evolution with the aging process, the tendency is not totally clear since it is highly dependent on the relative moisture content [61]. Some authors have found that the BDV of esters remains constant or even increases with the aging process, especially at the beginning [20, 33, 61]. Nevertheless, other works show that the BDV of esters decreases with the aging process, but this reduction is notably less than in mineral oil [44, 45]. The difference among the BDV of esters and mineral oil increases with the aging process [20], since the evolution of esters with aging is better, due to their higher saturation moisture. In [61], the BDV of different natural esters (sunflower, FR3, and Midel eN), a synthetic ester (M7131), and a mineral oil was measured. The natural ester FR3 had the higher BDV (88 kV new, 89 kV aged), followed by the synthetic ester Midel 7131(80 kV new, 85 kV aged). The natural ester Midel eN had the better evolution of BDV with the aging (68 vs 90 kV), whereas the sunflower oil remained constant (76 kV). The mineral oil BDV was always lower than that of the esters (71 kV when new) and, contrary to the esters, its BDV was noticeably decreased with the aging (decrease of 49%, up to 36 kV) [61].

Dielectric loss factor (tan δ) is one of the most important properties of the insulating liquids and an indicator of the aging status. Due to the higher number of dipolar particles in the esters, they have a higher loss factor than mineral oil [48, 51, 68]. Most authors have found that, during the aging process, tan δ of the esters is increased. At the beginning of the aging, the increase can be due to the increase in moisture content in the oil [51, 58]. At the end of the aging, when moisture is lower than that of new samples, the increase could be due to ester degradation [48]. However, other authors have found that tan δ decreases with the aging of the esters [39], concluding that this factor is more related to the moisture content of the fluid than with the aging [39, 40]. There is a big difference in the tan δ evolution with aging depending on the aging conditions. In [61], aging was carried out in air and sealed systems without additional materials (Tests 1 and 2) and in air and sealed systems with additional materials (Tests 3 and 4). When adding additional materials, changes in tan δ were higher than when aging without additional materials, since these materials are catalysts of the degradation processes. Results also showed that the loss factor of all the esters, both natural (sunflower, FR3, and Midel eN) and synthetic (M7131), is higher than that of mineral oil (10^{-3} vs 10^{-4}).

The dielectric permittivity (ε) of esters is higher than that of mineral oil and it is similar to that of cellulose materials [33, 51]. Whereas the relative permittivity of mineral oil is around 2, in the case of natural ester, this value is higher than 3 [33, 68]. This property is highly dependent on the moisture content of the insulating liquid. It increases when the moisture content increases, so its evolution with the aging of the esters depends on the changes in moisture. Some authors have found that the permittivity of esters remains almost constant during the aging process [40].

The resistivity of esters is lower than that of mineral oil ((10^6 vs 10^4 MΩm) [36, 48], which matches the results obtained with tan δ [48, 58]. Its evolution with the aging is like that of tan δ, decreasing when the moisture content increases and with the degradation of the liquid. Nevertheless, unlike tan δ, the resistivity of mineral oil decreases in a similar way to esters, caused by the aging products and a higher degradation, which leads to a higher concentration of polar substances. Moreover, an increase in the moisture in mineral oil is more harmful than in esters since its saturation moisture content is lower [61].

3.4 Assessment of Paper Degradation

Cellulose properties are modified due to the different processes that take place during the thermal aging. It is necessary to evaluate the evolution of the chemical, mechanical, and dielectric properties of the impregnated cellulose to ensure the feasibility of the insulation system.

3.4.1 Chemical Properties

As occurs in the esters, some chemical properties of the cellulose change with its degradation. The most important change in the cellulose itself is the polymerization degree reduction and the moisture content variation. Other chemical parameters that are used as indicators of paper and oil aging are the gases and furans produced as subproducts of the degradation processes.

3.4.1.1 Moisture Content

Moisture content of cellulose changes during the thermal aging. At the beginning, after the drying processes of both paper and impregnating liquids, the moisture content of cellulose is higher when using esters than with mineral oil. This is because the system reaches an equilibrium between the moisture content of the oil and paper and, as explained in Section 3.3.1, the moisture content of esters is higher than that of mineral oil. During aging, water is consumed through different processes, which have greater or lesser importance depending on the liquid used. Cellulose itself suffers a hydrolytic process, catalyzed by the acids, which produces breakage of the cellulose chains, as shown in Figure 3.3a. This process always occurs during the aging of the cellulose and causes the reduction of water content but, at the same time, the degradation of paper. If a high content of acids is dissolved in cellulose, this process is accelerated. Water from the cellulose and from the ester itself is consumed by the ester through another hydrolytic process, as shown in Figure 3.3b. Due to their hydrophilic characteristics, esters absorb more water than mineral oil [34, 43, 48]. Because of that, during the aging process, water from the cellulose is consumed by the ester, which reduces the hydrolytic reactions of the cellulose since there is little water available to react and, therefore, the degradation is slowed down. When aging with mineral oil, hydrolysis of the oil does not occur or it is scarce, so the cellulose increases its moisture content, since it absorbs the water produced by the oil during the aging [41, 43]. Moreover, the low molecular weight acids produced by the mineral oil are easily dissolved in the cellulose [33] and they catalyze the hydrolytic process of the cellulose itself, so the degradation is accelerated. The acids produced by the esters during hydrolysis are high molecular acids, as shown in Figure 3.3b. These acids interact with the cellulose O-H group through a transesterification process. After that, the cellulose is protected from degradation since a water barrier is produced. The water produced by the transesterification reaction can be consumed by the ester, producing more acids, and allowing more groups to be transesterified. Because of that, as will be explained in this section, the degradation of the cellulose is slowed down when using esters as the aging increases.

3.4.1.2 Degree of Polymerization

The DP has traditionally been used as an indicator of the condition of the insulating cellulose in power transformers [16, 70]. Insulation paper consists of cellulose, hemi-cellulose, and some residual lignin that has escaped complete removal during the insulation paper pulping process [52]. The cellulose consists of linear, polymer chains of cyclic, β-D-glucose ($C_6H_{12}O_6$) units. The number of such units per chain is called the polymerization degree. DP of the paper is measured indirectly through the viscosity of a mixture of paper, deionized water, and cupriethylenediamine hydroxide

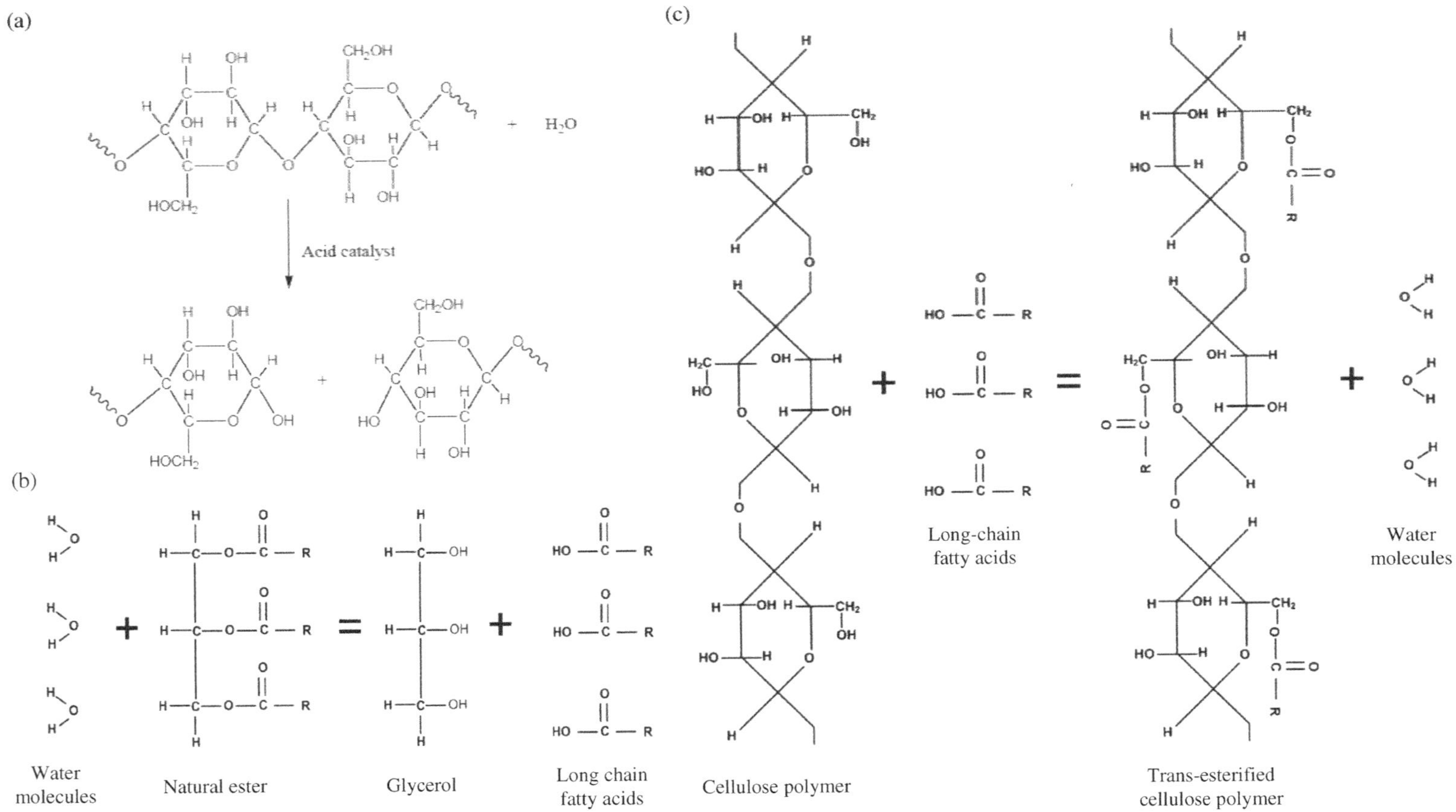

Figure 3.3 (a) Hydrolysis of cellulose. *Source:* Lelekakis et al. [15]. (b) hydrolysis of esters and (c) transesterification. *Source:* Frimpong et al. [69].

solvent (Cuen) [71]. These chains of cellulose break up during transformer operation by exposure to oxygen, moisture, and heat. As mentioned in Section 3.2, the aging conditions affect the degradation process. In [9], oxygen increased the rate of degradation of paper in oil containing 0.3–5% moisture by a factor of 2.5 and water increased the rate in simple proportion to its concentration. With the reduction of oxygen in the oil from the saturation level to less than 300 ppm, the aging rate was reduced by a factor of 16. When adding copper to the aging experiment, the oil is easily oxidized, so the oxygen content was higher, with the corresponding effect on the degradation [9]. Most of the accelerated thermal aging data can be fitted using the Ekenstam equation [16]:

$$\frac{1}{DP} - \frac{1}{DP_0} = kt \tag{3.3}$$

Where DP_0 is the initial DP of paper, at aging time 0 (usually ≈ 1300 [70]), DP is the DP after the aging time t, and k is the reaction rate constant. This equation is valid when DP is not close to the critical value (≈150–200 [70]), considered as the end of life of paper, since it loses its mechanical strength when DP is lower than that value, despite its dielectric properties not being greatly affected.

When aging with esters, DP of the paper is reduced, but it stabilizes at a higher value than when aging with mineral oils [48, 50]. All the authors agree with this, as the general conclusions indicate that the aging rate is lower with esters, since a longer aging time is needed to achieve the same DP value. However, at the beginning of the degradation, different results have been found in the bibliography. Some of the authors have observed a faster reduction of DP when using esters, since the moisture content of the cellulose is higher at the beginning of the aging [51, 52]. The polymer chains of the amorphous region of the cellulose are easily broken with temperature and the moisture content in this region is higher with esters, which accelerates the degradation. Other authors have found that the reduction of DP is similar when it is impregnated with mineral oil or esters [45, 48, 50]. After the first steps of the degradation, all the results show that the degradation process is slowed down in the paper impregnated with esters because, unlike mineral oil, esters absorbed water from the cellulose.

According to the results collected in [9], the evolution of DP during the aging is affected by the aging conditions. Considering the same paper, impregnating liquid, and aging temperature, the DP decreases slowly when aging under low water conditions. Whereas in high water and low or high oxygen conditions, the DP decreased from 1200 to 200 in 500 hours, it took 2000 hours to reach the same DP value when aging in low water and low or high oxygen conditions. It confirms the detrimental effect of moisture on paper degradation.

In [50], during the first 1280 hours of aging at 120 °C, the DP reduction of the paper impregnated in natural ester was 53 and 42% with mineral oil. However, from 1280 to 2800 hours of aging, the DP reduced by 6% in the natural ester-impregnated paper and 18% with the mineral oil. Vegetable oils delay the late degradation rate of Kraft paper, mainly due to the hydrophilic nature of natural esters and hydrolysis reaction. That effect is more significant at temperatures higher than 130 °C. In [72], after aging at 155 °C for 1500 hours, the DPs of the samples aged in vegetable oil were five times higher than those of the samples aged in mineral oil. With mineral oil, a DP = 200, generally considered as an end-of-life criterion, was reached after 850 hours, while with the vegetable ester, that limit was not reached before 1500 hours. The insulation lifetime seems to be, then, increased by using natural esters [43].

According to [39], the DP of the paper reached 200 when using mineral oil, while with esters, the DP was still around 400. This protective effect of esters is also enhanced under nitrogen atmosphere. In [41], higher DP values were reported when aging in natural ester than when using either synthetic ester or mineral oil, specifically with aging temperatures of 130, 150, and 190 °C under the same experimental conditions. In [38], the degradation rates in vegetable oil at 110 and 130 °C were reported to be slower than in mineral oil at the end of the aging process.

3.4.1.3 Fourier Transform Infrared Spectroscopy and X-ray Spectroscopy

FTIR can also be used to analyze the degradation of cellulose through the changes in the absorbance intensity or the appearance of new peaks. These changes are more evident in the regions that are affected by the aging of paper, such as the carbonyl region [51]. Some authors have also found that the transesterification of cellulose by the acids produced by esters can be seen in the FTIR through a new peak in $\approx 1746\,cm^{-1}$ [46, 51, 52] that does not appear with mineral oil or new paper. The intensity of this peak is higher as longer is the aging process, since more cellulose O–H groups are transesterified [46]. X-ray spectroscopy also gives information about the chemical structure of the cellulose and can be used to evaluate the degradation of the solid insulation [38, 46].

FTIR and X-ray can be used to estimate the DP of the paper, since there is a correlation between the intensity of the absorbance of the peaks of C–H and C=O functional group and the weight percent of the C element and between the intensity of the absorbance of O–H and the weight percent of the O element of transformer insulation paper. Both weight percent of C and O elements can be related to the average number of chain scission (DP) [46].

3.4.1.4 Furanic Compounds, Methanol Content, and Gases Production

Furanic compounds are a subproduct of the aging of the cellulose, since they cannot be found when aging the oil without paper [34, 51]. They can be measured through nonintrusive techniques, such as chromatography. Different furanic compounds can be found: 2-furfuraldehyde (2FAL), furfuril alcohol, 2-acetylfuran, 5-metil-furfuraldehyde, and 5-hydroxymethyl-2-furfuraldehyde (5HMF). The most significant are 2FAL and 5HMF [34]. The design of the aging process affects the results of furfural concentration since some authors have found that the higher the temperature, the higher the concentration of furanic compounds [34]. During the aging process, the furfural concentration increases, and it is higher when aging with mineral oil than with esters [39, 41]. Different authors have concluded that there is a relationship between furanic compounds and the DP, and they have developed a mathematical model to define this relationship [34, 35]. Equations that relate the logarithmic value of 2-FAL linearly to the DP of paper are commonly used in the industry to estimate the DP [1, 73], such as the Vuarchex equation [39]. A logarithmic relationship was established between them in [74], and the DP measured in mineral oil is comparable with the one calculated through 2-FAL content. A linear relationship was proposed by [35] between the log-scaled 2-FAL and the DP of paper with synthetic ester. According to [23], furfural analysis could become a powerful trouble-shooting tool for detecting severe insulation aging.

Methanol concentration is also used as an indicator of paper degradation. It is measured through gas chromatography and the results show that the methanol concentration increases with the aging and it is higher in esters than in mineral oil [35]. As occurs with furanic compounds, there is a relationship between the methanol concentration and the DP, as it increases linearly with the reduction in DP, and its concentration values are higher than 2-FAL during the early stage of paper aging [35], but its production is different with synthetic esters and mineral oil.

Finally, dissolved gas analysis has been widely used to identify faults during transformer operation and, also, to determine the aging of the cellulose. In the laboratory tests, these gases can be dissolved in the oil or in the air or the nitrogen gap of the test vial. Different gases appear during the aging process, as CH_4, C_2H_4, C_2H_6, H_2 and the most significant, CO and CO_2. The gas concentration increases with the degradation of cellulose [39, 51], but this increase depends on the aging conditions. When aging in nitrogen atmosphere, its protective effect makes less gases appear than when aging in air [39]. It also depends on the impregnating liquid. Ester-impregnated cellulose produces more CO and CO_2 than paper impregnated with mineral oil, as it was found by [41, 51], but the temperature needed to start the gases' production is higher when using esters. Differences in the gas production between different esters can also be found [41]. Synthetic esters produce more CO and CO_2 than natural esters.

3.4.2 Mechanical Properties

According to [75], several studies consider that the mechanical and electrical performances of cellulosic insulation are significantly related. The macroscopic mechanical behavior of paper can be accurately described by the assumption of orthotropy, with three main directions (MD, CD, and ZD). The cellulose fibers have a high tensile stiffness and strength along their axis, which are not as high in the direction across the fiber, causing the paper network to be stronger in the in-plane directions (MD and CD) than in the out-of-plane direction (ZD). The in-plane tensile failure strength in the MD is 2–4 times greater than in the CD [76, 77]. The mechanical behavior of different types of paper and board differs in detail, but some characteristics are common, such as highly anisotropic behavior, anisotropic hardening, nonlinear response, different behavior in tension and compression, viscoelasticity, strain rate sensitivity, damage evolution, and dependency on the moisture content [3]. Some of the most relevant studies on the mechanical behavior of paper materials were classified and reviewed in [78].

3.4.2.1 Tensile Strength

3.4.2.1.1 Measurement of the Tensile Mechanical Properties of Paper Insulation

The mechanical strength of the paper used as conductor insulation, as well as for pressboard used as structural component, plays a vital role for the functioning of power transformers. The tensile strength of the insulation can be influenced by their different raw materials and manufacturing processes [1]. As a consequence of aging and several chemical reactions, the cellulose fibers and the interfiber bonds are weakened [1, 23], and the solid insulation loses mechanical strength, which is generally quantified through the tensile strength. The use of mechanical strength for the determination of the deterioration rate of the paper insulation at different aging temperatures was proposed for the first time by [79].

The tensile strength can be measured using the wide-span tensile test, according to [80]. Due to the viscoelastic nature of paper, the rate at which the strain is applied in the test affects the results; thus, a defined constant rate of strain, relative humidity, and temperature must be maintained [23]. The length of the tested paper sample also affects the results. The ascending part of the stress–strain curves finishes when the tensile strength is reached, and then a localized failure zone is developed. According to [81], the failure mechanism in paper subjected to uniaxial tension is related to the length of the tested paper strip: strips whose length is higher than a particular critical length, L_{critical}, will show sudden fracture; if the length is lower than a particular value, L_{stable}, there will be a stable evolution of the failure zone; intermediate lengths of the paper strip between those two values will show a non-immediate rupture. This is relevant for the insulation of power transformers, because the length of the paper changes in its different orientations, and so the failure mode is expected to change [3].

Wide-span tensile strength is a component of the more complex bursting, folding, and tearing strengths, and represents the interfiber bonding of paper [23]. When the wide-span tensile test is performed, the testing machine records the load (*N*) by means of the load cell, and the elongation of the paper sample (mm) by the moving crosshead. The tests provide the maximum load at breakage (*N*) and the extension of the paper strip (mm). Other parameters can be obtained from them: the tensile strength, σ, per unit width, as defined in the paper industry, see Eq. (3.4), although its value in MPa is also commonly used, see Eq. (3.5); the deformation (%), see Eq. (3.5); the paper modulus in MN/m (or MPa) from the initial linear part of the plot of tensile stress against strain; and the tensile energy of absorption (J/m^2). The tensile index can be calculated from Eq. (3.7), and can be more convenient than the tensile strength, as it allows us to compare the mechanical response of papers with different thicknesses [23].

$$\sigma\left(\mathrm{kN/m}\right) = \frac{F(N)}{w(m)} \tag{3.4}$$

$$\sigma\left(\mathrm{MPa}\right) = \frac{F}{A} = \frac{F}{w\ t} \tag{3.5}$$

$$\varepsilon\left(\%\right) = \frac{L_f - L_0}{L_0}\ 100 \tag{3.6}$$

$$\text{Tensile index}\left(\mathrm{Nm/g}\right) = \frac{F(N)}{w(m)\ g\left(\mathrm{g/m^2}\right)} \tag{3.7}$$

Where:

F (*N*): Maximum force recorded by the tensile machine.
A (mm^2): Cross-sectional area of the paper strip, which is the product of the width, *w*, by the thickness, $t, A = w \cdot t$.
L_f (mm): Elongation recorded by the tensile machine.
L_0 (mm): Initial distance between jawns in the tensile test.
g (g/m^2): Grammage of the paper.

The zero-span tensile test measures the strength of the actual fibers within the paper [23], preferentially oriented in the MD. The measurement instrument has two jaws positioned so that the tensile force is across a plane through the thickness. In practice, there is a small gap between 0.2 and 2 mm and the zero-span tensile strength is derived by extrapolation from those measurements. The tensile force is calculated per unit area of sample and, therefore, generally quoted as a pressure. The pressure to break at zero gap of any paper is calculated from Eq. (3.8). The breaking length is calculated from Eq. (3.9), and the zero-span tensile index from Eq. (3.10). According to [23], this test is only useful with aged paper, as it is not applicable to new and slightly aged Kraft paper at zero gap, because the pressure to breakage is too large for the instrument. The degree of interfiber bonding is critical for the failure point of paper insulation, and can be calculated from the ratio of wide-span to zero-span tensile index, which remains constant until the paper begins to lose all mechanical strength, then drops rapidly to zero [23].

$$\text{Failure load} = K\left(P - P_0\right) \tag{3.8}$$

$$\mathrm{BL}\left(\mathrm{km}\right) = \frac{\text{Failure load}\left(\mathrm{kg}\right)}{w(m)\ g\left(\mathrm{g/m^2}\right)} \tag{3.9}$$

$$\text{Tensile index}\left(\frac{\text{Nm}}{\text{g}}\right) = 9.81 \ \text{BL}(m) \tag{3.10}$$

Where:

K: Instrument constant, which is 0.422 kg/psi.

P (psi): Observed pressure at break.

P_0 (psi): Pressure required to overcome the zero load which maintains the jaws in contact.

3.4.2.1.2 Limitations of the Tensile Strength as a Single Measure to Describe Mechanical Failure

Although almost all of the studies considering the mechanical performance of cellulosic insulation, as well as the specifications given by paper manufacturers, are focused on the tensile strength, it has been reported by [3] that the strength of the paper insulation against compressive and shear stresses and deformations could be equally or even more relevant than its resistance to tensile deformations. The paper insulation should have enough rigidity to withstand shear stresses between conductors and spacers, and compressive forces in the windings [1].

Mechanical FE simulations show that, depending on the kind of mechanical loads that affect the windings of power transformers, the shear and compressive stresses and strains over the paper or board can be more critical than the tensile ones. Although there are some publications studying the response of pressboard to loads in ZD, such as [82–90], little research has been done to obtain the shear and compressive mechanical properties of thin paper insulation. It is highly probable that mechanical failure of paper is led by interactive phenomena: a complex stress state induces tensile, compressive, and shear deformations, and it is the combination of all of them which causes the mechanical breakdown. An interactive mechanical failure model for ordinary Kraft and crepe insulation papers, inspired by the Norris criterion, was presented in [3].

3.4.2.1.3 Need for an Indirect Estimation of the Mechanical Properties of the Solid Insulation

Despite the undeniable relevance of the mechanical properties of paper and board for their performance as insulation, it is not possible to test them in operating transformers, as that would require large paper samples which would produce significant disruption to the transformer insulation [23]. Therefore, an indirect estimation of mechanical integrity is needed. The DP is correlated with the mechanical strength, and it is subsequently related to the concentration of degradation products in the oil. The DP is generally used as an indicator of the mechanical condition of the insulation because it is more easily measured [91].

However, the real conditions in an operating power transformer are complex: temperature and water content vary along the winding, leading to temperature and water content gradients affecting the different layers of paper insulation, which can be made of several different qualities of paper (such as thermally upgraded or normal Kraft paper) [1]. In [73], paper samples were extracted from the top, middle, and bottom of each phase and from the tapping, HV, LV, and tertiary windings of a failed transformer and the DP were measured. A considerable variation in the DP (approximately 25–45%) was noticed between the different windings in the same phase, which can be caused by lower temperatures associated with oil flow and greater cooling.

Furthermore, the different insulating liquids that can be used in a power transformer will influence the reduction of the mechanical strength, as the chemical reactions during degradation will differ. In [36], paper is aged impregnated in mineral oil and in different vegetable esters and, after 180 days at 90 °C, the maximum reduction in tensile strength was 20.66% in mineral oil and 16.16%

in corn oil. In [92], ordinary Kraft and Diamond Dotted papers (DDP) were aged in mineral oil and natural ester. The natural ester better preserved the mechanical properties of the Kraft paper during aging at 130–150 °C, with respect to the values measured when aged in mineral oil. However, the effect in the DDP, whose initial mechanical properties are better, seemed to be less pronounced. In [93], the rates of degradation in the mechanical properties of Kraft paper aged in mineral oil were around two times higher than the ones with the vegetable ester.

3.4.2.2 Relationship Between Degree of Polymerization (DP) and Mechanical Properties

The DP of the cellulosic insulation has been found to have a strong correlation with its mechanical properties [1]. The tensile strength, the burst strength (a biaxial tensile test), the elongation or the folding endurance show a correlation with the DP [94], see Figure 3.4. The embrittlement of the paper insulation due to aging reduces both the elongation at breakage and the DP, being these values minimum when the end of life is reached.

In [14], the nonlinear sigmoidal relationship between the DP and tensile strength at different aging states is presented for different aging temperatures. The initial reduction in tensile strength with the DP is slow, but at about a DP of 500, the reduction accelerates. According to [14], the tensile strength is significantly influenced by the final molecular properties: the high molecular weight chains and hydrogen bonding in the cellulosic insulation contribute most to the linking of crystalline regions, via the amorphous regions, and to the mechanical strength of the material. The results do not seem to depend on the temperature, but on the molecular properties.

The DP measurement is a "destructive" method, as it requires a piece of the insulation paper from the transformer under analysis, and that can cause local failure and is not a feasible method for an operating transformer. There are many nondestructive methods in the literature which measure other parameters that are directly linked to the DP and can be used in operating transformers (for instance, [95–99]), such as 2-FAL and DP relationship, explained in Section 3.4.1.4.

3.4.2.3 Scanning Electron Microscope (SEM)

The Scanning Electron Microscope (SEM) is commonly used for the observation of specimen surfaces. Its basic functioning principle is that, when the sample is irradiated with a fine electron beam,

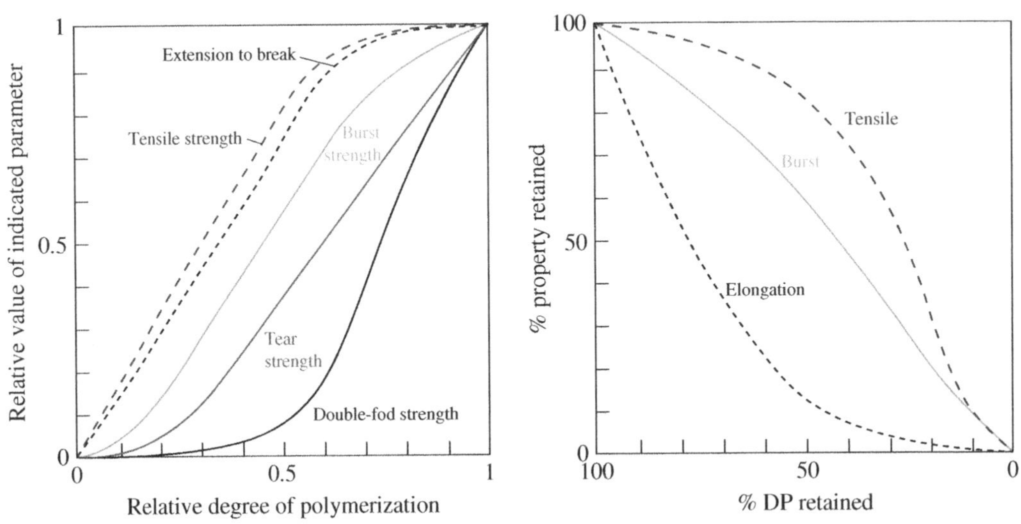

Figure 3.4 Different mechanical properties of the paper insulation as a function of the DP. *Source:* Based on Oommen and Arnold [94].

the specimen surface emits secondary electrons, making it possible to observe the topography of the surface by the two-dimensional scanning of the electron probe over the surface [3]. SEM makes it possible to appreciate the differences in microstructure for the various types of paper insulation. That microstructure affects the macroscopical mechanical response. For instance, the preferential orientation of the cellulose fibers, which is the aspect with the greatest impact on mechanical anisotropy, can be seen through the SEM. Besides, it is possible to detect local defects which frequently have an influence on mechanical fracture initiation and growth, which can lead to big cracks or total failure of the insulation.

In [52], the micro surface of insulation paper aged in different liquids was analyzed through SEM. The initial microstructure showed closely packed cellulose fibers with dense interwoven networks and no signs of deformation. After 175 days of thermal aging at 110 °C, the cellulose fibers have become thinner and the cracks from the flaws have expanded. That damage in the microstructure was less serious aging with natural ester than with mineral oil [52]. In [93] and [100], SEM was used to study the cellulose fibers in the rupture section, showing that the breaking happened due to the loosening of the interface between matrix and fiber when the insulation was not aged, and, on the other hand, when paper is heavily aged, the fiber and matrix share the same rupture plane. Some results of [93] are shown in Figure 3.5, where the deterioration of Kraft paper is greater when aging with mineral oil than with natural ester. The same DP (≈300) was reached after 261 hours in the case of mineral oil and 736 hours with natural ester. According to [8], SEM analysis over paper insulation which had suffered mechanical failure showed that microscopic structures are maintained in non-damaged areas of the insulation during the aging process, indicating that crack initiation and growth is mostly a local phenomenon.

3.4.2.4 Refractive Index of Cellulose Fibers (RI)

According to [101], the variations in refractive index (RI) are due to various packing densities of cellulose in the molecular structure. In [102], the refractive indices of cellulose fibers suspended in insulating oil were measured using a phase contrast dispersion microscope by the Dispersion Staining Method (DSM), and their relationship with the DP is described. In the DSM method, white light is dispersed at the boundary of the cellulose and the oil, the spectrum that does not refract is intercepted by an optical mask, and the cellulose fibers appear colored. That dispersion color indicates the RI of the cellulose and varies due to the extent of aging. When the insulation is not degraded, the fiber has a low RI, because it has an amorphous region to some degree, the wavelengths are long, and the dispersion colors are bluish. When the material is aged, the wavelengths are low, and the dispersion color is reddish. According to [102], it is possible to assess the DP of the paper insulation by measuring its dispersion colors.

(a) (b)

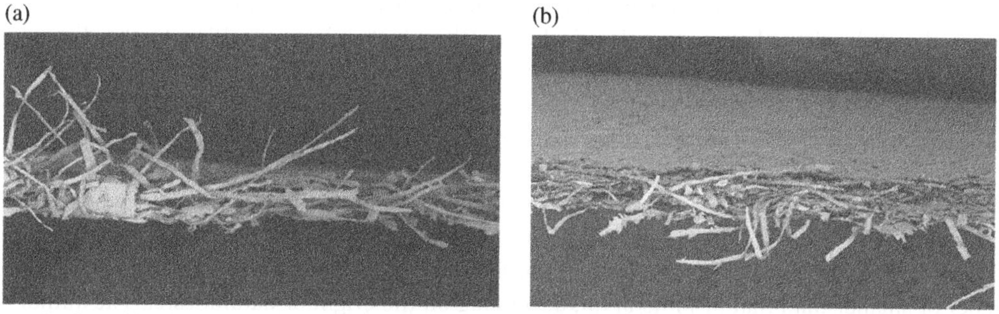

Kraft aged in mineral oil (130 °C, 261 h), DP = 300 | Kraft aged in natural ester (130 °C, 736 h) DP = 287

Figure 3.5 SEM of Kraft paper aged at 130 °C with (a) mineral oil for 261 hours (DP = 300) and (b) natural ester for 736 hours (DP = 287). *Source:* Fernández-Diego et al. [93].

In [103], the crystallinity of insulation paper was analyzed with X-ray diffraction (XRD) pattern. The crystallinity index firstly increases rapidly with aging duration, while it increases slowly in the late aging stage. Consequently, the RI increases with the thermal aging time. The dispersion staining colors (DSCs) showed an initial increase of the purple areas, and reduction of the blue areas. If aging continues, some areas of the cellulose become red or orange, corresponding to an increase in the refractive indices. However, the color distribution is not even for cellulose fibers at the same aging time. The primary color ratio-based ratio of blue to red, and the primary color ratio, showed a good correlation with the DP of insulation paper. That relationship was compared with the one with 2-furaldehyde (2-FAL) and carbon oxides. According to [103], the proposed aging markers show potential in aging state evaluation and residual life prediction of paper insulation in transformers.

3.4.3 Dielectric Properties

In transformer insulation, the dielectric properties of cellulosic materials are important to ensure the correct and safe operation of the machine. Different dielectric properties are measured to determine the viability to use a certain insulation system. Moreover, the evolution with aging of these properties must be analyzed. The main dielectric properties studied are BDV (kV) or electrical strength (E_d, kV/mm), PDIV and partial discharges pattern, dielectric losses (tan δ), dielectric permittivity (ε), resistivity (R), and polarization/depolarization current. The first two are usually measured under AC voltage whereas frequency domain and polarization/depolarization current methods are used to determine the other properties. Frequency domain spectroscopy (FDS) and polarization and depolarization currents (PDC) are considered two methods suitable for power transformers' diagnosis [104].

3.4.3.1 Breakdown Voltage

BDV represents the capacity of cellulose to withstand a high voltage and avoid the formation of an electric arc. It depends on different properties, such as the moisture content, the temperature [33, 55], the impregnating liquid [36, 59], and the cellulose material [54] or the cellulose type and thickness [37, 59]. Kraft paper and pressboard impregnated with esters have shown AC dielectric strengths equal to or higher than that of mineral oils [33, 59]. Despite the fact that several authors have studied the BDV of oil impregnated paper and pressboards, the results have revealed that the direct breakdown of the cellulose itself rarely occurs; it is the failure of the weakest component of oil/pressboard interface, which gradually damages the cellulose surface, known as creep discharge [33, 59], that finally causes the "breakdown" of pressboard.

Paper thickness has a great influence on the dielectric results. The greater the thickness, the higher the BDV. Nevertheless, the electrical strength, which relates the BDV to the distance between electrodes (or the insulation thickness) is reduced with the increase of paper thickness [59]. Moreover, this value is reduced with the paper and oil degradation.

As in the case of the BDV of esters, the evolution of BDV of impregnated paper with aging is not clear. Some authors have found that it decreases with aging [59] whereas other studies show an increase of BDV with aging [36], related to a reduction of moisture content of the paper when aging with esters.

According to the results obtained in [59], the BDV of paper impregnated with esters could be higher or smaller than that of mineral oil, depending on the origin of natural ester. The authors found that rice oil-impregnated paper had smaller BDV than mineral oil, whereas the BDV of corn oil and peanut oil-impregnated papers was higher than that of mineral oil. Nevertheless, it is evident that the aging process effect on BDV is more relevant in the case of mineral oil [55, 59]. The influence of the degradation of paper is also higher than that of the degradation of liquids, especially when using low-density papers [54].

3.4.3.2 Partial Discharges

Similar to BDV, PD behavior depends on different factors, such as the electrode configuration or the moisture content of the oils [105]. PDIV of esters is similar to that of mineral oil, but charge and repetition patterns are different [57].

3.4.3.3 Dielectric Loss Factor

As explained in Section 3.3.2, dielectric losses of esters are higher compared with mineral oil. As a result, dielectric losses of paper impregnated with the esters are also higher [37]. Evolution of tan δ with the aging process is not totally clear. Most authors have found that changes in the tan δ of solid insulation when aging with esters are more related to the decrease in moisture content than to the degradation of the paper [33, 40, 55]. In this way, considering that the cellulosic materials reduce their moisture content when aging with esters whereas it is increased when aging with mineral oil, as explained in Section 3.4.1.1, tan δ of the solid insulation is decreased with the aging process when using esters [37, 72].

Like BDV, dielectric losses depend on temperature. They increase with the increase of temperature [55].

To analyze the evolution of this property with the aging process or the changes of moisture, it is necessary to measure tan δ in the frequency domain. When moisture or degradation of paper changes, tan δ curves shift along the frequency axis and, if only the power frequency is studied, most of the information is lost and this can lead to an incorrect interpretation of what happens. The dissipation factor changes with the frequency and a maximum can be found at low frequencies. Dielectric losses are reduced with the increase of the frequency, reaching some minimums and maximums. This evolution is provoked by the changes of polarization processes that interfere with the dielectric behavior [37]. The effect of aging (or moisture) is more noticeable at low frequencies, where the influence of water is more relevant due to the conductivity effect.

3.4.3.4 Dielectric Permittivity

When paper is impregnated with esters, its dielectric permittivity (ε) is higher than when using traditional transformer oil. This is because the permittivity of natural ester is closer to that of insulating paper, and both are higher than that of mineral oil [33, 68].

$$\frac{\varepsilon_{\text{paper}}}{\varepsilon_{\text{natural ester}}} < \frac{\varepsilon_{\text{paper}}}{\varepsilon_{\text{mineral oil}}} \tag{3.11}$$

Far from being a disadvantage, the increase in dielectric permittivity allows a homogeneous distribution of electric field in the paper, which results in lower stress and a higher distribution of charge compared with low permittivity [68]. Also, the higher the permittivity, the lower the electrical stress on the edges of conductors, reducing the risk of creep discharges, which produces PD and damages the insulation, leading to the rupture of the cellulose [33].

In [68], the permittivity of mineral oil and ester-impregnated papers was studied. When analyzing the mineral oil and ester-based oil relative permittivity evolution with temperature, it was observed that ester's permittivity reduced more than mineral oil permittivity. However, solid insulation material permittivity increased with temperature rise. This result is in accordance with the tendency found by [106]. The difference between permittivity of ester and mineral oil-impregnated paper increased with the increase of temperature.

As occurred with the dielectric losses, this property depends on the moisture of both liquid and solid insulation and on the frequency. When studying the permittivity on frequency domain (AC), it becomes a complex permittivity. The real part represents the ability to polarize the sample and the imaginary part represents the dielectric losses. Most of the authors that have analyzed the permittivity have found that the real part remains almost constant with the increase of the frequency, whereas the imaginary part has a similar behavior to the dielectric losses (it decreases with the increase of the frequency). With the aging process, both parts are decreased due to the reduction of the moisture content of the paper when aging with esters [68, 107], whereas it is increased when aging with mineral oil.

3.4.3.5 Conductivity

The conductivity of oil-impregnated paper cannot be explained as a superposition of oil and paper conductivities. The charge carriers are transported along the pressboard fibers when it is not impregnated, so electrical conduction becomes dependent on the fiber network. When impregnating with oils, the cavities in the network are filled with oil, and also an additional charge carrier transport occurs in the oil volumes. Consequently, the conductivity of oil-impregnated paper is higher than the conductivity of the paper, as more charge carriers accumulate at the interfaces (polarization mechanisms) and can be transported through the structure (conduction processes). Moreover, conduction through the oil also affects the transport paths along the cellulose fibers, and the transport paths may change [54]. Considering the effect of the liquid in the insulation conductivity, it can be deduced that the ester-impregnated paper has a higher conductivity than the mineral oil-impregnated paper, since their conductivity is higher than that of mineral oil [20, 58]. During the aging process, the conductivity of paper impregnated with natural ester is decreased [53] due to the reduction of the moisture content, which reduced the presence of polar substances. Nevertheless, the conductivity of ester-impregnated paper remains higher than that of mineral oil-impregnated paper.

This property also depends on the temperature. The conductivity of the impregnated paper increases with the increase of temperature, which results in the increase of tan δ.

3.4.3.6 Polarization and Depolarization Currents

Polarization/depolarization currents are measured in time domain, applying a constant voltage for a certain time (e.g. 5000 seconds). In [42], the currents were studied in mineral oil and natural ester-impregnated paper. Their values did not change notably with the aging of the samples, but they depended on the time during which the voltage was applied. The currents of the papers impregnated with mineral oil are smaller than those of ester-impregnated paper.

3.5 Remaining Life of Transformer Insulation

The solid insulation aging is usually taken to be a good indicator of transformer loss of life. During its operation, a transformer suffers thermal, electrical, and mechanical stresses that deteriorate its insulation system. This insulation degradation can reduce the useful life of the transformer; for this reason, the assessment of the insulation of the transformer is critical in order to provide an estimation of its remaining life.

Different methods have been developed for estimating the remaining useful life of the transformer. The evaluation methods that have been more extensively applied include the IEEE/IEC models and the kinetic models.

The IEEE/IEC models estimate the loss-of-life of the transformer from the point of view of insulation degradation and are mainly based on temperature rise of transformers, their loading history, and ambient temperature. Otherwise, the kinetic models can evaluate the effect of temperature profile, the availability of water and oxygen using oil/insulation paper properties which show the chemical environment in the transformer. Generally, they estimate the loss-of-life using the DP or tensile strength values.

3.5.1 IEEE C57.91

Although moisture, oxygen, and temperature have a critical effect on the degradation of transformer insulation, IEEE guide [108] considers the temperature of the insulation as the controlling variable, assuming the impact of moisture and oxygen on insulation deterioration is considerably reduced through modern oil preservation systems. It has been established that the highest degradation of the insulation system is located where the highest or hottest-spot temperature is. Figure 3.6a shows the effect of the winding hottest-spot temperature on the per unit transformer insulation life as described by IEEE guide [108] taking into account the temperature as the principal variable affecting insulation life. Figure 3.6a gathers how an increase of temperature above reference temperature (110 °C) exponentially accelerates the aging rate, whereas an improvement of transformer insulation life is obtained if temperature is reduced below 110 °C. IEEE method assumes the insulation paper lasts for a certain number of hours (150 000 hours at 110 °C) specifying as an end-of-life criterion a DP of 200.

This relation of insulation aging to time and hottest-spot temperature can be defined as [108]:

$$\text{Per unit transformer insulation life} = A * e^{\left[\frac{B}{\Theta_H + 273}\right]} \tag{3.12}$$

where A, B are constants and Θ_H is the winding hottest-spot temperature, °C.

According to IEEE guide [108], it seems that a value of B of 15 000 and a value of A of $9.80 * 10^{-18}$ would be appropriate for cellulose aging in mineral oil. These values were based on many works that had measured the aging rates of different forms of cellulose, under controlled conditions.

The curve of Figure 3.6a can be used in two ways. Firstly, it could be used as the basis to estimate an aging acceleration factor (F_{AA}) for a given load and temperature or for a changing load and temperature over a 24-hour period:

$$F_{AA} = e^{\left[\frac{15\,000}{383} - \frac{15\,000}{\Theta_H + 273}\right]} \tag{3.13}$$

This F_{AA} is higher than 1 when hottest-spot temperature is superior to the reference temperature (110 °C) and lower than 1 for temperatures below 110 °C. This acceleratory factor could be used to determine an equivalent aging factor (F_{EQA}) for a period Δt_n over a total time period. The equivalent loss of life at a referential temperature that would be experienced during a defined time period for a given temperature cycle could be estimated as follows [108]:

$$F_{EQA} = \frac{\sum_{n=1}^{N} F_{AA,n} * \Delta t_n}{\sum_{n=1}^{N} \Delta t_n} \tag{3.14}$$

where F_{EQA} is the equivalent aging factor for the total time period; n is index of the time interval, t; N is total number of time intervals; $F_{AA,n}$ is the aging acceleration factor for the temperature which exists during the time interval Δt_n; and Δt_n is the time interval.

Secondly, the per unit transformer insulation life curve (Figure 3.6a) can also be used to calculate the percent loss of total life. This application requires the definition of the normal insulation life at the reference temperature in a given time period. The IEEE guide recommends that users select

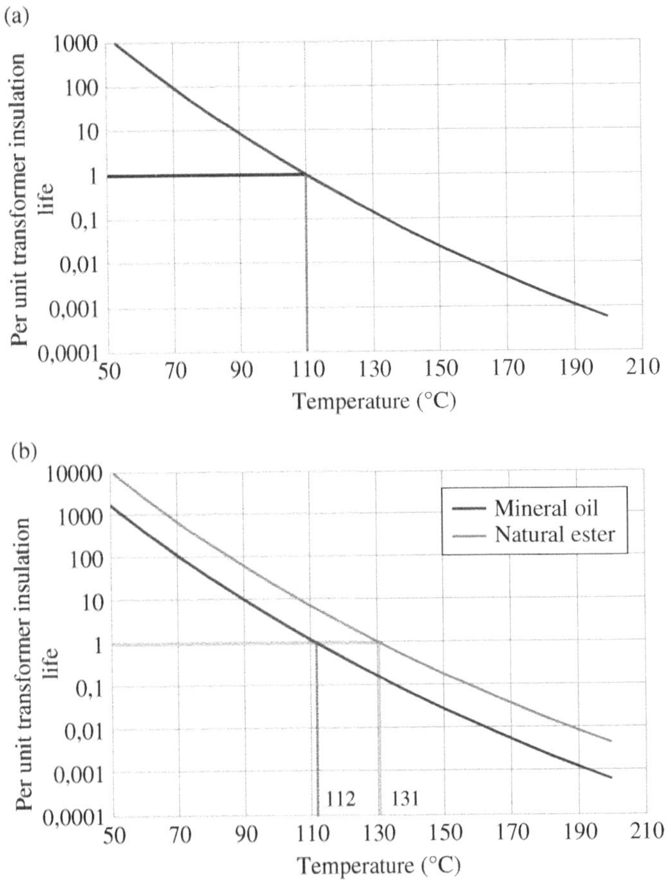

Figure 3.6 Transformer insulation life versus hottest-spot temperature, described by (a) IEEE guide. *Source:* IEEE [108]. (b) *Source:* Modified from Rapp and Luksich [109].

their own lifetime estimation, with 180 000 hours being used as a normal lifetime [108]. Then, the equivalent aging can be calculated (Eq. (3.15)) and it can be used to estimate the life lost in the total time period by multiplying it by the time:

$$\%\ \text{Loss of life} = \frac{F_{\text{EQA}} * t * 100}{\text{Normal insulation life}} \tag{3.15}$$

According to Biçen et al. [110], there are different experimental works about thermal-accelerated aging tests in which insulation papers are aged in natural esters taking 5–8 times longer to reach end of life in comparison to those immersed in mineral oil at the same temperature. Additionally, the hot-spot temperatures that are reached along the length of the windings of the transformers are different for natural ester and mineral oil-immersed transformers [111]. Consequently, these authors [110] proposed that aging acceleration factor should be revised due to the slowdown of the degradation process suffered by insulation papers in natural esters. Eq. (**3.13**) is rewritten and the aging accelerated factor can be calculated as:

$$F_{\text{AA}} = R * \left[e^{\left[\frac{15\,000}{383} - \frac{15\,000}{\Theta_H + 273}\right]} \right] \tag{3.16}$$

where R is a ratio of slow insulation aging for natural ester which can be selected between (1/5) and (1/8) depending on the properties of natural ester.

Furthermore, Biçen et al. [110] modeled two different 250 MVA power transformers whose technical characteristic features were the same for natural ester and mineral oil. They calculated the hot-spot temperature used in loss of life calculation, through different procedures for natural ester and mineral oil transformers. They estimated the hot-spot temperatures and aging parameters (insulation loss-life and remaining life) taking into account the transient characteristics of the load, as well as the ambient temperature. They obtained that the life loss of insulation for mineral oil-immersed transformer was approximately 2.4 times higher than the one of the natural ester-immersed power transformer. Moreover, they also found that the natural ester-immersed transformer was very advantageous in the nominal operating conditions; however, this advantage decreased with increasing load values.

Rapp and Luksich [109] also tried to adapt the relationship of insulation aging to time and hottest-spot temperature to other insulation materials. They analyzed the results obtained in different studies that had used sealed tube aging systems to compare paper degradation rates of Kraft paper, cotton/Kraft, and thermally upgraded Kraft paper (TUK) in mineral oil and natural ester fluid insulation. In order to compare data obtained at different temperatures, these authors [109] converted the time for each temperature to a unit life basis using Eq. (3.12) and the end-of-life criterion of DP = 200. This allowed them to estimate the A constant in Eq. (3.12) for natural ester (Table 3.5), giving a unit life temperature of approximately 131 °C (Figure 3.6b). These results support the ratio of slow insulation aging for natural ester proposed by Biçen et al. [110], because an increase of 19 °C in unit life temperature in natural ester in comparison with mineral oil [109] represents an insulation aging for natural ester of approximately (1/7) of the one obtained for mineral oil.

Similar results to those obtained in [109] were found by Zhang et al. [112] for Nomex® T910 paper aged in natural ester (FR3) and Kraft paper aged in mineral oil. They estimated a unit life temperature of 112 °C for Kraft paper in mineral oil and 133 °C for T910 paper in natural ester.

3.5.2 IEC 60076-7

As in the IEEE guide [108], the aging model included in the IEC 60076-7 standard [113] is based only on the insulation temperature as the controlling parameter of the insulation paper degradation. Similarly to IEEE guide [108], in the IEC standard [113], the rate of aging is referred to the winding hot-spot temperature since the temperature distribution is not uniform. It is the part that is operating at the highest temperature which will normally undergo the greatest deterioration.

In the IEC standard [113], the relative aging rate (V) is defined in two different ways depending on the type of paper that is considered:

$$V_{\text{non-thermally upgraded paper}} = 2^{\frac{\Theta_H - 98}{6}} \tag{3.17}$$

Table 3.5 Constants to determine per unit transformer insulation life.

Reference	Insulation fluid	A	B
[108]	Mineral oil	$9.80 * 10^{-18}$	15 000
[109]	Natural ester	$7.82 * 10^{-17}$	15 000

$$V_{\text{thermally upgraded paper}} = e^{\left[\frac{15000}{383} - \frac{15000}{\Theta_H + 273}\right]} \tag{3.18}$$

where Θ_H is the winding hottest-spot temperature, °C.

The aging rate (V) for non-thermally upgraded paper is based on the fact that the hot-spot temperature at which there is a normal aging rate equal to 98 °C at nominal load and ambient temperature of 20 °C. It is also based on the fact that a 6 °C increase in a transformer's hot-spot temperature (Θ_H) can double its insulation aging rate. Additionally, it can be observed that for thermally upgraded paper, V is equivalent to the F_{AA} defined in the IEEE guide [108].

The loss of life over a time period is determined according to IEC [113] as follows:

$$L = \int_{t_1}^{t_2} V * dt \quad \text{or} \quad L = \sum_{n=1}^{N} V_n * t_n \tag{3.19}$$

where V_n is the relative aging rate during Interval n, obtained according to Eq. (**3.17**) or Eq. (3.18); t_n is the nth time Interval; n is the number of each time Interval, and N is the total number of invervals during the period considered.

Unlike the IEEE guide [108] in which a normal lifetime of the power transformers equal to 180 000 hours (20.55 years) has been defined, in the IEC standard [113], a total life has not been established, although it is mentioned that the transformer life can be 30 years (262 800 hours) depending on the aging rate determined by Θ_H.

Biçen et al. [110] presented that the relative aging rate for non-thermally upgraded paper of IEC [113] should be revised similarly to F_{AA} for natural esters due to the differences in the deterioration process in comparison with mineral oil:

$$V_{\text{non-thermally upgraded paper}} = R * \left(2^{\frac{\Theta_H - 98}{6}}\right) \tag{3.20}$$

where R is a ratio of slow insulation aging for natural ester which can be selected between (1/5) and (1/8) depending on the properties of natural ester.

Once the methods for determination of loss of life described in IEEE and IEC have been explained, it can be concluded that there are terminological differences when certain quantities are defined as summarized in Table 3.6.

Table 3.6 Loss of life data according to [108] and [113].

IEEE		IEC	
Θ_H (°C)	**110**	**Θ_H (°C)**	**98**
Normal transformer life (years)	20.55	Normal transformer life (years)	Mentioned 30
Aging acceleration factor	F_{AA}	Relative aging rate	V
Equivalent aging factor	F_{EQA}		
Equivalent aging	%Loss of life	Loss of life	L

Sources: Based on IEEE [108] and IEC [113].

3.5.3 Kinetic Approach to Modeling

The purpose of the kinetic approach with power transformers is to understand the degradation of the cellulosic insulation and to roughly predict the lifetime of the electrical machine based on its present state, the hot-spot temperature, and the operational conditions [35]. There are two possible schemes for aging assessment [113]: prediction of reduced mechanical strength using equations for strength development or predictions of the tensile strength based on algorithms for aging kinetics as a function of the DP-value of the paper insulation. However, some authors claim that the choices of DP or tensile-strength value as the end-of-life criterion, which significantly affect the predicted life expectancy, can sometimes be arbitrary [23]. According to [91], insulation samples have been removed from transformers, which were still working perfectly satisfactorily, with DPs well below 200. Thus, an improvement in the understanding of the underlying mechanisms of aging and loss of strength is needed to optimize the management and replacement of transformers.

3.5.3.1 Polymerization Degree

As previously explained, one of the most used means of determination of paper deterioration and the remaining life is the DP of cellulose insulation.

One of the first formulas established for the kinetics of cellulose degradation was a pseudo zero-order kinetic Eq. (3.21) [17] which is based on Ekenstam's work [114] and the Arrhenius equation:

$$\ln\left(1-\frac{1}{\mathrm{DP}_0}\right)-\ln\left(1-\frac{1}{\mathrm{DP}}\right)=k*t \tag{3.21}$$

If DP_0 and DP are large, then the Eq. (3.21) can be expressed as:

$$\frac{1}{\mathrm{DP}}-\frac{1}{\mathrm{DP}_0}=k*t \tag{3.22}$$

where DP_0 is the number average degree of polymerization at the time of 0; DP is the number average degree of polymerization at the time of t; k is the constant rate of reaction; and t is the aging time (h).

The Arrhenius equation gives the dependence of the rate constant of a chemical reaction on the temperature:

$$k=A*\mathrm{e}^{-E_a/R*T} \tag{3.23}$$

where A is pre-exponential or the frequency factor (h^{-1}); E_a is the activation energy (J/mol); R is the ideal gas constant (8.314 J/mol/K); T is the absolute temperature (K).

Therefore, the expression known as the Ekenstam equation results in:

$$\frac{1}{\mathrm{DP}}-\frac{1}{\mathrm{DP}_0}=A*\mathrm{e}^{-E_a/R*T}*t \tag{3.24}$$

Subsequently, this formula was modified by Emsley et al. [91] because they observed that as long as the DP value of insulation paper decreases to a certain value, its behavior gradually deviates from the pseudo zero-order kinetic. In that case, the reaction rate is no longer a constant, which is also known as the leveling-off degree of polymerization (LODP). Consequently, these authors [91] proposed the DP would be better estimated as:

$$\frac{1}{\mathrm{DP}}-\frac{1}{\mathrm{DP}_0}=\frac{k_{1_0}}{k_2}*\left(1-\mathrm{e}^{-k_2*t}\right) \tag{3.25}$$

where k_{1_0} and k_2 are temperature-dependent constants and are related by:

$$k = k_{1_0} * e^{-k_2 * t} \tag{3.26}$$

Being k (Eqs. (3.21)–(3.23)) the rate at which the DP decreases and k_2 the rate at which k changes.

In 2005, Zervos and Moropoulou [115] defined a kinetic model to describe the depolymerization of pure cotton cellulose in sealed vessels:

$$\left(\frac{1}{DP} - \frac{1}{DP_0}\right) * 100 = a * \left(2^{k*t} - 1\right) \tag{3.27}$$

where k and a are constants.

The following year, Calvini and Gorassini [116] considered that due to experimental uncertainty about the actual values of DP, it was preferable to express the kinetic models in terms of scissions per chain $S = \left(\frac{DP_0}{DP} - 1\right)$ rather than scissions per monomer $S = \left(\frac{1}{DP} - \frac{1}{DP_0}\right)$:

$$\left(\frac{DP_0}{DP} - 1\right) = \left(\frac{DP_0}{LODP}\right) * \left(1 - e^{-kt}\right) \tag{3.28}$$

Ding and Wang [117] defined a new kinetic model in which DP = $f(t)$ unlike the models defined by Ekenstam, Emsley, Zervos, and Calvini where (1/DP) = $f(t)$:

$$1 - \frac{DP}{DP_0} = w_{DP}^{*} * \left(1 - e^{-kt}\right) \tag{3.29}$$

where w_{DP}^{*} is the capacity of the DP degradation resorvoir. This value can be obtained by introducing the constraint condition $w_{DP}(t = t_f) = 1$, being t_f the time to failure.

Additionally, some authors have used Arrhenius's law to calculate lifetime of power transformers in hours instead of per unit transformer life as in IEEE guide [108]:

$$\ln L(T) = A + \frac{B}{T} \quad \text{or} \quad L(T) = A * e^{B/T} \tag{3.30}$$

where ln$L(T)$ is the lifetime of power transformer; A and B are constants.

The lifetime estimation of a power transformer using the Arrhenius approximation requires the definition of an end-of-life criterion, for instance, when DP value reaches 200.

All the previously kinetic models described were defined to be applied to insulation papers aged in mineral oil, nevertheless, in recent decades different authors have used them to quantify the deterioration suffered when insulation fluids are esters (Table 3.7).

Tables 3.8 and 3.9 include the results of thermal aging obtained by some authors for esters and mineral oil.

Although it is known from operational experience that insulation paper deterioration is caused by many factors that influence the cumulative effect of temperature over time, the kinetic models based on esters evaluated here have only considered the effect of temperature. The works analyzed in this section provide a constant value of the pre-exponential factor instead of defining an expression which includes the impact of variables such as moisture or oxygen content. Consequently, more research is needed to define kinetic models that incorporate the synergistic effect of temperature, moisture, and oxygen on degradation rate of insulation papers.

Table 3.7 Estimated parameters of Eq. (3.24).

Reference	Insulation paper	Insulation fluid	k	A (h^{-1})	E_a (kJ/mol)
[118]	Thermally upgraded paper	Mineral oil		$2.41 * 10^8$	93.44
	Thermally upgraded paper	Natural ester		$3.99 * 10^6$	81.08
[35]	Thermally non-upgraded Kraft paper	Gemini X		$4.96 * 10^6$	111.9
	Thermally non-upgraded Kraft paper	MIDEL 7131		$1.49 * 10^3$	102.0
[119]	Cellulose paper + pressboard	Mineral oil	$3.95 * 10^{-5}$ (130 °C)		
	Cellulose paper + pressboard	Mixed oil (with natural esters)	$2.07 * 10^{-5}$ (130 °C)		

Similar to the DP results, the decrease of tensile strength is also influenced by different variables. For this reason, there are works in which kinetic models based on tensile strength have been applied to model paper degradation as explained in the following subsection.

3.5.3.2 Tensile Strength

The critical parameter that conditions the life expectancy of power transformers is believed to be the mechanical strength of the paper insulation, which reduces as a result of the aging process. If the paper cannot withstand the shear stresses in the winding during an electric fault, such as a short circuit or large inrush currents, there will be displacements in the conductors [1]. If breakdown does not immediately occur, a change in the geometry will increase future stresses and the probability of development of defects.

A tensile strength reduction due to aging of 50% of its original value is established in [6] as criterion for the end of life. However, according to [23], mechanical testing of aged insulation paper whose tensile strength is 20% of the initial value reports that it still retains some mechanical integrity. According to the IEEE loading guide [108], a 50%-reduction criterion is probably far too conservative, as many transformers operate well with far lower strength of the paper. The life expectancy will change radically if we consider 50% of the initial tensile strength as the end-of-life criterion or a different value (such as 30 or 20%). Furthermore, the service conditions in real power transformers may differ from the ones considered in laboratory aging tests, affecting degradation processes. The key to defining a satisfactory end-of-life point may lie in defining a probability function of failure, based on experimental measurements [23].

According to [14], the tensile strength of paper at any moment during the aging process, σ_t, can be expressed as a function of the initial tensile strength, σ_0, the number of polymer chains, N_C, by Eq. (3.31), where A is a constant independent of the degradation temperature over the range of temperatures for which $\ln \sigma$ versus N_C is linear.

$$\sigma_t = \sigma_0 e^{AN_C} \tag{3.31}$$

The degradation of the tensile strength of the paper insulation can be described by a first-order kinetic model [14], see Eq. (3.32), and the chain scission reaction can be represented by a zero-order kinetic equation, see Eq. (3.33). Combining and integrating Eqs. (3.32) and (3.33), (3.34) is obtained. That can also be expressed as in Eq. (3.35), which gives a linear plot whose slope is

Table 3.8 Estimated parameters of Eq. (3.25).

Reference	Insulation paper	Insulation fluid	k_{1_0}			k_2		
				A (h^{-1})	E_a (kJ/mol)		A (h^{-1})	E_a (kJ/mol)
[120]	Kraft paper	Mineral oil	$8.12*10^{-6}$ (90 °C)			$4.31*10^{-3}$ (90 °C)		
			$29.70*10^{-6}$ (110 °C)			$1.87*10^{-3}$ (110 °C)		
			$56.50*10^{-6}$ (130 °C)			$0.25*10^{-3}$ (130 °C)		
	Thermally upgraded paper	Mineral oil	$3.76*10^{-6}$ (90 °C)			$1.56*10^{-3}$ (90 °C)		
			$2.06*10^{-6}$ (110 °C)			$3.22*10^{-3}$ (110 °C)		
			$10.90*10^{-6}$ (130 °C)			$4.70*10^{-3}$ (130 °C)		
	Kraft paper	Biotemp	$3.51*10^{-6}$ (90 °C)			$4.04*10^{-3}$ (90 °C)		
			$29.70*10^{-6}$ (110 °C)			$14.20*10^{-3}$ (110 °C)		
			$186*10^{-6}$ (130 °C)			$44.1*10^{-3}$ (130 °C)		
	Thermally upgraded paper	Biotemp	$1.68*10^{-6}$ (90 °C)			$5.03*10^{-3}$ (90 °C)		
			$7.53*10^{-6}$ (110 °C)			$14.10*10^{-3}$ (110 °C)		
			$20.60*10^{-6}$ (130 °C)			$21.60*10^{-3}$ (130 °C)		
[38]	Conventional paper	Mineral oil	$5.40*10^{-6} \pm 7.00*10^{-7}$			$(5.35 \pm 1.87)*10^{-4}$		86.94 ± 18.50
		Biotemp	$1.20*10^{-6} \pm 1.75*10^{-7}$			$2.77*10^{-4} \pm 6.67*10^{-7}$		151.24 ± 8.28
[56]	Kraft paper	Mineral oil		$1.23*10^{11}$	122.51		$1.57*10^{13}$	120.64
		Natural ester 1		$3.73*10^{10}$	119.08		$5.13*10^{10}$	99.23
		Natural ester 2		$2.04*10^{8}$	101.31		$1.79*10^{9}$	87.33
[121]	Kraft paper	Mineral oil	$18.16*10^{-6}$ (110 °C)			$2.21*10^{-3}$ (110 °C)		
		Mixed oil (mineral oil + natural ester)	$8.24*10^{-6}$ (110 °C)			$1.79*10^{-3}$ (110 °C)		
[51]	Pressboard	Mineral oil	$4.49*10^{-7}$ (120 °C)			$1.11*10^{-4}$ (120 °C)		
	Pressboard	Natural ester	$7.40*10^{-7}$ (120 °C)			$5.35*10^{-4}$ (120 °C)		

Table 3.9 Estimated parameters of Eq. (3.30), applying different end-of-life criteria.

Reference	Insulation paper	Insulation fluid	A	B
DP = 450				
[45]	High-density pressboard	Mineral oil	$2.70*10^{-17}$	18 659
	Thermally upgraded paper	Mineral oil	$1.66*10^{-14}$	16 273
	High-density pressboard	Palm fatty acid ester	$6.22*10^{-18}$	19 472
	Thermally upgraded paper	Palm fatty acid ester	$4.25*10^{-18}$	20 132
DP = 200				
[122]	Thermally upgraded Kraft paper (nondrying sample) at 120 °C	Natural ester	$4.35*10^{-12}$	13 110
	Thermally upgraded Kraft paper (drying sample) at 120 °C	Natural ester	$7.32*10^{-12}$	13 110

proportional to the zero-order rate constant of the chain scission. Some plots of $\ln \sigma_t$ versus time are shown in [14].

$$-\frac{d\sigma_t}{dN_C} = A\sigma_t \tag{3.32}$$

$$\frac{dN_C}{dt} = k \tag{3.33}$$

$$\sigma_t = \sigma_0 e^{-Akt} \tag{3.34}$$

$$\ln \sigma_t = \ln \sigma_0 - Akt \tag{3.35}$$

However, some laboratory tests over aged paper revealed that the tensile strength can increase during the first stages of aging, while the DP reduces [91]. It is not possible to model that phenomenon of tensile strength is a function of bond scission, as suggested by [14], and that would overestimate the change of tensile strength in the later stages. In [91], that experimental observation was interpreted as a process in which part of the loss of strength due to bond scission is compensated by relaxation of the broken fragments into a more stable configuration, which would allow new hydrogen bonds to develop within and between fibers. The authors proposed a model in which the tensile strength, σ, is related to the DP according to Eqs. (3.36) and (3.37).

$$\frac{d\sigma}{dt} = -k_1 k_3 + k_4 \mathrm{DP} \tag{3.36}$$

$$\sigma_t - \sigma_0 = K_1 e^{-k_2 t} + K_2\left[\ln\left(e^{k_2 t} - K_3\right)\right] - K_4 \tag{3.37}$$

where K_1, K_2, K_3, K_4 are constants which incorporate k_1, k_2, k_3, k_4 and the initial DP value, DP_0, and σ_t and σ_0 are the final and initial tensile strength.

Table 3.10 Correlation of DP and retained tensile strength (%).

Retained σ (%) = $C*DP+D$ [45] where: σ is tensile strength, C and D are constants			
Insulation paper	**Insulation fluid**	**C**	**D**
High-density pressboard	Mineral oil	0.1406	0.9774
High-density pressboard	Palm fatty acid ester	0.1279	18.566
Retained σ (%) = $C*\ln(DP)+D$ [45]			
Thermally upgraded paper	Mineral oil	104.23	−585.43
Thermally upgraded paper	Palm fatty acid ester	83.337	−447.95
$\sigma = C*DP+D$ [123]			
Kraft paper	Natural ester	$2.70*10^{-2}$ (7.59% moisture, 85 °C)	−17.04
Kraft paper	Natural ester	$3.56*10^{-2}$ (2.00% moisture, 85 °C)	−39.50
Kraft paper	Natural ester	$1.97*10^{-2}$ (7.59% moisture, 120 °C)	9.19
Kraft paper	Natural ester	$2.48*10^{-3}$ (2.00% moisture, 120 °C)	−5.90

According to [91], the previous equations can be rearranged and their integration gives:

$$\frac{d\sigma}{dDP} = \frac{K_a}{DP^2} - \frac{K_b}{DP} \tag{3.38}$$

where constants K_a and K_b depend on the aging mechanisms, which are affected by different variables (insulating liquid, aging temperature, water content, etc.). In [23], values of K_a and K_b are provided for the zero-span and wide-span tensile test, aged in air and in oil.

Besides tensile strength, some kinetic models can be developed based on other mechanical properties. For instance, in [93], a kinetic damage model for lifetime prediction was proposed, and the mechanical property that better represented the aging process was the energy consumed per unit volume of the failure zone.

It was found by different authors that in insulation papers aged in mineral oil, there is a profound correlation between chain scission in cellulose with the reduction of mechanical strength. Similar behavior has been obtained for solids aged in esters which has allowed to propose different relations as can be seen in Table 3.10.

3.6 Conclusions

In this chapter, the types of cellulosic materials used in power transformers have been exposed. The most commonly used materials are paper and board, due to their good thermal, mechanical, and dielectric properties. Their high availability and low price also make them the preferred materials for transformer manufacturers. The degradation processes that affect the cellulose include oxidation, hydrolysis, and pyrolysis, which are different when using esters and traditional oil. The aging of the cellulose may cause changes in its dielectric and mechanical properties, which can lead to some failures that can even produce the end of life of the transformer. In fact, the most common failure locations are the windings, and mechanical and dielectric types of failure are the most typical failure modes.

When using new liquids as cooling and insulating materials in power transformers, as occurs with esters, it is necessary to study the effect of the liquid on the properties of cellulose and on the degradation process. Accelerated thermal aging tests are carried out in the laboratories to prove the suitability of the esters. Despite the available standards concerning thermal aging procedures, several differences can be found in the temperature, atmosphere, initial moisture content, and materials used by each author. This affects the obtained results, since the degradation processes and the aging rate change with these factors, making it difficult to compare results from different studies.

Despite the differences between the thermal aging procedures, some general conclusions about the evolution of the properties of the esters can be drawn. The esters have a higher viscosity than traditional mineral oils, except for palm oil. The viscosity of esters is slightly increased with aging and this increase is larger when the aging process is carried out considering the cellulose. One of the advantages of esters is their high flash and fire points, always higher than those of mineral oil. They are reduced with the degradation of the liquid but remain higher than in mineral oil. The acidity of esters is also higher than that of mineral oil but far from being a disadvantage, the acids produced by the esters protect the cellulose. The acidity increases with the aging since acids are products of the degradation processes. At the beginning of the aging, the moisture content of esters is increased since they absorb water from cellulose. After that, esters consume water through a hydrolysis process. This affects the dielectric response of these liquids, and the esters increase their BDV with aging, contrary to traditional mineral oil. In the first steps of aging, the dielectric losses of esters and their conductivity increases due to the absorption of water from cellulose. Esters have a higher permittivity than traditional oils and it is close to that of the cellulose.

When impregnating with esters, the cellulose properties are affected by the liquid. During the aging, cellulose suffers a hydrolytic process in which it reacts with the water and the paper is weathered. Esters absorb water from cellulose during the aging, which reduces the hydrolysis of cellulose, reducing its degradation. Moreover, esters produce high molecular acids during the aging through the hydrolysis process and these acids react with cellulose through transesterification, protecting the cellulose from water. The transesterification is confirmed through FTIR analysis. The acids produced by mineral oil are harmful to cellulose since they act as catalysts of the hydrolysis of cellulose. Due to this, cellulose reduces its moisture content when aging with esters, which reduces the aging rate and the degradation of cellulose. This is evident when analyzing the depolymerization degree of cellulose when aged with esters. The degradation is slower in comparison with that of mineral oil and it takes more time to reach the same DP level when aging with esters. Different degradation products are produced during the aging of cellulose. Furanic compounds, methanol and gases concentrations are used as indicator of the degradation of paper. Esters produce less furanic compounds but more methanol, CO, and CO_2 than mineral oil. Mechanical properties are reduced during the aging of cellulose. Tensile strength is essential to ensure the safe operation of the transformer. When aging with esters, cellulose maintains its tensile strength during more time than when aging with mineral oil. Tensile strength is highly related to DP, and one can be estimated from data about the other. Also, there is a relationship between DP, tensile strength, and furanic and methanol concentrations. Dielectric properties of cellulose are also affected by the impregnating liquid and the aging. BDV, dielectric losses, permittivity, and conductivity of ester-impregnated paper are higher than those of mineral oil-impregnated cellulose. These properties depend highly on the moisture, so the evolution of the dielectric properties of the cellulosic materials is better when aging with the esters than with the mineral oil, since esters reduce the moisture content of the paper.

The degradation of the cellulose can be used to estimate the loss of life of the transformer. There are two standards that indicate the conditions, equations, and parameters to be used when estimating the life remaining of the transformer insulation. These guides and constants have been adapted by some authors to be used with materials different from mineral oil, as natural esters. The remaining life can be estimated using the depolymerization degree reduction of solid insulation or its tensile strength. The parameters used in the equations change depending on the solid insulation (e.g. kraft, pressboard, thermally upgraded paper, DDP), the dielectric liquid (e.g. mineral oil, natural ester, synthetic ester), or the aging conditions (e.g. temperature, oxygen, moisture). The results show the lower aging rate of ester compared with mineral oil, but more research is needed to develop valid models under all the conditions and for different insulating materials.

References

1 Lundgaard, L.E., Lundgaard, L.E., Allan, D. et al. (2007). Ageing of cellulose in mineral-oil insulated transformers. *CIGRE*.

2 IEC (2014). IEC 60076-14:2014. Power transformers - Part 14: Liquid-immersed power transformers using high-temperature insulation materials.

3 Oria, C., Ortiz, A., and Ferreño, D. (2020). *An Experimental, Analytical and Numerical Study of the Mechanical Failure of the Cellulosic Insulation of Continuously Transposed Conductors in Power Transformers Under the Influence of Short Circuits and Thermal Ageing*. University of Cantabria.

4 IEC (2010). IEC 60763 Specification for laminated pressboard.

5 IEC (2006). IEC 61061 Non-impregnated densified laminated wood for electrical purposes.

6 IEC (2013). IEC 60216-1:2013 - Electrical insulating materials - Thermal endurance properties - Part 1: Ageing procedures and evaluation of test results.

7 IEC (2005). IEC 60216-2:2005 Electrical insulating materials - Thermal endurance properties - Part 2: Determination of thermal endurance properties of electrical insulating materials - Choice of test criteria.

8 Oria, C., Carrascal, I., Ortiz, A. et al. (2020). Experimental and numerical analysis of cellulosic insulation failures of continuously transposed conductors under short circuits and thermal ageing in power transformers. *IEEE Trans. Dielectr. Electr. Insul.* 27 (1): 325–333.

9 Emsley, A.M., Xiao, X., Heywood, R.J., and Aii, M. (2000). Degradation of cellulosic insulation in power transformers. Part 3: effects of oxygen and water on ageing in oil. *IEE Proceedings – Science Measurement and Technology*, vol. 147, No. 3, May 2000.

10 Rapp, K.J., McShane, C.P., and Luksich, J. (2005). Interaction mechanisms of natural ester dielectric fluid and Kraft paper, *2005 IEEE International Conference on Dielectric Liquids ICDL 2005*, Coimbra, Portugal (26 June–1 July 2005), pp. 393–396.

11 Lampe, W. and Spicar, E. (1976). The oxygen-free transformer, Reduced ageing by continuous degassing. *CIGRE* 12 (05).

12 Fabre, J. and Pichon, A. (1960). Deteriorating processes and products of paper in oil. Application to transformers. *CIGRE* 137: 1–18.

13 Lundgaard, L., Hansen, W., Linhjell, D., and Painter, T. (2004). Aging of oil-impregnated paper in power transformers. *IEEE Trans. Power Deliv.* 19 (1): 230.

14 Hill, D.J.T., Le, T.T., Darvenizab, M., and Saha, T. (1995). A study of degradation of cellulosic insulation materials in a power transformer. Part 2: tensile strength of cellulose insulation paper. *Polym. Degrad. Stab.* 49: 429–435.

15 Lelekakis, N., Martin, D., and Wijaya, J. (2012). Ageing rate of paper insulation used in power transformers part 2: oil/paper system with medium and high oxygen concentration. *IEEE Trans. Dielectr. Electr. Insul.* 19 (6): 2009–2018.

16 Emsley, A.M. and Stevens, G.C. (1994). Review of chemical indicators of degradation of cellulosic electrical paper insulation in oil-filled transformers. *IEE Proc. Sci. Meas. Technol.* 141 (5): 324–334.

17 Emsley, A.M. and Stevens, G.C. (1994). Kinetics and mechanism of the low temperature degradation of cellulose. *Cellulose* 1: 26–56.

18 Emsley, A.M. (1994). The kinetics and mechanism of degradation of cellulosic insulation in power transformers. *Polym. Degrad. Stab.* 44: 343–349.

19 Hill, D.J.T., Le, T.T., Darveniza, M., and Saha, T. (1995). A study of degradation of cellulosic insulation materials in a power transformer, part 1. Molecular weight study of cellulose insulation paper. *Polym. Degrad. Stab.* 48: 79–87.

20 Gasser, H.P., Krause, C., Lashbrook, M., and Martin, R. (2011). Aging of pressboard in different insulating liquids. *Proceedings of IEEE International Conference on Dielectric Liquids*. Trondheim, Norway (26–30 June 2011).

21 IEC (1993). IEC 61198:1993. Mineral insulating oils - Methods for the determination of 2-furfural and related compounds.

22 Martins, M.A.G. (2010). Vegetable oils, an alternative to mineral oil for power transformers—experimental study of paper aging in vegetable oil versus mineral oil. *DEIS* 26 (6): 7–13.

23 Emsley, A.M., Heywood, R.J., Ali, M., and Xiao, X. (2000). Degradation of cellulosic insulation in power transformers. Part 4: effects of ageing on the tensile strength of paper. *IEE Proc. - Sci. Meas. Technol.* 147 (6): 4–9.

24 Krause, C. (2012). Power transformer insulation - history, technology and design. *IEEE Trans. Dielectr. Electr. Insul.* 19 (6): 1941–1947.

25 Lesaint, O., Saker, A., Gournay, P. et al. (1998). Streamer propagation and breakdown under ac voltage in very large oil gaps. *IEEE Trans. Dielectr. Electr. Insul.* 5 (3): 351–359.

26 Lundgaard, L., Herstad, K., Anker, M.U., and Sletbak, J. (1986). Flashover along solid surfaces parallel to the electric field in liquid insulation at 50 Hz. *International Conference on Large High Voltage Electric System, CIGRE*, Paris (27th August–4th September of 1986), pp. 1–7

27 Tenbohlen, S., Jagers, J., Bastos, G. et al. (2015). Transformer reliability survey. *CIGRÉ*.

28 IEEE (2012). IEEE Standard C57.100 Test procedure for thermal evaluation of insulation systems for liquid-immersed distribution and power transformers, vol. 2011.

29 IEC (2011). IEC TS 62332-1:2011 - Electrical insulation systems (EIS) - Thermal evaluation of combined liquid and solid components - Part 1: General requirements.

30 NEMA MW-1000, Magnet Wire (2008). Transformer Oil Resistance and Hydrolytic Stability, Long Term Exposure Test.

31 IEC, IEC 60216-3:2006 Electrical insulating materials - Thermal endurance properties - Part 3: Instructions for calculating thermal endurance characteristics. 2006.

32 IEC (2006). IEC 60216-4-1:2006 Electrical insulating materials - Thermal endurance properties - Part 4-1: Ageing ovens - Single-chamber ovens.

33 Liao, R., Hao, J., Chen, G. et al. (2011). A comparative study of physicochemical, dielectric and thermal properties of pressboard insulation impregnated with natural ester and mineral oil. *IEEE Trans. Dielectr. Electr. Insul.* 18 (5): 1626–1637.

34 Martins, M.A.G. (2005). Experimental study of paper aging in vegetable oil versus mineral oil. *IEEE Electr. Insul. Mag.* 26 (6): 7–13.

35 Matharage, S.Y., Liu, Q., Wang, Z.D. et al. (2018). Aging assessment of synthetic ester impregnated thermally non-upgraded kraft paper through chemical markers in oil. *IEEE Trans. Dielectr. Electr. Insul.* 25 (2): 507–515.

36 Ariffin, M.M., Ishak, M.T., Hamid, M.H.A. et al. (2017). Ageing effect of vegetable oils impregnated paper in transformer application. *International Conference on High Voltage Engineering and Power Systems, ICHVEPS 2017*, vol. 2017, Denpasar, Indonesia (2–5 October 2017), pp. 183–187.

37 Lison, L., Kolcunova, I., and Pavlik, M., (2015). The influence of accelerated ageing on the oil impregnated paper. *Proceedings of 2015 16th International Scientific Conference on Electric Power Engineering EPE 2015*, no. 1, Kouty nad Desnou, Czech Republic (20–22 May 2015), pp. 743–747.

38 Yang, L., Liao, R., Sun, C. et al. (2010). Influence of vegetable oil on the thermal aging rate of Kraft paper and its mechanism. *2010 International Conference on High Voltage Engineering and Application*, New Orleans, LA, USA (11–14 October 2010).

39 Coulibaly, M.-L., Perrier, C., Marugan, M., and Beroual, A. (2013). Aging behavior of cellulosic materials in presence of mineral oil and ester liquids under various conditions. *IEEE Trans. Dielectr. Electr. Insul.* 20 (6): 1971–1976.

40 Trnka, P., Hornak, J., Prosr, P. et al. (2020). Various aging processes in a paper-natural ester insulation system in the presence of copper and moisture. *IEEE Access* 8: 61989–61998.

41 Martins, M.A.G. and Gomes, A.R. (2012). Comparative study of the thermal degradation of synthetic and natural esters and mineral oil: effect of oil type in the thermal degradation of insulating kraft paper. *IEEE Electr. Insul. Mag.* 28 (2): 22–28.

42 Hao, J., Liao, R., Chen, G. et al. (2012). Quantitative analysis ageing status of natural ester-paper insulation and mineral oil-paper insulation by polarization/depolarization current. *IEEE Trans. Dielectr. Electr. Insul.* 19 (1): 188–199.

43 García, B., García, T., Burgos, J.C., and Urquiza, D. (2017). Studying the loss of life of natural-ester-filled transformer insulation: impact of moisture on the aging rate of paper. *IEEE Electr. Insul. Mag.* 33 (1): 15–23.

44 Rapp, K.J. and Luksich, J. (2014). Accelerated aging of aramid insulation in various dielectric liquids. *Proceedings of 2014 IEEE 18th International Conference on Dielectric Liquids ICDL 2014*, Bled, Slovenia (29 June–3 July 2014), pp. 41–44.

45 Tokunaga, J., Koide, H., Mogami, K. et al. (2017). Lifetime estimation of cellulosic insulation material in palm fatty acid ester for power transformers. *IEEE Trans. Dielectr. Electr. Insul.* 24 (2): 1010–1017.

46 Munajad, A., Subroto, C., and Suwarno (2017). Study on the effects of thermal aging on insulating paper for high voltage transformer composite with natural ester from palm oil using fourier transform infrared spectroscopy (ftir) and energy dispersive x-ray spectroscopy (EDS). *Energies* 10 (11).

47 Fofana, I., Wasserberg, V., Borsi, H., and Gockenbach, E. (2002). Challenge of mixed insulating liquids for use in high-voltage transformers, part 2: investigations of mixed liquid impregnated paper insulation. *IEEE Electr. Insul. Mag.* 18 (4): 5–16.

48 Carcedo, J., Fernández, I., Ortiz, A. et al. (2015). Aging assessment of dielectric vegetable oils. *IEEE Electr. Insul. Mag.* 31 (6): 13–21.

49 Ortiz, A., Delgado, F., Ortiz, F. et al. (2018). The aging impact on the cooling capacity of a natural ester used in power transformers. *Appl. Therm. Eng.* 144: 797–803.

50 Fernández, I., Delgado, F., Ortiz, F. et al. (2016). Thermal degradation assessment of Kraft paper in power transformers insulated with natural esters. *Appl. Therm. Eng.* 104: 129–138.

51 Bandara, K., Ekanayake, C., Saha, T.K., and Annamalai, P.K. (2016). Understanding the ageing aspects of natural ester based insulation liquid in power transformer. *IEEE Trans. Dielectr. Electr. Insul.* 23 (1): 246–257.

52 Liao, R., Liang, S., Sun, C. et al. (2009). A comparative study of thermal aging of transformer insulation paper impregnated in natural ester and in mineral oil. *Eur. Trans. Electr. Power* 20: 518–533.

53 Ciuriuc, A., Dumitran, L.M., and Notingher, P.V. (2013). Dielectric properties of paper insulation aged in the presence of vegetable transformer oil. *Proceedings of IEEE International Conference on Solid Dielectrics ICSD*, Bologna, Italy (30 June–4 July 2013), pp. 549–552.

54 Schober, F., Harrer, S., Kuchler, A. et al. (2016). Transient and steady-state dc behavior of oil-impregnated pressboard. *IEEE Electr. Insul. Mag.* 32 (4): 8–14.

55 Fofana, I., Wasserberg, V., Borsi, H., and Gockenbach, E. (2002). Challenge of mixed insulating liquids for use in high-voltage transformers.1. Investigation of mixed liquids. *IEEE Electr. Insul. Mag.* 18 (3): 18–31.

56 Carcedo, J., Fernandez, I., Ortiz, A. et al. (2016). Quantitative study on the aging of kraft paper in vegetable oils. *IEEE Electr. Insul. Mag.* 32 (6): 29–35.

57 Cui, L., Chen, W., Vaughan, A.S. et al. (2017). Comparative analysis of air-gap PD characteristics: vegetable oil/pressboard and mineral oil/pressboard. *IEEE Trans. Dielectr. Electr. Insul.* 24 (1): 137–146.

58 Hemmer, M., Badent, R., and Leibfried, T. (2003). Electrical properties of vegetable oil-impregnated paper insulation, *Conference on Electrical Insulation and Dielectric Phenomena (CEIDP), Annual Report*, Albuquerque, NM, USA (19–22 October 2003), pp. 60–63.

59 Van Nguyen, D., Nguyen, L.P., and Quach, T.N. (2019). Investigation of AC breakdown properties of paper insulators and enamel insulation impregnated with rice oil, corn oil and peanut oil for transformers. *IET Sci. Meas. Technol.* 13 (9): 1352–1361.

60 Patra, M.R. and Burjupati, N.R. (2018). Effect of moisture and geometrical parameter on dielectric behavior of ester oil impregnated insulation using dielectric spectroscopy measurement. *Proc. IEEE Int. Conf. Prop. Appl. Dielectr. Mater.* 2018: 423–428.

61 Tenbohlen, S. and Koch, M. (2010). Aging performance and moisture solubility of vegetable oils for power transformers. *IEEE Trans. Power Deliv.* 25 (2): 825–830.

62 Wilhelm, H.M., Franch, V., Tulio, L., and Franch, A.F. (2015). Compatibility of transformer construction materials with natural ester-based insulating fluids. *IEEE Trans. Dielectr. Electr. Insul.* 22 (5): 2703–2708.

63 Gomez, A.M.D., Prakash, N.B., and Vigneshwaran, B. (2015). Investigation on electrical characteristics of dielectric fluids under aging phenomena. *2015 International Conference on Communications and Signal Processing ICCSP 2015*, Melmaruvathur, India (2–4 April 2015), pp. 455–459.

64 Ö. (Best) Kaymaz (2014). Transformers – Midel 7131 Test Results Report. BEST Transformers.

65 Sancibrian, O., Vila, C., Fernandez, I. et al. (2019). Analysis of the impregnation process of cellulosic materials by ester-based insulating liquids. *ARWtr 2019 – Proceedings of 2019 6th Advanced Research Workshop on Transformers*, Córdoba, Spain (7–9 October 2019), pp. 59–63.

66 Bandara, K., Ekanayake, C., Saha, T., and Ma, H. (2015). Investigation of moisture influence on dielectric response of ester oil impregnated pressboard. *IEEE Power Energy Soc. Gen. Meet.* 2015: 1.

67 Liu, Q., Wang, Z.D., and Perrot, F. (2009). Impulse breakdown voltages of ester-based transformer oils determined by using different test methods. *Annual Report - Conference on Electrical Insulation and Dielectric Phenomena, CEIDP*, no. 1, Virginia Beach, VA, USA (18–21 October 2009), pp. 608–612.

68 Prevost, T.A. (2009). Dielectric properties of natural esters and their influence on transformer insulation system design and performance - An update, *2009 IEEE Power Energy Society General Meeting PES'09*, Calgary, AB, Canada (26–30 July 2009), pp. 1–7.

69 Frimpong, G.K., Oommen, T.V., and Asano, R. (2011). A survey of aging characteristics of cellulose insulation in natural ester and mineral oil. *IEEE Electr. Insul. Mag.* 27 (5): 36–48.

70 Heywood, R.J., Emsley, A.M., and Ali, M. (2000). Degradation of cellulosic insulation in power transformers. Part 1: Factors affecting the measurement of the average viscometric degree of polymerisation of new and aged electrical papers. *IEE Proc. - Sci. Meas. Technol.* 147 (2): 86–90.

71 IEC (2004). IEC 60450:2004 Measurement of the average viscometric degree of polymerization of new and aged cellulosic electrically insulating materials.

72 Ciuriuc, A., Dumitran, L.M., Notingher, P.V. et al. (2016). Lifetime estimation of vegetable and mineral oil impregnated paper for power transformers. *Proc. 2016 IEEE Int. Conf. Dielectr. ICD 2016* 2: 720–723.

73 Lelekakis, N., Guo, W., Martin, D. et al. (2012). A field study of aging in paper-oil insulation systems. *IEEE Electr. Insul. Mag.* 28 (1): 12–19.

74 Shroff, D.H. and Stannett, A.W. (1985). A review of paper ageing in power transformers. *IEE Proc.* 132 (6): 312–319.

75 Karsai, K., Kerényi, D., and Kiss, L. (1987). *Large Power Transformers*. Budapest: Elsevier.

76 Stenberg, N. (2002). *On the Out-of-Plane Mechanical Behaviour of Paper Materials*. Stockholm: Department of Solid Mechanics, Royal Institute of Technology.

77 Persson, K. (1991). Material Model for Paper: Experimental and Theoretical Aspects. Diploma Report, *Diploma report*. Lund Institute of Technology, Lund, Sweden.

78 Oria, C., Ortiz, A., Ferreno, D. et al. (2019). State-of-the-art review on the performance of cellulosic dielectric materials in power transformers: mechanical response and ageing. *IEEE Trans. Dielectr. Electr. Insul.* 26 (3): 939–954.

79 Montsinger, V.M. (1930). Loading transformers by temperature. *Trans. Am. Inst. Electr. Eng.* 49 (2): 776–790.

80 ISO 1924-2:2008 (2008). Paper and board. Determination of tensile properties. Part 2: Constant rate of elongation method (20 mm/min).

81 Tryding, J. (1996). *In-Plane Fracture of Paper*. Lund University.

82 Stenberg, N., Fellers, C., and Östlund, S. (2001). Measuring the stress–strain properties of paperboard in the thickness direction. *J. Pulp Pap. Sci.* 27 (6): 213.

83 Stenberg, N. and Fellers, C. (2003). The out-of-plane Poissons ratios of paper and paperboard. *Nord. Pulp Pap. Res. J.* 17 (4): 387.

84 Nygårds, M. (2009). Modelling the out-of-plane behaviour of paperboard. *Nord. Pulp Pap. Res. J.* 24 (1): 72.

85 Girlanda, O., Wei, K., Brattberg, T., and Schmidt, L.E. (2012). Influence of density on the out-of-plane mechanical properties of pressboard. *Conference on Electrical Insulation and Dielectric Phenomena (CEIDP) Proceeding*, Montreal, Canada (14–17 October 2012), pp. 247–250.

86 Girlanda, O., Tjahjanto, D.D., Östlund, S. et al. (2013). Characterization and modelling of the mechanical properties of pressboard. *Annual Report Conference on Electrical Insulation and Dielectric Phenomena*, Chenzhen, China (20–23 October 2013), p. 563.

87 Girlanda, O., Tjahjanto, D.D., Östlund, S. et al. (2014). Modeling and experimental validation of the mechanical behavior of pressboard. *2014 IEEE Electrical Insulation Conference (EIC)*, Philadelphia, PA, USA (8–11 June 2014).

88 O. Girlanda., Sahlén, F., Schmidt, L. E. et al. (2015). Analysis of the micromechanical deformation in pressboard performed by X-ray microtomography. *2015 IEEE Electrical Insulation Conference (EIC)*, Seattle, WA, USA (7–10 June 2015), pp. 89–92.

89 Joffre, T., Girlanda, O., Forsberg, F. et al. (2015). A 3D in-situ investigation of the deformation in compressive loading in the thickness direction of cellulose fiber mats. *Cellulose* 22 (5): 2993–3001.

90 Girlanda, O., Tjahjanto, D.D., Östlund, S., and Schmidt, L.E. (2016). On the transient out-of-plane behaviour of high-density cellulose-based fibre mats. *J. Mater. Sci.* 51: 8131–8138.

91 Emsley, A.M., Heywood, R.J., Ali, M., and Eley, C.M. (1997). On the kinetics of degradation of cellulose. *Cellulose* 4: 1–5.

92 Fernández, C., Fernandez, I., Ortiz, A. et al. (2019). Kraft and diamond dotted paper thermally aged in mineral oil and natural ester: mechanical characterisation. *2019 IEEE 20th International Conference on Dielectric Liquids*, Roma, Italy (23–27 June 2019).

93 Fernández-Diego, C., Ortiz, A., Carrascal, I. et al. (2019). Damage assessment of transformer Kraft paper insulation aged in mineral and vegetable oils. *Cellulose* 26 (4): 2653–2672.

94 Oommen, T.V. and Arnold, L.N. (1981). Cellulose insulation materials evaluated by degree of polymerization measurements. *Proceedings of the 15th Electrical Electronics Conference*, New York, USA (19–22 October 1981), pp. 257–261.

95 Scheirs, J., Camino, G., Avidano, M., and Tumiatti, W. (1998). Origin of furanic compounds in thermal degradation of cellulosic insulating paper. *J. Appl. Polym. Sci.* 69: 2541–2547.

96 Baird, P.J., Herman, H., and Stevens, G.C. (2006). Non-destructive measurement of the degradation of transformer insulating paper. *IEEE Trans. Dielectr. Electr. Insul.* 13 (1): 309.

97 Mirzaie, M., Gholami, A., and Tayyebi, H. (2007). Thermal degradation of cellulose paper insulation in power transformers. *2007 International Conference on Solid Dielectrics*, Winchester, UK (8–13 July 2007), p. 673.

98 Jalbert, J., Gilbert, R., Tétreault, P. et al. (2007). Identification of a chemical indicator of the rupture of 1,4-?-glycosidic bonds of cellulose in an oil-impregnated insulating paper system. *Cellulose* 14 (4): 295–309.

99 Matharage, S.Y., Liu, Q., and Wang, Z.D. (2016). Aging assessment of Kraft paper insulation through methanol in oil measurement. *IEEE Trans. Dielectr. Electr. Insul.* 23 (3): 1589–1596.

100 Carrascal, I., Fernández, C., Casado, J.A. et al. (2018). Quantification of Kraft paper ageing in mineral oil impregnated insulation systems through mechanical characterization. *Cellulose* 25: 3583–3594.

101 Cranston, E.D. and Gray, D.G. (2008). Birefringence in spin-coated films containing cellulose nanocrystals. *Colloids Surf. A Physicochem. Eng. Asp.* 325: 44–51.

102 Yoshida, M. and Konishi, Y. (2009). Diagnosis for aging degradation of insulating paper in power transformers by measuring the refractive index of cellulose fibers. *2009 IEEE Bucharest Power Techology Conference*, Bucharest, Romania (28 June–2 July 2009).

103 Zhang, J., Wang, F., Li, J. et al. (2018). Ageing characterization of oil impregnated paper insulation based on dispersion staining colors. *IEEE Trans. Dielectr. Electr. Insul.* 25 (5): 1587–1597.

104 Gubanski, S., Boss, P., Cseépes, G. et al. (2003). Dielectric response methods for diagnostics of power transformers. *IEEE Electr. Insul. Mag.* 19 (3): 12–18.

105 Cavallini, A., Montanari, G.C., and Ciani, F. (2005). Analysis of partial discharge phenomena in paper-oil insulation systems as a basis for risk assessment evaluation, *2005 IEEE International Conference on Dielectric Liquids ICDL 2005*, Coimbra, Portugal (26 June–1 July 2005), pp. 241–244.

106 Li, J., Grzybowski, S., Sun, Y., and Chen, X. (2007). Dielectric properties of rapeseed oil paper insulation. *Annual Report - Conference on Electrical Insulation and Dielectric Phenomena, CEIDP*, Vancouver, BC, Canada (14–17 October 2007), pp. 500–503.

107 Bandara, K., Ekanayake, C., and Saha, T.K. (2016). Analysis of frequency domain dielectric response of pressboard insulation impregnated with different insulating liquids. *IEEE Trans. Dielectr. Electr. Insul.* 23: 2042–2050.

108 IEEE (2011). IEEE C57.91-2011 - IEEE Guide for Loading Mineral-Oil-Immersed Transformers and Step-Voltage Regulators.

109 Rapp, K.J. and Luksich, J. (2011). Review of Kraft paper/natural ester fluid insulation system aging. *Proc. – IEEE Int. Conf. Dielectr. Liq.*, Trondheim, Norway (26–30 June 2011), pp. 12–15.

110 Biçen, Y., Çilliyüz, Y., Aras, F., and Aydugan, G. (2011). An assessment on aging model of IEEE/IEC standards for natural and mineral oil-immersed transformer. *Proc. - IEEE Int. Conf. Dielectr. Liq.*, Trondheim, Norway (26–30 June 2011), pp. 20–23.

111 Santisteban, A., Delgado, F., Ortiz, A. et al. (2017). Numerical analysis of the hot-spot temperature of a power transformer with alternative dielectric liquids. *IEEE Trans. Dielectr. Electr. Insul.* 24 (5): 3226–3235.

112 Zhang, X., Ren, L., Yu, H. et al. (2018). Dual-temperature evaluation of a high-temperature insulation system for liquid-immersed transformer. *Energies* 11 (8): 1957.

113 IEC (2018). IEC 60076-7. Power transformers – Part 7: Loading guide for mineral-oil-immersed power transformers.

114 Ekenstam, A. (1936). The behaviour of cellulose in mineral acid solutions: Kinetic study of the decomposition of cellulose in acid solutions. *BER* 69: 553.

115 Zervos, S. and Moropoulou, A. (2005). Cotton cellulose ageing in sealed vessels. Kinetic model of autocatalytic depolymerization. *Cellulose* 12 (5): 485–496.

116 Calvini, P. and Gorassini, A. (2006). On the rate of paper degradation: lessons from the past. *Restaurator* 27 (4): 275–290.

117 Ding, H.Z. and Wang, Z.D. (2008). On the degradation evolution equations of cellulose. *Cellulose* 15 (2): 205–224.

118 Rapp, K.J. and Luksich, J. (2001). Reaction rates of paper aged in natural ester dielectric fluid. *Conference on Electrical Insulation and Dielectric Phenomena (CEIDP), Annual Report*, Kitchener, ON, Canada (14–17 October 2001), pp. 209–212.

119 Feng, D., Hao, J., Liao, R. et al. (2019). Comparative study on the thermal-aging characteristics of cellulose insulation polymer immersed in new three-element mixed oil and mineral oil. *Polymers (Basel)* 11 (8): 1–16.

120 Lijun, Y., Ruijin, L., Caixin, S., and Sun, H. (2008). Study on the influence of natural ester on thermal aging characteristics of oil-paper in power transformers. *International Conference on High Voltage Engineering and Application*, Chongqing, China (9–12 November 2008), pp. 437–440.

121 Liao, R., Hao, J., Yang, L. et al. (2010). Improvement on the anti-aging properties of power transformers by using mixed insulating oil. *2010 International Conference on High Voltage Engineering and Application ICHVE 2010*, New Orleans, LA, USA (11–14 October 2010), pp. 588–591.

122 Subroto, C., Munajad, A., and Suwarno, S., Lifetime estimation of cellulose paper in natural ester dielectric fluid. *International Conference on High Voltage Engineering and Power Systems ICHVEPS 2017 - Proceeding*, vol. 2017, Denpasar, Indonesia (2–5 October 2007), pp. 307–311, 2017.

123 Sari, Y.E., Al-Ghifary, F., Dewi, T.I.D.K., and Suprijadi, S. (2018). Study on characteristics of paper degradation in natural ester from palm oil-based insulation liquid. *4th IEEE Conference Power Engineering and Renewable Energy, ICPERE 2018 – Proceedings*, Solo, Indonesia (29–31 October 2018).

4

Degradation Assessment of Ester Liquids

A. J. Amalanathan[1], R. Sarathi[1], N. Harid[2], and H. Griffiths[2]

[1] Department of Electrical Engineering, Indian Institute of Technology Madras, Chennai, India
[2] HVDM Research Group, APEC Research Centre, Khalifa University, Abu Dhabi, PO, UAE

4.1 Introduction

The ester fluids have been gradually introduced for use in transformer insulation since 1998 because of their better environmental characteristics compared with mineral oils. Concurrently, research on the properties and behavior of ester fluid insulation has been progressing worldwide, focusing on performance under different electrical and thermal stresses, as well as on chemical reactions occurring in service such as oxidation and hydrolysis. [1, 2]. However, their relatively higher cost compared with mineral oil limited their use to areas where additional measures are needed for the environmental protection. Nevertheless, the higher thermal class of ester fluids makes them viable in densely populated areas or commercial buildings such as malls and airports. Further, the transformer-filled ester insulation can be placed closer to buildings, which provides the additional benefits of reduced space requirements as specified in the National Electrical Code (NFPA 70), and shorter low-voltage (LV) cable runs and hence lower power losses. The ester fluids being hygroscopic in nature, precautionary measures must be taken for storage and during handling to prevent the fluid from reacting with atmospheric moisture. This makes them recommended for use in non-breathing transformers. Thus, with the increasing demand for environmentally friendly power transformer for voltages greater than 100 kV, the dielectric fluid needs to be replaced, considering some design changes that can accommodate the ester fluids inside the transformer [3]. With the aforementioned benefits on alternate dielectric fluids compared to petroleum-based oil, many transmission lines greater than 400 kV are now being operated with ester fluids taking into consideration the overall lifetime cost of transformer installation. Ester fluids are classified into two groups depending on their source and their adaptability to transformer applications: synthetic ester and natural ester. Their chemical structure and their desirable properties such as flash point, breakdown voltage, and oxidation stability are discussed in the next sections.

4.1.1 Types of Ester Fluids

Based on the process of manufacturing, the ester-based fluids are broadly classified into two types: natural esters produced from vegetable oils extracted from plant crops and synthetic esters derived

Alternative Liquid Dielectrics for High Voltage Transformer Insulation Systems: Performance Analysis and Applications, First Edition. Edited by U. Mohan Rao, I. Fofana, and R. Sarathi.

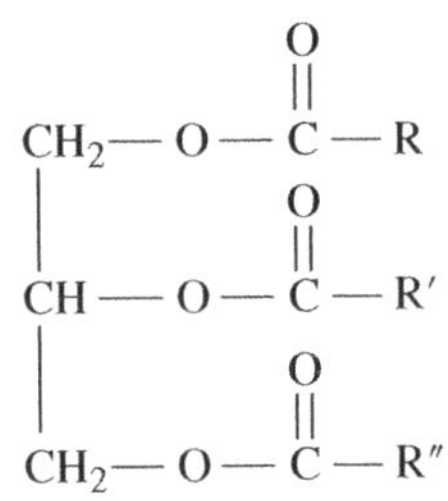

Figure 4.1 Structure of natural ester.

Table 4.1 Fatty acid composition of vegetable oils.

	Saturated fatty acid		Monounsaturated	Polyunsaturated	
Oil fatty acid	**Palmitic acid (C16 : 0)**	**Stearic acid (C18 : 0)**	**Oleic (C18 : 1)**	**Linoleic (C18 : 2)**	**Linolenic (C18 : 3)**
Palm	45	4	40	10	—
Olive	11	3	71	10	1
Rapeseed	4	2	62	22	10
Sunflower	7	5	19	68	1
Soybean	11	4	24	54	7

from chemical reactions. Natural ester consists of glycerol backbone with different fatty acid groups that can be saturated, monounsaturated, and polyunsaturated. The general schematic structure of natural ester is shown in Figure 4.1.

The R, R′ and R″ groups refer to the different alkyl groups with a major variation in the number of C=C double bonds. The chemical composition of different fatty acid groups shown in Table 4.1 determines the physical properties of the fluid. The higher the amount of saturated acids, in the ester group, the more stable is the fluid toward oxidation and less viscous compared to unsaturated acids which make the fluid unstable [4]. Although natural esters are produced from a wide variety of plant crops, their application to transformers is still now limited to those produced from soybean, rapeseed, and sunflower due to their more acceptable insulating characteristics.

Synthetic esters are usually derived from the combination of more than one alcohol functional group (polyol) with carboxylic acids to produce structures with several acids leading to a central polyol structure. Most preferably, the synthetic ester is defined as a family of polyol esters, and in particular these fluids are pentaerythritol tetra ester as shown in Figure 4.2.

In this figure, R represents an alkyl group ranging from C_5H_{11} to C_9H_{19}. Currently, many commercial transformer-grade ester fluids are available, examples include MIDEL 7131, ProEco TR3746, Envirotemp 200 for synthetic esters and BIOTEMP, Envirotemp FR3, MIDEL 1204, MIDEL 1215 for natural esters.

4.1.2 Properties of Ester Fluids

The ester fluids providing a higher fire safety enhances the transformer insulation performance through various other properties such as moisture content, dielectric permittivity, and breakdown strength which will be briefly discussed in this section.

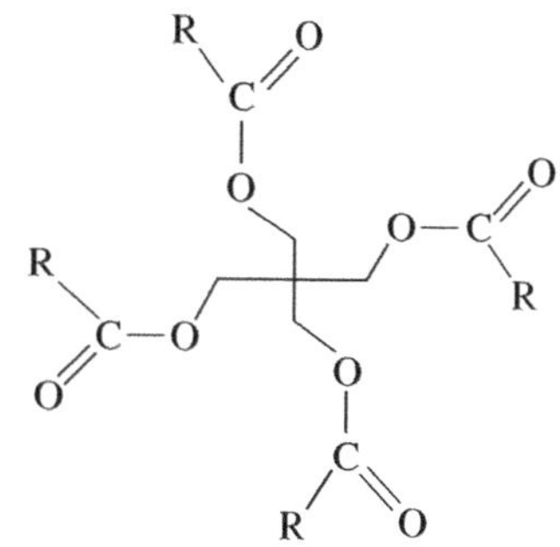

Figure 4.2 Structure of synthetic ester.

4.1.2.1 Breakdown Voltage

Ester fluid-filled power transformers need to withstand over voltages that can be much higher than their nominal voltage ratings. Under abnormal conditions, such over voltages of AC, DC, lightning, or switching types can sometimes exceed the withstand levels and cause fluid breakdown [5, 6]. With the increase in transmission voltage levels, the withstand levels of insulating fluids become higher, which means that the insulation costs increase. This calls for significant effort from transformer designers and researchers to optimize the quantity and quality of insulating liquids, taking into account environmental constraints. Technical standards such as IEC 60156 and ASTM D877 provide recommendations for measuring the breakdown voltage of small liquid samples in homogeneous electric field in the laboratory. These tests are used as a quality check of insulating fluids by the utilities before filling the power transformers. Due to the more viscous nature of alternate fluids compared to mineral oil, a longer rest time before carrying out the tests and time interval between two consecutive breakdowns must be allowed to disperse any gas bubbles formed during the process. In practice, however, the breakdown voltage depends on a number of factors including temperature, impurities, moisture content, pressure, surface condition of metallic conductors and magnitude, and duration of applied voltage stress. These aspects make the breakdown voltage statistical in nature depending on the presence of such various influencing factors and the combined mechanisms associated with them. Under the influence of an external electric field, solid impurities align along the direction of field lines with moisture absorbed from the atmosphere forming a bridge between the high-voltage and low-voltage electrodes, causing electrical breakdown. Table 4.2 shows the AC and DC breakdown voltages of ester fluids, with natural ester showing a higher breakdown voltage compared to synthetic ester [7]. Under DC voltage, the electric field distribution under DC voltage depends on the rise in its humidity, which could have a significant influence on the resistivity of the fluid [9]. The rate of change in applied voltage waveform is responsible for the

Table 4.2 Breakdown voltage of ester fluids under AC and DC voltages.

		DC voltage (kV)	
Type of oil	**AC voltage (kV)**	**Positive polarity**	**Negative polarity**
Mineral oil	78	53	55
Natural ester	87	58	84
Synthetic ester	61	47	57

Source: Based on Reffas et al. [7]; Beroual et al. [8].

reduction in AC breakdown voltage compared to DC breakdown voltage. Also, it is known that the voltage polarity has a major effect on the breakdown voltage of all insulating fluids. Breakdown under negative polarity occurs at a higher voltage than the positive polarity. In a process similar to what occurs in gases, the free electrons generated in the fluid are attracted nearer to the anode leaving the positive ions behind. Further, these ions move toward the ground electrode, which is assumed as an extension of needle electrode, causing a reduction in the breakdown voltage [10]. Under negative polarity, the injected electrons act as a space charge shield weakening the applied voltage gradient and thereby increasing the breakdown voltage [11].

During the breakdown mechanism, the streamer propagation forms an important factor affecting the breakdown of insulating fluids. The propagation of streamers depends on the localized electric field and parameters of the applied voltage such as magnitude, rise time, shape, and polarity. Also, it is necessary to understand the ability of such insulating fluids to resist against short duration transients such as standard lightning impulse (1.2/50 μs) with its specified tolerance limit as specified in IEC 60060-2 standard for high-voltage test techniques [12]. Streamer propagation is analyzed nowadays using ultrahigh-speed cameras and a beam splitter. Figure 4.3 shows typical photographs of negative streamers for synthetic ester (MIDEL 7131) and natural ester (FR3) insulating fluids. One of the streamer's characteristic that is important for liquid insulation design is the stopping length, defined as the length of longest branch moving toward the opposite electrode. It is known to increase with the applied voltage. Although the streamer stopping lengths are close in both ester fluids as shown in Figure 4.3, the shape characteristics such as channel width and number of branches are found to be different. There are various modes of propagation of streamers: the slow mode, where the streamer has more branches and propagates with a constant speed of less than 3 km/s, and the fast mode, in which the streamers travel faster at speeds ranging between 5 and 50 km/s, with a more filamentary branch structure and no offshoots.

Lightning impulse breakdown voltages of different ester fluids measured in a point–plane electrode configuration with different gap spacing are shown in Figure 4.4. At the smallest gap distance of 25 mm, all fluids show the same breakdown voltage under positive polarity, increasing with the gap spacing at a rate that is dependent on the fluid type. The breakdown voltage under negative polarity shows a similar behavior but with higher magnitudes than positive polarity for

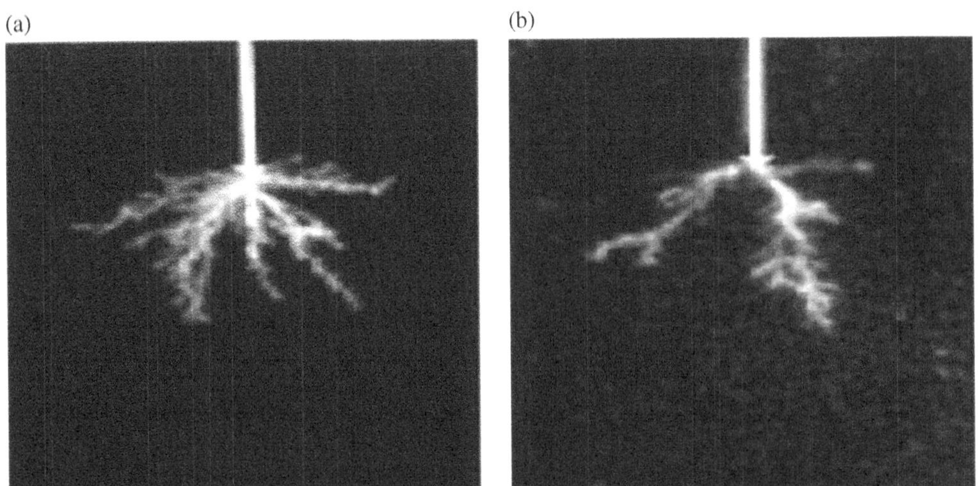

Figure 4.3 Typical photographs of negative streamers in (a) synthetic ester and (b) natural ester.

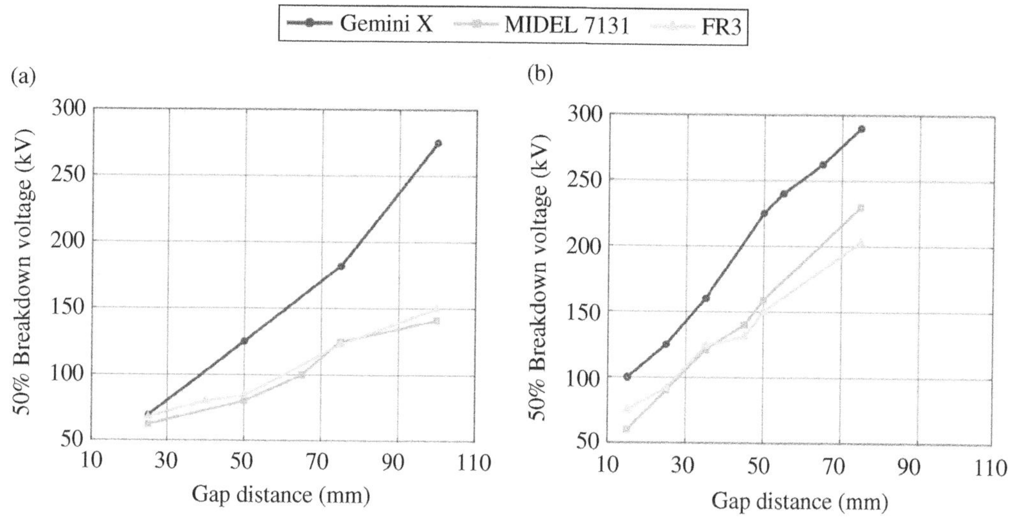

Figure 4.4 Fifty percent breakdown voltage of ester fluids versus gap distances for (a) positive polarity and (b) negative polarity. *Source:* Based on Dang et al. [13].

the same gap distance. The positive streamers are more filamentary than negative streamers. The latter have less branches, which requires a higher impulse breakdown voltage compared to positive lightning for the same stopping length irrespective of the electrode gap [13].

4.1.2.2 Moisture Content

Moisture ingress into the insulating fluid is known to have a substantial effect on the breakdown voltage and hence a better understanding of the impact of moisture content and the tolerance limits allowed for ester fluids is needed. Figure 4.5 shows the influence of moisture on the AC breakdown voltage of ester fluids, mineral oil, and silicone oil. As can be seen in the figure, while the breakdown voltage of mineral oil and silicone oil decrease sharply with moisture content over

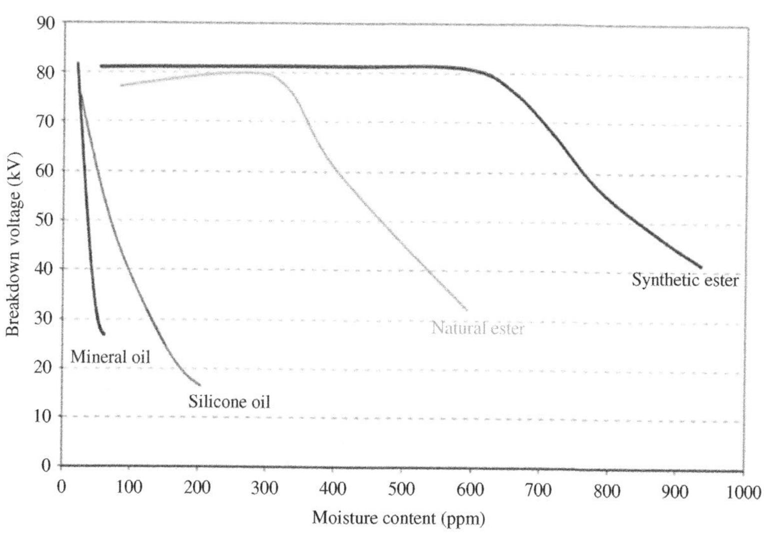

Figure 4.5 Effect of moisture on the AC breakdown voltage of ester fluids.

the 0–200 ppm range, the ester fluids can absorb a much larger amount of water without compromising their breakdown voltage. This is due to the polar nature of ester fluids, which gives them the ability to attract hydrogen bonds present in water molecules. The moisture content of mineral and silicone oil at normal operating temperature (90 °C) is 65 and 220 ppm, whereas for natural and synthetic ester fluids it is around 300 and 700 ppm, respectively. Natural ester consists of three ester linkages per molecule, whilst synthetic ester contains around four ester linkages per molecule, and these differences in chemical structure could be the reason for higher amount of water dissolved in synthetic ester fluids. When the transformer tends to cool down to its ambient temperature (shut down process), the mineral oil is saturated to 118% and only 26% is saturated in case of ester fluids. This indicates that no free water exists in ester fluids, but mineral oil could show a much lower breakdown voltage when the transformer is re-energized, increasing the risk of failure. Further, the rate of cellulosic paper aging is also related to moisture content, where the lifetime of paper reduces by a factor of 10 for each 1% increase in water content. Consequently, it is necessary to keep the cellulosic pressboard/paper as dry as possible and, to this end, ester fluids outperform mineral oil in trapping more water content from the solid insulation [14, 15].

4.1.2.3 Flash Point and Fire Point

Transformer fires are prone to occur more frequently making fire as one of the major safety concerns in power transformers. Table 4.3 shows the flash point and fire point of ester fluids in comparison to silicone oil and mineral oil. In transformers, the oil temperature can rise to more than 160 °C due to hotspot formation, which can cause severe damage to the solid and liquid insulants [16]. If such localized hotspots are not detected at an early stage, they can cause flashover of the cellulosic pressboard material inside the transformer, leading to fire hazards. The flash point is the lowest temperature at which vapor pressure forms an ignitable mixture with air near the surface of the fluid. The fire point is the lowest temperature at which an insulating liquid in an open container attains a vapor pressure that is sustainable for continuous burning once ignited.

The ester fluids and silicone oil come under a higher thermal class K, in comparison, mineral oil with a lower thermal class O is more easily exposed to fire hazards in transformers. The amount of energy produced by the complete combustion of insulating liquid is defined as the net calorific value. The difference between the calorific value of ester fluid and mineral oil could be associated with the variation in their chemical structure, where each chemical bond results in a different energy [17].

4.1.2.4 Viscosity

The viscosity is one of key properties of insulating fluids that determines the heat transfer mechanism occurring inside the transformer. It also defines the resistance to fluids in motion; a high viscosity slows down the flow velocity within the winding ducts, which rises the operating temperature of the transformer. At ambient temperature, a large discrepancy in the viscosity is found between ester fluids and mineral oil as shown in Table 4.4 but only a slight change is seen at higher

Table 4.3 Flash point and fire point of ester fluids.

Parameter	Test method	Mineral oil	Silicone oil	MIDEL 7131	MIDEL 1215
Flash point	ISO 2719/ASTM D93	150 °C	300 °C	260 °C	>315 °C
Fire point	ISO 2592/ASTM D92	170 °C	343 °C	316 °C	>350 °C
Net calorific value	ASTM D240-02	46 MJ/kg	≥32 MJ/kg	30.8 MJ/kg	37.5 MJ/kg

Table 4.4 Dynamic viscosity of ester fluids.

	Dynamic viscosity (mPa. s)	
Type of oil	**30 °C**	**90 °C**
Mineral oil	10.2	3.4
Silicone oil	45.6	15
Natural ester	41.3	5.6
Synthetic ester	42.1	8.1

temperature. In general, the viscosity of silicone oil and ester fluids are higher than that of mineral oil [18, 19], indicating lower heat transfer involved in the convection process, and this issue should be taken care while designing the cooling system of ester-filled power transformers.

4.1.2.5 Oxidation Stability

The oxidation stability is an important factor affecting the performance of insulating fluids. Oxygen can be present in the fluid naturally or may ingress from the external environment. The acidity and dissipation factor are the most important factors used to evaluate the oxidation stability of insulating fluids where the silicone oil based on polydimethylsiloxane is the most resistant fluid toward oxidation [20]. Petroleum-based oils oxidize at temperature as low as 105 °C, leading to the formation of degrading products such as organic acids and sludge that reduces the dielectric properties of the insulating oil. Synthetic ester fluids start to oxidize slowly at 125 °C depending on the content of saturated fatty acids, but natural esters are more susceptible to oxidation due to their chemical structure containing more fatty acid chains. The presence of carbon–carbon double bond is the main site for oxidation reaction, which increases the viscosity of the fluid along with the formation of oxygenated by-products such as alcohols, aldehydes, acids, and ketones [21]. In terms of oxidation stability, all natural esters are not chemically unstable, i.e. polyunsaturated fatty acid chains with two or three double bonds are oxidized more easily than monounsaturated single chain molecules. The lower oxidation stability of natural esters compared to silicone oil, synthetic esters, and mineral oil makes them more feasible for sealed-type transformers with nitrogen in the headspace [22]. Recently, more studies have been formulated on the addition of antioxidants (propyl gallate, butylated hydroxy toluene (BHT), ascorbic acid) with natural esters to improve their oxidation stability [23].

4.1.2.6 Dielectric Constant and Dissipation Factor

The polarizability of an insulating fluid upon application of an external electric field determines the permittivity or dielectric constant of the fluid. In transformers operating with mixed solid (cellulosic pressboard/paper) and liquid, the liquid insulation should possess high dielectric permittivity and low losses. Table 4.5 shows typical values of the permittivity and dissipation factor of ester fluids and mineral oil. The dielectric constant of ester fluids being closer to that of pressboard insulation reduces the electric field stress at the solid/liquid interface as opposed to silicone and mineral oil where high field intensities can take place, reducing the breakdown strength [26]. The dielectric constant is also temperature dependent where at high temperature, the fluid molecules become more agitated and the tendency to orient in the direction of the field decreases, which causes a decrease in the dielectric constant [24]. The dissipation factor of ester fluids does not match the very

Table 4.5 Dielectric permittivity and dissipation factor of ester fluids [17, 20, 24, 25].

	ε_r		tan δ	
Type of fluid	**20 °C**	**90 °C**	**20 °C**	**90 °C**
Mineral oil	2.20	2.10	0.00012	0.0011
Silicone oil	2.70	2.5	0.01	—
Natural ester	3.20	2.93	0.0020	0.0163
Synthetic ester	3.30	2.95	0.0008	0.0212

Source: Based on Perrier et al. [20]; Hamid et al. [24]; Mehta et al. [25].

low values of mineral oil. The resistivity of the insulating fluid influences its dissipation factor, and the presence of various additives in the oil's hydrocarbon structure help in retarding the contamination of oil.

4.1.2.7 Biodegradability

Transformer oil spillage occurs more often in a power system network and it could constitute a major environmental hazard if not properly handled. The extent to which the fluid is metabolized by naturally occurring microbes present in soil or any other water resource is termed as biodegradability. Such oil spills must be dealt with quickly to reduce the cost involved in the cleaning process. In accordance with the OECD 301 guideline, the ester fluids are classified as 95% biodegradable liquid, whereas mineral and silicone oils show more resistance to biodegradation [27, 28]. Thus, less cost remediation is involved in case of oil spillage from ester-filled transformer.

4.1.3 Fluid Maintenance and Storage Issues

Considering the above discussions on the advantages of ester fluids over mineral oil, fluid maintenance and storage issues need to be handled carefully to avoid any system failure resulting from the use of incorrect mineral limits for ester fluids. To overcome this problem, separate standards have been setup for both natural and synthetic ester fluids [29, 30]. The usage of vegetable oil exposed to cold weather conditions is a major issue where it freezes at subzero temperatures and its pour point is below −30 °C even after the addition of depressants [31]. A significant number of antioxidants should be present in the sealed unit of ester transformers in order to avoid the entry of air and moisture during its operational lifetime. The ester fluid is found to be incompatible with the standard gasket material due to oil leakage in the on-load tap changers (OLTCs), and as an alternative, fluorosilicone rubber could be used without any adverse effects [32]. Since corrosive sulfur present in mineral oil has resulted in transformer failures [33], ester fluids which is noncorrosive in nature can absorb certain sulfur concentrations during its reclamation process with various adsorbents [34], which needs to be monitored. The thermal evaluation of mineral oil showed a lower top oil temperature (TOT) and winding hot spot temperature (HST) [27] but from the aspect of aging, the ester fluids observed a better thermal performance under varying loading conditions [2]. Thus, the thermal parameters of the insulating fluid before filling the transformer and its loading power factor should be considered at their planning stage. In order to aid the impregnation of cellulose with insulating fluids, it is preferred to heat between 60 and 80 °C during its filling inside the power transformer and left around 24 hours prior to the energization. Practically, the time taken by the ester fluids to completely fill the cellulose pores is longer than the mineral oil impregnation time [35]. Also, for a storage tank that was previously filled with mineral oil, it is recommended to carry out a thorough cleaning and rinsing of the bottom pipe lines, valves, and pumps of transformers with ester fluids before refilling with the ester fluid. Existing storage tanks used for mineral oil can be used for ester fluids but care should be taken on the selection of pumps with required horsepower and capacity due to the fluid's varying viscous nature. Further, an additional vacuum processing is recommended for ester fluids before filling the transformer tank in order to sufficiently degasify the liquid and to avoid excessive foaming [30].

4.2 Procedure of Accelerated Thermal Aging

During its operational lifetime, both solid and liquid transformer insulation undergoes multiple stresses as a result of exposure to high electric fields, high temperature, and temperature variations depending on the loading cycle, in addition to chemical reactions affecting their dielectric properties. Before introducing any new insulating liquid into the transformer, it is necessary to understand how its dielectric properties (Section 4.1.2) are affected by different operating temperatures under laboratory conditions. This is because thermal aging is the most predominant factor influencing the transformers performance among other stresses and causes more deterioration to both the solid and liquid insulants [36]. For the evaluation of the thermal behavior of insulating fluids in power transformers, accelerated thermal aging tests are usually performed in laboratories considering the effect of various materials such as copper and pressboard [37, 38]. This section reviews the standards listed to perform the aging of insulating fluids based on their dielectric properties, thermal class, and moisture content.

4.2.1 ASTM D1934-95

ASTM D1934-95 standard is used to evaluate the oxidative aging of petroleum-based fluids using open beaker method [39]. It involves two procedures for the aging process, the first is the aging without metal catalyst (procedure A) and the second corresponds to the aging in the presence of metal catalyst (procedure B). Either fixed shelf stage oven or rotating shelf oven can be used for procedure A, whereas in case of procedure B, due to addition of metal catalyst such as copper, the rate of oxidation will be very high and hence slowly rotating shelf oven is adapted for uniformity. Both procedures use 300 ml of test fluid to be tested in 400 ml beaker, and it is aged for 96 hours in a circulating air-controlled oven at 115 °C. Mostly this test method is applicable for oils used as impregnating medium in electrical power transmission cables.

4.2.2 IEC 62332-2

IEC 62332-2 standard is used for thermal evaluation of electrical insulation system (EIS) containing both solid and liquid components, without restriction to its voltage class [40]. There are two types of processes involved in determining the thermal endurance of insulation system for liquid immersed transformers. The first is the sealed tube aging process and the second corresponds to the aging of full-scale models. The sealed tube aging is not practical for the components with different thermal properties, and the full-scale model is also impractical for large-scale power transformers. The simultaneous accelerated aging test performed with both liquid and solid insulants cannot be performed efficiently due to their different thermal indexes, where the solid insulation aging process takes a longer time. To account for this, a model has been introduced in this standard where solid materials are allowed to be aged at a separate temperature from that of the bulk liquid considering them under the same apparatus. This is made possible by providing an independent temperature control for both the solid and liquid insulants using separate circuits. The test system should contain an aging vessel, power supply, control measures, safety, and sampling system. Before the start of the experiment, the solid materials are allowed to dry at 110 °C for two days while the insulating liquid is allowed for degassing [41]. Once the above process is complete, the solid samples are allowed to vacuum impregnate with the liquid at 60 °C for 12 hours and then finally subjected to aging inside the test vessel.

4.2.3 Temperature

The aging temperature of the test is chosen based on the thermal class of the insulating fluid as shown in Table 4.6. The choice of liquid aging temperature should be at least 15% higher than its operating range and should not be more than 10% below its flash point. The duration of aging is selected such that there is more sampling between the end points. At each subsequent period, both the solid and liquid sampling are removed and evaluated using the diagnostic testing procedures.

4.2.4 Atmosphere

Depending on the oxidation stability of insulating fluids, the atmosphere under which the aging is done has to be selected. For the ester fluids, natural esters have to be aged only under nitrogen blanketing since its application has been limited only to sealed type transformers [42] whereas in case of synthetic ester, aging can be performed either in air or any other inert ambience.

4.2.5 Moisture

Both the solid and liquid insulants are tested for their initial moisture content prior to aging and after each sampling duration [43]. The moisture content before the start of the experiment indicates whether the solid materials have dried up to its full extent. Also, the changes between the initial and final moisture content can determine the amount of degradation occurring during its aging cycle. The moisture content of paper/oil samples are determined using capacitive probes and Karl Fischer titration through equilibrium curves [44]. Since the moisture saturation limits of ester fluids are different from mineral oil, the relative moisture content (W_{rel}) is determined using the relation [45],

$$W_{rel} = \frac{W_{abs}}{W_L} \tag{4.1}$$

Table 4.6 Recommended aging temperature and duration of insulating fluids.

	Expected thermal class			
Duration of aging period (hours)	**90**	**120**	**180**	**220**
1500/3000/4500/6000	110 °C	140 °C	200 °C	240 °C
500/1000/1500/3000	125 °C	155 °C	215 °C	255 °C
250/500/750/1000	140 °C	170 °C	230 °C	270 °C

where W_{abs} is the absolute moisture content in the liquid and W_L refers to the moisture saturation limit in the liquid. Recently, fiber brag grating-based optical sensors were developed for online moisture measurements in transformer insulation and were found to be cost-effective during the manufacturing stage, but retrofitting these sensors in existing transformer is not possible because of the requirements to access to winding [46].

4.3 Assessment of Liquid Degradation

The insulating fluid is subject to very high temperatures depending on the transformer's loadability, and this combined with various other degrading factors such as oxygen, moisture, contaminants and gas bubbles can reduce its operational lifetime significantly, which is a major concern for aged power transformers. Problems that arise as a result of one or more of these factors include partial discharge within the bulk liquid insulation or at the solid–liquid interface and static electrification. In response to this, continuous assessment of the performance of insulating fluids [47] using condition monitoring techniques has become a priority task that most power utilities adopt in their asset management plans. The following sections describe experimental approaches used for measuring the partial discharge inception voltage and the static electrification current of ester fluids, with a discussion on the effect of thermal aging on these parameters.

4.3.1 Partial Discharge Inception Voltage

Partial discharge inception voltage (PDIV) is the voltage at which localized discharge takes place within the voids or any gaseous pockets in the fluid without bridging the gap between the high-voltage conductors and grounded metallic parts. One of the major causes of failure in the liquid insulation is corona discharge, a type of partial discharge, which may occur due to protrusions in the winding or the intensification of electric field at weak links and edges inside the transformer [48]. Identification of such incipient discharges is a real challenge where there has not been any specific relevant standard for testing related to the evaluation of PDIV in transformer insulating liquids. The existing specifications should be reviewed to include new PDIV detection methods that are more sensitive under laboratory conditions than under field conditions. As per IEC 61294 standard [49], the PDIV is defined as the voltage at which apparent charge is greater than or equal to 100 pC when the voltage is increased at the rate of 1 kV/s. Further it was re-defined as the voltage at which PD pulses are repetitive in nature as per IEC 60270 standard [50]. In addition to the method of detection, the rate of increase of the test voltage recommended by the standard may also need to be reviewed and complemented, as shown in a study comparing the electrical properties of mineral oil and ester fluids, where the PDIV measured using a stepwise increase of the applied voltage was found to be much lower than that measured according to the ramp increase recommended by IEC 61294 [51]. Later on, various other techniques were used to identify partial discharges such as acoustic emission techniques and radio frequency (RF) techniques under laboratory environment [52–54]. PD current pulses are characterized by a rise time in the nanosecond range and emit electromagnetic radiation in the ultrahigh frequency (UHF) range of 300–3000 MHz [55]. In recent times, UHF sensors exploiting this feature have been widely used for measuring PDIV. These sensors provide enhanced sensitivity and better signal to noise ratio (SNR), and have the advantage of being noninvasive and nondestructive [56].

4.3.1.1 Measurement of PDIV Under AC and DC Voltage

The IEC 61294 recommends the use of a needle-plate electrode system for measuring the PDIV of dielectric liquids. The PD inception and propagation are dependent on the radius of curvature of needle electrode, moisture content, and contaminated particles in the oil. Assuming the needle to be a perfectly conductive hyperboloid and neglecting space charge effects, the electric field at the tip of the needle inside the liquid insulation [26] is given by

$$E_{\max} = \frac{2V}{r.\ln\left(1+\frac{d}{r}\right)} \tag{4.2}$$

where V is the instantaneous applied voltage, r is the tip radius, and d is the gap spacing between the electrodes.

The needle tip radius should be inspected regularly using an optical microscope and should be replaced if any signs of degradation are observed after every 10 PDIV measurements.

The PDIV of natural ester (MIDEL 1215) under AC and DC voltage of both polarities for different aging temperature and aging duration is shown in the Figure 4.6. The results show that for all

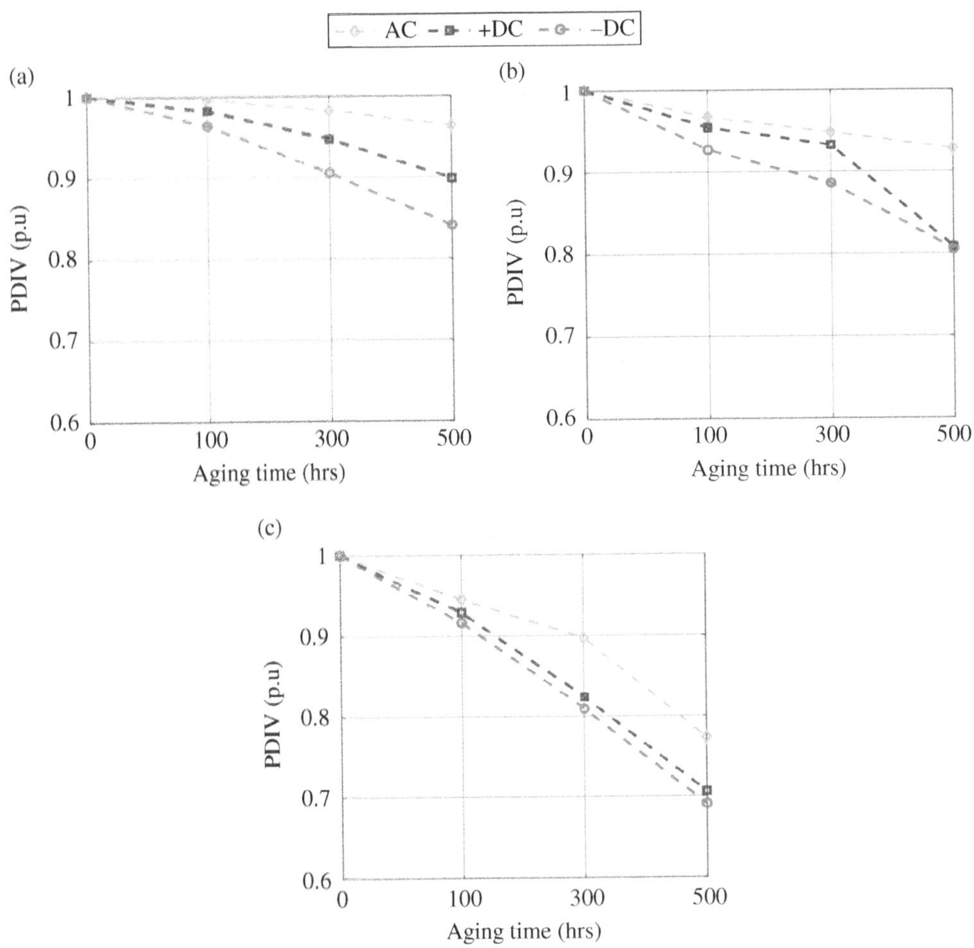

Figure 4.6 PDIV of ester fluids aged at different temperatures under different voltage profiles. (a) 90 °C, (b) 140 °C, and (c) 160 °C.

aging temperatures used in this test, the PDIV of ester decreases with aging duration. The standard deviation also narrows with increasing aging temperature and aging duration. The rate of decrease of PDIV is dependent on the aging temperature and the type and polarity of the applied voltage. The higher the aging temperature the lower the PDIV. The fall in PDIV at higher aging temperature (160 °C) over the 500 hours aging duration is 30.91, 29.33, and 22.68% for the negative DC, positive DC, and AC voltage, respectively. The ionization of the needle tip under positive polarity produces an aiding electric field initiating discharges at a lower voltage compared to negative polarity [26]. Under positive DC voltage, the intensity of light and number of pulses during PD activity are more compared to negative DC voltage. This phenomenon indicates that a higher voltage is required under negative DC voltage for substantial PD activity to occur [57]. The change in PDIV with aging temperature and aging duration can be a useful quantitative parameter for assessing the fluid degradation.

4.3.1.2 Measurement of PDIV Under Harmonic Voltage

With the advancement of power electronic devices and their growing introduction in the transmission network, it has become essential to understand the partial discharge activity in ester fluids under AC harmonic frequencies and different total harmonic distortions (THD) [58] contained in the applied voltage. One way to explore this is to measure the PDIV with application of AC voltage at harmonic frequencies, or with application of an AC voltage with superimposed harmonics containing different THD content. Tables 4.7 and 4.8 show the results of such tests for natural ester fluid aged at temperatures of 90 °C, 140 °C, and 160 °C, and for aging durations of 100 and 500 hours. As shown in Table 4.7, the PDIV decreases moderately by 4–6% harmonic frequency for both aging temperatures. A similar observation can be made from the results of Table 4.8. For example, after 100 hours of aging at 90 °C, the PDIV reduces by about 3% when the harmonic frequency superimposed on the fundamental increases from the third to the seventh harmonic at the same THD content of 4%. However, an increase in the THD content to 40% causes a larger reduction (about 12%) in the PDIV at this aging temperature. Like the results shown in Figure 4.6 for DC and AC voltages, the aging temperature and duration also cause a reduction in the PDIV under harmonic voltages. The statistical information on the PD characteristic features can be used as a diagnostic tool for in-situ PD measurements on transformers. PDIV tests on ester fluid at different gap distances and aging conditions reported in [59] confirmed that the results showed a perfect fit with the Weibull distribution function. Also, it was found in [60] that the scale parameter (α) giving the characteristic lifetime (63.2%) of insulation failure decreases with aging temperature.

Table 4.7 PDIV of thermally aged ester fluids under AC harmonic frequencies.

	PDIV (kV)							
Sample	**2f**		**3f**		**5f**		**7f**	
Aging temp/ aging time	**100 hours aged**	**500 hours aged**	**100 hours aged**	**500 hours aged**	**100 hours aged**	**500 hours aged**	**100 hours aged**	**500 hours aged**
90 °C	11.67	11.39	11.58	11.26	11.48	10.95	11.17	10.76
140 °C	11.53	10.86	11.42	10.69	11.26	10.43	10.96	10.35
160 °C	11.51	10.52	11.22	10.24	11.12	10.21	10.45	9.43

Table 4.8 PDIV of thermally aged ester fluids under AC harmonic frequencies with THDs.

Sample	PDIV (kV)							
	f+3f (4% THD)		f+3f (40% THD)		f+7f (4% THD)		f+7f (40% THD)	
Aging temp/ aging time	100 hours aged	500 hours aged	100 hours aged	500 hours aged	100 hours aged	500 hours aged	100 hours aged	500 hours aged
90 °C	11.48	11.26	10.32	10.01	11.08	10.61	10.19	9.76
140 °C	11.12	10.43	10.13	9.54	10.83	10.07	9.87	9.13
160 °C	10.57	9.75	9.96	9.02	10.05	9.22	9.23	8.23

4.3.2 Flow Electrification

Static electrification is a major concern occurring in oil-filled power transformers that needs attention. The forced oil convection used for heat transfer can result in internal flashover, thereby creating a fire hazard [61]. The oil circulation inside the transformer can cause surface charge accumulation on the pressboard material raising its surface potential, and this phenomenon could lead to incipient discharges over a longer period of time. When the insulating fluid comes in contact with pressboard/metallic interfaces, the solid–liquid couple, which were initially neutral, become polarized due to physico-chemical reactions and then leads to the formation of charges with opposite polarity at the interfacial region. The major source of charge generation at such interfaces is caused by intermolecular collisions. Thus, the electrostatic charging can be explained by collision and impulse process, which could lead to space charge distribution in both the solid and liquid insulation forming the electrical double layer (EDL) [62].

The first layer nearer to the solid surface holds either positive or negative charges depending on its chemical interactions and is known as fixed or compact layer. The second layer is composed of ions that are loosely packed, providing an electrical screening to the first layer. The ions present in this second layer are free to move into the fluid and is known as diffuse layer. The mechanism of charge formation of the liquid with the metallic/pressboard insulation is shown in Figure 4.7. The excess electrons present in the outer metallic structure attracts the positive ions and accumulate on the surface, leaving behind the negative ions to be carried by the insulating fluid (Figure 4.7a) away from the compact layer. The streaming electrification of insulating fluid on the pressboard/paper interfaces in power transformers causes the positive charges to be diffused into the fluid from the EDL (Figure 4.7b). This can be attributed to the molecular structure of cellulose as shown in Figure 4.8, which consists of six hydroxyl groups on its outer layer. The positively charged hydrogen atoms in the –OH groups have a higher affinity toward negative ions present in the fluid and thus negative ions accumulate on the paper/pressboard surface.

4.3.2.1 Flow Electrification Measurement Methods

Depending on the simulation of fluid flow such as planar flow or centrifugal motion, there are different methods [63] used to measure the electrostatic charging tendency of insulating fluids under laboratory condition. The flow loop device [64, 65] was initially developed to estimate the accumulated charge by the planar flow of the fluid on the pressboard using a capacitive sensor. It consists of a pump, a heat regulator, a flow meter, a fluid tank, a relaxation vessel, and a capacitive sensor (Figure 4.9) to perform the charge exchange and movements at the fluid/pressboard interfaces.

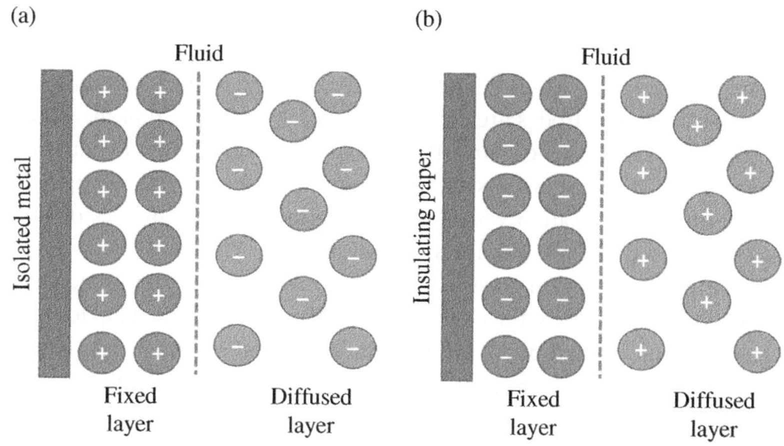

Figure 4.7 Electrical double-layer formation and ion distribution at (a) oil/metallic interface and (b) oil/paper interface.

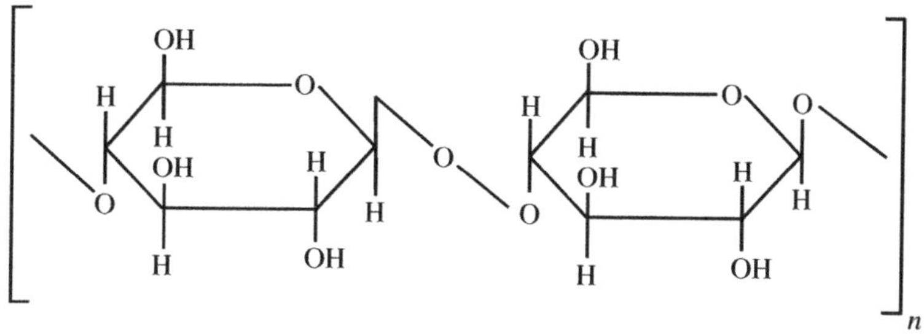

Figure 4.8 Molecular structure of cellulose.

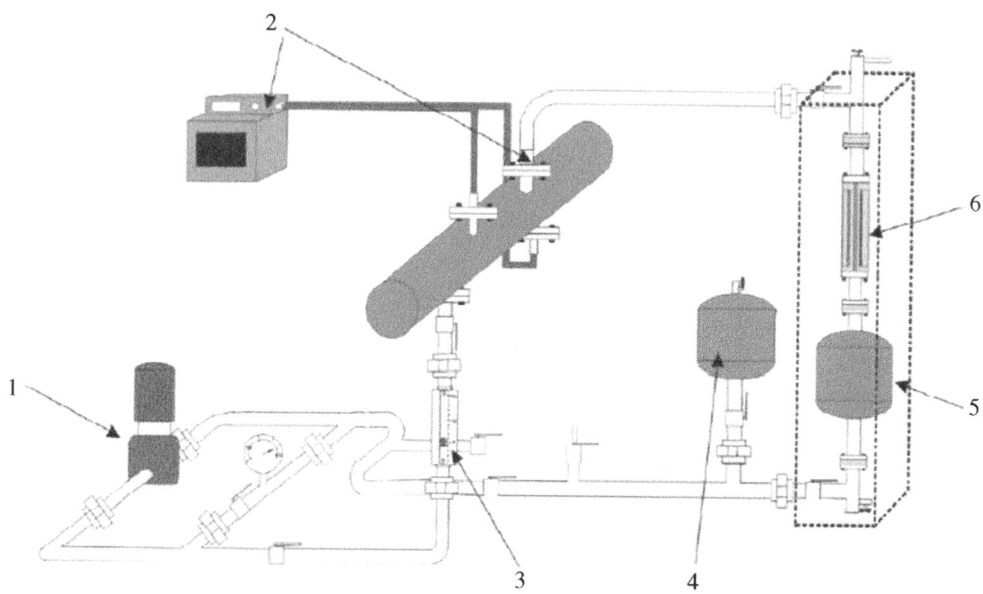

Figure 4.9 Test loop to simulate oil flow (1: pump, 2: heat regulation, 3: flow meter, 4: fluid tank, 5: relaxation vessel, 6: capacitive sensor). *Source:* Based on Paillat et al. [65].

The fluid flow induced by the pump is regulated through a flow meter to the heat exchanger to acquire the desired temperature. It is then fed to the sensor which consists of a rectangular pressboard channel with pressboard sheets stuck together and inserted in a PTFE frame to insulate the pressboard from the electrodes. These electrodes are connected to a picoammeter that measures the accumulation current which is related to the charge that is trapped on the pressboard surface. Both the upstream and downstream leakage currents are measured, while the remnant charges in the fluid are allowed to flow into the relaxation vessel. One of the major disadvantages of this method is that it requires a large volume of oil (20 l) for this analysis.

Before the experiment is started, the loop and pressboard duct are dried through a nitrogen gas flow in vacuum ambience for 24 hours. Further, the oil is allowed to flow through the loop for two hours in order to impregnate the pressboard duct and a relaxation period of 24 hours is maintained. Hence, a relatively long time is needed for the pre-processing as well as for the conditioning of both the solid and liquid components. These drawbacks led to the introduction of filter techniques for measuring the charging tendency of fluid. The Ministatic charge tester [66] was developed to measure electrostatics involved in the jet fuels which was then adapted for application to insulating fluids (Figure 4.10). In this method, the oil sample from a 50 ml plastic syringe is forced through a filter sheet where the charge separation occurs on the surface of the cellulosic paper. This filter paper contains numerous pores with a larger surface area than the pressboard ducts or paper used in transformers. The charges carried away by the filtered oil are measured by the picoammeter, and this filtered current is divided by the volume flow rate of the fluid to obtain the flow current in terms of charge density. There are various factors on which charge density depends, such as oil velocity, filter medium, and temperature, and therefore its value cannot be considered as an absolute value for comparison between different oil samples [63]. Since the measured currents are from the filter holder, the conventional current would be negative and it should be assigned an opposite sign to obtain the charging current of the fluid.

To overcome the drawbacks of the methods discussed above, techniques based on centrifugal motion of the fluid or rotating model were introduced. The Couette charger [67] consists of aluminum co-axial cylinders with the fluid filling the annulus between inner and outer cylinder as shown in Figure 4.11. The fluid-impregnated pressboard material is rolled in a manner similar to

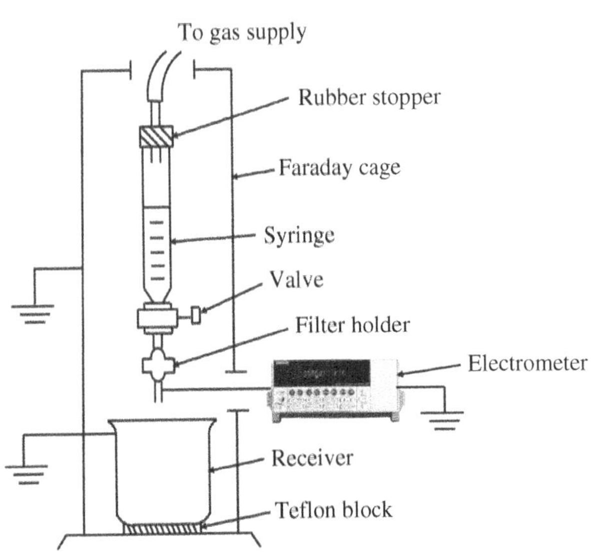

Figure 4.10 Ministatic charge tester. *Source:* Based on Sierota, A. and Rungis [63].

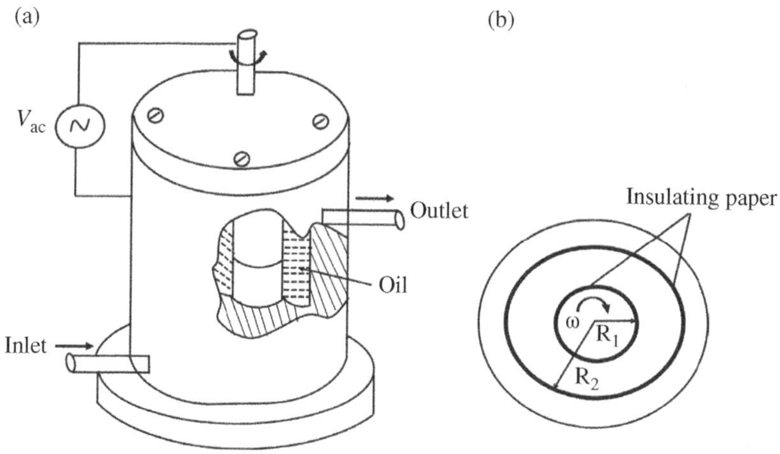

Figure 4.11 Couette charging apparatus with (a) front view on inner cylinder rotation and (b) top view of metallic cylinder. *Source:* Lyon et al. [67] / with permission of IEEE.

that of a real transformer. This method can perform the current measurements under both laminar and turbulent flow. The flow regime is changed by rotating the inner cylinder at speeds up to 2000 rpm. Unlike the previous methods, this method allows for the impact of external electric field on the charging tendency of the insulating fluid under different flow regimes [68]. In a more advanced version of the Couette charger, the absolute charge sensor (ACS) is used for direct measurement of charge densities independent of fluid properties within the instrument. Although this method provides better accuracy, the long time taken for conditioning of process and the large quantity of fluid required make it more complex.

The large-scale models used for measuring the static electrification phenomenon was complemented by the introduction of small models such as the spinning disk system and the CIGRE Test Cell [69]. The charging tendency of the fluid should be measured taking into account the effect of temperature, moisture and type, thickness, and surface roughness pressboard material. The spinning disk technique adopted by CIGRE [69] is based on the rotation of a disc coated with solid pressboard material inside a metallic aluminum vessel filled with insulating liquid, as shown in Figure 4.12. The torque is provided with a variable speed motor, which permits adjustment of the immersion depth of the disc into the liquid. The charge separation at the interfaces cause charges to migrate toward the walls of the aluminum vessel, which are then measured by an electrometer. The experiment can be performed with different pressboard thicknesses, disc diameters, and disc velocities. Also, stray currents are minimized by enclosing the spinning disc system in a Faraday cage. A comparative evaluation of the different methods used for measuring the electrostatic charging tendency (ECT) of insulating fluids is summarized in Table 4.9.

4.3.2.1.1 *Effect of Spinning Velocity*

The change in the streaming current with disc velocity can influence the streaming current by varying the hydrodynamic flow conditions within the fluid. The increase in disc velocity will cause increased streaming current due to shear stress and change in its diffusion sublayer thickness [70]. This phenomenon may also be partially attributed to the frictional stress existing between the pressboard and the insulating fluid. An increase in the degree of aging can have a major impact on the charging tendency of the fluid due to the formation of various oxidation by-products and may even cause the current to flow in the opposite direction [71]. Thus, the fluid flow velocity in the pipeline of the cooler in transformers are usually limited to 3 m/s or less [72].

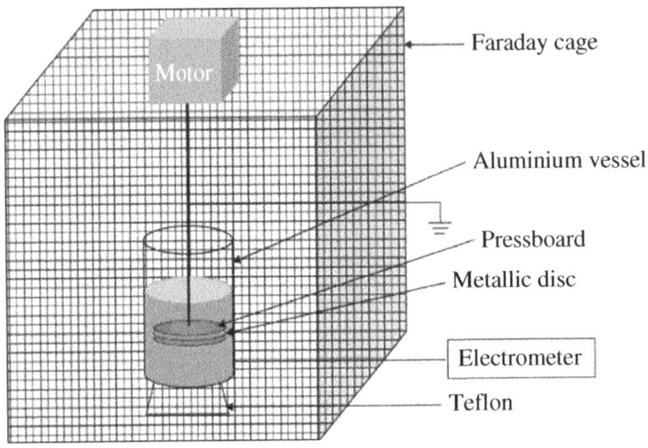

Figure 4.12 Spinning disc system used for streaming electrification of insulating liquids. *Source:* C.J. W. Group [69].

Table 4.9 Evaluation of different methods used for ECT of fluids.

Characteristics	Planar flow	Rotating flow	Filter methods
Type of flow	Laminar/turbulent	3-dim rotating flow	Laminar
Sample component	Fluid/pressboard (duct or pipe)	Fluid/pressboard (CTC, CC), oil (metallic disc)	Fluid
Reference component	Fluid/pressboard simultaneously	Fluid/pressboard or fluid/ metallic disc simultaneously	Paper and steel
Sample size of the liquid component	High	Low (medium)	Low
Sample size of the solid component	High	Low	Low
Reproducibility	Medium (low)	Medium (low)	High
Applicability to transformers	High	Medium	Low
Standardization	Not suitable	Less suitable	Suitable

4.3.2.1.2 *Effect of Temperature*

Transformers can reach very high temperatures during operation, and this impacts negatively on the charging tendency of the paper/fluid insulation. Indeed, the magnitude of the streaming current increases exponentially with increasing temperature, and this behavior may be related to the diffusion coefficient and the ionic mobility within the fluid [73]. Moreover, at high temperatures the motion of the fluid becomes turbulent causing more charge carriers to be accumulated on the pressboard material. Compared with mineral oils, ester fluids experience a higher charging current with aging, with noticeable difference observable starting around 80 °C.

4.3.3 Spectroscopic Studies

Spectroscopic analysis is the mostly widely used method for both quantitative and qualitative analysis of insulating liquids. The determination and identification of spectra produced due to the interaction of electromagnetic radiation within the fluids depends on their structural and molecular composition. Depending on the spectral region of electromagnetic radiation employed for the analysis, these methods are classified as ultraviolet (UV), visible (Vis), infrared (IR), and radio (nuclear magnetic resonance). The most commonly used spectroscopic methods are UV-Vis absorption, fluorescence, and chemiluminescence spectroscopy, and these are discussed in the following sections.

4.3.3.1 UV-Visible Spectroscopy

The UV-Visible region of electromagnetic radiation lies in the range of 200–700 nm and absorption in this wavelength range is related to the electronic transition within the molecules. The major transitions occur from the bonding or nonbonding orbitals with the antibonding orbitals where the interaction within the fluid can lead to absorbance, fluorescence, and reflection. The UV-visible spectroscopy is the most sensitive nondestructive technique [74] that measures the amount of absorption based on the chromophores present in the insulating liquid. According to Beer-Lambert's law, the amount of light absorbed by the solute in an isotropic homogeneous medium is proportional to the concentration of solute in the fluid and it is given by

$$A = \varepsilon l c \tag{4.3}$$

where ε is the absorptivity, l is the path length of light through the solution, and c is the concentration of species absorbing the light. The schematic experimental setup used for UV-Vis absorption spectroscopy is shown in Figure 4.13. It consists of a light source that is of deuterium and tungsten halogen lamp providing radiation in the UV-Visible spectrum range. This light source is fed to the monochromator to disperse the beams to the sample fluid and the reference fluid, and then measure them through a detector.

The results of UV-Vis absorption spectroscopy carried out on thermally aged ester fluids is shown in Figure 4.14. A substantial shift in the wavelength is possible for increasingly higher aging temperatures and longer aging durations. This may be due to the new chromophoric moieties that are formed during the darkening of the fluid's color. Also, when the fluid is highly concentrated, there occurs a saturation in its absorption value and hence a derived absorption plot is used at the inset to provide a quantitative indication of the shift of wavelength. Furthermore, the dissolved decay products could be identified by the evaluation of area under the absorbance curve [75]. Usually, higher absorption peaks are absorbed at 275 and 315 nm, which corresponds to copper ions that are dissolved during the aging experiments on the insulating liquid. Accordingly, it can

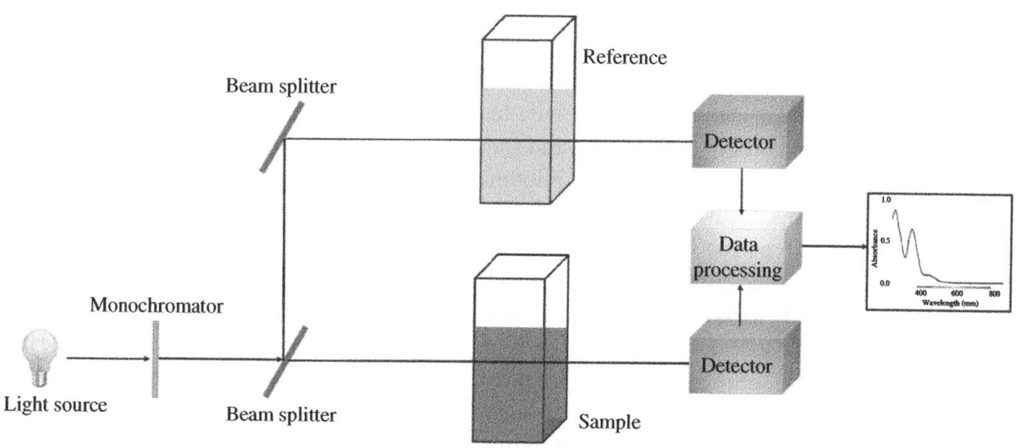

Figure 4.13 Schematic diagram of UV-Visible spectroscopy measurement.

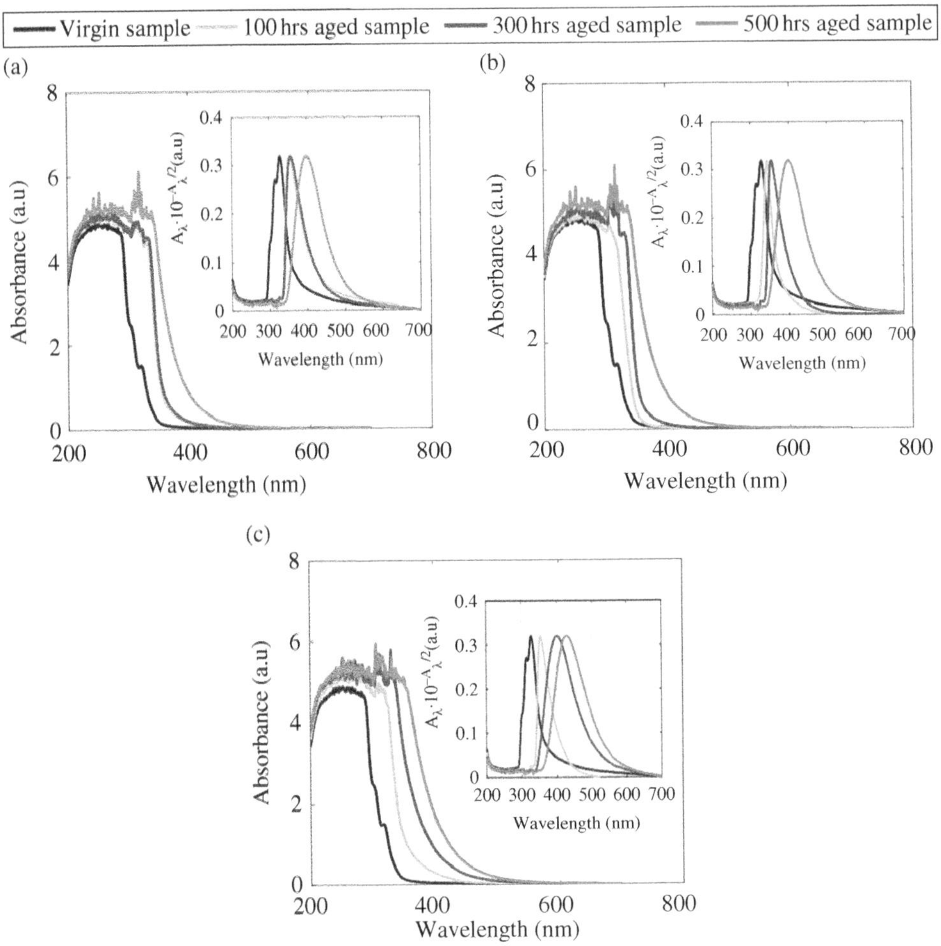

Figure 4.14 UV-visible absorption spectrum of ester fluid aged at (a) 90 °C, (b) 140 °C and (c) 160 °C.

be said that the UV-Vis absorption method could provide information on the degradation status of the insulating fluid and on the new elements formed during the fluid's pigmentation.

4.3.3.2 Fluorescence Spectroscopy

Fluorescence is a type of luminescence that occurs due to the emission of light by a substance that has absorbed light or any other electromagnetic radiation. During this process, the excitation of molecules occurs, taking them from the ground state to the excited state. Subsequently, the molecules return to the ground state again losing photons in the form of fluorescence [76]. The emitted light always has a longer wavelength (lower energy) than the absorbed radiation. Transformer fluid insulation, which consists of a large number of hydrocarbons, shows an intrinsic fluorescence property, and during aging, these bulky hydrocarbon molecules can break leading to certain oxygenated compounds with a loss of fluorescence. Also, vegetable oils contain a rich source of fluorescent substances such as unsaturated fatty acids, vitamin E, and its derivatives [77]. The schematic experimental setup used for fluorescence measurement shown in Figure 4.15 consists of monochromator both at the excitation and emission side unlike the UV-Vis absorption spectroscopy where it is placed only at the excitation source side. The emission spectrum can be obtained

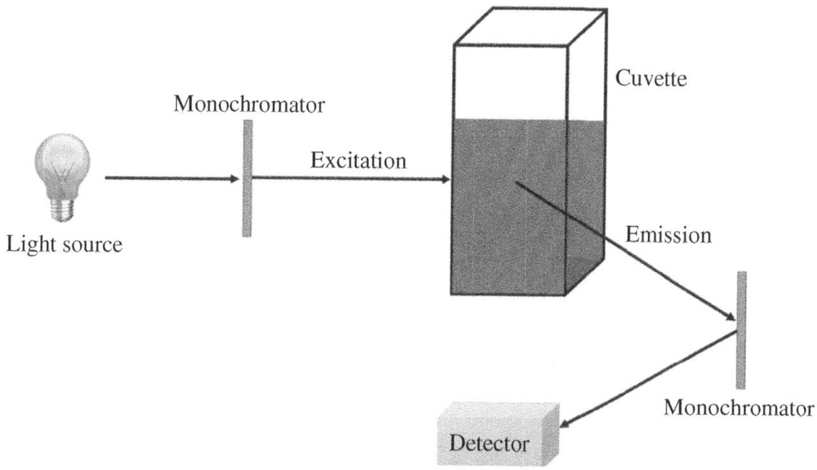

Figure 4.15 Schematic diagram used for fluorescence measurement.

by varying the emission intensity as a function of wavelength at a constant excitation wavelength. Also, the excitation emission matrix (EEM) in 3D form could be retrieved by varying the emission spectra at different excitation wavelengths.

The degradation of ester fluid with aging leads to the formation of new carbonaceous particles with a reduction in fluorescence intensity due to the loss of fluorophore [78]. The excitation-emission matrix (EEM) of unaged and aged ester fluid for different time periods is shown in Figures 4.16 and 4.17. The excitation maxima of unaged ester fluid lie in the range of 300–415 nm and a prominent Stokes shift occurs for the fluid aged for a longer duration [60]. The excitation maxima from the EEM plot are in close correspondence with the derived absorption wavelength from the UV-Vis absorption spectroscopy. Thus, an understanding of the characteristic changes in the EEM plot of insulating fluids and its correlation with other dielectric properties can be used as a quantitative indication for the condition monitoring of transformers.

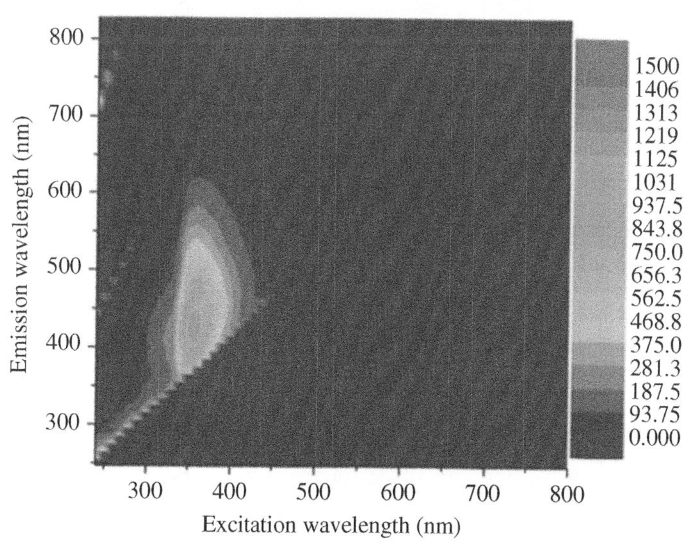

Figure 4.16 EEM spectra of unaged ester fluid.

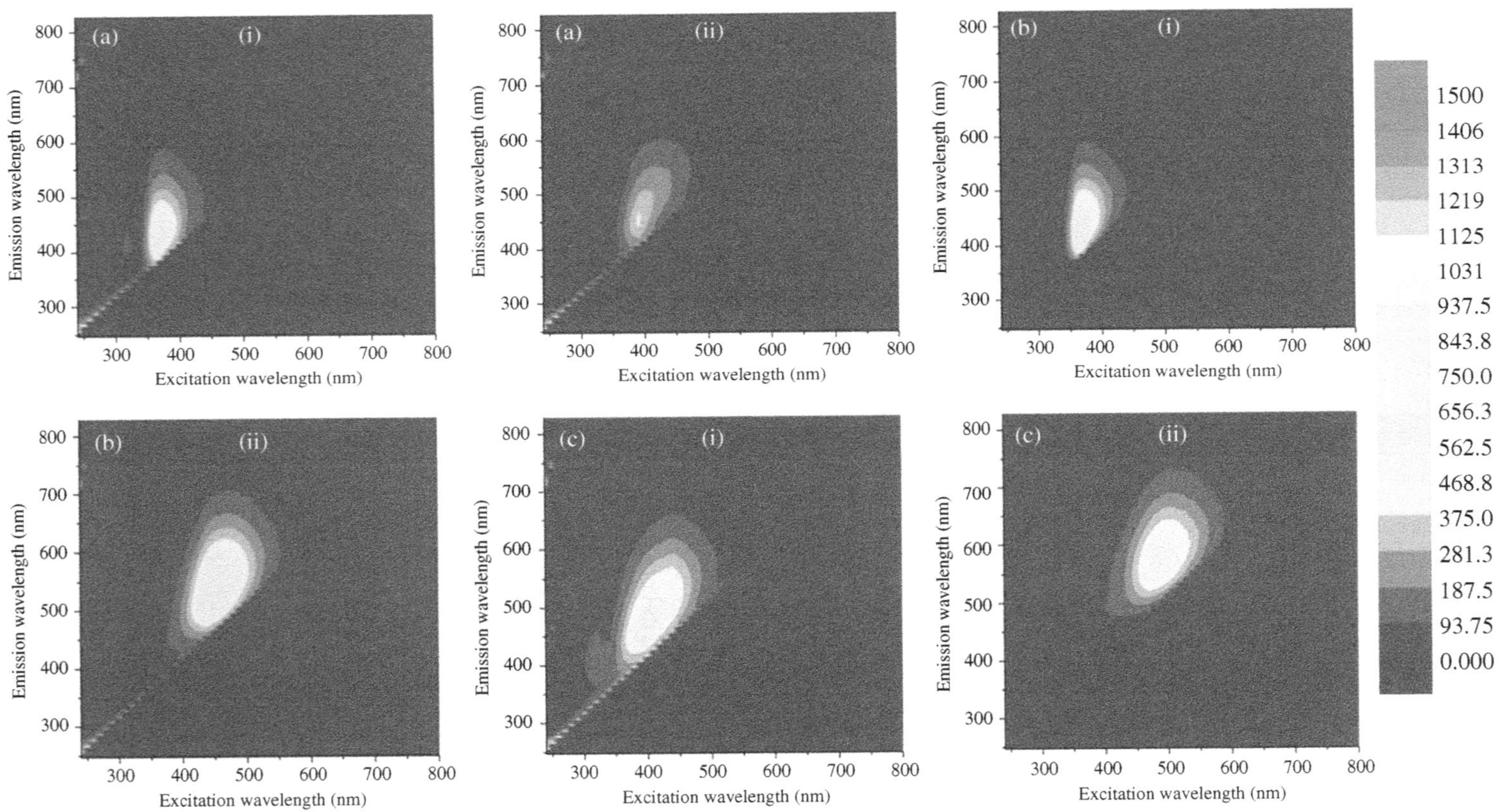

Figure 4.17 EEM spectra of ester fluids aged at (a) 90 °C, (b) 140 °C and (c) 160 °C for (i) 100 hours and (ii) 500 hours aging duration.

4.3.4 Dielectric Response Spectroscopy

The interaction of an external electric field with the dipole moment of insulating fluids is assessed in the dielectric response measurements. The experimental setup consists of a three terminal test cell as shown in Figure 4.18 to contain the fluid samples according to IEC 60247 [79]. The desired sinusoidal supply voltage with variable frequency is fed using an OMICRON-DIRANA unit. The maximum temperature allowed in the test cell is around 110 °C, and the samples are tested at different temperatures by maintaining equilibrium conditions during the measurement period.

The aging of insulating fluid inside the transformer can lead to the formation of various acidic and polar substances affecting its permittivity and dissipation factor [80]. The oxidative reaction of ester fluids at high temperature may cause the degradation of unsaturated fatty acids and the free radical mechanism could further lead to the increase in the fluid's conductivity. The variation of the loss factor tan δ is investigated over a wide frequency range (0.1–1 kHz) as shown in Figure 4.19. A high tan δ is usually observed at low frequencies due to the interfacial polarization that causes charge formation at the interface between the dipole molecules of fluids and particles formed with aging. At high frequencies, the loss factor reduces substantially. The thermal aging at high temperature causes an increase in the loss factor, which is indicative of an increase in its ionic mobility due to hopping of charge carriers [81]. Compared with mineral oil, the ester fluid has a higher tan δ because of its higher dielectric constant and relatively higher conductivity. Also, an inverse relation between the dielectric constant and coulombic attraction of oppositely charged ions could be responsible for the higher dissipation factor. The frequency domain spectroscopy (FDS) analysis could help identify the different polarization mechanisms and the loss factor, thus giving an indication on the degradation status of fluid.

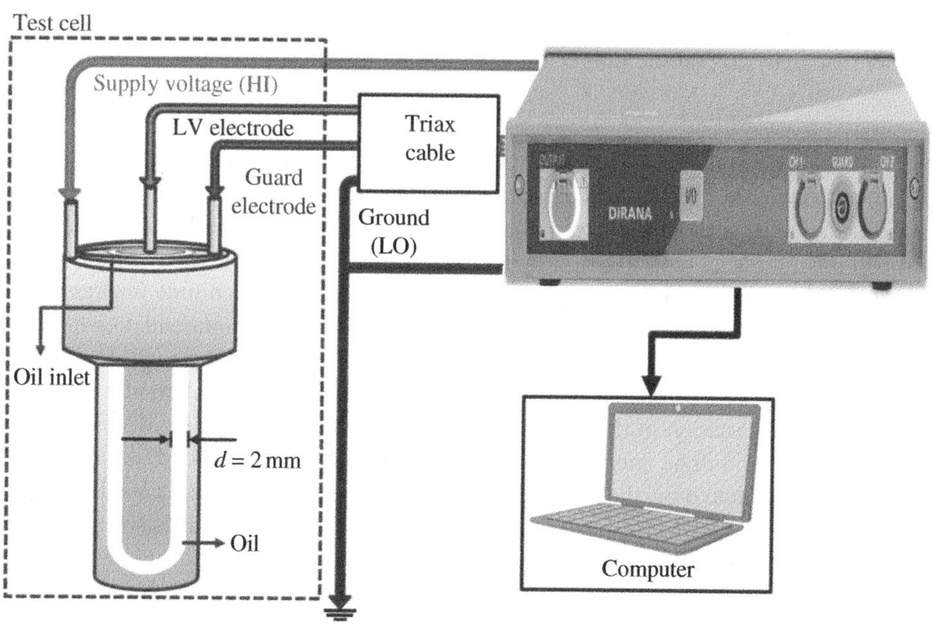

Figure 4.18 Experimental setup used for dielectric response spectroscopy. *Source:* IEC 60247 [79].

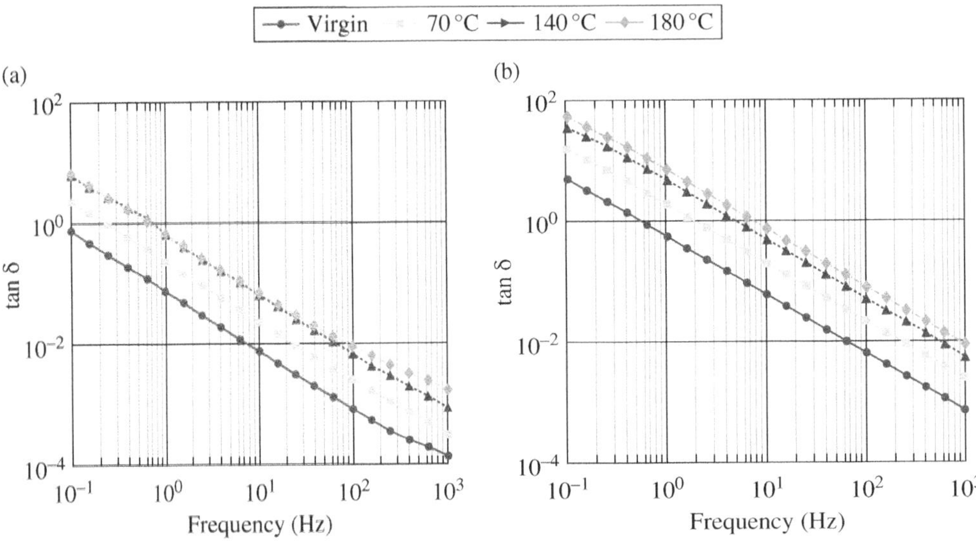

Figure 4.19 Variation of dissipation factor with frequency for aged ester fluid at temperatures of (a) 30 °C and b) 90 °C.

4.3.5 Physico-Chemical Studies

Apart from the above discussions on dielectric properties of ester fluid, various reactions such as oxidation and hydrolysis inside the transformer can cause the formation of free radicals, accelerating the aging process [82]. Also, there is a possibility for solid particle (pressboard and copper) contaminants to intrude into the fluid, which can lead to localized partial discharge and then eventually breakdown of the insulation. A better understanding of the physico-chemical changes occurring during operation can help in the condition assessment of the fluid. Different methods that can be used for analyzing these changes are discussed in the following sections.

4.3.5.1 Interfacial Tension

The interfacial tension between electrical insulating liquids and water is an indirect measure of the impurities contained in either liquid, which migrate under charge attraction into the water at the interface, creating an associated tensile force [30]. An increase in contaminants due to degradation causes a drop in the interfacial tension between fluid and water. Thus, measurement of the IFT is an effective tool to monitor the fluid condition, and a reduction of its magnitude (expressed in mN/m) indicates increased content of polar contaminants in the liquid. Recommendation for testing transformer oil IFT is available in the ASTM D971 [83] and IEC 62961 [84] standards. Natural ester fluids have a lower IFT compared with mineral oils because of their inherently different chemical structure. The variation in the interfacial tension (IFT) with aging time of ester fluids under thermal aging conditions measured according to ASTM D971standard is shown in Figure 4.20. The time taken to draw the ring through the interface was around 30 seconds and the total measurement was completed at 60 seconds. Upon aging, the moisture diffusion from the pressboard material to the ester fluids occurs, and it is influenced during its suspension with water molecule, decreasing its IFT. For natural ester, an IFT value of less than 22 mN/m indicates an uncertain condition inside the transformer [85] and if it reaches below 15 mN/m, then complete resampling of fluid is to be done [86]. In general, a reduction of more than 40% of the IFT value

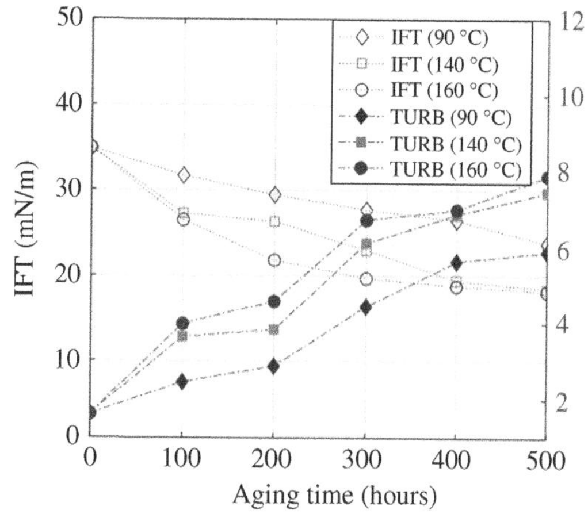

Figure 4.20 IFT and turbidity of ester fluid during thermal aging.

Table 4.10 Limits on IFT based on voltage class of transformer.

Test method	Value for voltage class		
	≤69 kV	69–288 kV	>345 kV
Interfacial tension (mN/m)-minimum limit	24	26	30

Source: IEEE C57.106 [87].

from the value measured with new fluid should trigger further investigative checks, as per IEEE C57.147. Also, depending on the voltage class of transformer, ASTM D971 describes a minimum limit on the IFT for better operation of insulating fluid inside the power transformer (Table 4.10).

4.3.5.2 Turbidity

Turbidity is a measure of transparency, which indicates the amount of suspension of solid particles in the fluid. The optical-based laboratory technique quantitatively determines the amount of microscopic solid suspension formed in insulating liquids [88]. Turbidity increases with aging as shown in Figure 4.20 and it could be reduced by removing the suspended or colloidal particles by a centrifugation process [89]. However, some tiny particles may still be present in noncolloidal form that gets mixed with insulating fluid causing an increase in the turbidity even after the centrifugation process. The insoluble colloidal suspensions are mainly from the copper and aged pressboard along with the fluid, which can further cause a darkening of the fluid's color.

4.3.5.3 Viscosity

The viscosity of insulating fluids determines the cooling effectiveness of power transformers during normal operating regime and therefore their temperature rise. For virgin and aged ester fluid samples, Figure 4.21 shows that the viscosity decreases with increase in the measuring temperature due to the weak cohesion forces occurring between molecules, while only a slight variation in the viscosity occurs at the lowest aging temperature of 90 °C. A significant increase occurs at the higher aging temperatures of 140 and 160 °C and at longer aging durations. The molecular structure of synthetic ester being of pentaerythritol type does not allow much impact on thermal stress

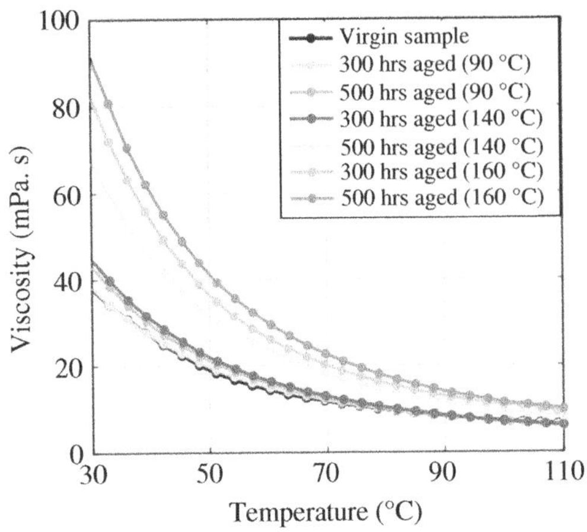

Figure 4.21 Viscosity of ester fluid aged at different temperatures.

Table 4.11 Composition of thermal aged ester fluid (160 °C) based on GC-MS data.

	Relative composition (%)			
Organic groups/compounds	**Unaged**	**100 hours aged**	**300 hours aged**	**500 hours aged**
Carboxylic acids & anhydrides	4.20	16.60	42.10	47.10
9, 12 Octadecadienoic acid ($C_{18}H_{32}O_2$)	4.2	14.20	19.00	22.40
Carbonyl compounds	2.80	4.30	3.80	11.20
Antioxidants (Tocopherol)	3.20	1.30	2.82	0

compared to natural ester which is of triglyceride structure [90]. This difference in structure between natural and synthetic ester fluid causes a variation in their visco-elastic behavior. The rotational Reynolds number being a function of viscosity defines the rheology of ester fluids, and this parameter could be used to relate the laminar and turbulent flow conditions responsible for the static charges inside power transformers.

4.3.5.4 Organic Composition of Oil Using GC-MS

Gas chromatography/mass spectrometry (GC-MS) technique is used nowadays to analyze the organic composition present in the insulating fluid. An inert gas such as helium is used as a carrier and, based on its volatility, the different compounds can be identified. Table 4.11 gives the relative composition of different functional groups of an ester fluid sample aged at 160 °C for different durations. The functional groups are identified using the NIST MS database. The thermal oxidation of ester fluid leads to carboxylic acids, ketones and alcohols accompanied with a reduction in its ester composition, indicating its interconversion via thermohydrolysis. The acids and ketones are usually formed by alkoxy radical β-scission, hydrogen transfer, and disproportion [91].

4.4 Assessment of Paper Degradation

The cellulosic pressboard/paper insulation used in transformers contains 90% cellulose, 6% hemicellulose, and 3% lignin. With prolonged exposure to operational stresses, degradation of these materials takes place and hence understanding the kinetics involved in its deterioration is important in predicting the paper/pressboard insulation lifetime. The different methods used to diagnose the cellulosic material are given in the following sections.

4.4.1 Surface Discharge Analysis

Surface discharge is one of the important phenomena that can lead to a catastrophic failure in power transformers under normal operating conditions. Different types of surface discharge such as sliding discharge, creeping discharge, and surface tracking can form only when the tangential electric field at the fluid/pressboard interface is greater than 1 kV/mm [92]. These kinds of discharges start from the fluid breakdown at the outer coil winding and then traverse toward the pressboard barrier, where the plasma may ignite forming white marks following moisture evaporation from the cellulosic material [93]. At a later stage, carbonized traces are also visible leading to a conductive path that facilitates breakdown. It is general practice to match the permittivity of both the fluid and pressboard so as to avoid the appearance of discharges. Ester fluids offer a better matching of their permittivity with the pressboard insulation compared to mineral oils with a lower stress exhibited at the interface [94]. Further, the surface discharge also depends on the impurities such as bubbles or copper particles affecting the discharge characteristics. Using similar techniques as PDIV, the surface discharge inception voltage (SDIV) can be measured. Table 4.12 shows the SDIV of thermally aged ester impregnated pressboard measured under AC voltages and harmonic frequencies with different THD contents. Under AC voltage, the charges remain on the pressboard surface even after the first discharge, increasing the localized electric field required for the initiation of the discharge [95]. Also, the effect of harmonics shows an increase in the SDIV compared to fundamental frequency since the time required for the discharge to occur is much longer than the time needed for the voltage to reach its inception value.

Table 4.13 shows the SDIV of thermally aged ester-impregnated pressboard under DC voltages with ripple content. The inception voltage under pure negative DC is higher than that under positive DC and AC voltage. The discharges under positive polarity are more radially oriented and more conducting compared to negative polarity [96]. With the advancement in the HVDC power transmission and other grid-connected renewable sources, the insulating material is subjected to pulsating DC voltages [97]. For both polarities, the SDIV decreases with the increase in the percentage ripple of the applied voltage. This is attributed to an increased rate of change of the applied field under ripple DC voltage causing the discharge inception to start at a lower voltage compared to pure DC voltage.

Table 4.12 Surface discharge inception voltage (SDIV) of thermally aged ester-impregnated pressboard under AC voltages at harmonic frequencies with THDs.

	Voltage profile	Harmonic frequencies			Harmonic frequencies with THD	
Sample	**AC**	**3f**	**5f**	**7f**	**f+3f (40%)**	**f+7f (40%)**
Unaged	10.70	11.56	11.92	12.85	11.21	10.60
90 °C aged	10.20	11.35	11.65	12.36	10.95	10.25
140 °C aged	9.34	10.73	11.27	11.73	10.46	9.85
160 °C aged	8.74	10.25	10.48	11.61	10.03	9.14

Table 4.13 Surface discharge inception voltage (SDIV) of thermally aged ester-impregnated pressboard under DC voltages with ripple content.

	Voltage profile		+DC with ripple		−DC with ripple	
Sample	**+DC**	**−DC**	**3%**	**5%**	**3%**	**5%**
Unaged	22.53	28.31	20.30	19.34	25.34	24.21
90 °C aged	21.12	27.27	18.76	17.48	23.36	22.24
140 °C aged	18.69	26.68	15.45	14.23	19.51	18.23
160 °C aged	18.27	24.34	14.69	13.31	17.56	16.12

4.4.2 Surface Potential Measurement

Surface potential on the pressboard material can cause serious insulation failure in converter transformers creating more risks for its safe and reliable operation [98]. The charges accumulate on the surface of pressboard and can become trapped in the voids or defecting parts of the insulating material. Conventionally, there are different techniques used for charge injection when measuring the surface potential such as the sharp point emitter and external corona source. The corona generated using the needle-plane electrode configuration is shown in Figure 4.22. A DC voltage of required magnitude is supplied through a high-voltage amplifier from the signal generator for a specific time duration. The charge deposition is based on the corona discharge process when the sample is at one position (Position 1), which is moved to Position 2 where the potential decay is measured using the electrostatic voltmeter.

The magnitude of charge deposition depends on the resistivity of the material, and these charges lead to build-up the potential on the surface. The potential decay associated with the insulating material is given as

$$V = V_0 . e^{-\lambda t} \tag{4.4}$$

where V_0 represents the initial potential, $\lambda = 1/\tau$ is the decay rate, and τ is the mean lifetime of potential decay process.

The potential decay characteristics of virgin and thermally aged ester-impregnated pressboard samples is shown in Figure 4.23 for positive and negative applied DC voltages. Clearly, all samples experience a surface potential that increases with increasing aging temperature. From the figure, the initial (t = 0) surface potential under negative polarity is higher than that under positive

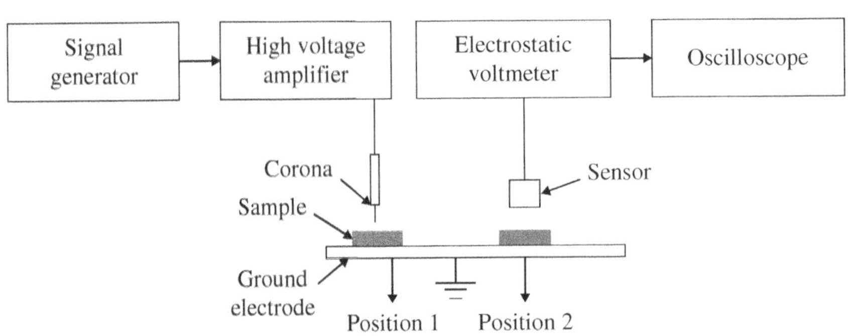

Figure 4.22 Surface potential measurement setup.

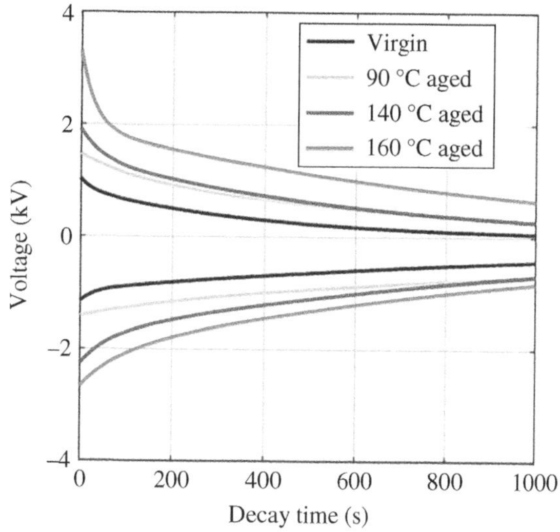

Figure 4.23 Surface potential decay characteristics of thermally aged ester-impregnated pressboard material.

polarity for all pressboard samples. However, the surface potential decays at a higher rate under positive polarity than the negative polarity [99]. More charges are trapped on the surface of aged pressboard material under negative polarity causing an increased mean lifetime. After a certain decay time, the decay rate tends to an approximately constant value depending on the aging temperature, indicating the absence of remnant charges in the material. The conduction and diffusion of ester fluid into the pressboard material is closely related to the trap energy distribution of the insulating polymers [100]. The trap density proportional to the product of time and voltage gradient $t(\mathrm{d}V/\mathrm{d}t)$ [101] and the energy of demarcation associated with the decay process can be calculated according to Simmon and Tam's theory as:

$$N_t = \frac{4\varepsilon_0\varepsilon_r}{eL^2kT}\left|t\frac{\mathrm{d}V}{\mathrm{d}t}\right| \tag{4.5}$$

$$E_t = kT\ln\left(\nu t\right) \tag{4.6}$$

where ε_0 is permittivity of free space (8.854×10^{-12} F/m), ε_r is the dielectric constant of the material, e is the charge of the electron (1.602×10^{-19} C), L is the thickness of the specimen, k is the Boltzmann constant (1.38×10^{-23} m^2 kg/s^2/K^1), ν is the escape frequency of electron (10^{12} Hz), and T is the temperature. The first peak in the trap distribution characteristic shown in Figure 4.24 indicates the shallow traps and the second peak reflects the deep traps formed on the pressboard material. The transfer of charges from the surface to bulk is more severe in the case of aged material compared to unaged one, leading to deep trap formation. For negative polarity, the shallow trap density increases with aging temperature and reaches peak values at trap energy levels between 0.87 and 0.89 eV, as shown in Figure 4.24a. The same observation holds for positive polarity corona, but the peak densities of aged samples appear to occur at about the same trap energy of 0.89 eV (Figure 4.24b). These results indicate that the trap density and trap depth can be used to grade the aging status of pressboard material in the transformer.

4.4.3 Impedance Spectroscopy

Impedance spectroscopy is a relatively new technique that utilizes an AC signal to excite the system and measures its current and voltage response. The Nyquist plot, which consists of plotting the

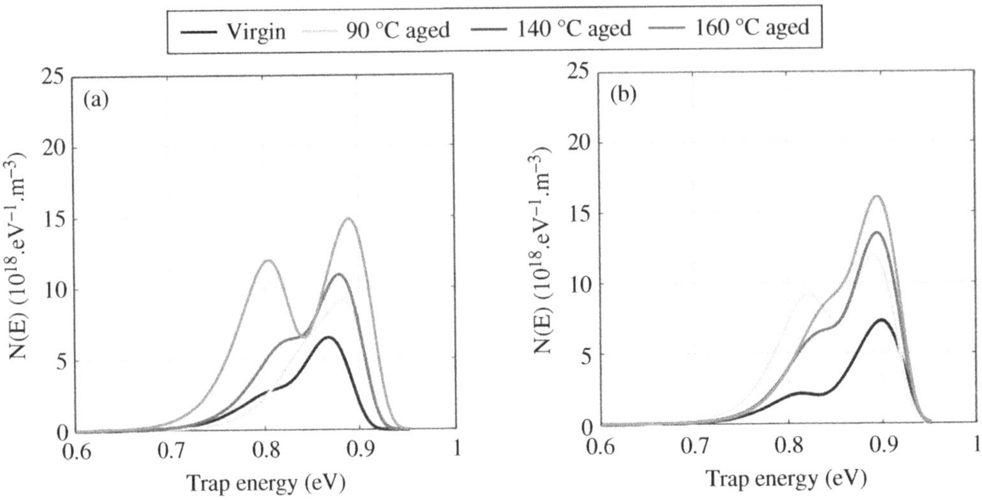

Figure 4.24 Trap distribution characteristics under (a) positive corona and (b) negative corona.

real and imaginary parts of the measured complex impedance data at different frequencies to analyze the dielectric behavior of the insulating material. The complex Nyquist plot and the frequency explicit plots (Bode plots) are used nowadays to determine electrical conductivity and dipole relaxation in the cellulosic pressboard material [102]. Figure 4.25a shows that for thermally aged ester impregnated pressboard, the Nyquist plot does not follow a full semicircle indicating a non-Debye type relaxation. The high impedance value for the virgin, unaged sample reflects the pressboard material acting as a barrier. The first semicircular arc seen for all samples at high frequencies in the inset of Figure 4.25a corresponds to the first relaxation frequency and shows a marginal reduction in the impedance with aging temperature. The second arc defines the interfacial polarization effect and charges accumulating at the surface of the dielectric at low frequencies, corresponding to further relaxations. As can be seen, the charge accumulation increases with aging temperature. The intercept of the semicircle with the real axis (Z') determines the bulk resistivity of the material. An increase in degradation of the pressboard material due to thermal aging results in the rupture of polymeric chains, which in turn reduces the grain boundary resistance between the glucose units of cellulose.

Figure 4.25b shows the variation of the real part of impedance Z' with frequency at different aging temperatures. The applied electric field and local distortions in the pressboard with aging causes a reduction in Z' indicating a negative temperature coefficient of resistance at low frequencies. A reduction in the peak value of the loss factor (Z'') and an increase in the corresponding frequency is observed with aging temperature (Figure 4.25c). Figure 4.25d shows the variation with frequency of the loss factor (Z'') normalized to the first peak value $\left(Z''_{\max}\right)$ of the virgin and thermally aged samples. The shift in peak toward higher frequencies indicates a reduction in resistivity and an increase in the loss factor of the material [103]. The Nyquist plot was fitted using a double RC parallel circuit in series as shown in Figure 4.26, where the bulk resistances (R_1, R_2) and capacitances (C_1, C_2) of the pressboard material can be retrieved from the results of impedance spectroscopy (Table 4.14). The first RC circuit (R_1C_1) represents the pressboard insulation, whereas second RC circuit (R_2C_2) indicates the insulating fluid impregnated on the surface of pressboard material.

The dielectric relaxation time calculated based on the frequency of lossy peak maxima ($Z''_{\max}$) is given by:

$$2\pi f_p \tau = 1 \tag{4.7}$$

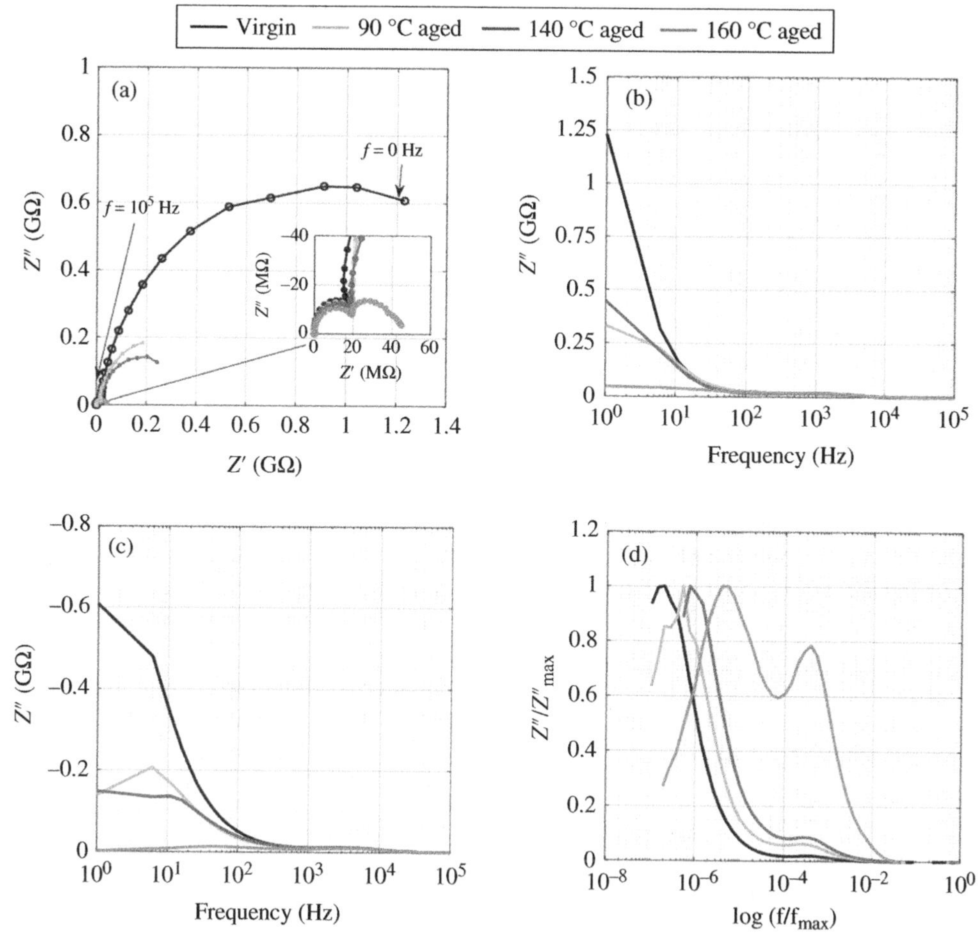

Figure 4.25 (a) Nyquist plot of thermally aged samples, (b) real part of impedance with frequency, (c) imaginary part of impedance with frequency, (d) normalized part of impedance with frequency.

Figure 4.26 Equivalent circuit model.

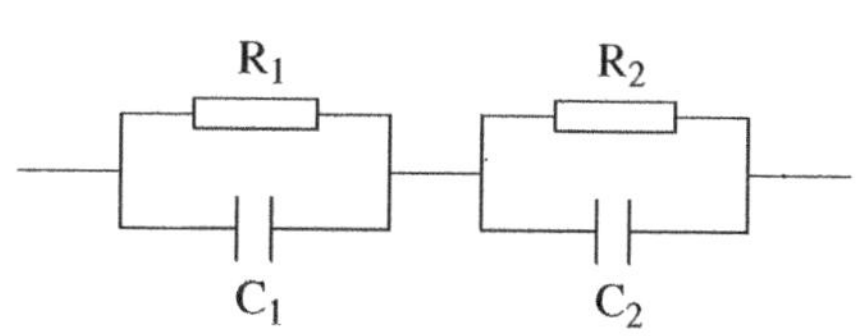

Table 4.14 Equivalent electrical circuit parameters.

Sample	R_1 (GΩ)	R_2 (GΩ)	C_1(pF)	C_2(pF)
Unaged	1.16	0.42	87.06	49.14
90 °C aged	0.36	0.02	58.52	3.25
140 °C aged	0.27	0.02	51.02	2.61
160 °C aged	0.04	0.01	45.32	2.06

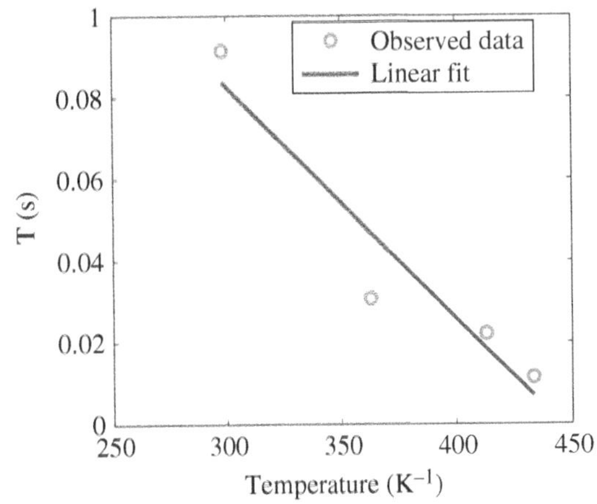

Figure 4.27 Relaxation time with aging temperature.

where f_p corresponds to maximum frequency of Z''_{max}, and τ is the relaxation time of the dielectric. The bulk resistance of pressboard material decreases with increasing aging temperature, whereas an inverse linear relation is followed between relaxation time and aging temperature (Figure 4.27). Consequently, the characteristic variation in relaxation time may be used as reference in industry to determine aging level of insulating pressboard material.

Recently, electric modulus has been used to provide an alternative tool to analyze the charge transfer mechanism and mobility of the pressboard material. Electric modulus (M^*) is defined as the inverse of complex permittivity (ε^*).

$$M^* = \frac{1}{\varepsilon^*} = \frac{\varepsilon'}{\varepsilon'^2 + \varepsilon''^2} + i\left(\frac{\varepsilon''}{\varepsilon'^2 + \varepsilon''^2}\right) = M' + jM'' \tag{4.8}$$

$$M^* = M' + jM'' = j\omega C_O Z^* \tag{4.9}$$

where $\omega = 2\pi f$ is the angular frequency, C_O is the geometrical capacitance of empty test cell, and Z^* is the measured complex impedance of the sample. The transformed impedance data can provide localized relaxation as compared to conductive or resistive analysis.

The modulus spectrum shown in Figure 4.28a reveals asymmetry with reduced semicircular arcs, which indicates that a temperature-dependent hopping mechanism is involved in the process of charge transfer and a non-Debye type of relaxation. Figure 4.28b represents the variation with frequency of the real part of electric modulus (M') for thermally aged material. No change in the short-range mobility was observed for thermally aged material due to its strong restoring force toward the electric field. The low frequency part of the peak values in the modulus represents the long-distance range of ion mobility, whereas the high-frequency part indicates only localized motion. There is no shift in the frequency of the maximum value M''_{max} as shown in Figure 4.28c, which also confirms the static nature of ionic charges even at high aging temperature of the pressboard samples. Thus, the transition from long range to short range mobility can be determined by observing the peak of the imaginary part of the electric modulus M''.

4.4.4 Py-GC/MS

The pyrolysate composition is a fingerprint of the structure of the material, as the organic groups evolve via the cleavage of covalent bonds present in the material [99]. This property is exploited by performing analytical pyrolysis-gas chromatography/mass spectrometry (Py-GCMS) that

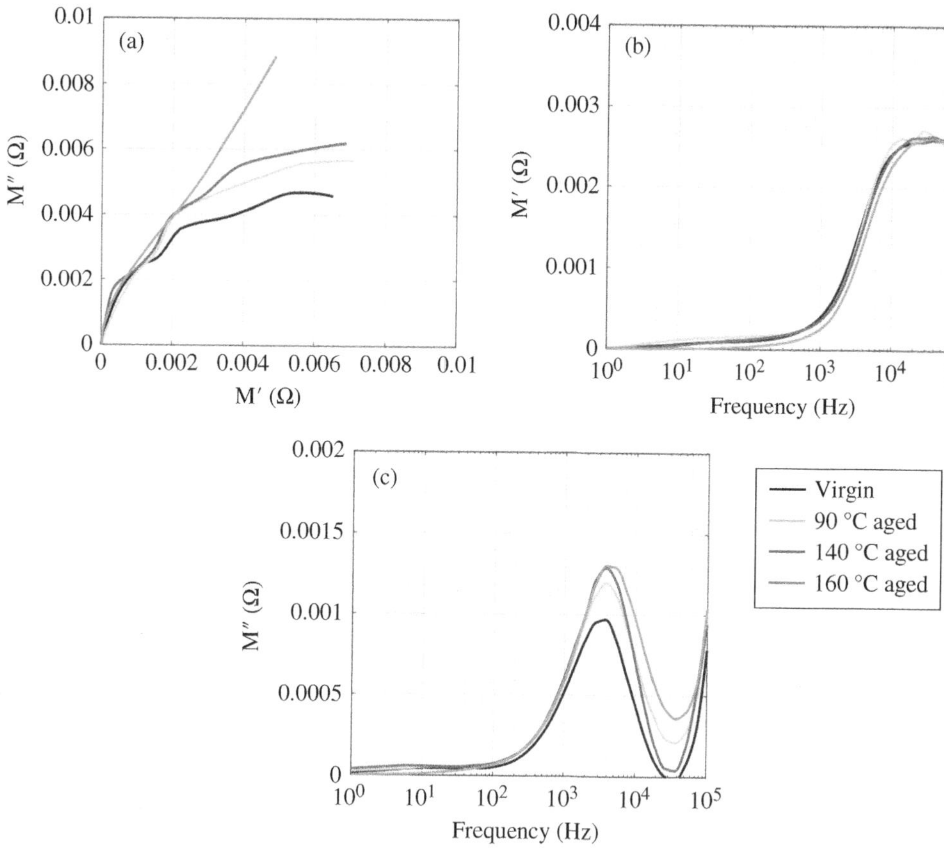

Figure 4.28 (a) Electrical modulus spectrum, (b) real part of modulus with frequency, (c) imaginary part of modulus with frequency.

investigates the composition of vapors from a material during pyrolysis to find a characteristic variation of the material due to thermal aging. Figure 4.29 shows the variation of organic functional groups in pressboard samples aged at different temperatures. The major compounds formed due to the thermal decomposition of natural-ester-impregnated pressboard were anhydrosugars, furanic derivatives, and aliphatic hydrocarbons. An increase in the concentration of anhydrosugars like 1,6-Anhydro-β-d-glucopyranose (levoglucosan) and 2,3 anhydro-D-mannosan is related to the decrease in the degree of polymerization of cellulosic pressboard material. The weakening of the ester C–O bond due to thermal aging can lead to the formation of carboxylic acids during analytical fast pyrolysis. Thus, acidic residues present in the pressboard material may contribute to accelerate the aging process, resulting in a reduced lifetime of the pressboard insulation.

4.4.5 Laser-Induced Breakdown Spectroscopy

Optical methods are also used to characterize the surface property of insulating material and a similar approach could be used to help understand the degradation of thermally aged ester-fluid-impregnated paper. Recently, laser-induced breakdown spectroscopy (LIBS) has gained a prominence for understanding the composition of both the liquid and solid material [104]. Typically, the experimental setup used for the LIBS is shown in Figure 4.30. A Q-switched Nd^{3+}: YAG laser (LAB-150-10-S2K, Quanta-Ray LAB series, Spectra Physics) produces a beam with a pulse duration of 10 ns and at a repetition rate

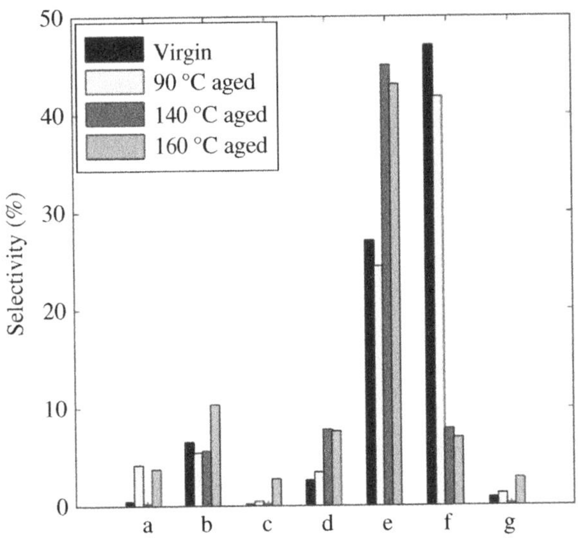

Figure 4.29 Variation of organic functional groups in pressboards aged at different temperatures. (a) Alcohols, (b) aliphatic hydrocarbons, (c) anhydrosugars, (d) carbonyl compounds, (e) carboxylic acids, (f) esters, and (g) furanic derivatives.

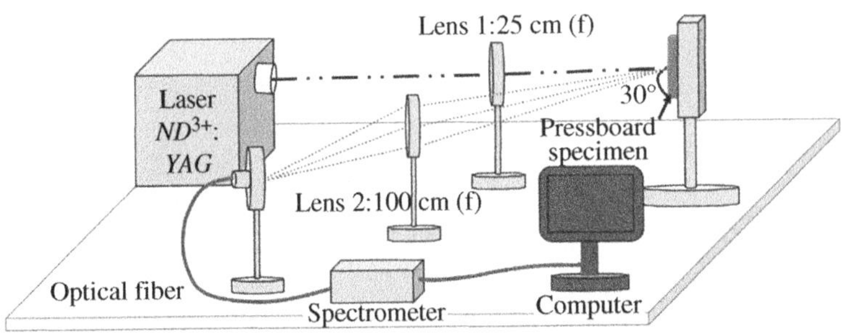

Figure 4.30 Experimental setup used for LIBS studies. *Source:* Noll [104].

of 10 Hz. The optically induced plasma ignites once the energy of the laser is greater than the threshold fluence of the material. The laser and the spectrometer are synchronized with the help of a control unit to be able to acquire the emission of plasma yielding the atomic spectra at the time of ablation.

The emission spectra obtained for thermally aged ester-impregnated pressboard (EIP) is shown in Figure 4.31a. The intensity of optical emission due to the Cu I (324.754 nm) peak is found to increase with the aged samples' temperature. This indicates that the surface conductivity of pressboard material increases as a result of aging compared with the virgin pressboard sample [105]. The presence of N II (500.415, 567.956 nm) and O II (313.472, 407.586 nm) peaks observed in the emission spectra could be due to external atmospheric conditions such as ionization of air surrounding the pressboard surface, which is involved in the discharge process [106]. The carbonaceous reaction in the pressboard material is identified by the C II (657.805 nm) peaks in the visible region of the emission spectrum. Since the number of collisions between the particles in the plasma species is very large, a local thermal equilibrium (LTE) can be assumed in a significantly small region of the plasma plume. Based on the above assumptions, the Boltzmann-Saha equation was used to calculate the plasma temperature (*Te*) and the emission lines were identified by comparing them with the NIST database [107] as

$$T_e = 1.44\left[\frac{E_2 - E_1}{\ln\left\{\frac{I_1\lambda_1 A_2 g_2}{I_2\lambda_2 A_1 g_1}\right\}}\right] \tag{4.10}$$

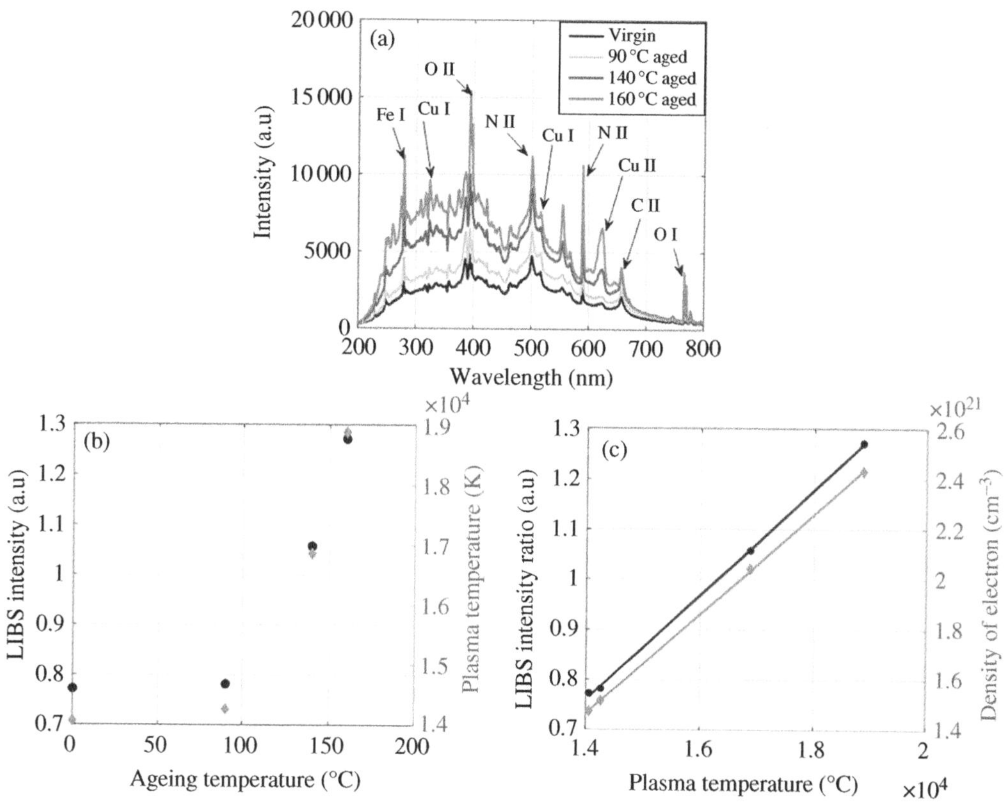

Figure 4.31 (a) LIBS emission spectra of aged pressboard, (b) LIBS intensity ratio and plasma temperature with respect to aging temperature, and (c) correlation of plasma temperature with intensity ratio and electron density.

where E_1 and E_2 are the excited energy levels, λ_1 and λ_2 are the wavelengths corresponding to intensities of atomic species I_1 and I_2, respectively, g_1 and g_2 are the statistical weights of the excited energy level, A_1 and A_2 are the transition probabilities of states. The LIBS intensity ratio and the temperature of the plasma formed during laser ablation are calculated based on the ratio of Cu I emission peak at 324 nm to the Cu I emission peak at 515.3 nm. Thus,

$$n_e = \frac{I_a}{I_i} * 6.6 * 10^{21} * \frac{A_i g_i}{A_a g_a} * \exp\left(\frac{E^{\text{ion}} + E_i + E_a}{T_e}\right) \tag{4.11}$$

where n_e is the electron density, E_i and E_a are the energies of upper states corresponding to Cu I peaks with their intensities as I_i and I_a, E^{ion} is the Ist ionization potential of Cu (7.7264 eV).

The parametric variation of LIBS intensity ratio and plasma temperature with respect to aging temperature of EIP material is shown in Figure 4.31b. Both quantities increase with aging temperature, which indicates the degradation of the EIP material. Also, a linear correlation was obtained between plasma temperature and intensity ratio and electron density. The calculated electron density increases with plasma temperature (Figure 4.31c), which indicates an increased number of electrons involved in the plasma discharge.

4.5 Conclusions and Future Scope

This chapter provided a discussion on the important aspects that must be considered when ester insulating fluid is to be used as an alternative to petroleum-based oil having different properties. The main properties of esters are compared with those of mineral oils and accordingly the transformer applications in which ester fluids outperform mineral oils are highlighted. The test methods for characterizing the fluid's properties and the diagnostic techniques for assessing the degradation of ester fluids and the pressboard/paper insulation were summarized. From the above discussions of the results obtained under the laboratory conditions, some of the techniques may be applicable to real-time transformer applications for the condition monitoring of the solid and liquid insulation. Considering the future scope of expansion in the transformer-related networks, the use of ester fluids with different added nanoparticles may offer a promising alternative for improving the dielectric properties while retaining the same insulation levels. This topic is currently the subject of extensive research, which is expected to grow in the foreseeable future.

References

1 N'cho, J.S., Fofana, I., Hadjadj, Y., and Beroual, A. (2016). Review of physicochemical-based diagnostic techniques for assessing insulation condition in aged transformers. *Energies* 9 (5): 367.
2 Rao, U.M., Sood, Y.R., and Jarial, R.K. (2017). Ester dielectrics: current perspectives and future challenges. *IETE Tech. Rev.* 34 (4): 448–459.
3 Lashbrook, M. (2014). Ester fluids for power transformers at > 100kV. *Transformers Magazine* 1 (2): 14–19.
4 CIGRE brochure 436 (2010). Experience in service with new insulating liquids.
5 Rozga, P. and Stanek, M. (2016). Characteristics of streamers developing at inception voltage in small gaps of natural ester, synthetic ester and mineral oil under lightning impulse. *IET Sci. Meas. Technol.* 10 (1): 50–57.
6 Beroual, A., Khaled, U., Mbolo Noah, P.S., and Sitorus, H. (2017). Comparative study of breakdown voltage of mineral, synthetic and natural oils and based mineral oil mixtures under AC and DC voltages. *Energies* 10 (4): 511.
7 Reffas, A., Beroual, A., and Moulai, H. (2020). Comparison of breakdown voltage of vegetable olive with mineral oil, natural and synthetic ester liquids under DC voltage. *IEEE Trans. Dielectr. Electr. Insul.* 27 (5): 1691–1697.
8 Beroual, A., Sitorus, H.B.H., Setiabudy, R., and Bismo, S. (2018). Comparative study of AC and DC breakdown voltages in Jatropha methyl ester oil, mineral oil, and their mixtures. *IEEE Trans. Dielectr. Electr. Insul.* 25 (5): 1831–1836.
9 Tenbohlen, S. and Koch, M. (2010). Aging performance and moisture solubility of vegetable oils for power transformers. *IEEE Trans. Power Del.* 25 (2): 825–830.
10 Bartnikas, R. (1994). *Electrical Insulating Liquids, Engineering Dielectrics*, vol. 3, 284–301. Philadelphia, USA: American Society for Testing and Materials.
11 Xiang, J., Liu, Q., and Wang, Z.D. (2018). Streamer characteristic and breakdown in a mineral oil and a synthetic ester liquid under DC voltage. *IEEE Trans. Dielectr. Electr. Insul.* 25 (5): 1636–1643.
12 IEC Standard 60060-2 (2010). High-voltage test techniques—part2: measuring systems.
13 Dang, V.H., Beroual, A., and Perrier, C. (2012). Investigations on streamers phenomena in mineral, synthetic and natural ester oils under lightning impulse voltage. *IEEE Trans. Dielectr. Electr. Insul.* 19 (5): 1521–1527.

14 Przybylek, P. (2017). Drying transformer cellulose insulation by means of synthetic ester. *IEEE Trans. Dielectr. Electr. Insul.* 24 (4): 2643–2648.
15 Fernandez-Diego, C., Ortiz, A., Carrascal, I.A. et al. (2019). Damage assessment of transformer Kraft paper insulation aged in mineral and vegetable oils. *Cellulose* 26 (4): 2653–2672.
16 Fernandez, F.O., Ortiz, A., Delgado, F. et al. (2017). Transformer health indices calculation considering hot-spot temperature and load index. *IEEE Electr. Insul. Mag.* 33 (2): 35–43.
17 Dombek, G. and Gielniak, J. (2018). Fire safety and electrical properties of mixtures of synthetic ester/mineral oil and synthetic ester/natural ester. *IEEE Trans. Dielectr. Electr. Insul.* 25 (5): 1846–1852.
18 Martin, D., Khan, I., Dai, J., and Wang, Z.D. (2006). An overview of the suitability of vegetable oil dielectrics for use in large power transformers. *Euro TechCon.*: 4–23.
19 Salama, M.M., Mansour, D.E.A., Daghrah, M. et al. (2020). Thermal performance of transformers filled with environmentally friendly oils under various loading conditions. *Int. J. Elec. Power.* 118: 105743.
20 Perrier, C., Beroual, A., and Bessede, J.-L. (2006). Improvement of power transformers by using mixtures of mineral oil with synthetic esters. *IEEE Trans. Dielectr. Electr. Insul.* 13 (3): 556–564.
21 Mehta, D.M., Kundu, P., Chowdhury, A. et al. (2016). A review on critical evaluation of natural ester vis-a-vis mineral oil insulating liquid for use in transformers: Part 1. *IEEE Trans. Dielectr. Electr. Insul.* 23 (2): 873–880.
22 IEC 62770 (2013). Fluids for electrotechnical applications-unused natural esters for transformers and similar electrical equipment.
23 Raymon, A., Pakianathan, P.S., Rajamani, M.P.E., and Karthik, R. (2013). Enhancing the critical characteristics of natural esters with antioxidants for power transformer applications. *IEEE Trans. Dielectr. Electr. Insul.* 20 (3): 899–912.
24 Hamid, M.H.A., Ishak, M.T., Din, M.M. et al. (2016). Dielectric properties of natural ester oils used for transformer application under temperature variation. *IEEE International Conference on Power and Energy (PECon)*, Melaka, Malaysia (28–29 November 2016), pp. 54–57.
25 Mehta, D.M., Kundu, P., Chowdhury, A. et al. (2016). A review of critical evaluation of natural ester vis-a-vis mineral oil insulating liquid for use in transformers: Part II. *IEEE Trans. Dielectr. Electr. Insul.* 23 (3): 1705–1712.
26 Azcarraga, C.G., Cavallini, A., and Piovan, U. (2014). A comparison of the voltage withstand properties of ester and mineral oils. *IEEE Electr. Insul. Mag.* 30 (5): 6–14.
27 Tokunaga, J., Nikaido, M., Koide, H., and Hikosaka, T. (2019). Palm fatty acid ester as biodegradable dielectric fluid in transformers: a review. *IEEE Electr. Insul. Mag.* 35 (2): 34–46.
28 Oommen, T.V. (2002). Vegetable oils for liquid-filled transformers. *IEEE Electr. Insul. Mag.* 18 (1): 6–11.
29 IEC 61203 (1995). Synthetic organic esters for electrical purposes-guide for maintenance of transformer esters in equipment.
30 IEEE C57.147 (2018). Guide for acceptance and maintenance of natural ester insulating liquid in transformers, pp. 1–85.
31 Rafiq, M., Lv, Y.Z., Zhou, Y. et al. (2015). Use of vegetable oils as transformer oils–a review. *Renew. Sust. Energ. Rev.* 52: 308–324.
32 Wilhelm, H.M., Franch, V., and Tulio, L. (2015). Compatibility of transformer construction materials with natural ester-based insulating fluids. *IEEE Trans. Dielectr. Electr. Insul.* 22: 2703–2708.
33 Rehman, S., Alhems, L.M., Jadim, R. et al. (2017). Experimental investigation of temperature effect on corrosive sulfur formation in transformers. *IEEE Trans. Dielectr. Electr. Insul.* 24 (5): 3201–3206.

34 Holt, A.F., Facciotti, M., Amaro, P. et al. (2013). An initial study into silver corrosion in transformers following oil reclamation. *IEEE Electrical Insulation Conference (EIC)*, Ottawa, pp. 469–472.

35 Dai, J. and Wang, Z.D. (2008). A comparison of the impregnation of cellulose insulation by ester and mineral oil. *IEEE Trans. Dielectr. Electr. Insul.* 15 (2): 374–381.

36 Loiselle, L., Fofana, I., Sabau, J. et al. (2015). Comparative studies of the stability of various fluids under electrical discharge and thermal stresses. *IEEE Trans. Dielectr. Electr. Insul.* 22 (5): 2491–2499.

37 Yuliastuti, E. (2010). Analysis of dielectric properties comparison between mineral oil and synthetic ester oil. M.S dissertation. Department of Electrical Sustainable Energy, Delft University of Technology.

38 Thakur, S. and Sarathi, R. (2019). Investigation on insulation performance of thermally aged natural ester oil impregnated pressboard. *IET Sci. Meas. Technol.* 13 (8): 1194–1202.

39 ASTM D1934-95 (2012). Standard test method for oxidative aging of electrical insulating liquids by open-beaker method.

40 IEC TS 62332-2 (2014). Electrical insulation systems (EIS) – thermal evaluation of combined liquid and solid components – part 2: simplified test.

41 Tokunaga, J., Koide, H., Mogami, K., and Hikosaka, T. (2016). Comparative studies on the aging of thermally upgraded paper insulation in palm fatty acid ester, mineral oil, and natural ester. *IEEE Trans. Dielectr. Electr. Insul.* 23 (1): 258–265.

42 Coulibaly, M.L., Perrier, C., Marugan, M., and Beroual, A. (2013). Aging behavior of cellulosic materials in presence of mineral oil and ester liquids under various conditions. *IEEE Trans. Dielectr. Electr. Insul.* 20 (6): 1971–1976.

43 Singha, S., Asano, R., Frimpong, G. et al. (2014). Comparative aging characteristics between a high oleic natural ester dielectric liquid and mineral oil. *IEEE Trans. Dielectr. Electr. Insul.* 21 (1): 149–158.

44 IEC 60814 (1997). Insulating liquids—oil-impregnated paper and pressboard—determination of water by automatic coulometric Karl Fischer titration.

45 Liao, R., Hao, J., Chen, G. et al. (2011). A comparative study of physicochemical, dielectric and thermal properties of pressboard insulation impregnated with natural ester and mineral oil. *IEEE Trans. Dielectr. Electr. Insul.* 18 (5): 1626–1637.

46 Ansari, M.A., Martin, D., and Saha, T.K. (2020). Advanced online moisture measurements in transformer insulation using optical sensors. *IEEE Trans. Dielectr. Electr. Insul.* 27 (6): 1803–1810.

47 Saha, T.K. (2003). Review of modern diagnostic techniques for assessing insulation condition in aged transformers. *IEEE Trans. Dielectr. Electr. Insul.* 10 (5): 903–917.

48 Ganeshan, S., Murugesan, I., Cavallini, A. et al. (2017). Identification of partial discharges in power transformers: An approach driven by practical experience. *IEEE Electr. Insul. Mag.* 33 (5): 23–31.

49 IEC 61294, 1993-10-07 (1993). Insulating liquids – determination of partial discharge inception voltage (PDIV) – test procedure.

50 IEC 60270 (2000) High-voltage test techniques – partial discharge measurement.

51 Wang, Z.D., Liu, Q., Wang, X. et al. (2011). Discussion on possible additions to IEC 60897 and IEC 61294 for insulating liquid tests. *IET Electr. Power. App.* 5 (6): 486–493.

52 Baker, P.C., Judd, M.D., and McArthur, S.D.J. (2010). A frequency-based RF partial discharge detector for low-power wireless sensing. *IEEE Trans. Dielectr. Electr. Insul.* 17 (1): 133–140.

53 Sarathi, R., Singh, P.D., and Danikas, M.G. (2007). Characterization of partial discharges in transformer oil insulation under AC and DC voltage using acoustic emission technique. *J. Electr. Eng.* 58 (2): 91.

54 Bartnikas, R. (2002). Partial discharges. Their mechanism, detection and measurement. *IEEE Trans. Dielectr. Electr. Insul.* 9 (5): 763–808.

55 Sinaga, H.H. (2012). Detection, identification and localization of partial discharges in power transformers using UHF techniques. Ph. D dissertation. University of New South Wales Australia.

56 Siegel, M., Beltle, M., Tenbohlen, S., and Coenen, S. (2017). Application of UHF sensors for PD measurement at power transformers. *IEEE Trans. Dielectr. Electr. Insul.* 24 (1): 331–339.

57 Xiang, J., Liu, Q., and Wang, Z. D. (2016). Inception and breakdown voltages of insulating liquids under DC stress. *IEEE International Conference on High Voltage Engineering and Application (ICHVE)*, Chengdu, China (19–22 September 2016), pp. 1–4.

58 Sonerud, B., Bengtsson, T., Blennow, J., and Gubanski, S.M. (2009). Dielectric heating in insulating materials subjected to voltage waveforms with high harmonic content. *IEEE Trans. Dielectr. Electr. Insul.* 16 (4): 926–933.

59 Chandrasekar, S. and Montanari, G.C. (2014). Analysis of partial discharge characteristics of natural esters as dielectric fluid for electric power apparatus applications. *IEEE Trans. Dielectr. Electr. Insul.* 21 (3): 1251–1259.

60 Amalanathan, A.J., Sarathi, R., Prakash, S. et al. (2019). Investigation on thermally aged natural ester oil for real-time monitoring and analysis of transformer insulation. *High Voltage* 5 (2): 209–217.

61 Crofts, D.W. (1988). The electrification phenomena in power transformers. *IEEE Trans. Electr. Insul.* 23 (1): 137–146.

62 Metwally, I.A. (1996). Characterization of static electrification in power transformers. *IEEE Trans. Dielectr. Electr. Insul.* 3 (2): 307–315.

63 Sierota, A. and Rungis, J. (1994). Electrostatic charging in transformer oils. Testing and assessment. *IEEE Trans. Electr. Insul.* 1 (5): 840–870.

64 Moreau, O., Paillat, T., and Touchard, G. (2003). Flow electrification in transformers: Sensor propotype for electrostatic hazard. *Electrostat* 178: 31–36.

65 Paillat, T., Touchard, G., and Bertrand, Y. (2012). Capacitive sensor to measure flow electrification and prevent electrostatic hazards. *Sensors* 12 (11): 14315–14326.

66 Bourgeois, A., Mortha, G., Paillat, T. et al. (2006). Flow electrification in power transformers: study of a potential remedy. *IEEE Trans. Dielectr. Electr. Insul.* 13 (3): 650–656.

67 Lyon, D.J., Melcher, J.R., and Zahn, M. (1988). Couette charger for measurement of equilibrium and energization flow electrification parameters: application to transformer insulation. *IEEE Trans. Dielectr. Electr. Insul.* 23 (1): 159–176.

68 Washabaugh, A.P. and Zahn, M. (1996). Flow electrification measurements of transformer insulation using a Couette flow facility. *IEEE Trans. Dielectr. Electr. Insul.* 3 (2): 161–181.

69 C.J. W. Group (2000). Static electrification in power transformers, CIGRE, General Session, paper 12/15.13, Paris.

70 Kedzia, J. and Willner, B. (1994). Electrification current in the spinning disk system. *IEEE Trans. Dielectr. Electr. Insul.* 1 (1): 58–62.

71 Talhi, M., Fofana, I., and Flazi, S. (2013). Comparative study of the electrostatic charging tendency between synthetic ester and mineral oil. *IEEE Trans. Dielectr. Electr. Insul.* 20 (5): 1598–1606.

72 Brubaker, M.A. and Nelson, J.K. (2000). A parametric study of streaming electrification in a full-scale core-form transformer winding using a network-based model. *IEEE Trans. Power Deliv.* 15 (4): 1188–1192.

73 Liu, D., Du, B., Liu, F., and Wang, S. (2016). Effects of multiple parameters on static electrification and breakdown strength of transformer oil. *IET Sci. Meas. Technol.* 10 (6): 597–601.

74 Nollet, L.M. and Toldra, F. (2015). UV–Visible absorption, fluorescence, and chemiluminescence spectroscopy. In: *Handbook of Food Analysis-Two Volume Set*, 1417–1432. CRC Press, Taylor and Francis Group.

75 ASTM D6802 (2002). Test method for determination of the relative content of dissolved decay products in mineral insulating oils by spectrophotometer.

76 Sikorska, E., Khmelinskii, I., and Sikorski, M. (2012). Analysis of olive oils by fluorescence spectroscopy: methods and applications, *Olive oil-constituents, quality, health properties and bioconversions*, pp. 63–88.

77 Choe, E. and Min, D.B. (2007). Chemistry of deep-fat frying oils. *J. Food Sci.* 72 (5): 77–86.

78 Panigrahi, S.K., Thakur, S., Sarathi, R., and Mishra, A.K. (2017). Understanding the physico-chemical properties of thermally aged natural ester oil adopting fluorescent technique. *IEEE Trans. Dielectr. Electr. Insul.* 24 (6): 3460–3470.

79 IEC 60247 (2004). Insulating liquids - measurement of relative permittivity, dielectric dissipation factor (tan d) and d.c. resistivity.

80 Bandara, K., Ekanayake, C., Saha, T.K., and Annamalai, P.K. (2016). Understanding the ageing aspects of natural ester-based insulation liquid in power transformer. *IEEE Trans. Dielectr. Electr. Insul.* 23 (1): 246–257.

81 Thakur, S., Sarathi, R., and Danikas, M.G. (2019). Investigation on thermal ageing impact on dielectric properties of natural ester oil. *Electr. Eng.* 101 (3): 1007–1018.

82 Wilhelm, H.M., Tulio, L., Jasinski, R., and Almeida, G. (2011). Aging markers for in-service natural ester-based insulating fluids. *IEEE Trans. Dielectr. Electr. Insul.* 18 (3): 714–719.

83 ASTM D971 (2012). Standard test method for interfacial tension of oil against water by the ring method.

84 IEC 62961 (2018) Insulating liquids-test methods for the determination of interfacial tension of insulating liquids- determination with the ring method.

85 Zhang, X. and Gockenbach, E. (2008). Asset-management of transformers based on condition monitoring and standard diagnosis. *IEEE Electr. Insul. Mag.* 24 (4): 26–40.

86 Hamrick, L. (2009). Transformer oil sampling. https://studylib.net/doc/18113398/transformer-oil-sampling (accessed 4 January 2011).

87 IEEE C57.106 (2007) IEEE guide for acceptance and maintenance of insulating oil in equipment.

88 ASTM D6181 (2003) Standard test method for measurement of turbidity in mineral insulating oil of petroleum origin.

89 Loiselle, L., Rao, U.M., Fofana, I., and Jaya, T. (2020). Monitoring colloidal and dissolved decay particles in ester dielectric fluids. *IEEE Trans. Dielectr. Electr. Insul.* 27 (5): 1516–1524.

90 Dombek, G. and Nadolny, Z. (2016). Liquid kind, temperature, moisture, and ageing as an operating parameters conditioning reliability of transformer cooling system. *Maint. Reliab.* 18 (3): 413–417.

91 Pfaendtner, J. and Broadbelt, L.J. (2008). Mechanistic modeling of lubricant degradation. 1. Structure– reactivity relationships for free-radical oxidation. *Ind. Eng. Chem. Res.* 47 (9): 2886–2896.

92 Sikorski, W. (2019). Development of acoustic emission sensor optimized for partial discharge monitoring in power transformers. *Sensors* 19 (8): 1865.

93 Mitchinson, P.M., Lewin, P.L., Strawbridge, B.D., and Jarman, P. (2010). Tracking and surface discharge at the oil/pressboard interface. *IEEE Electr. Insul. Mag. 26* (2): 35–41.

94 Thirumurugan, C., Kumbhar, G.B., and Oruganti, R. (2019). Effects of impurities on surface discharges at synthetic ester/cellulose board. *IEEE Trans. Dielectr. Electr. Insul.* 26 (1): 64–71.

95 Yi, X. and Wang, Z.D. (2015). The influences of solid surface on the propagation of creepage discharge in insulating liquids. *IEEE Trans. Dielectr. Electr. Insul.* 22 (1): 303–312.

96 Beroual, A., Dang, V.H., Coulibaly, M.L., and Perrier, C. (2013). Investigation on creeping discharges propagating over pressboard immersed in mineral and vegetable oils under AC, DC and lightning impulse voltages. *IEEE Trans. Dielectr. Electr. Insul.* 20 (5): 1635–1640.

97 Bao, L., Li, J., Zhang, J. et al. (2016). Influences of temperature on partial discharge behavior in oil-paper bounded gas cavity under pulsating DC voltage. *IEEE Trans. Dielectr. Electr. Insul.* 23 (3): 1482–1490.

98 Nakane, R., Okubo, H., and Kato, K. (2019). HVDC electrical insulation performance based on charge activity in oil-pressboard composite insulation structures. *IEEE Trans. Dielectr. Electr. Insul.* 26 (2): 576–583.

99 Amalanathan, A.J., Sarathi, R., Harid, N. et al. (2020). Investigation on the performance of thermally aged natural ester fluid impregnated pressboard material. *IEEE Trans. Dielectr. Electr. Insul.* 27 (5): 1578–1586.

100 Shen, W.W., Mu, H.B., Zhang, G.J. et al. (2013). Identification of electron and hole trap based on isothermal surface potential decay model. *J. Appl. Phys.* 113 (8): 083706.

101 Simmons, J.G. and Tam, M.C. (1973). Isothermal currents in semiconductors and insulators containing arbitrary trap distributions. *Phys. Rev. B* 7: 3706–3713.

102 Liu, J., Fan, X., Zhang, Y. et al. (2020). Quantitative evaluation for moisture content of cellulose insulation material in paper/oil system based on frequency dielectric modulus technique. *Cellulose* 27 (4): 2343–2356.

103 Barsoukov, E. and Macdonald, J.R. (2018). *Impedance Spectroscopy: Theory, Experiment, and Applications*. John Wiley & Sons.

104 Noll, R. (2012). *Laser-Induced Breakdown Spectroscopy: Fundamentals and Applications*, 484–488. Berlin- Heidelberg, Germany: Springer-Verlag.

105 Aparna, N., Vasa, N.J., Sarathi, R., and Rajan, J.S. (2014). Feasibility study for detecting copper contaminants in transformer insulation using laser-induced breakdown spectroscopy. *Appl. Phys. A Mater. Sci. Process.* 117 (1): 281–288.

106 Aparna, N., Vasa, N.J., and Sarathi, R. (2018). Analysis of copper contamination in transformer insulating material with nanosecond-and femtosecond-laser-induced breakdown spectroscopy. *J. Phys. D. Appl. Phys.* 51 (23): 235601.

107 Sansonetti, J.E. and Martin, W.C. (2005). Handbook of basic atomic spectroscopic data. *J. Phys. Chem. Ref. Data* 34 (4): 1559–2259.

5

End Life Behavior of Ester Liquids in High-Voltage Transformers

U. Mohan Rao, I. Fofana, L. Loiselle, and T. Jayasree

Research Chair on the Aging of Power Network Infrastructure (ViAHT), University of Quebec at Chicoutimi, Chicoutimi, QC, Canada

5.1 Introduction

Insulation system in oil-filled transformers will be generally designed to last for several decades in successful operation. Owing to several degradation mechanisms like oxidation, hydrolysis, and cellulose paralysis, designed life of the insulation system varies on a large scale. These degradation mechanisms are complex chemical reactions that are highly interrelated and involves in various external factors [1]. Hence, utility engineers perform various conditioning monitoring activities in ensuring the effective functioning of the insulation system. These condition-monitoring activities involve oil characterization through various diagnostic tests and oil treatment [2]. Diagnostic characterizations of insulation oil are performed on a routine practice while oil treatment is performed in extreme situations and at the end of life. The intent of oil treatment is to remove the decay products in the insulation oil and maintain the oil properties close to the pristine conditions. It is understood that degradation of oil-paper insulation is unavoidable and hence is essential to understand the degradation process and its effect on the behavior of oil-paper insulation.

Meanwhile, shortfalls of mineral insulating oils like low dielectric strength, low flash point, and poor biodegradability made the engineers to search for alternative insulating fluids [3]. Ester dielectric fluids is a topic of high attention for utility engineers and researchers interested in alternative dielectric fluids. Several researchers confirmed that ester fluids are potential alternative to mineral insulating oils for use in high-voltage transformers [4–6]. Degradation aspects of ester fluids in comparison to mineral oils (MOs) were investigated and are reported in [7–9]. It is inferred that degradation of ester fluids is slower than MOs and confirmed the longevity of insulation paper in ester fluids.

Degradation of insulating liquid causes dissolved gasses, polar solvents, water, free radicles, acids, etc. Similarly, degradation of insulating paper causes decay cellulose particles, acids, gases, water, methanol, and furanic compounds. It is to be understood that degradation of oil-paper insulation causes byproducts in solid, liquid, and water phases. However, the influence of solid-phase particles will generally be more detrimental to the insulating integrity. Thus, in this chapter, the focus will be on decay particles in solid phase. Therefore, degradation of oil-paper insulation is evident by production of decay products that may or may not be soluble in the insulation oil. The soluble decay particles are very minute in size and are almost found soluble/dissolved in insulation

Alternative Liquid Dielectrics for High Voltage Transformer Insulation Systems: Performance Analysis and Applications, First Edition. Edited by U. Mohan Rao, I. Fofana, and R. Sarathi.

oil. Similarly, colloidal decay products are of different sizes and are distributed within the oil in the transformer. Soluble and colloidal particles act as local conducting elements in the oil and adversely affect the physiochemical and dielectric properties of the insulation oil. As per CIGRE 157 (technical brochure), large portion of the transformer failures is observed due to defects in insulation system, and decay particles contribute more toward the failure of the insulation system [10]. ASTM developed a standard test measurement to estimate the number of particles in the insulating fluid that also quantifies the size (diameter) of the particles [11]. This method is found to be potential in quantifying the number of colloidal particles, but the actual traces of soluble (dissolved in oil) particles are questionable. The critical evolution and impact of these decay products on insulation system has been discussed in the subsequent sections of this chapter.

It is to be recalled that ester dielectric fluids are found to be a potential alternative to MO for use in oil-filled transformers. Explicit studies on degradation aspects of these alternative oils improve the existing knowledge on workability and lifetime performance. Until date, majority of the research on ester fluids is focused on the performance testing and understanding the degradation behavior of these new liquids. However, there is significantly less literature on treatment of these new liquids and understanding the influence of soluble and colloidal degradation particles in esters [12] [13]. Hence, there is a great scope for research on understanding the influence of soluble and colloidal particles on new insulating fluids.

In addition to this, regeneration avenues of ester liquids are still a challenging and interesting aspect in the communities of transformers. Traditionally, fuller's earth is widely accepted for regeneration of mineral insulating oils. Therefore, as a first step it is essential to understand the suitability of fuller's earth for regeneration of ester liquids. Therefore, in this chapter, the treatment of ester liquids with fuller's earth at different quantities and at different filtration temperatures is reported. After detailed observations and experimental measurements, it is understood that the scope for fuller's earth for regeneration of ester liquids is too low. At an outset, the contents in this chapter cover the influence of colloids and soluble particles on the liquid properties. The discussions also span to the impact of fuller's earth treatment on ester liquids at selected conditions.

5.2 Evolution of Colloidal and Soluble Decay Particles

5.2.1 Perspective of Decay Particles

As discussed in the previous chapters, insulation system in the liquid-filled power transformer comprises of insulating liquid and insulation paper. Insulating liquid may be mostly MO (traditional) or ester liquids (alternative candidates). Similarly, solid insulation is mostly cellulose in nature. It is to be seen that solid in conjunction with liquid forms a composite insulation system for the transformer. Therefore, a combined and collaborative degradation happens to the system while serving together. Degradation of liquid hampers the performance of the solid, and premature aging happens to the system and vice versa. Thus, the degradation mechanisms (oxidation, hydrolysis, thermal, and electrical) explained in Chapter 1 are highly interrelated. Owing to electrical, thermal, oxidation, and electrical stresses, the degradation mechanisms expedites with stressing times and leads to generation of aging byproducts. These products may be in solid, liquid, and gas phases. However, the byproducts in solid phase are more detrimental to the insulation system. The liquid and gas phase products are not vigorous to the insulation system, but their presence for a long-term affects the system. However, the presence of liquid and gaseous phase products contribute comparably to that of the solid byproducts to the insulation premature aging.

Generally, decay content (sludge), acids, moisture, gases, and free radicles are evolved with aging times. The generation of acids and water precedes the evidence of the sludge in the oil. This sludge is generally seen as an end byproduct of oxidation and hydrolysis. Theoretically, acids and water arises with insulation (both oil and paper) degradation, which further agglomerates to form colloids. These colloids will contribute largely to the sludge formation in liquid and also adsorb onto the large surface of the cellulose insulants. In addition to this, it is to be mentioned that with the promotion of oxidation, peroxides are initially evident as primary oxidation products. Further, the secondary oxidation products including nonvolatile, volatile, high-molecular-weight acids, and free fatty acidic compounds are derived from hydroperoxides. These will contribute to soluble particles too. The direct detection of soluble and colloidal particles at an early stage is challenging. Thus, it is customary to monitor the parameters that will be influenced with the evidence of soluble and colloidal particles. Colloids can generally be understood by the monitoring of acidity and interfacial tension, while soluble decay particles may be monitored by dissolved decay products and turbidity measurements. It is to mention that both soluble and colloidal particles are highly interrelated, and it is difficult to specify the exact influencing parameter. The above classification on monitoring parameters is indicative toward the major contributing factors.

5.2.2 Size And Influence of Decay Particles

Insulation system in oil-filled transformers is subjected to high electrical and thermal stresses through core-winding assembly. Further, these electrical and thermal stresses are variable and are highly dependent on the operating conditions of the transformer. Along with these stresses, oxidation and hydrolysis of oil-paper insulation remain proactive. Also, with increase in temperature, pyrolysis of cellulose insulants will aid the above mentioned reactions. All these reactions are highly interrelated and are also dependent on external factors like moisture, temperature, oxygen, loading conditions, etc. The major byproducts of these reactions are acids, water, gasses, cellulose fibers, furfurals, polar solvents, and metal oxides that contaminate the oil. Generally, cellulose fibers and metallic particles contribute more to the decay particles in oil-filled transformer. Presence of metallic particles is a high indication of damage to the insulation system than that of the cellulose fibers.

The geometry (shape and size) of these particles play a critical role in deciding the degree of oil contamination and vulnerability. The size of the decay products introduced into the oil may vary from few microns to hundreds of microns. The distribution of particles size is as follows: (i) tiny particles: thousands of particles will be less than 25 microns, (ii) small particles: hundreds of particles will be in the range of 25–100 microns (iii) high-grade particles: significantly less particles over 100 microns. However, some metallic particles that wear out of core-winding assembly and rubber scraps from gaskets/sealing may be even bigger in size. The shape of these particles is not definite and is highly irregular with rough surfaces at the micro level [10]. The presence of these particles in the insulating fluid adversely affects the performance of transformer and properties of the oil. The phenomenal effects of soluble and colloidal particles evolved in insulating fluids with aging depend on their geometry. Tiny particles (<25 microns) are almost soluble in oil, and these particles affect the oil fluidity and influence all the physicochemical properties of oil. Presence of these soluble particles majorly influences properties like dielectric strength, acidity, interfacial tension, and color.

Small particles (25–100 microns) aid to the effects of tiny particles and lead to oil contamination, blocking of ducts, hampering the oil flow through cooling tubes, and sludge depositions. These particles are almost colloidal particles that may turn to soluble particles with operating times due to high temperatures. These cellulose fibers may not be directly soluble with operating times, but they may further split into multiple particles known as sister particles. These sister particles may also further split into multiple particles, and this process continues with time. This splitting process is mainly attributable to operating time, internal temperature, and pressure that act in physical reduction of particle size. There is a high probability that the size of these sister particles may fall less than 25 microns and act as soluble particles. High-grade particles (>100 microns) further intensifies the consequences of particles less than 100 microns. High-grade particles act as conductive particles and initiate local electrical stress. These local fields weaken the overall insulation integrity by initiating local partial discharges, conducting channels, and hotspots [14]. The details of the soluble and colloidal particles in transformer insulating oil is illustrated in Figure 5.1.

The illustration of the influence of different particles on insulation is indicative approach while considering the significant effects of particles. However, presence of all the particles influences the oil properties accompanied by ruining the effectiveness and functionality of the insulation system. The arrowheads in the above figure indicate particle splitting and formation of sister particles that are having size less than 25 microns. It is established that presence of these particles is witnessed with aging of the insulation accompanied by change in color of the oil.

5.3 Colloidal Particles – Centrifugal Treatment (ASTM D1698)

This section reports the behavior and influence of colloidal particles in MO, natural ester (NE), synthetic ester (SEA), and other synthetic ester (SEB). All the insulating liquids have been subjected to accelerated thermal aging as per the modified ASTM D1934 at different temperatures. Thermal aging is performed in presence of the insulating papers and copper catalyst. The oil-paper ratio adopted in this study is 1:10 to demonstrate the nominal ratio adopted in a typical transformer. As

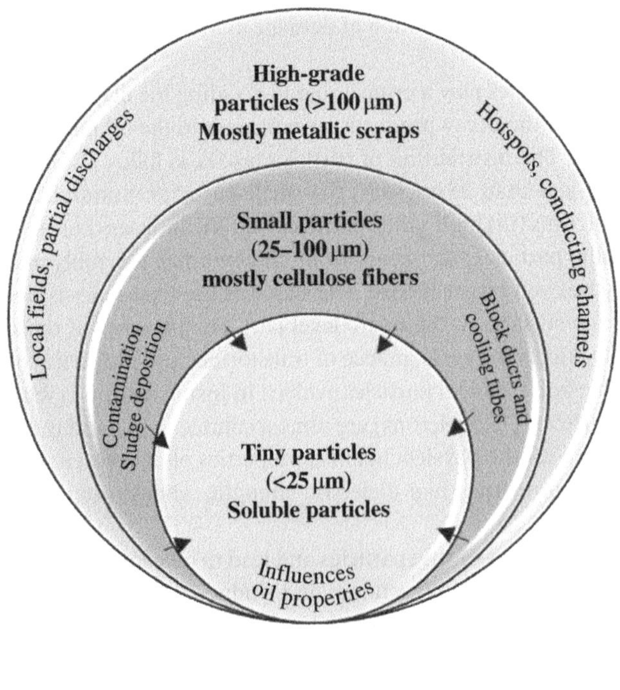

Figure 5.1 Illustration of decay particles and their influence on transformer insulating liquid. *Source:* Loiselle et al. [15] / with permission of IEEE.

mentioned earlier, thermal aging is controlled one with an aging history of the liquid samples at different temperatures (115, 150, 175, and 195 °C). Each temperature is subjected for a duration of two weeks each totaling to eight weeks of aging duration. In order to simulate moisture migrations and provide the most of hydrolysis influence on degradation, heating is performed for 12 hours and cooling (to room temperature) is done for 12 hours. Only heating hours are considered for the eight weeks of thermal aging. The degraded liquids at various aging conditions are subjected to centrifuge treatment as per ASTM D1698. The centrifuged and aged oils are tested for ultraviolet visible spectroscopy, turbidity, and particle count as per ASTM D6802, ASTM D6181, and ASTM D6786, respectively. The experimental outline is illustrated in Figure 5.2.

It is to be mentioned that ester liquids (all aging conditions) and MO (115 °C) are evidenced with no colloidal particles after the centrifuge treatment. The visualization indicating the absence of colloidal particles in aged ester liquids after the centrifuge treatment is shown in Figure 5.3.

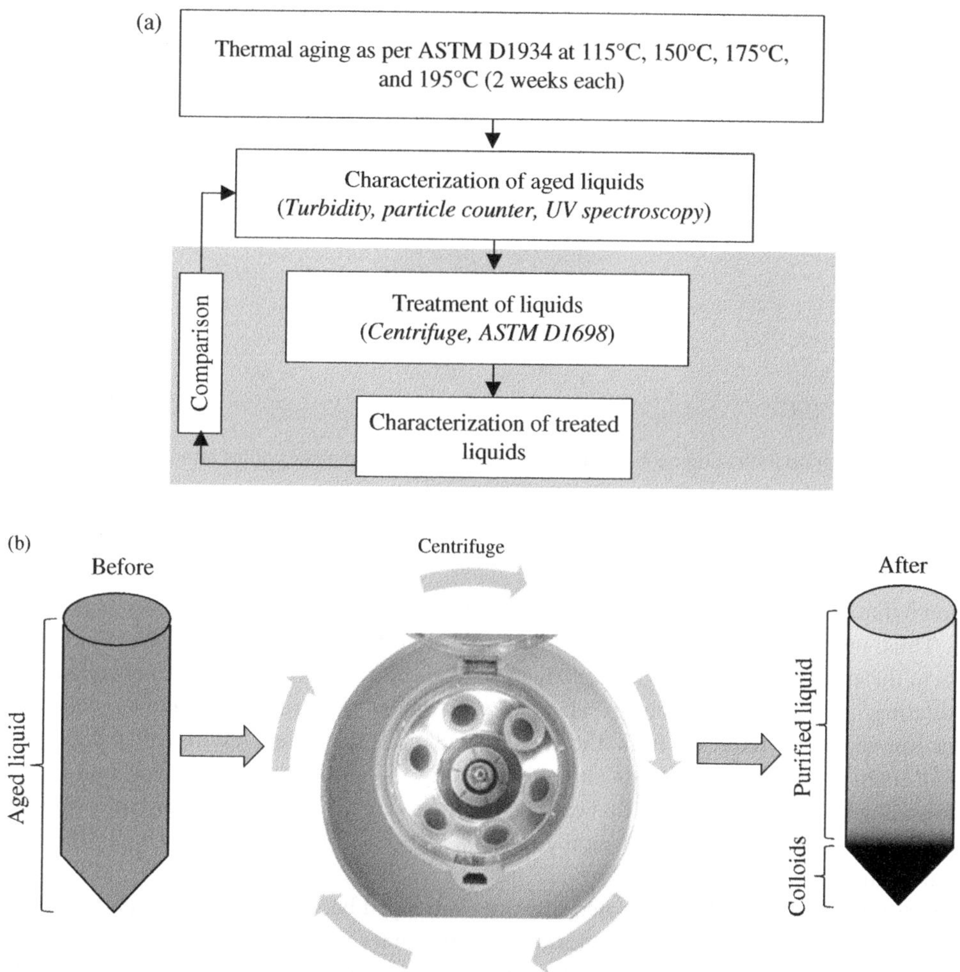

Figure 5.2 Illustration of the experimental and centrifuge treatment. (a). Illustration of the experimental for colloidal particles monitoring. (b). Illustration of the centrifuge treatment.

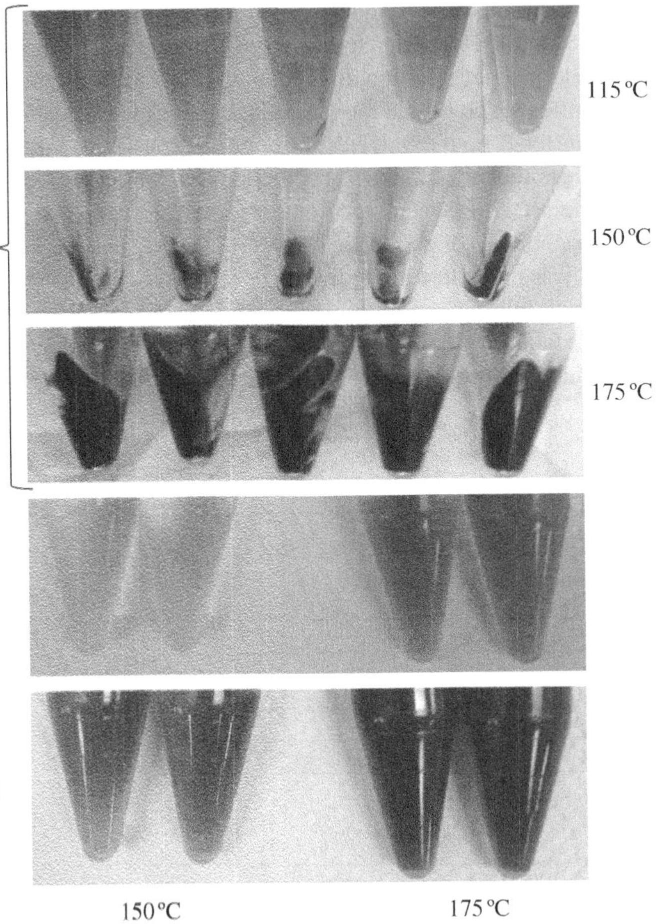

Figure 5.3 Visualization indicating no evidence of colloidal particles in ester liquids after centrifuging.

5.3.1 UV Spectroscopy

It is known that colloidal particles increase with increase in thermal degradation of the insulating liquids. Thus, increasing the absorbance of the liquid to ultraviolet visible light. Therefore, the increase in the absorbance curves to higher wavelength is controlled by the presence of colloids in the insulating liquid. In the present experiment, with centrifugation of aged liquids, colloidal particles are separated/removed from the insulating liquids. This separation will reduce the absorbing (light) nature of the liquids and subsequently the absorbance curves will move to a lower wavelengths. It is to be recalled that colloidal particles are not noticed in case of ester liquids and MOs at 115 °C. Therefore, the spectral curves of MO, SEA, SEB, NE, and treated MOs are shown in Figure 5.4.

In Figure 5.4a, the aged MO and treated MO are overlapped because no colloidal particles are seen at this stage of aging. However, the later stages of thermal aging are noticed and the influence is depicted in Figure 5.4b–d. Considering the aging factors associated with 150, 175, and 195 °C, colloidal particles are absent for SEA, SEB, and NE, while they are evident for MO. In Figure 5.4c and d, MO is rigorously degraded, thus shifting the absorbance curves beyond the wavelength limit

(360–600 nm) as per the ASTM D6802. To comment on the degradation behavior, it is to be observed that the aged ester liquids are better than that of the centrifuged MO.

5.3.2 Turbidity

Turbidity of the liquid demonstrates the ability of the liquid to be transparent to the light. Therefore, owing to aging by products, turbidity of the transformer insulating liquid increases with increase in degradation. Therefore, one may understand that removal of colloidal particles partially rejuvenates the turbidity of the insulating liquid. The variation in the turbidity of aged MO, SEA, NE, and SEB along with centrifuged MO is shown in Figure 5.5.

It is seen that turbidity of the MO increased tremendously with thermal aging. The same is reduced with the centrifugal treatment. The turbidity of the ester liquids is also increased with thermal degradation. However, the turbidity of the aged ester liquids and centrifuged MO is comparable.

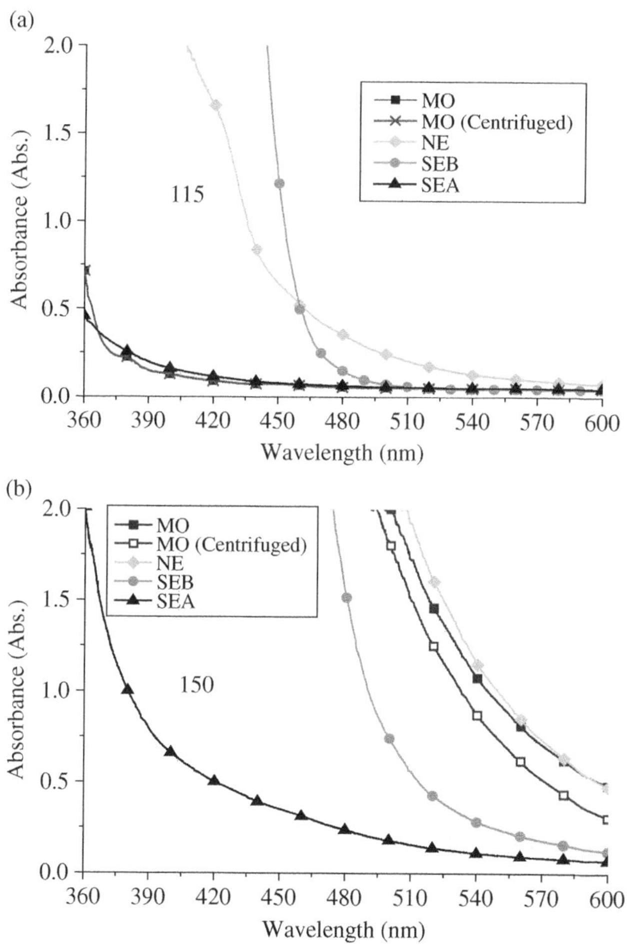

Figure 5.4 UV/Vis curves for aged and treated liquids (a) at 115 °C, (b) at 150 °C, (c) at 175 °C, and (d) at 195 °C.

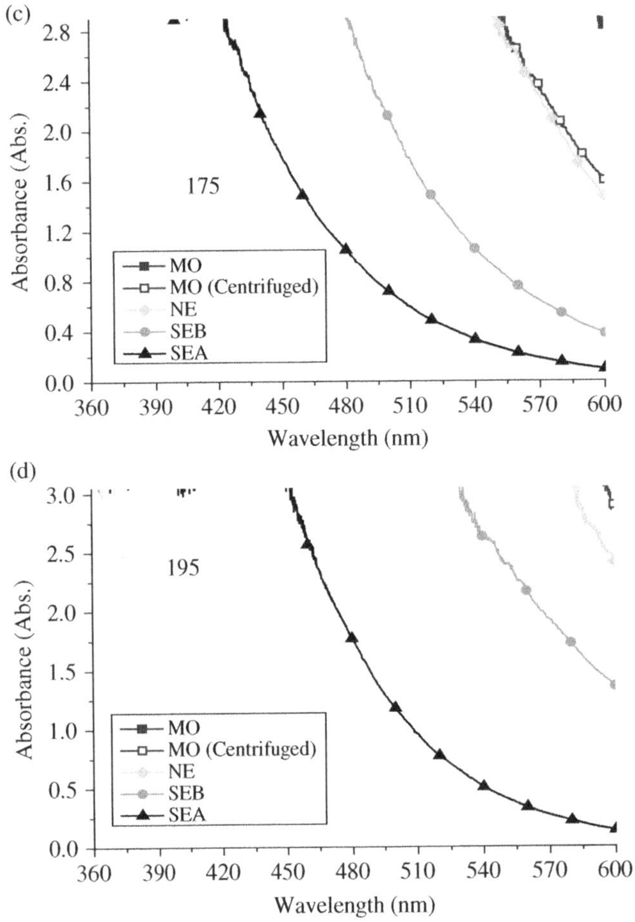

Figure 5.4 (Continued)

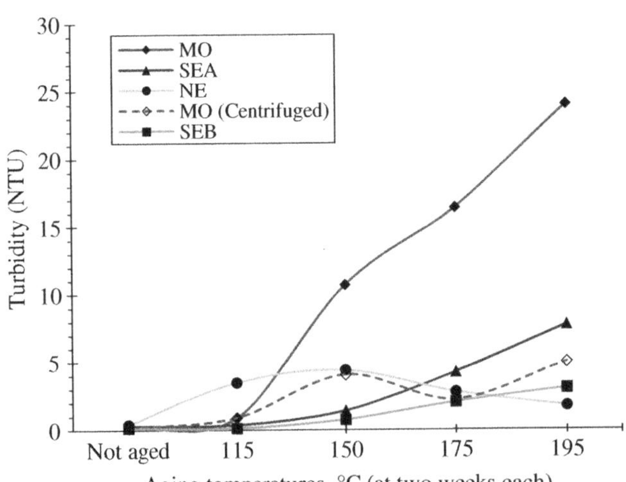

Figure 5.5 Turbidity of the aged liquids and centrifuged mineral oil. *Source:* Loiselle et al. [15] / with permission of IEEE.

5.3.3 Particle Counter

Because centrifuged liquid is involved in the removal of colloidal particles that are separable by the centrifuged treatment, the number of decay particles reduces. As centrifugal removal of particle is based on mechanical agitation, this removal is attributable to the size of the particle. Thus, to understand the change in the number of particles with centrifugal treatment, particle count measurements are carried out for aged MO, SEA, SEB, NE, and centrifuged MO. The results of the same are presented in Figure 5.6.

It is clearly noticed from Figure 5.6 that a few decay particles are retailed in the insulating liquid even after centrifugal treatment. These decay particles could be tiny colloids or major dissolved particles. In addition, on observing the aged and centrifuged MO, a phenomenal crossover is noticed in the

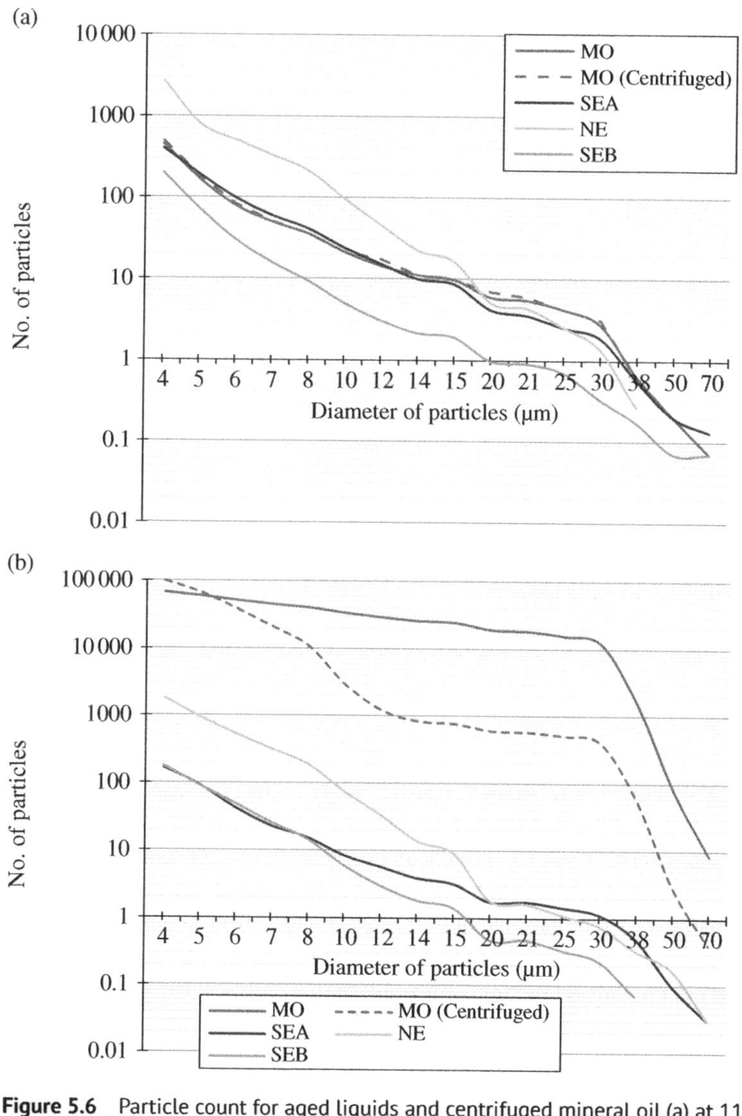

Figure 5.6 Particle count for aged liquids and centrifuged mineral oil (a) at 115 °C, (b) at 150 °C, (c) at 175 °C, and (d) at 195 °C.

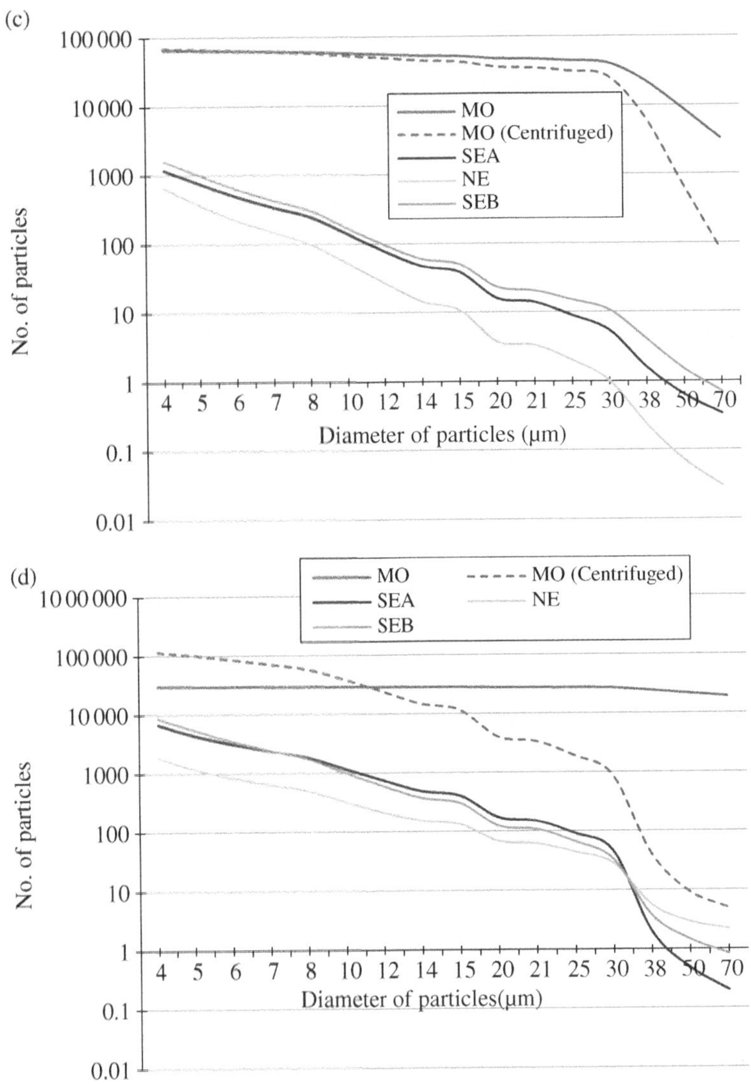

Figure 5.6 (Continued)

curves. This crossover demonstrated that higher diameter particles (above the crossover) are removed with centrifugal treatment, while the smaller diameter (below the crossover) are retained in the liquid.

One may assume that there exists a critical size for the colloidal particles that are generated with thermal degradation in the insulating liquids. Further detailed analytical measurements are to be performed to comment on the précised size (diameter) of the colloids in the liquid. This critical size may again vary for different insulating liquids and for different aging factors. Aiding the results of UV curves and the turbidity, particle count of aged ester liquids is lesser or comparable to that of the centrifuged MO.

At an outset, a few general remarks may be drawn based on the experimental observations. The same are listed below:

- The increase in the smaller diameter particles with centrifugal treatment may also be due to the possible splitting of the large particles into few smaller particles due to mechanical forces.

- On observing the aged MO and treated MO, in particle count results, a trending crossover is noticed with thermal aging. This indicated the possibility of increasing the smaller diameter particles with thermal degradation. This is because more splitting of large particles is possible with higher degradation, and the brittleness of the particles increases with increase in the thermal degradation.
- The hydrolytic behavior of ester liquids (due to hydrophilic nature of ester group) keeps the paper dry and thus may be a possible contribution toward the fact that colloids are not seen in ester liquids.
- Colloidal particles govern the thermal degradation of mineral insulating oils, while colloidal particle have least contribution toward the degradation of the ester liquids.
- Finally, centrifugal treatment of rejuvenating the insulating liquids is not a feasible solution. This is due to the fact that centrifugal treatment failed to remove tiny and smaller diameter particles.

5.4 Soluble Particles – Fuller's Earth Filtration (ASTM D7150)

To understand the influence of soluble particles on degradation of various insulating liquids, aged liquids are filtered using fuller's earth treatment. The insulating liquids, accelerated aging conditions, and aging history are followed the same that are similar to the Section 5.3. The only difference in the experimental being the treatment process, in this section fuller's earth filtration is adopted. The filtration setup (laboratory-based model) and experimental illustration are presented in Figure 5.7a and the experiment flowchart is presented in Figure 5.7b. One may understand that filtration of aged liquids directly also removes the colloidal particles. The influence of colloidal particles is individually discussed in the previous section. Therefore, the discussions in this section are intended to aim only the influence of soluble particles as per ASTM D7150. It is to be mentioned that filtering is performed at 80°C for single pass.

5.4.1 UV Spectroscopy

To monitor the influence of soluble particles, UV characterizations are performed before (aged) and after fuller's earth filtration. The UV curves of aged (MO, SEA, SEB, NE) and filtered (MO, SEA, SEB, NE) are presented in Figure 5.7.

From Figure 5.8, it is verified that dissolved decay concentration in the liquids is reduced with filtration. In addition, in case of MO the reduction in decay concentration due to filtering is increased with aging. This is due to the fact that higher aging causes higher concentration of decay contents [15, 16]. It is also noticed that generation of decay concentration in ester liquids is lesser than that in the mineral insulating oil. In Figure 5.8d, mineral oils have moved beyond the spectral limits due to its higher degradation. The overall average absorbance (average of the absorbance's at different wave lengths from 360–600 nm) as a function of thermal aging is shown in Figure 5.9.

It is noticed that at lower aging factors, there is no significant difference with filtration in case of ester liquids. After 150 °C, the difference in absorbance with filtration is significantly noticed. In case of mineral insulation oils, this difference is seen at an earlier stage compared to that of the ester liquids. As per the UV studies, the improvement seen in spectral curves and absorbance indicates a positive feasibility of fuller's earth filtration for ester liquids. However, further studies are required to conclude and verify the same.

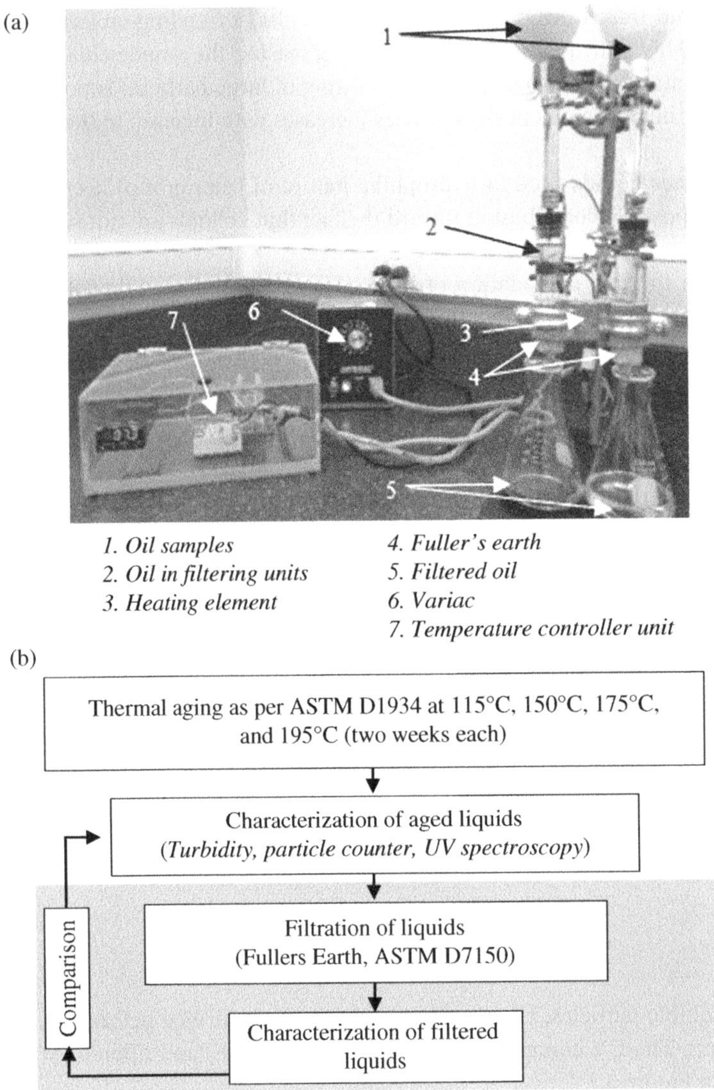

Figure 5.7 Illustration of the experimental and view of filtering setup. (a). View of the laboratory model for fuller's earth filtration. (b). Illustration of the experimental for soluble particles monitoring.

5.4.2 Turbidity

Turbidity of thermally aged and filtered (fuller's earth, 80 C, one pass) liquids is measured as per ASTM D6181 and the results of the same are discussed in this section. The details of the turbidity for aged and filtered MO, SEA, SEB, and NE are shown in Figure 5.10.

From the turbidity results, it is noted that turbidity of mineral oil and ester liquids is increasing with fuller's earth filtration. Theoretically, it is expected that removal of soluble particles with filtration should improve the liquid properties (reduce the turbidity here). However, it is rather increasing in the present case. This may understood that pars of fuller's earth may be carried to the filtrate during the filtration process. To avoid this, a bed of cotton has been introduced at the bottom of the filtering units.

5.4.3 Particle Counter

Particle count in the insulating liquids is measured before and after filtration (aged and filtered). The details of the number of particles for MO, SEA, SEB, and NE at different thermal aging conditions have been presented in Figure 5.11.

It is noticed that the reduction in number of larger diameter particles reduced in case of MO and NE. This may be noticed by the crossover noticed with aged and filtered liquids. The accumulation of smaller diameter particles with aging times in case of MO may be because higher aging causes more degraded particles with thermal degradation of cellulose. In case of NE and SEA, the accumulation of the number of smaller diameter particles is almost comparable. The same is highlighted in red circles in Figure 5.11. This may be due to the introduction of additional particles

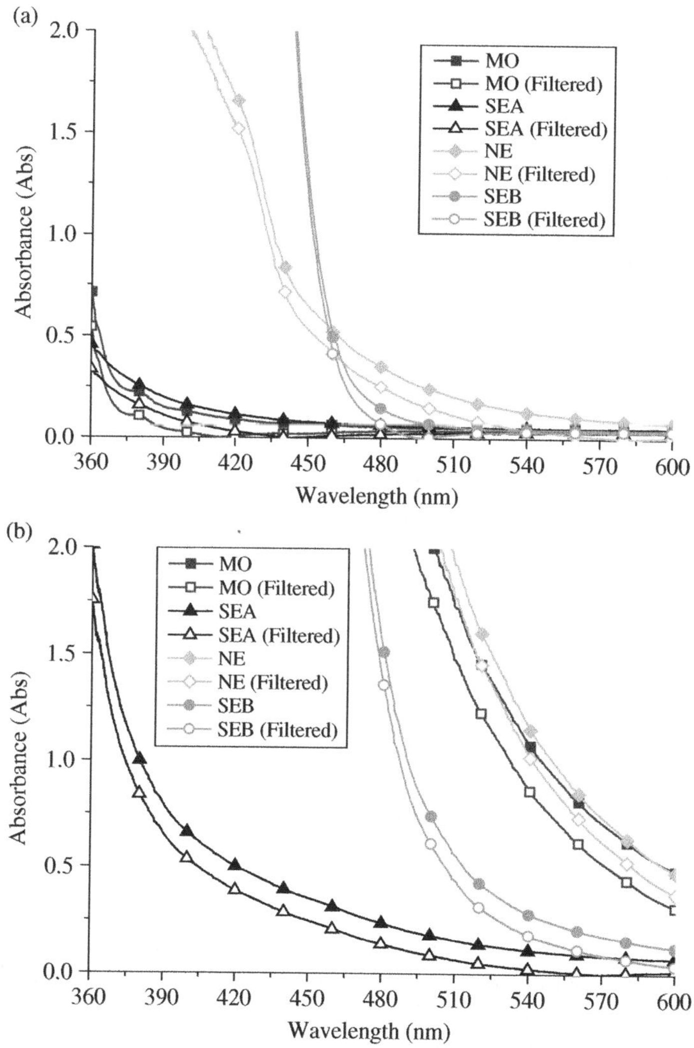

Figure 5.8 UV/Vis spectral curves for aged oils and filtered oils (a) at 115 °C, (b) at 150 °C, (c) at 175 °C, and (d) at 195 °C. *Source:* Based on Loiselle et al. [15].

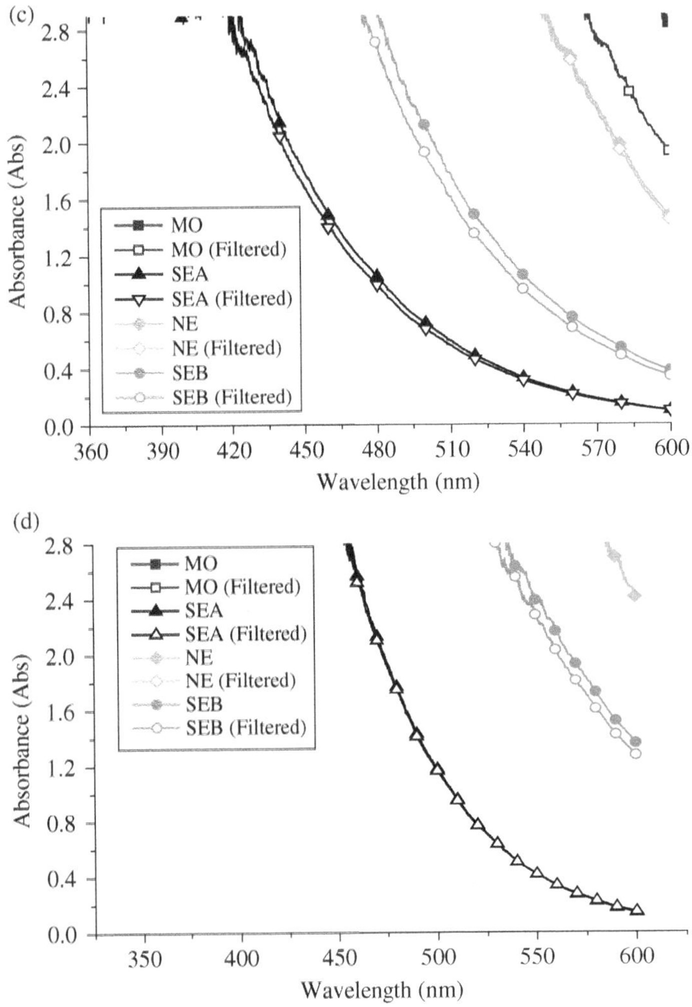

Figure 5.8 (Continued)

during filtration from the adsorbent (fuller's earth). The reaction of ester with hydrogen bonds of the rough surfaces of the hydrous aluminum silicates can introduce tiny particles.

5.5 Feasibility of Fuller's Earth Filtration for Ester Liquids

Following the ambiguity on the feasibility of the fuller's earth treatment for ester liquids in the above section, further detailed investigations are carried out. In this section, MO, SEA, and NE are thermally aged at 115 °C. Thermal aging is performed in presence of the solid insulation (cellulose) in the ratio of 1:20 and copper catalyst. Thermal aging is performed for 1500 hours with an aging history at 50, 1000, and 1500 hours. The discussions in this sections aim to understand the potential of fuller's earth filtration. This is accomplished by considering different liquid adsorbent ratios and different preheating temperatures. It is to be recalled that in the previous section, 35 g/ml of fuller's earth and preheating temperature of 80 °C are adopted.

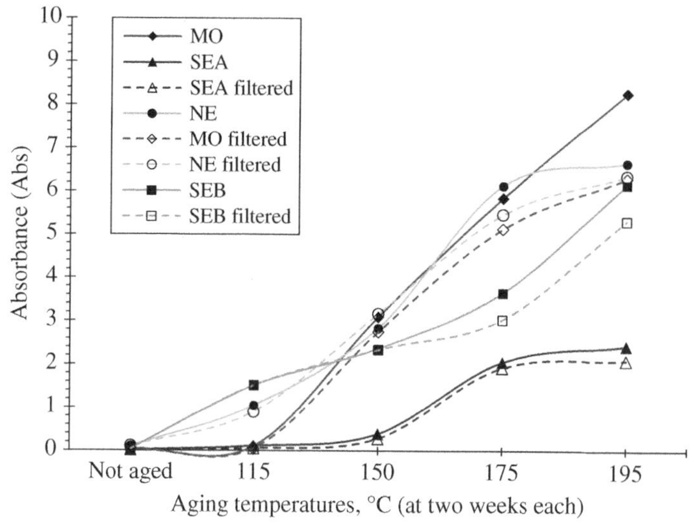

Figure 5.9 Absorbance of aged and filtered liquids.

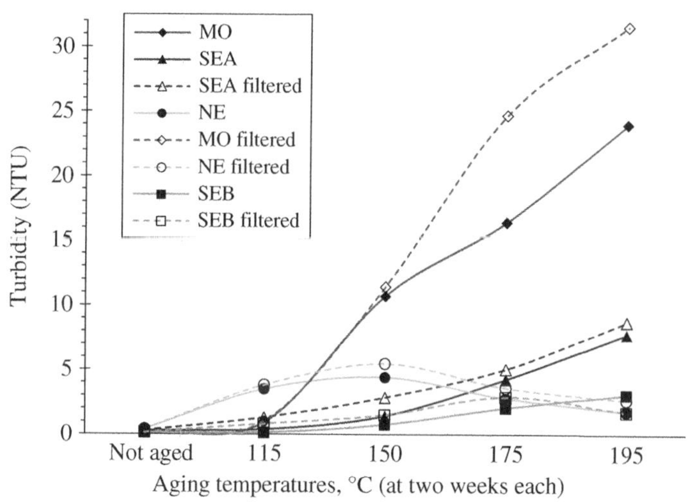

Figure 5.10 Turbidity of aged and filtered liquids.

5.5.1 On Ratio of Fuller's Earth to Liquid

In the present section, the ratio of fuller's earth to liquid is varied from 30 to 40 g/ml for filtering the aged MO, NE, and SEA. The filtration is performed at 30, 40, and 40 g/ml at 80 °C and the results of change in dissolved decay products and turbidity are discussed in this section. The changes in the turbidity of aged and filtered liquids at different oil-adsorbent ratios are presented in Figure 5.12a.

It is observed from Figure 5.12a that at 1 g/30 ml for MO, the turbidity is improved (all the aging factors), while at other ratios it exhibited a fluctuating behavior at different aging times. In case of NE, none of the ratios worked and with increase in the fuller's earth ratio, the turbidity increased. Similarly, SEA exhibited a fluctuating behavior for 1 g/35 ml and 1 g/40 ml, while it is reduced for

1 g/30 ml. The changes in the concentration of dissolved decay products with filtration are presented in Figure 5.12b. It is seen that decay content concentration has been increased for NEs and exhibited a fluctuating trend in case of SE. The increase in concentration of decay content and turbidity in NEs indicated a possible addition of fuller's earth particles in the filtrate. Based on the above results, it is inferred that fullers earth has a low potential to be a good adsorbent for ester dielectric fluids.

5.5.2 On Treatment Temperature

In the present section, the ratio of fuller's earth to liquid is fixed at 35 g/ml and filtration temperature is varied. The filtration temperatures considered are 70, 80, and 90 °C for MO, NE, and SEA under similar aging conditions as described in the previous section. The fuller's earth filtration is performed at different temperatures, and the results of change in dissolved decay products and

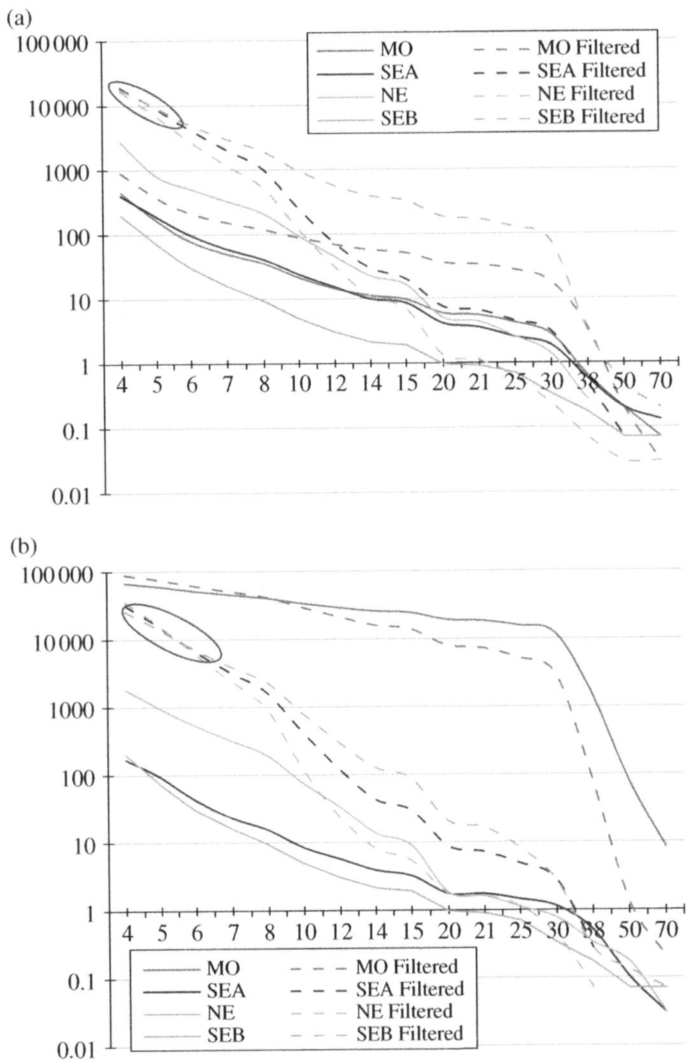

Figure 5.11 Particle count for aged oils and filtered oils (a) at 115 °C, (b) at 155 °C, (c). at 175 °C, and (d) at 195 °C.

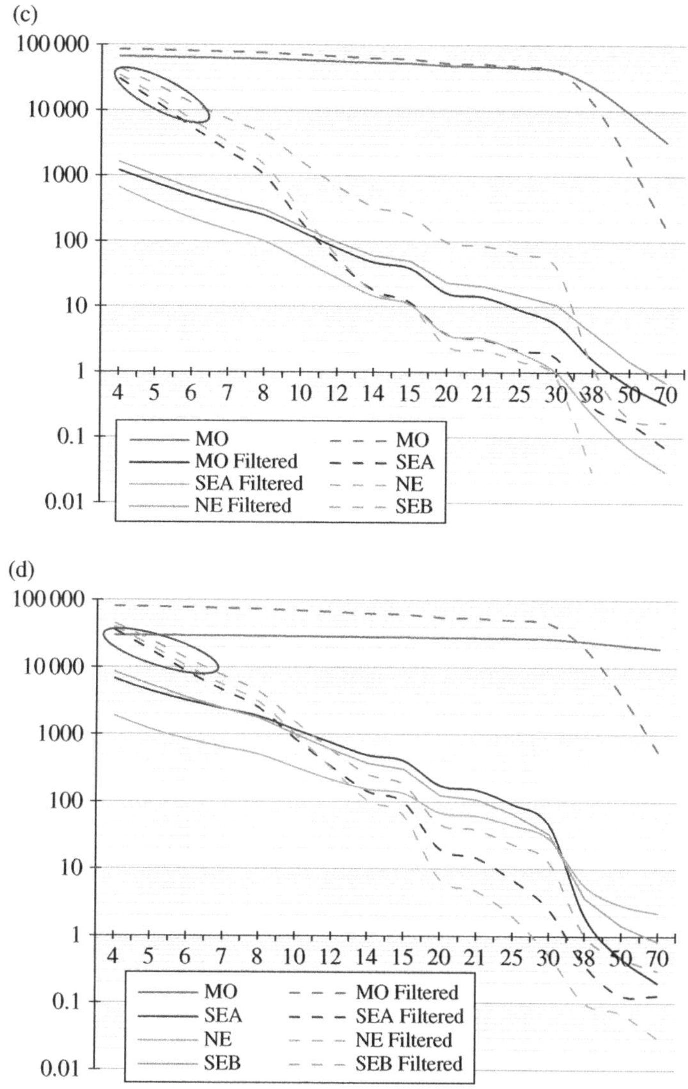

Figure 5.11 (Continued)

turbidity are discussed in this section. The changes in the turbidity of aged and filtered liquids at different filtering temperatures are presented in Figure 5.13a.

It is observed from Figure 5.12a that at 1 g/30 ml for MO, the turbidity is improved (all the aging factors), while at other ratios it exhibited a fluctuating behavior at different aging times. In case of NE, none of the ratios worked and with increase in the fuller's earth ratio, the turbidity increased. Similarly, SEA exhibited a fluctuating behavior for 1 g/35 ml and 1 g/40 ml, while it is reduced for 1 g/30 ml. The changes in the concentration of dissolved decay products with filtration are presented in Figure 5.12b. It is seen that decay content concentration has been increased for NEs and exhibited a fluctuating trend in case of SEA. The increase in concentration of decay content and turbidity in NEs indicated a possible addition of fuller's earth particles in the filtrate. Based on the above results, it is inferred that fuller's earth has a low potential to be a good adsorbent for ester dielectric fluids and potential alternatives need to be investigated [17].

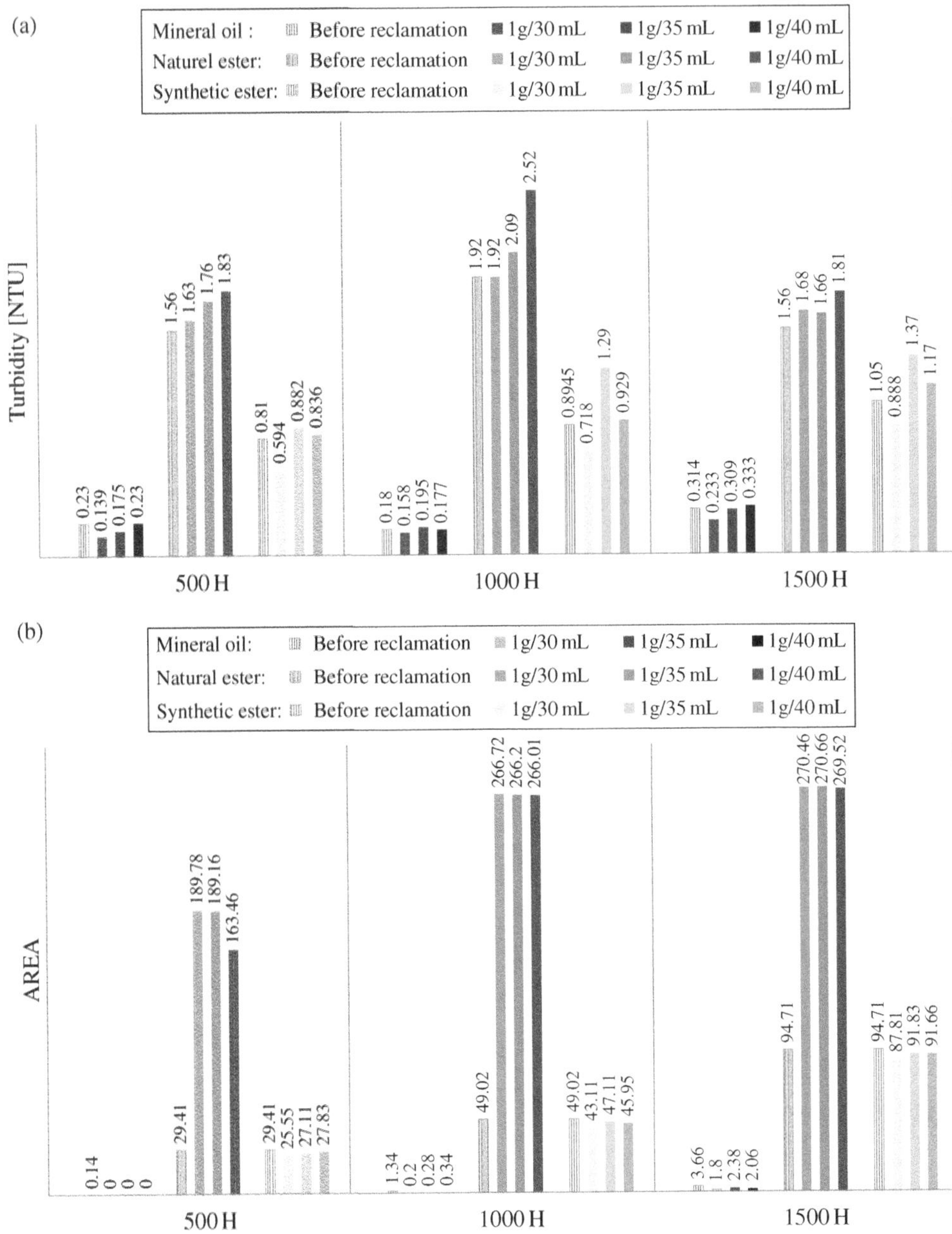

Figure 5.12 Variations of turbidity and dissolved decay content with filtration ratios. (a) Changes in turbidity. (b). Changes in dissolved decay products.

5.6 Conclusions and Future Scope

The influence of colloidal and soluble decay particles on the state of the liquid (aging) has been reported for ester liquids in comparison to the mineral insulating oil. It is understood that influence of particle size (diameter) is a critical parameter for the liquids to have soluble and colloidal particles. It was also found that ester fluid has not noticed with the presence of colloidal particles with aging. In fact, soluble decay contents are involved with the deterioration of ester fluids, thus the influence of X-wax needs to be investigated under different fault conditions. It is revealed that fuller's earth has

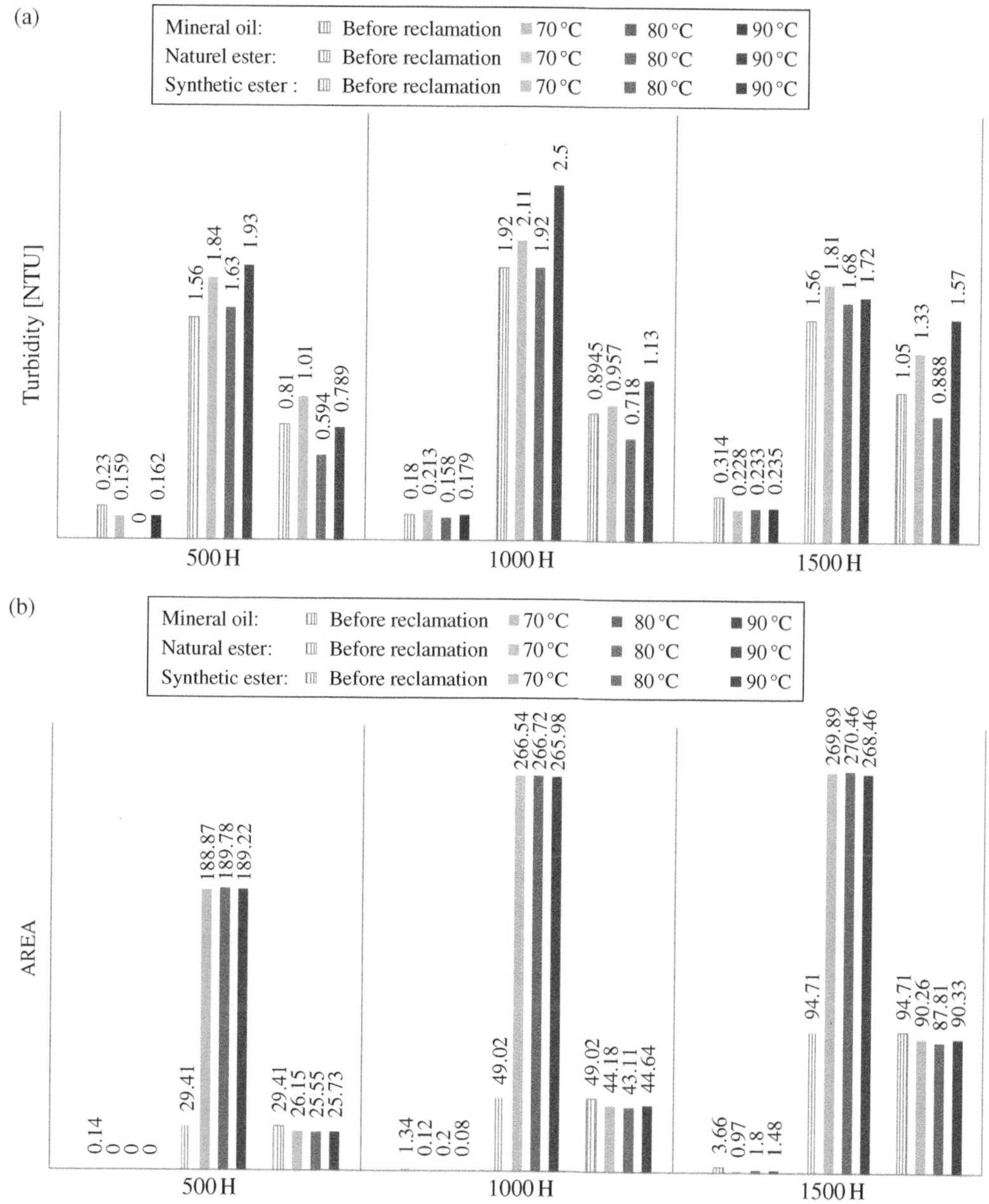

Figure 5.13 Variations of turbidity and dissolved decay content with filtration temperatures. (a) Changes in turbidity. (b) Changes in dissolved decay products.

a low potential to be a good adsorbent for ester dielectric fluids. Therefore, a detailed investigation of the suitable adsorbents has to be carried out. However, ester liquids need to be verified for the efficiency of regeneration at different ratios and pretreatment temperatures with fuller's earth.

References

1 Fofana, I. (2013). 50 years in the development of insulating liquids. *IEEE Electr. Insul. Mag.* 29 (5): 13–25.

2 Wang, M., Vandermaar, A.J., and Srivastava, K.D. (2002). Review of condition assessment of power transformers in service. *IEEE Electr. Insul. Mag.* 18 (6): 12–25.

3 Beroual, A., Sitorus, H.B.H., Setiabudy, R., and Bismo, S. (2018). Comparative study of AC and DC breakdown voltages in Jatropha methyl ester oil, mineral oil, and their mixtures. *IEEE Trans. Dielectr. Electr. Insul.* 25 (5): 1831–1836.

4 Bakrutheen, M., Willjuice Iruthayarajan, M., and Narayani, A. (2018). Influence of ultrasonic waves on viscosity of edible natural esters based liquid insulation. *IEEE Trans. on Dielectr. Electr. Insul.* 25 (5): 1628–1635.

5 Rozga, P., Stanek, M., and Pasternak, B. (2018). Characteristics of negative streamer development in ester liquids and mineral oil in a point-to-sphere electrode system with a pressboard barrier. *Energies* 11: 1088.

6 Mohan Rao, U. and Jarial, R.K. (2019). Measurement of transformer solid insulation degradation using dilatometry and x-ray diffraction analysis. *Measurements, Elsevier* 131: 701–705.

7 Fernández, I., Delgado, F., Ortiz, F. et al. (2016). Thermal degradation assessment of kraft paper in power transformers insulated with natural esters. *Appl. Therm. Eng.* 104: 129–138.

8 Li, Y., Zhong, L., Yu, Q. et al. (2019). Influence of moisture content on cellulose structure and breakdown strength of vegetable oil-impregnated paper. *IEEE Trans. on Dielectr. Electr. Insul.* 26 (4): 1245–1252.

9 Bandara, K., Ekanayake, C., Saha, T.K., and Annamalai, P.K. (2016). Understanding the ageing aspects of natural ester based insulation liquid in power transformer. *IEEE Trans. on Dielectr. Electr. Insul.* 23 (1): 246–257.

10 CIGRE 157 Technical brochure (2000). Effect of particles on transformer dielectric strength. *WG* 12: 17.

11 ASTM D6786-08 Standard (2008). *Test Method for Particle Count in Mineral Insulating Oil Using Automatic Optical Particle Counters*. USA: ASTM.

12 Mohan Rao, U., Fofana, I., and Loiselle, L. (2019). Preliminary studies on soluble and colloidal decomposition products in ester filled transformers. *Proceedings of 20th IEEE International Conference on Dielectric Liquids (ICDL)*, Roma, Italy (23–27 June 2019), pp. 1–5.

13 Hao, J., Dan, M., Li, Y. et al. (2017). Cellulosic particles accumulation and conductivity characteristics of natural ester used in power transformer. *Elsevire Procedia Eng.* 202: 168–175.

14 Degeratu, S., Rotaru, P., Rizescu, S. et al. (2015). Condition monitoring of transformer oil using thermal analysis and other techniques. *J. Therm. Anal. Calorim.* 119 (3): 1679–1692.

15 Loiselle, L., Rao, U.M., Fofana, I., and Jaya, T. (2020). Monitoring colloidal and dissolved decay particles in ester dielectric fluids. *IEEE Trans. Dielectr. Electr. Insul.* 27 (5): 1516–1524.

16 Mohan Rao, U., Fofana, I., Bétié, A. et al. (2019). Condition monitoring of in-service oil-filled transformers: case studies and experience. *IEEE Electr. Insul. Mag.* 35 (6): 33–42.

17 Leila, S., Zafour, H., Mohan Rao, U., and Fofana, I. (2019). Regeneration of transformer insulating fluids using membrane separation technology. *Energies* 12 (3): 368.

6

Prebreakdown and Breakdown Phenomena in Ester Dielectric Liquids

Pawel Rozga[1], T. Jayasree[2], U. Mohan Rao[2], I. Fofana[2], and P. Picher[3]

[1] *Institute of Electrical Power Engineering, Lodz University of Technology, Lodz, Poland*
[2] *Research Chair on the Aging of Power Network Infrastructure (ViAHT) University of Quebec at Chicoutimi, Chicoutimi, QC, Canada*
[3] *Institut de recherche d'Hydro-Québec, Varennes, Canada*

6.1 Introduction

Prebreakdown phenomena in dielectric liquids have been the issues presented in research works for many years, wherein a particular increase was observed at the beginning of the second half of the twentieth century [1–5]. The studies on prebreakdown phenomena were conducted mainly at the voltages of impulse character that to the greatest extent reflecting the electrical nature of the physical phenomena preceding the breakdown of the liquid. Together with an increase of measurement techniques, the knowledge on prebreakdown phenomena has become more comprehensive. This has been mainly due to the introduction of photo-optical techniques, which made possible to follow the spatiotemporal development of the discharges in liquids [6–16]. The methods based on shadowgraph or streak photographs have enabled to recognize the spatial shape of discharge forms. In turn, when the measurement of propagation time has been added the propagation velocity – one of the fundamental parameters describing character of electrical discharges in liquids – has been possible to determine. The photomultiplier tube (PMT) has become an indispensable measuring instrument in the experimental research on electric discharges in dielectric liquids. The results obtained with its use (sequence of discrete light pulses of nanosecond risetimes observed both during initiation and propagation of the discharges) have confirmed the assumptions from previous works that defined propagation of the discharges as of step character [6, 9, 11, 12, 14–16].

Most of research works cited above have been carried out with the use of bare electrodes of point-to-plane configuration, which have created the nonuniform electric field distribution. Applying such electrode system resulted from few reasons. First, the certainty about the place of discharge initiation has been obtained, because discharges always have started to develop from the area around the high-voltage (HV)-point electrode. This has been a positive fact from the point of view of the optical methods usage – observed space might be limited to the space around the HV electrode. Second, the point-to-plane electrode systems have not required the high inception voltages in contrast to the systems of uniform electrical field distribution and the same gap distances that are characterized much higher inception voltages. Additionally, using the systems with HV-point electrode has given the possibility to observe the discharges, which have not caused the

Alternative Liquid Dielectrics for High Voltage Transformer Insulation Systems: Performance Analysis and Applications, First Edition. Edited by U. Mohan Rao, I. Fofana, and R. Sarathi.

breakdown (the so-called stopping length discharges) and which have been an excellent source of data on discharge initiation and initial phase of propagation of the electrical discharges in liquids. Finally, the systems of point-to-plane configuration have imitated, to some extent, the defects of real insulating systems, where existing microscopic field inhomogeneities may cause the local increase of electrical field stress and initiate of the discharges when overvoltage takes place.

Some variation of the point-to-plane electrode system is the system with pressboard plate placed on the flat grounded electrode. Such solution adopted, for example, in [17–19], stops the development of the discharges causing the breakdown, which allows for performing the measurements and analyses of behavior of the different liquids in wider range of testing voltages than in the case of the systems without insulating plate. Additionally, such types of the systems may simulate to a greater extent of the real situations, which may occur in insulating systems of electrical power devices. When inhomogeneity occurs and discharge is initiated, it does not develop between two conductive elements; the element of solid insulation always exists, which stops the discharge propagation.

In terms of prebreakdown and breakdown phenomena in dielectric liquids, the theories have been developed, which are aimed on explanation of the mechanisms of initiation, propagation, and breakdown of dielectric liquids [20–31]. They have been formulated from the very beginning of the research on the electrical properties of liquids. These theories have been firstly developed on the basis of the studies on pure hydrocarbon liquids, their mixtures, and transformer mineral oils of various origins (naphthenic and paraffinic). The latest part of the research have been the studies on dielectric liquids that are alternative to mineral oil including natural and synthetic esters [17–19, 32–39]. The studies on prebreakdown phenomena in ester liquids has become extremely important from a cognitive and practical point of view. The ester liquids have been used more and more often as insulating-cooling medium in power transformers, including those of highest voltage class. The approach for the studies concerning esters has been based on comparison of the results obtained for them with the results concerning mineral oil received at the same testing conditions. This allows the identification of common features and differences between the distinctive liquids in terms of the initiation and propagation of electrical discharges, as well as breakdown, and on this basis to present some practical recommendations for manufacturers and users of transformers.

6.2 Research Methods in Assessment of Prebreakdown Phenomena in Ester Liquids

As was mentioned above, the prebreakdown and breakdown phenomena in dielectric liquids have been conducted mainly under lightning impulse voltages. Two approaches may be identified in such studies:

- Those based on standards allowing to determine the lightning impulse breakdown voltage (LIBV) of a given dielectric liquid
- Those based on advanced experimental techniques allowing to assess the nature of the discharges that is a way of their propagation from initiation to breakdown

In the first case, apart from the determination on the basis of a statistical approach, the breakdown voltage of the liquid at lightning stress, the physical nature of the discharges is impossible to recognize. However, the relatively simple methodology applied allows for a quick comparison of liquids of different physico-chemical properties as, for example, ester liquids with mineral oils [40–43]. In the second case, the studies needed to conduct are time-consuming and the equipment needed to perform the measurements requires from the researcher a wide knowledge in the field of not only electricity but also electronics, optics, or spectroscopic techniques. Nevertheless, the results obtained in this way provide a broad spectrum of data on the mechanisms determining the initiation and propagation of the discharges in liquids [33–36, 44–46].

6.2.1 Standard-Based Approach

There are two standards that describe the procedures of determining the lightning impulse breakdown voltage of dielectric liquids:

- ASTM D3300 "Dielectric breakdown voltage of insulating oils of petroleum origin under impulse conditions" [41]
- IEC 60897 "Methods for the determination of the lightning breakdown voltage of insulating liquids" [40]

The ASTM Standard suggests to do the standard lightning impulse voltage 1.2/50 μs of negative polarity. The electrode system should consist of two sphere electrodes of 3.8 mm gap distance or point-to-sphere electrode system of 25.4 mm gap distance. The voltage should be supplied in a step manner with 5 or 10 kV step. Each level of voltage includes three impulses with minimum 30 second interval between them. Measurement procedure is repeated five times and average value from these five measurements is treated as breakdown voltage. Additionally, standard deviation is calculated to show data spread. In the case of IEC Standard, the electrode system suggested consists of point HV electrode of radius of curvature of its tip in the range between 40 and 70 μm and grounded sphere of 12.5–13 mm in diameter. The gap distance is 10, 15, or 25 mm depending on the electrical strength of the liquid tested. In contrast to the ASTM Standard, the measurement is performed with one impulse per step (typically 5 kV) wherein the tested sample should withstand minimum three voltage levels before breakdown is noticed. Similarly as in the case of ASTM Standard 5 measurements are done and average values with standard deviation are given.

6.2.2 Experimental Approach

The experimental approach aimed on the assessment of streamers behavior in a given liquid under lightning impulse voltage is not standardized. It has been developed as a result of many years of studies conducted in scientific centers all over the world. Nevertheless, there are several universal assumptions without which a comprehensive assessment of the initiation and propagation mechanisms of electrical discharges in liquids is impossible. These assumptions are:

- having a controlled source of lightning impulse voltage with a measurement setup for measuring the peak value of voltage waveform (most often in the form of a resistive voltage divider, a peak value meter and an oscilloscope recording the voltage waveform), which simultaneously gives the possibility to determine the time to streamer initiation or the time to breakdown;
- having a system that allows for tracking the spatiotemporal development of the streamers based on one of the methods of photographic recording of discharges (shadowgraph method, refractive method, static or streak photography method);
- having a system for recording the streamer current, charge or the sequence of light pulses emitted by the streamers developing in a given liquid, which gives information about the degree of ionization of the discharge channels and the propagation mode.

Without the above three components, a comprehensive evaluation of the prebreakdown mechanisms in dielectric liquids is hardly probable. Additionally, it is possible to use other methods, which may complement these fundamental ones. One of them is spectrophotometer-based measurement of the spectra emitted by the streamers developed in liquid or by the breakdown channel. This method enables to identify the characteristic frequency radiation (wavelengths) for the specific discharge phenomenon [47–50]. The experimental approach focuses on evaluation of characteristic parameters describing electrical discharges in a given type of liquid. These parameters are then used in comparative assessment of ester liquids and mineral oil, taking into account the fact that mechanisms of initiation and propagation of the discharges in mineral oil are quite well recognized. The parameters considered in the liquids assessment are:

- inception voltage of the streamers, which is defined as the lowest value of voltage at which the streamers appear for the first time in a given testing conditions;
- stopping length of the streamers, which is defined as the final length of discharge channels that are reached during streamer propagation at given testing conditions; it is important to point out that stopping length accompanies only the discharges developing below the breakdown voltage;
- propagation velocity, which is defined as velocity of the streamers developed under electric field resulting from applied voltage and electrode configuration; typically, propagation velocity is given as the average velocity for given testing conditions; for the stopping length streamers, propagation velocity is evaluated as the ratio of stopping length of the streamers to time of streamer propagation, while for the streamers causing the breakdown, propagation velocity is calculated as ratio of gap distance to time to breakdown. However, when the streamer photo is taken out, propagation velocity may be determined as ratio of streamer length measured from photo collected to time of taking the photo.
- acceleration voltage, which is defined as a voltage at which sudden increase of propagation velocity is observed; in general this value defines transition of the streamers to the next mode (from slow to fast), mainly from second to third but some literature data reports also the transition from second to fourth mode;
- streamer shape, which is classified on the basis of collected streamer photos and depends on the polarity and magnitude of applied voltage; it is accompanied with above mentioned streamer mode.

6.3 Initiation of Streamers in Dielectric Liquids

The above quoted term "streamer" is a well-known and strictly defined term in relation to the theory of electrical discharges in gases. It means a stable electron avalanche of 10^9 electrons inside it was developed, which is based on collision ionization and photo ionization. The streamer channel in gases is a weakly conductive (low-temperature) plasma [51, 52]. Streamers in gases play a major role in flashover process. However, the final phase of the flashover is a result of a "leader" discharge development. The streamer in air is characterized by a charge density of about 0.003 μC/cm and a leader channel of more than 1 μC/cm.

In the case of dielectric liquids, the term "streamer" is used for description of the series of prebreakdown phenomena, which occur in liquids after discharge initiation and not always must finish themselves by breakdown. Thus, "streamer" in liquids, however, does not have a physical equivalent to "streamer" in gases. First, it is characterized by higher electron density of 10^{17} to 10^{18}/cm, which allows for evaluation of a charge density along the ionized gas channel in the order of 100 μC/cm. Thus, streamers in liquids possess another refractive index than the surrounded area. This fact is used in application of optical techniques enabling in taking the streamer photographs during propagation. The refractive index mentioned differs, however, depending on the streamer mode [6–12, 14, 15, 17, 27–35, 38, 39]. The term "streamer," as a name of the discharge phenomenon in liquids, is also connected with its similarity to the lightning corona discharge in gases. Photographically registered positive lightning corona discharge in air is, in its spatial form, very similar to the positive streamers developing in mineral oil at lightning impulse voltage.

Therefore, the term "streamer" in relation to prebreakdown phenomena in liquids should be used very carefully, being aware of the clear differences between the definition of this concept in terms of the theory of electrical discharges in gases and attempts to apply it in the nomenclature of discharge phenomena observed in liquids. Hence, in the literature, regarding the electrical discharges in liquids, the term "streamer" is sometimes omitted in favor of the classical naming such as "electrical discharge," "prebreakdown event," "predischarge phenomena," "partial discharge," or "prebreakdown phenomena." There is also a group of researchers, who are against the use of the term "streamer" to describe discharge phenomena in liquids. They suggest to use of the term "leader" [53], which in relation to large air gaps, means the development of a step discharge, which is closer in its nature to the nature of electrical discharges in liquids. However, taking into account that the term "streamer" has been commonly accepted in most of the research works published, it will be used further in this chapter to tell about electrical discharges developing in different kind of dielectric liquids.

6.3.1 Influence of Tip Radius on Streamer Initiation

Independent of the fact, how type of mechanism determines streamer initiation in dielectric liquids and how voltage polarity is, electrical field stress resulting from electrode geometry and applied voltage seems to be one of the most important factor used in assessment of streamer initiation mechanisms in dielectric liquids. In the systems of nonuniform electrical field distribution (as, for example, point-to-plane configuration), electric field distribution depends mainly on the radius of curvature of HV-point electrode and to a lesser extent on the distance between the electrodes (the so-called free liquid gap). Having the experimentally evaluated inception voltages, it is possible to determine, in an approximate way, the maximum value of electrical field stress in the vicinity of point electrode, which contributes to the streamer initiation [6–11, 18, 25, 28, 34, 54–56]. Many years of studies allowed for elaborating the formula Equation (6.1) for point-to-plane electrode system in oil, which enables the calculation of the inception electrical field stress without using the numerical methods.

$$E_i = \frac{2V_i}{r \cdot \ln\left(\frac{4d}{r}\right)} \tag{6.1}$$

In this formula V_i represents inception voltage from experimental studies, *r* is a radius of curvature of HV-point electrode, and *d* is a gap distance.

The experimental studies mentioned allowed for determination the typical values of inception electrical field stress, at which streamers are initiated under impulse voltage. These values are in the range between 1 and 10 MV/cm, wherein in the individual cases the values causing initiation may be lower and reach the values from the range between 0.5 and 1 MV/cm. Such a discrepancy in the values of inception electrical field stress results from few facts. First, generally lower values of inception electrical field stress concern the systems of uniform and quasi-uniform electrical field distribution, where the influence of well-known surface effect and volume effect on the electrical strength of such systems immersed in oil has been observed. If the nonuniform electrical field distribution is considered, the lower values of inception electrical field stress have been observed at higher values of radius of curvature of HV point, which is explained as increase of surface of this electrode, thus, as in the case of the systems of quasi-uniform fields, by the influence of surface effect. The higher values are, however, typical for the systems with a very small radius of curvature of the tip of the HV electrode. However, still it is important to remember that increase of radius of curvature brings increase of inception voltage when keeping the same gap distance between the electrodes. The relationship between the radius of curvature of the HV electrode and inception voltage of the streamers and inception electrical field stress was studied widely in [28] for mineral oil under HV lightning impulse 0.4/1400 μs. The radius of curvature of point electrode was, in the cited studies publication, changed in a wide range from 1 μm to 20 mm. The initiated streamers were of the stopping length type, thus they do not reach the grounded electrode. Therefore, they may be treated as developed at minimum value of testing voltage, for which electrical field stress was so low that initiated streamers discharge did not develop up to the form of breakdown channel.

Returning to the presented formula Equation (6.1), Figure 6.1, borrowed from [28], presents two relationships: inception voltage versus radius of curvature of HV point and inception electrical field stress versus radius of curvature of HV point.

The figure shows the influence of surface effect causing decrease of inception electrical field stress with increase of radius of curvature of HV point. The values from the Figure 6.1b are the results of calculations using above quoted formula Equation (6.1) based on the results of the measurements presented in the Figure 6.1a. As was mentioned above, the values from the range of 1–10 MV/cm are typical for inception electrical field stress and the radiuses of curvature lower than 1 mm.

In the given testing conditions, streamer initiation in liquids may, however, took place at lower values of electrical field stress. This may be concluded on the basis of the studies concerning lightning strength of selected fresh hydrocarbons. The values of electrical field stress, obtained from such studies, at which in the given conditions breakdown occurred, were assessed to be 1–2 MV/cm. Thus, initiation of the streamers, which do not transform into the form leading to breakdown, should take place at much lower values of electrical field stress than this corresponding with breakdown. Obviously, such considerations depend, in a large extent, on the electrode configurations and the gap distances. In the case of small gaps (up to 5 mm) and bare electrodes, streamer initiation is connected always with breakdown, thus electrical strength expressed by electrical field stress is equivalent to inception electrical field stress. Similar conclusion concerns the systems of uniform electrical field distribution.

Some publications [6, 9, 10, 25, 26, 28, 54, 57] report that, based on the theoretical considerations concerning the physico-chemical nature of liquids, the threshold value of inception electrical field stress is connected directly with phenomenon of electrostatic emission or field ionization in the liquid volume. In such case for field emission, these values, at which discharge current was registered, were calculated as 7–20 MV/cm, wherein these values depend on the radius of curvature of point electrode. At higher radius and lower electrical field stress, the phenomenon of critical volume was noticed as the phenomenon deciding on streamer initiation. Concerning field ionization, the values, at which this ionization is possible, were assessed as 10 to several MV/cm, wherein they depend on the chemical structure of a given liquid. Exemplary, higher inception electrical field stress characterizes cyclohexane, for which ionization energy is equal to 8.75 eV, while for mineral oil with dominance of aromatic compounds the inception electrical field stress is lower due to ionization energy of benzene, which is equal to 7 eV. It is, however, important to point out that the above mentioned values for electrostatic emission and field emission concern the liquids that are free of moisture and contaminations; for such liquids, the possibility of the appearance in the

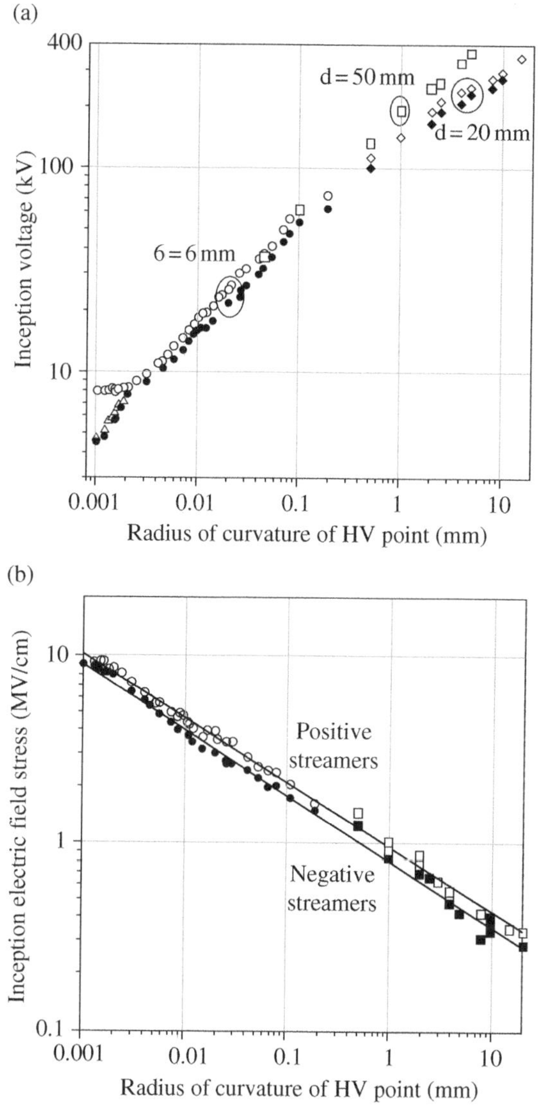

Figure 6.1 Dependence inception voltage (a) and inception electrical field stress (b) of positive (open symbols) and negative (black dots) streamers versus radius of curvature of HV-point electrode: for d = 6 mm radius of curvature was changed from 1 to 200 μm, for d = 20 and 50 mm radius of curvature was changed from 0.5 to 20 mm – electrical field stress was calculated on the basis of Equation (6.1) using the data of inception voltage. *Source:* Based on Denat et al. [35].

liquid volume contaminations or gas bubbles is not considered. From the above, it can be concluded that the quality of the liquid and its chemical composition (esters in composition differ to a large extent from mineral oils) may affect the streamer initiation, but, in non-uniform electric field (where formula (1) applies), the influence should be minimal or negligible.

6.3.2 Streamer Initiation Mechanisms

Streamer initiation in dielectric liquids, independently of the liquid kind (mineral oil or ester liquids) is based on the phenomenon of gas bubble formation in the vicinity of HV electrode. Mostly, this phenomenon is called cavitation mechanism [22–29, 54, 57]. There are three main mechanisms considered in the literature: formation of gas bubble as a result of liquid detachment from the surface of the initiating electrode (impact of electromechanical forces), bubble formation as a result of Joule heat (high-density current appearance), and bubble formation as a result of electrostatic emission. Mechanism of initiating bubble concerns both voltage polarities (the forces causing bubble appearance do not depend on whether HV point is anode or cathode), however, high-density currents may come from both cathode (electrostatic emission of thermal emission) and anode (field ionization, hole conduction). Nevertheless, inception of positive and negative streamers is described in another way.

In the case of positive polarity, bubble formation may takes place due to two possible reasons: as a result of impact of Joule heat or as a result of cavitation from acting the electro-mechanical forces on liquids. The process deciding about the initiation of negative streamers in liquids is field emission called also an electrostatic emission.

The experimental studies described in [9, 22, 57] indicated that before reaching the impulse peak value, the streamer is initiated, it is possible to record the current pulses of very small values (circa 10^{-5} A). Together with these pulses, the shock waves are noticed as a result of forming the cavitation spaces around the HV-point electrode. The recorded current emitted from the small dot of the point electrode in a very small period of time (from 10 to 100 ns) may give enough heat (energy) that may cause evaporation of a small volume of liquid in the vicinity of this point. It is assumed that this process may take place when local electrical field stress resulting from electrode geometry is around 10 MV/cm and the value of current emitted from mentioned dot of surface equal to $0.1\ \mu m^2$ is in the range of 100 nA. An increase of the temperature in such process is enough, in period of time that is typical for streamer initiation delay, to cause evaporation of liquid. This theory was confirmed by the studies on the influence of pressure on the streamer initiation, for example in cyclohexane. The studies showed that together with increase of the pressure up to 7 MPa recorded, discharge current was always represented by the continuous component connected with the current responsible for liquid evaporation. However, the current pulses connected with streamer development decreased with an increase of the pressure and disappeared at the pressure equal to 7 MPa. Additionally, increase of the pressure caused an increase of the inception voltage. In such case, energy needed for liquid evaporation had to be higher.

The second theory relates to the positive streamer initiation and says that streamers start to develop in micro-bubbles filled with vapor. The bubbles are created as a result of cavitation process occurring as the impact of electro-mechanical forces on the liquid layer adhering to the metal surface of electrode of high curvature or in the vicinity of micro-inequalities on quasi-plate electrode. In both cases, the local electrical field stress may reach even 1–10 MV/cm, thus enough value for streamer initiation. The possibility of appearance of local cavitation disturbances on the boundary of metal–liquid in the form of initiating bubbles was considered as directly proportional to the ratio εE^2, where E is the value of electrical field stress resulting from the voltage applied, and ε is the electrical permittivity of the liquid [26, 27].

In the case of negative polarity, mechanism of initiation is quite better recognized. The main process deciding about initiation is field (electrostatic) emission. When electrical field stress exceeds a given threshold value (in the case of negative polarity it is said that this value is above 2.5 MV/cm), fast electron avalanches are generated from the HV-point electrode, which then induce shockwaves and their final effect is creation of bubbles. The development of main discharge channel takes place solely in these bubbles. This process is independent of the pressure. Connection of negative discharge initiation with the generated shockwaves was confirmed by the photos of streamers where during the initiation phase these shockwaves were clearly visible even if the channel itself was difficult to recognize [22, 23, 27].

6.3.3 Research Progress on Streamer Initiation in Esters vs. Mineral Oil

The works devoted to streamer initiation in ester liquids versus mineral oil were carried out as independent studies or more often as a part of the studies concerning general comparison of breakdown properties of dielectric liquids selected [18, 33, 34, 37–39, 44, 46, 58–60]. In general, the results have indicated that in nonuniform fields the streamers in ester liquids are initiated at the similar values of testing voltage as in the case of mineral oil. The observations in this field are true for wide range of gap distances from few mm up to few cm. Exemplary, in [18], streamer inception voltages in three liquids tested, namely, synthetic ester, natural ester, and mineral oil, were compared. The nonuniform field was created by point-insulating plate electrode configuration at four gap distances 5, 10, 15, and 20 mm. For each gap and for both voltage polarities of standard LI voltage obtained, results were practically identical. This conclusion concerns 50% inception voltages ($V_{i50\%}$) evaluated from Weibull distribution plots and to a slightly lesser extent the so-called zero percent inception voltage ($V_{i0\%}$) determined as a location parameter of three-parameter Weibull distribution function. From the experimental studies, this was also noticed that at inception voltages, independently of the liquids considered, the spatial form of negative discharges resembled the compact bushy structure with many branches propagating both paraxial and lateral directions, while for positive polarity the structure of streamer spatial shapes formed a bush of a spherical boundary of rising radius as the streamer developed.

In turn, in [34] both the influence of gap distance and radius of curvature of high voltage point were considered in relation to mineral oil and different natural esters (vegetable oils). The increase in the radius of curvature caused increase in the inception voltage of the streamers independently of the liquids under test. In general, this was stated that no differences appear between esters and mineral oil in terms of streamer inception or these differences are negligible. Figure 6.2 confirms it through presentation of inception electric field E_i for 2 mm gap distance and different radius of curvatures changed from 10 to 90 μm. The value of E_i was calculated using Equation (6.1) taking into account the experimentally obtained inception voltages.

The authors confirmed also similarity between the liquids in terms of the influence of gap distance on streamer inception. The results obtained for the gaps changed from 2 to 15 mm showed increase of inception voltage in a very small magnitude and the differences in a very narrow range.

Inception voltage of the streamers in esters and mineral oil was compared also in [33] for both voltage polarities and 50 mm gap distance of point-to-plane electrode arrangement. The authors

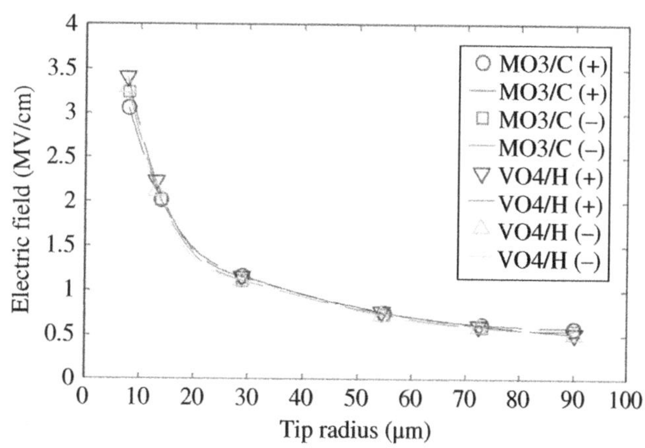

Figure 6.2 Influence of radius of curvature of high-voltage point (with d = 2 mm) on the inception electric field of streamers. The polarity is indicated in brackets. *Source:* Dang et al. [34] / with permission of IEEE.

found the same regularities as described above. The differences noticed between the ester liquids under and mineral oil were minimal form statistical point of view. They changed from 50 to 55 kV and from 45 to 50 kV for positive and negative polarity of lightning impulse voltage, respectively.

From above, this may be stated that due to the similar values of inception voltages in esters and mineral oil, it can be clearly indicated that the physical nature of the phenomenon of streamer initiation in esters and in mineral oil is similar. The cavitation (bubble) initiation mechanism is correct for both esters and traditional mineral oil.

6.4 Streamer Propagation

Regardless of the kind of dielectric liquid we deal with (esters or mineral oil), there are certain universal relationships in the propagation of streamers that are true for each of them. These relationships have been developed over many years, and on the basis of elaborated observations, classification of the individual phases of discharge development was proposed. The spatiotemporal observation of the streamers, as well as recorded streamer current or emitted light, were taken into account at classification of streamer propagation modes. However, the streamer propagation velocity was assessed as the main indicator, which, with the highest reliability, indicate the propagation mode independently of the liquid under consideration.

6.4.1 Overview of Propagation Modes

In general, classification of the streamer modes in dielectric liquids is connected with appearance of the subsequent development phases with increase of a testing voltage related to the inception voltage (more often) or breakdown voltage (seldom) in the given testing conditions. First, the classification of the streamers was proposed by Hebner [21] giving, for the distinctive phases of streamer development, the subsequent numbers. Development of Hebner's approach is numbering proposed by Lesaint and his co-researchers [12], who divided the propagation phases, independently of the voltage polarities, into slow (first and second mode) and fast (third and fourth mode). Using the adjectives "slow" and "fast" is related to the above mentioned propagation velocity, which significantly distinguishes these two groups of the discharges.

On the basis of the literature data available, Table 6.1 presents the proposal of general division of streamer propagation modes taking into account the polarity of testing voltage. The given propagation modes correspond to the characteristic propagation velocities. Table 6.1 shows the values that are true for mineral oils and hydrocarbon liquids; however, the ester-related values follow a similar trend with slightly higher values in each category.

Table 6.1 Propagation velocities for different streamer modes.

Propagation mode	Positive streamers	Negative streamers
First mode	0.1–1 km/s	0.1–1 km/s
Second mode	1–6 km/s	0.5–3 km/s
Third mode	20–50 km/s	10–20 km/s
Fourth mode	100 km/s	100 km/s

Not only propagation velocity is a characteristic feature distinguishing the modes of streamers developing in liquids. There is a close correlation between streamer mode, its shape, propagation velocity, intensity of light emitted by the streamer channels, and its current or charge. In particular, a clear distinction between the streamer modes is visible at transition from second to third mode. Together with propagation velocity, we can observe the change in the structure of streamers recorded photographically, as well as the change in the light waveforms corresponding with propagation. Figure 6.3 shows general, possible to observe in real testing conditions, idealized relationship between propagation velocity and testing voltage. Such relationship enables indication of characteristic points and values of voltage correlated with the distinctive modes of streamer development.

There are two typical relationships for positive lightning impulse voltage: solid line (no. 2) concerns classical mineral transformer oils, dotted line (no. 1) concerns ester liquids (both synthetic and natural). At the negative voltage polarity transition into the area of fast streamers is smoother, and the propagation velocities of the third mode streamers are lower than that corresponding with positive streamers. It is seen from the figure quoted that at the inception voltage the first or second mode streamers may be initiated, wherein the first mode streamers are very difficult to observe, and their investigations have been practically omitted. This results from the fact that they may develop in the specific testing conditions (the point electrode of a very small radius of curvature <10 μm) and in the liquids of very low boiling point. In such liquids, the evaporation and formation of gas bubbles require quite low energy. In the case of commercially applied dielectric liquids, such situation is not possible. Thus, the most often case is a direct initiation of second mode streamers. The first observed form of the second mode streamers is the above mentioned "stopping length" streamer. It is a kind of streamer, that although initiated, disappears in the inter-electrode space after reaching some distance from the HV point. Propagation of such streamer is stopped because of a low value of electrical field stress existing in some distance from HV-point electrode and simultaneously due to too low impact of space charge coming from the streamer itself, which is not able to keep the ionization processes inside the discharge channel. Further increase of the testing voltages causes that the second mode streamers propagate up to the grounded electrode

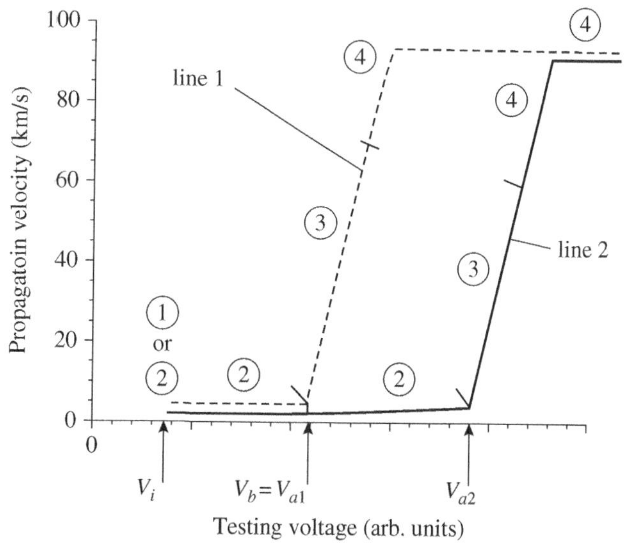

Figure 6.3 Relationship between propagation velocity of the streamers in dielectric liquids versus testing voltage: V_i – inception voltage, V_b – breakdown voltage, V_{a1}, V_{a2} – acceleration voltage for given case; 1,2,3,4 – subsequent propagation modes, made on the basis of [12, 21, 32]. *Source:* Based on Lesaint and Massala [12]; Hebner [21]; Duy et al. [32].

causing development of a breakdown channel (V_b in Figure 6.3). Such situation takes place usually in the case of transformer mineral oils and most of liquid hydrocarbons [7, 9–12, 14, 16, 21, 27]. An alternative situation is the appearance of third mode streamers after exceeding the threshold value of testing voltage; breakdown in such case is a result of the development of fast streamers (dotted line in Figure 6.3). Such case is characteristic mainly for dielectric esters and a small number of hydrocarbon liquids [19, 32–35, 39].

The characteristic value of testing voltage, for which, as was mentioned above, fast streamers start to develop, is called acceleration voltage and marked in the literature as V_a. As can be seen in Figure 6.3, the moment of acceleration, in relation to the testing voltage, may be different for different liquids. It may happen that acceleration voltage is equal to breakdown voltage in the given electrode system and for given dielectric liquid (as in the case of line 1 from Figure 6.3) or acceleration voltage may be much higher than the breakdown voltage (line 2 from Figure 6.3). The influence on the acceleration voltage and indirectly on the breakdown voltage has also the gap distance of the point-to-plane arrangement. The undeniable fact is also the influence of the type of liquid on the acceleration voltage. Figure 6.4 presents the literature-based example [12], where the relationship between propagation velocity and testing voltage for selected gap distances in mineral oil at positive lightning impulse voltage 0.4/1400 µs was shown. The breakpoint of the presented lines indicates the value of acceleration voltage.

Figure 6.5 shows, however, the case of different dielectric liquids tested, while the same measurement conditions are applied. The data presented concern the specific point-insulating plate electrode system and negative LI voltage tested, for example, in [19, 46]. The studies were focused on three dielectric liquids: mineral oil, synthetic ester, and natural ester. As it is clearly visible, for both esters tested, in the condition of experiment (voltage was increased up to the 2.4 of inception voltage multiply), the appearance of third mode negative streamers was noticed, and this was done for different values of testing voltage. However, at the same time, only slow developing streamers were noticed in mineral oil considered.

A possibility of the development of fourth mode streamers was also indicated in Figure 6.3. However, the appearance of this streamer mode is possible only when the voltage applied to the tested point-to-plane electrode system exceeds the inception voltage from 3 to 5 times depending on the liquid kind and field nonuniformity. The propagation velocity of fourth mode streamers reaches

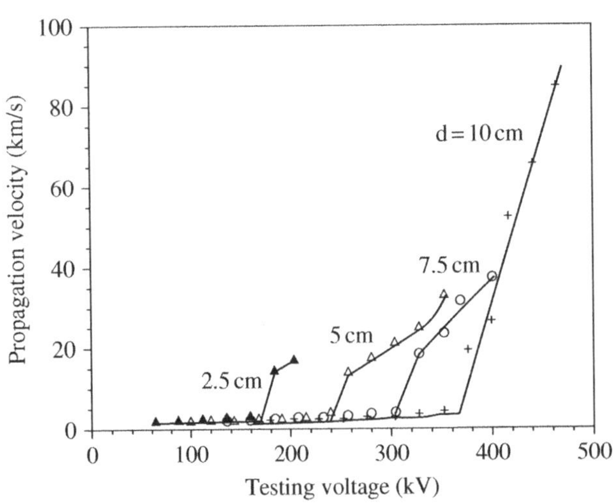

Figure 6.4 Relationship between propagation velocity and testing voltage for selected gap distances in the system of point-to-plane electrode configuration in mineral oil. *Source:* Lesaint and Massala [12] / with permission of IEEE.

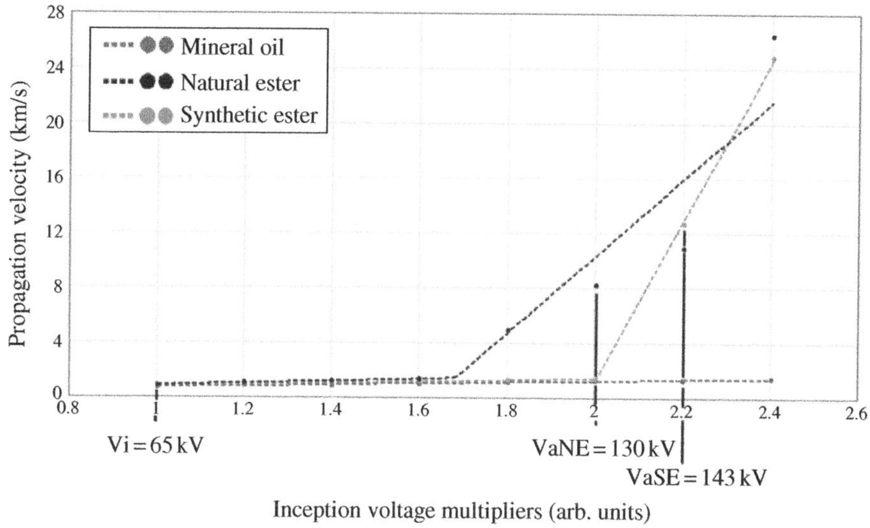

Figure 6.5 Dependence of propagation velocity of negative streamers versus inception voltage multipliers at 20 mm gap distance of point-insulating plate electrode system: Vi – inception voltage (the same for all three liquids), VaNE – acceleration voltage for natural ester tested, VaSE – acceleration voltage for synthetic ester tested. *Source:* Based on Rozga [19]; Rozga [46].

even 100 km/s, but the studies on such higher voltages has been conducted very seldom due to very short times of development of these streamers (below 1 μs) and high energy connected with breakdown process. As was mentioned in the beginning of this section, each propagation mode is characterized not only by specific propagation velocity but also by different properties in the field of streamer shape registered photographically, light waveforms registered, and current measured. Additionally, these features depend also on voltage polarity at which the streamers develop.

Figure 6.6 presents the sequence of streamer growth in the case of second mode streamers in point-to-plane electrode system at positive and negative polarity of lightning impulse voltage, respectively. The subsequent shadowgraph photos concern the increasingly longer propagation time. As it is seen from the figure, the forms of positive and negative second mode streamers are easy to recognize. The slow second mode streamers form a bush of a quasi-spherical boundary, which is filled with thin threadlike channels that propagate and penetrate the inter-electrode space with rising radius. The symmetry of registered shape results from the fact of directing the streamers from the space between the electrodes into the area of increasing electrical field stress. In the case of second mode negative streamers, they resemble the compact bushy structure with many branches propagating mainly lateral but also paraxial directions. In both cases, the development of the streamers is accompanied by shockwaves that are concurrent with the tips of the thin discharge channels. However, the shockwaves are much more visible in the photographs taken out for negative streamers.

The shape of streamers change when acceleration voltage is reached. In other words, the streamers change themselves when changing the propagation mode. Figure 6.7 presents the examples of shadowgraph photos of the third mode streamers. They were registered at specific electrode system: insulated HV electrode – insulated plate [17], in which due to the use of two insulated electrodes the fast developing streamers did not cause the breakdown and the photographs could be registered with satisfactory resolution. The presented photos include, however, the so-called return channel (the red one on the photos). It arouse as a result of the capacitive coupling between the electrodes of the system. The recorded streamer shapes of third mode streamers differ from these

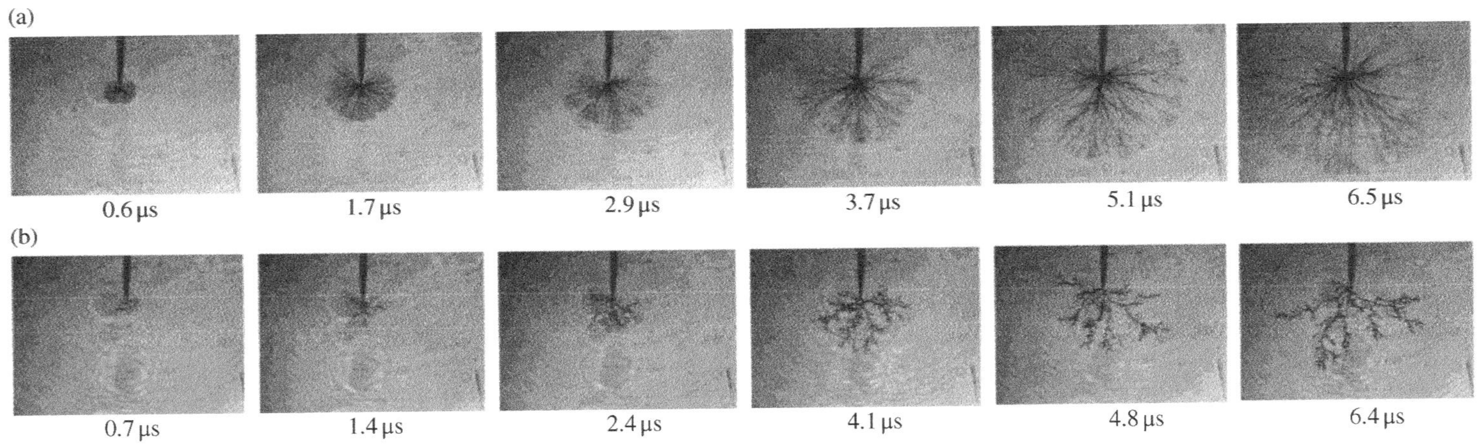

Figure 6.6 The sequence of shadowgraph photos of the streamers developing in mineral oil at 20 mm gap distance of point-to-plane electrode system, the values below the photos mean propagation time: (a) positive polarity of LI voltage, (b) negative polarity of LI voltage.

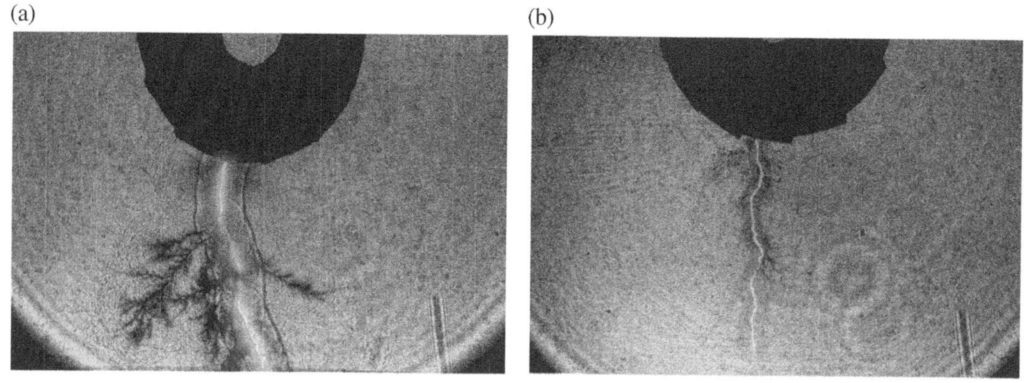

Figure 6.7 Examples of shadowgraph photos of third mode streamers developing in electrode system consisting paper-insulated HV electrode and pressboard-insulated grounded plate, $d = 20\,mm$: (a) positive polarity of LI voltage, (b) negative polarity of LI voltage.

characterizing second mode ones. The third mode streamers develop mainly along the axis defined by the tip of point electrode with a much smaller number of lateral branches, which are, however, thicker in relation to the second mode threadlike discharge channels. It is especially visible in the case of positive polarity (Figure 6.7a). In general, together with increase of testing voltage, number of side branches decrease and in the case of negative polarity leads practically to the "mono-streamer" (Figure 6.7b) as the one main channel connecting both the electrodes.

Among the characteristics of the streamers the special attention should be put on the streamer current or streamer light, which significantly differ when comparing slow and fast streamers. On the basis of the registered streamer current or courses of light emitted by them, it is easy to assess that both positive and negative streamers are characterized by step nature of propagation. Both current and light pulses follow each other with certain time intervals, increasing their peak values with time. This means that with each successive step the light is emitted with a greater intensity. Each successive step entails a step elongation of the streamer and in such case the streamer, as a coherent structure, must emit more light than, for example, in the initial stage. Figure 6.8 presents example of the typical courses of light corresponding with second and third mode streamers, respectively. Both subfigures concern positive lightning impulse voltage in the system of point-insulating plate electrode system with 20 mm gap. The light waveforms were completed with voltage waveforms.

The differences, which are noticed between the slow and fast streamers in the spatial shape of streamers, are reflected in courses of light. The third mode streamers have higher peak values of light pulses, which with the same PMT sensitivities in both cases, clearly shows the PMT saturation when recording light from fast streamers of the third mode type. Additional difference is also the frequency of the light pulses – for the third mode streamers this frequency is significantly higher. Moreover, the oscillograms quoted shows that the streamers form Figure 6.8b reach the insulating plate placed on the opposite, grounded electrode. It is depicted by wide light pulse finishing the series of discrete pulses. Their propagation velocity is easy to determine as a quotient of gap distance (20 mm) to time at which the streamer reaches the insulated plate ($< 1\,\mu s$). This gives the values higher than 20 km/s as the value characteristic for fast, third mode positive streamers. Concerning Figure 6.8a, this propagation lasts minimum 13 μs, thus propagation velocity must be less than 1.5 km/s, which is typical for slow, second mode streamers. Concluding the above presented general characteristic of streamer propagation in dielectric liquids, Table 6.2 sets some kind of systematizing of the typical phenomena possible to occur during streamer development in dielectric liquids under nonuniform electrical field and increasing voltage related to inception voltage of the streamers. Summary concerns rather small gap distances, however, may be true also for the

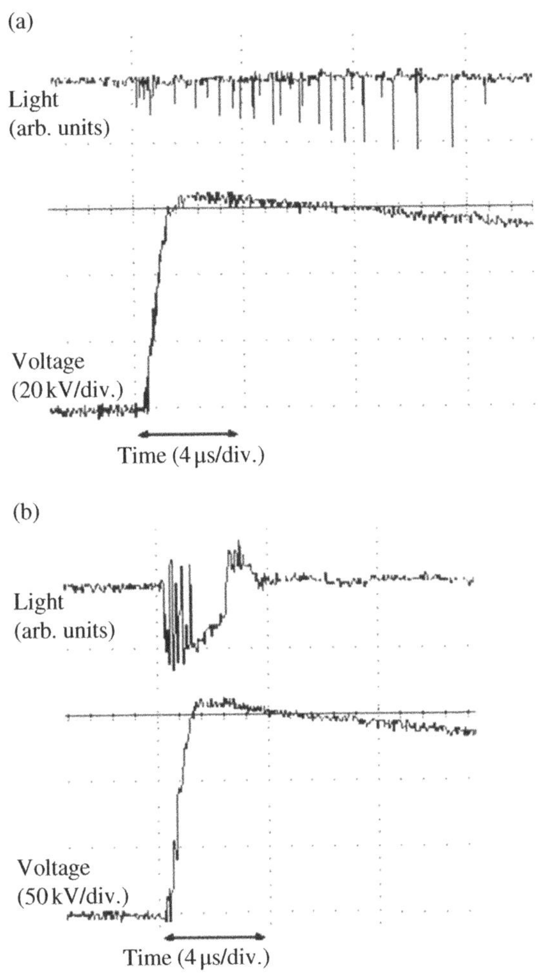

Figure 6.8 Examples of light oscillograms registered under positive LI voltage in point-insulating plate electrode system in synthetic ester: (a) $V = 65\,kV$ and corresponds with the inception voltage V_i, (b) $V = 156\,kV$ and corresponds with the 2.4 V_i, data are based on [61]. *Source:* Based on Rozga [61].

discharges in larger gaps. It is important to point out herein that the proposed systematics is connected with mineral oils and may change within the inception voltage multipliers when considering other types of dielectric liquids. In such case (when considering ester liquids), the values of inception voltage multipliers will be always lower, which will be further explained when focusing on comparison of prebreakdown phenomena in esters and mineral oil.

6.4.2 Streamer Development Theories

The differences between the structures of slow and fast discharges described above resulting from the differences in the physical processes determining the phenomena occurring during streamer development. The other processes are responsible for the development of slow first and second mode streamers and other for fast mode streamers. In the first case, the development is considered on the basis of the mechanisms taking into account the fact that the streamer channels propagate as a result of ionization processes inside them, which are filled with ionized gas (bubble/streamer theory or ionization in the gaseous phase). In turn, fast streamers are identified with the so-called ionization of hard liquid molecules when the streamer channels consist of ionized liquid.

Table 6.2 Sequence of phenomena that occur in the point-to-plane electrode system at lightning impulse voltage in mineral oil.

No.	Type of phenomena	Relative value of voltage related to inception voltage V_i
1.	Streamer initiation (first or second mode streamers)	$V = Vi$
2.	Propagation of the "stopping length" streamers reaching maximum 80% of a given gap distance	$V = 1.2–1.6\,V_i$
3.	Propagation of the slow second mode streamers up to the grounded electrode without breakdown (threshold phenomenon registered very seldom)	$V = 1.4–1.6\,V_i$
4.	Breakdown as a result of slow, second mode streamers (with long times to breakdown > 10 μs)	$V = 1.8–2.4\,V_i$
5.	Breakdown as a result of fast, third mode streamers (with short times to breakdown < 2 μs)	$V = 2.4–3.2\,V_i$
6.	Stabilization of the phenomena in the form of breakdown as a result of fast, third mode streamers, but characterized by very short times to breakdown (<1 μs) and high energy of breakdown channel	$V > 3\,V_i$
7.	Breakdown as a result of fourth mode streamers	$V > 3,5\,V_i$

Streamer development as a result of the so-called gaseous phase ionization may be explained as follows. After discharge initiation in the gas bubble in a given moment of time t_n, which is related to the moment of supplying the lightning impulse voltage to the electrode system, the streamer development follows as a result of ionization processes directly inside the streamer channels. In other words, the streamer channels are filled with ionized gas and propagation happens within the subsequent micro-discharges inside the channels through the collision ionization of the molecules of low ionization potential [1–8, 11, 12, 15, 21–24]. The result of the micro-discharges mentioned is a gas generation and increase of the pressure at the tip of the streamer channel. It leads to its step lengthening. A gaseous channel is filled mainly by H_2 (hydrogen), which is a main product of the discharges in oil (80%). This statement has been confirmed, for example, by the DGA studies that indicate on the partial discharges in the transformer oil when hydrogen concentration in tested sample is high. Additionally, the fact of gaseous nature of discharge channels has been proved by the spectroscopic measurements of the streamer spectra in oil, where mainly Hα Balmera lines, connected with the process of excitation of hydrogen atoms and their returning to the lower level, have been recognized as the most characteristic lines for the developing streamers [47–50]. When the micro-discharge reaches the end of the channel, positive ions (in the case when potential of HV electrode is positive) hit from the inside in its front wall causing liquid evaporation and collision dissociation of liquid molecules. Within these processes at the tip of the channel a gas is formed, the pressure increases rapidly and consequently step lengthening of the channels follows with simultaneous generation of the shockwaves (they are clearly visible in the above quoted Figure 6.6, especially in the case of negative streamers). Discharge development corresponds also with the flashes that are synchronous with current pulses. Since the majority of the gas inside streamer channel is hydrogen, its ionization causes formulation of positive ions, whose diameters are much lower than the diameter of H_2 molecules, hence their mean free path is almost as long as corresponding mean free path of the electrons. Positive hydrogen ions are therefore the perfect bullets in the discharge process being able to dissociate the oil molecules at the tip of the gas channel. The alternative way is dissociation resulting from collisions between liquid molecules and electrons from the depths of liquid when channel has the potential of HV electrode (channel is then treated as a kind of extension of point electrode). When the process at the end of channel finishes discharge goes out. The next elementary lengthening of streamer, which corresponds with next moment of time t_{n+1}, takes place when re-ignition happens. It is possible when electrical field stress in the place of streamer initiation has the value that is similar to the inception electrical field stress. This becomes when the space charge accumulated around the HV electrode after previous micro-discharge is dispersed. This dispersion may occur due to diffusion or impact of the electrostatic forces. When HV electrode is negative, the role of bullets has of course electrons.

There are a large number of arguments for the theory of gaseous phase ionization, which is higher than the number of arguments against. This fact may cause recognition that it is a right theory. The most important argument is experimentally documented influence of pressure on streamer propagation in liquid hydrocarbons, especially influence on propagation velocity and registered light pulses. The studies conducted in cyclohexane and transformer oil [44, 51, 62] clearly indicated that with an increase of the pressure, the number and amplitude of current and light pulses decrease, wherein for negative impulse voltage, the peak values of these pulses saturated themselves at the pressure equal to 20 bars [2 MPa] and above this value any pulses were registered. Analogical phenomenon was observed for positive streamers, but the threshold pressure was 30 bars [3 MPa]. An increase of the pressure, at the constant testing voltage, caused, however, the change of the streamer shapes in terms of intensity of branching. Among the latest studies, the influence of pressure on the streamer branching in long gaps (80 mm) was confirmed in [16], where the authors noticed that the degree of branching of the streamers is reduced with increasing pressure. Figure 6.9, borrowed solely from [16], shows this dependence for positive polarity of LI voltage.

The gaseous phase ionization, due to relatively low voltages, to which it corresponds to (low electrical field stresses), is treated as the theory described the way of development of slow, first and second mode streamers. A transformation of the slow streamers into fast is connected with the changes in the ionization processes, so with an acceleration of the propagation. The theory describing the processes of fast streamer propagation is called sometimes liquid phase ionization. According to this theory, ionization processes cover liquid molecules and streamer channels are solely filled with ionized liquid. Hence, the ionization processes concern not only the molecules of low ionization potential but also these of higher potentials, which may be excited within vibrations and emit then a quantum of energy connected with a characteristic radiation frequency. The values of electrical field stress, which correspond with the fast streamers development, have been assessed on higher than 30 MV/cm, and in individual cases may reach even 100 MV/cm. This

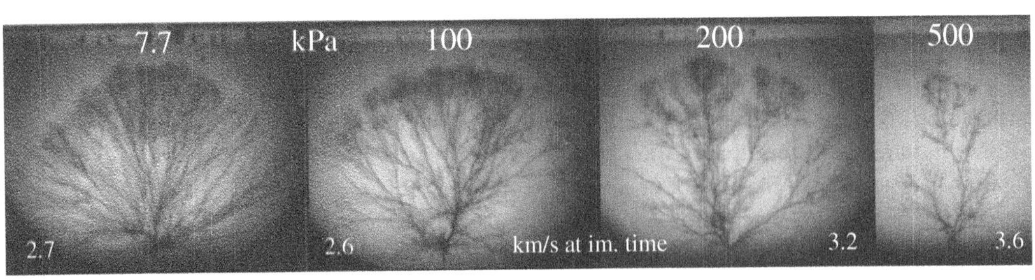

Figure 6.9 Shape of long, positive second mode streamers at 25 μs, shortly before breakdown, at 277 kV. Long second mode streamers did not exist at higher pressures. The branching is reduced with increasing pressure. The entire gap is shown; velocity at image time is indicated. *Source:* Linhjell et al. [16].

results, among others, from higher voltages at which the fast streamers start to appear. However, these values are enough high to excite or ionize the so-called hard molecules of a given liquid [2, 3, 5, 8, 11–16, 50]. Figure 6.10 shows schematically the development of individual discharge channel as a result of the process of liquid phase ionization on the example of positive streamer development. It is important to remember that few similar channels may develop simultaneously in different directions of interelectrode space.

Figure 6.10a symbolizes idealized positive discharge channel in the moment of propagation break that is between two successive step lengthening. The channel, in such situation, has a potential of HV electrode because it is some kind of its extension. Taking into account the fact that this channel is very thin (in most cases even thinner than the tip of HV point), the electrical field stress in front of its head may reach the value of electrical field stress exceeding significantly 10 MV/cm. At such high electric field, a field ionization in front of the head of streamer channel may cause ionization of neutral particles. Hence, in the electrode space in the vicinity of streamer head, the electrons separated from ionized neutral particles may appear. Fragment (b) shows, however, schematically how the electrons that appear in previous step move. They are heading toward the head of streamer channel. The alternative phenomenon for presented situation may be collision ionization. In such case, the electrons are detached from the acceptors in high electrical field around the streamer head and ionize the neutral molecules.

Figure 6.10b shows, in turn, next propagation stage when electrons penetrated into the main channel and within the ionization processes inside the channel they created the plasma together with positive charge existing there. Positive space charge, which was left outside the channel after penetration of electrons into the channel, creates now a new channel head extended of elementary section. In the case of negative streamers, the electrons must move of course from the head into the liquid and farther from the head the lower kinetic energy of the electrons is. These electrons are caught by the so-called acceptors (the molecules of electron affinity) and create negative space charge, which becomes a new head. In the thin layer around this new head, where electrical field stress is enough high, the field ionization may happen. Due to much higher, than in the case of positive discharges, the radius of curvature of negative channel head, the liquid volume under the

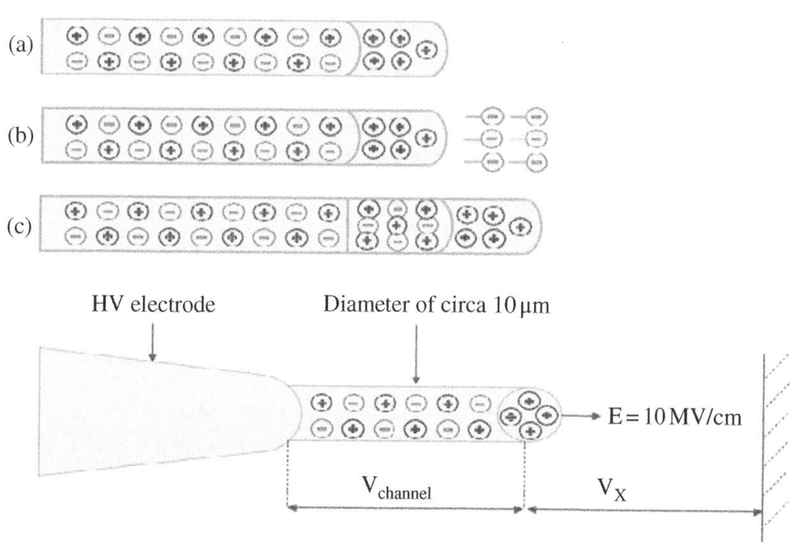

Figure 6.10 Streamer development according to the liquid phase ionization.

high electrical field is much lower, and hence are the shorter length and lower propagation velocity of the negative streamers. In the literature, we can notice the series of arguments for the theory of liquid phase ionization, but the arguments against are also quoted. The stronger argument results from the studies focused on the hydrocarbon liquids, which concerned the influence of the substances of low ionization potential or the so-called electron scavengers on the streamer development. From these studies, this result that adding the substances of low ionization potential accelerates the positive streamers, while the electron scavengers accelerate the negative ones [6–8, 10, 11]. The examples of the results in this field borrowed from [6, 7] are presented in Figure 6.11.

Apart from the influence on propagation velocity, it was noticed also the influence of the additives mentioned on the length of the streamer channels, what is partially connected with each other. The above quoted substance of low ionization potential (DMA) caused clearly increase of the length of positive streamers and practically did not change the length of negative streamers. However, electron scavengers changed the shape of negative streamers making them similar to the positive ones. The role of electron scavengers is to capture the electrons just after the channel head causing that in the thin layer around the streamer head the space charge of high density arises. Hence, some kind of "new head" behaves as a small radius. In this way, the electrical field stress on the surface of the head reminds on the enough high level such as that it may maintain intense field ionization. The second important argument for the rightness of the theory of liquid phase ionization is possibility of precise calculation of electrical strength of the hydrocarbons of simple and branched chain considering the assumption that loss of energy of the electrons follows as a result of excitation for vibrations of the C-C bonds in carbon skeleton or analogically C-H bonds. From the arguments against this theory, one seems to be the most important; it is difficult to explain the influence of pressure on propagation velocity of the streamers, as well as on breakdown voltage [2–8, 11, 21, 26].

Concluding, the above presented considerations concerning the theories of streamer development in dielectric liquids, the following statements may be formed:

- At relative low testing voltages in relation to the inception voltage (lower values of electrical field stress), the streamer channels are filled with gas (the influence of pressure on channel length is observed), and the velocities of the streamers are characteristic for first or second propagation mode.

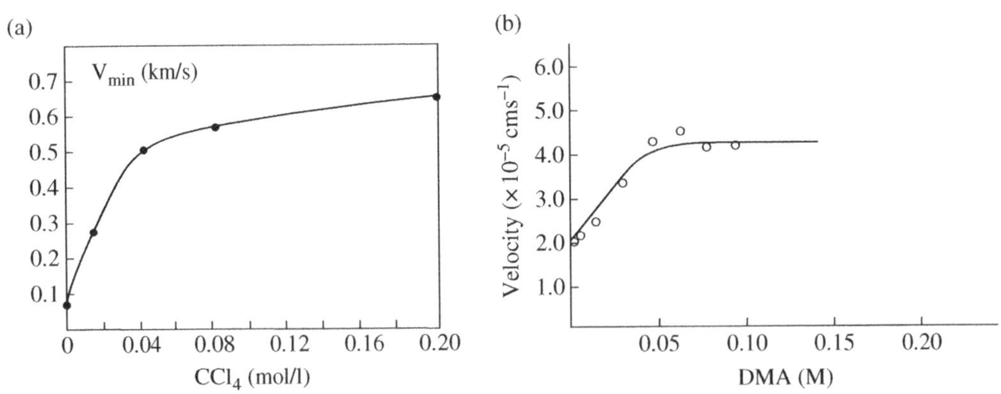

Figure 6.11 Literature examples showing the influence of additives on propagation velocity of the streamers: (a) influence of concentration of CCl4 (electron scavenger) on propagation velocity of negative streamers developing in cyclohexane under lightning impulse, $d = 2\,\text{mm}$, $r = 3\,\mu\text{m}$, $U = 33\,\text{kV}$, (b) influence of concentration of DMA (additive of low ionization potential) on propagation velocity of positive streamers developing in 2,2,4-trimethylopentane. *Source:* Based on Sharbaugh et al. [6]; Beroual and Tobazeon [7].

- At higher testing voltages related to the inception voltage (higher values of electrical field stress), the streamer channels are filled with liquid (influence of pressure is not observed), and the velocities of the streamers are in the range from several to dozen of kilometers per second, which are typical for third or fourth propagation modes.
- At the intermediate values of testing voltage, streamers may constitute the channel filled with a mixture of gas and liquid, or during the development, the streamers may change its character from slow to fast or vice versa.

Thus, both the theories described above seem to be true, but gaseous phase ionization concerns the streamers developing at lower testing voltages (lower electrical field stresses), while at higher voltages (higher electrical field stresses) the liquid phase ionization is the prevailing theory.

6.4.3 Streamer Propagation and Breakdown in Esters vs. Mineral Oils

Since natural and synthetic ester liquids appeared on transformer market, the studies on streamer propagation and breakdown in esters started to be carried out in order to explain the differences in discharge characteristics between esters and mineral oils. Hence, the literature concerning pre-breakdown and breakdown phenomena in esters under this type of stress is quite numerous and have increased year by year. The nonuniform field has been typically applied in the lightning tests of the esters, as it has been done commonly in the laboratory studies of mineral oils and other hydrocarbons [18, 19, 32–39, 44–46, 58–61, 63–66].

First, the propagation of the streamers in ester liquids is very often assessed on the basis of the "stopping length" (L) parameter. L means typically the distance from the tip of the point electrode where the streamers are initiated up the end of the farthest streamer channel recorded photographically. The stopping length, as was mentioned above, are connected only with the phenomena which did not cause the breakdown in the tested electrode system. In general, the stopping length increases with applied voltage, and this is true independently of the liquid kind. In comparative assessment of streamers developing in esters and mineral oil, the general view based on experimental studies is that the stopping length of streamers propagating in esters is longer than that of streamers in mineral oil when considering the same condition of experiment. However, this observation is true mainly for longer gap distances (>30 mm). For shorter gaps, the influence of liquid kind on stopping length is less clear and practically noticed only in the case of negative polarity of testing voltage. Exemplary, the work [37] compared three vegetable oils and mineral oil in terms of stopping length for two gap distances 25 and 50 mm. The positive streamers penetrate the 25 mm gap with similar length independently of the liquid type. When the gap is longer (50 mm), the differences appeared and stopping length of the streamers in esters was greater. For negative polarity, both for 25 and 50 mm, the streamers have reached longer way when developed in esters than in mineral oil. Additionally, this was also indicated that at the given voltage level applied to the electrode system, the streamers in ester liquids of natural origin propagate up to the opposite electrode leading to breakdown, while in mineral oil at the same voltage level the streamers are still of stopping length type. This is, without any doubt, the advantage of the mineral oil in terms of streamer propagation characteristics.

Similar insights were noticed in [65] in relation to the palm oil and 50 mm gap. The authors gave in their studies the data on the so-called streamer increasing rate (expressed in [mm/kV]) calculated on the basis of recorded photos. In the case of two types of palm oil, the values obtained for positive polarity were 0.19 and 0.16 mm/kV, while in mineral oil this value was equal to 0.16 mm/kV. In the case of negative polarity, the differences were more visible in favor of mineral oil, for which this parameter was assessed as 0.12 mm/kV, while for palm oils tested they were 0.18 and 0.16 mm/kV, respectively. The vegetable oil (based *Jatropha curcas* seeds) and mineral oil were compared to each other also in [64] for small gaps (20 and 30 mm) and in terms of both polarities of LI voltage. The findings were that for positive polarity, the streamers developed in a similar way at the values of voltage below breakdown voltage. In other words, no specific differences were noticed in terms of stopping length of positive streamers. However, as in previously cited researches, negative streamers in vegetable oil reached greater distance at the same voltage level than the streamers in mineral oil. The studies for longer gap distances presented in [32, 66] to a greater extent revealed the discrepancies in stopping length of the streamers in natural ester and mineral oil. With gap distances increased, stopping lengths of the streamers compared between natural ester and mineral oil showed significant differences, unfavorable for the natural ester. Especially in [40], where data concerned the comparative assessment of streamer behavior in rapeseed oil and mineral oil at gap distances 50 and 100 mm, the results unambiguously showed longer range of the streamers that did not cause the breakdown in the case of vegetable oil than mineral oil. This observation is equally true for both voltage polarities. Complex data on stopping length of the streamers in synthetic ester liquid was presented in turn in [33, 34, 38]. The number of works devoted to the synthetic esters is smaller as smaller number of synthetic ester liquids are available on transformer market. Anyway, the work [33] compared among others the stopping length of the streamers between synthetic ester, natural ester, and mineral oil. The 50 mm gap distance as mostly considered in the cited paper clearly showed the best lightning performance of mineral oil in terms of stopping length. It was the shortest among the liquids considered, both for negative and positive polarity. However, as above quoted, for positive polarity this was confirmed that the stopping length of the streamers in natural ester was on a similar level as in the case of mineral oil. Synthetic ester, together with natural esters and mineral oils, was tested at 20 mm gap distance and both lightning impulse voltage polarities in [34]. Small gaps of synthetic esters also indicated similarities to the observations noticed for natural esters (vegetable oils). The data collected at positive polarity overlapped for all liquids tested; however, these concerned negative polarity indicated that mineral oil has the best performance in terms of stopping length of the streamers.

In turn, new synthetic ester, namely, trimethylolpropane trimester (TME), was compared with other liquids at 25 mm gap distance in [38] where stopping length of the streamers were investigated as one of the most important parameters. The study once again confirmed what was found earlier for other ester liquids (both natural and synthetic) that streamer propagation at the voltages lower than breakdown voltage is faster in esters, especially under negative polarity of lightning impulse voltage. Streamers in esters at this voltage polarity stop their propagation after reaching longer distance than the streamers in mineral oil at the same voltage level. TME behaved, however, similar to natural ester tested at both polarities of testing voltage. Figure 6.12, as reproduced from [38], summarizes to a certain extent all above presented considerations showing differences between three types of liquids at both voltage polarities in terms of stopping length of the streamers.

With the analysis of streamer stopping length, the analysis of streamer propagation velocity is strictly connected. Having the data on streamer stopping length and simultaneously knowing the time at which the streamers reach a given distance or disappear in the interelectrode space, it is possible to determine propagation velocity of the streamers of stopping length type. Typically, average value of propagation velocity is a factor used in comparative assessment. What is obvious and results from direct analysis of streamer length, the faster are the streamers in esters because at the same time they reach longer distance. In general, propagation velocity increases with the increase of testing voltage. Such finding equally concerns esters and mineral oil with trend noticed at the voltages that did not cause the breakdown. It means that ester liquids are typically characterized

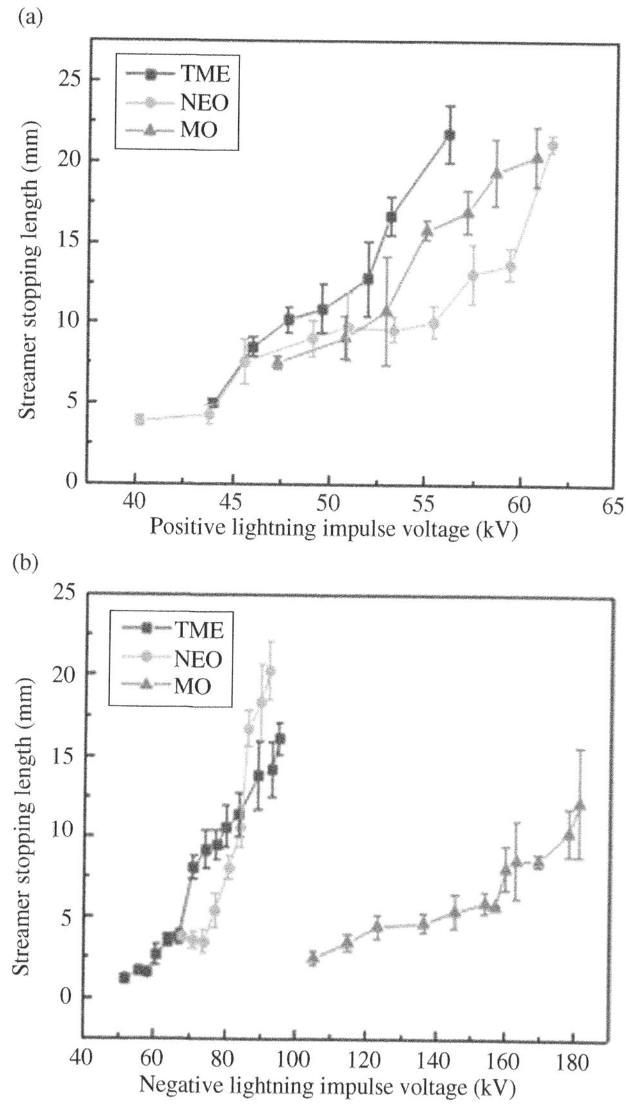

Figure 6.12 Streamer stopping length at various LI voltage (gap distance = 25 mm): TME – synthetic trimethylolpropane triesters, NEO – natural ester oil, MO – mineral oil, (a) positive polarity, (b) negative polarity. *Source:* Wang et al. [38] / with permission of IEEE.

by slightly higher propagation velocities or velocities almost equal to these corresponding with streamers in mineral oil. This last situation is observed when considering small gaps (up to 25–30 mm [33, 34, 36, 59, 60]. At higher gaps, the differences in the propagation velocities of slow mode streamers become greater to the disadvantage of esters [33, 35, 66]. In esters, similarly as for mineral oils, the differences in propagation velocity of positive and negative streamers exist and the positive ones are faster.

In terms of assessment of streamers on the basis of propagation velocity, the best indicator seems to be mentioned earlier acceleration voltage, meaning simply transition of the streamers from slow propagating to fast propagating. The conclusion found in this field by all researchers studying streamer phenomena in esters is that, independently of the gap distance, the acceleration voltage in ester liquids is always lower than in the case of mineral oil (look at the Figure 6.5 quoted earlier). Simultaneously, acceleration voltage in esters is, in most cases, identical with breakdown voltage or at least very close to this value. In opposite to this situation, acceleration voltage in mineral oil is always far from breakdown voltage. This means that the esters are more susceptible to the development of fast streamers. Above finding is stronger when increasing the gap distance – the differences in propagation velocities of streamers between esters and mineral oils become then bigger, and also bigger is the differences between acceleration voltages in esters and mineral oil. The data in this field are described in details in [33].

Since the breakdown voltage has been already mentioned it is worth to present some data concerning this quantity. For ester liquids, comprehensive breakdown tests have been conducted both in relation to the standards [33, 36, 42, 43] and to assess the relationship between the LI breakdown voltage and the gap distance [18, 19, 32–39, 43, 46–50, 58–61, 63–66]. The studies based on IEC 897 standard (25 mm gap of point-to-sphere electrode system) showed that ester liquids of different origin are characterized by similar LI breakdown voltage as mineral oil [36, 43]. The results were true independently of voltage polarity with small deviation in favor of the esters at positive polarity and in favor of the mineral oil at negative polarity (approx. 4–5 kV in relative values).

In contrast to the tests based on IEC 897 standard, the results concerning method suggested by the ASTM D3300 standard (3.8 mm gap of sphere-to-sphere electrode system) showed the higher values of breakdown voltage for mineral oil tested [42]. Additionally, the authors of cited work presented the results of a comparison of LI breakdown voltage and switching impulse (SI) breakdown voltage of esters and mineral oil in a comprehensive way. Using the same electrode system as for the standard-based tests, the authors have analyzed the influence of voltage polarity and testing methodology in determination of LI and SI breakdown voltage. They found that mineral oil is characterized by slightly higher values of breakdown voltage independently of the voltage waveform. The reduction of mean breakdown voltage of esters was 15 under LI and 8% under SI. The analysis of the voltage polarity shows, however, that lower LI breakdown voltage characterizes synthetic ester and natural ester, and this is true for both polarities. On the other hand, four different testing methods such as rising-voltage method with one shot per step, rising-voltage method with three shots per step, up-and-down method, and multiple-level method were applied to compare the 50% breakdown voltage. Each method clearly indicated the mineral oil as the liquid of the best lightning performance. However, when 1% breakdown probability (called as withstand voltage) is compared, the results are more favorable for both esters. From the studies, this was obtained that the 1% LI breakdown voltage from Weibull plots were 153.1 and 151.0 kV for synthetic ester and natural ester respectively, which were only slightly lower than the corresponding withstand voltage of mineral oil (156.7 kV).

In turn, in [60] the authors compared positive LI breakdown voltage of synthetic ester, olive oil, rapeseed oil, and mineral oil and obtained more promising results in relation to esters of natural origin. The tests were performed for quasi-uniform electric field represented by sphere-to-sphere electrode system with a 2.5 mm gap distance. Applying up-and-down method, the 1, 10, 50, and 90% breakdown probabilities were read, and they indicated a lower lightning strength of synthetic ester among the other considered liquids. Exemplary, a breakdown voltage median in the case of synthetic ester was similar to the median concerning mineral oil, as well as mineral oil–olive oil mixture (20–80%). However, when comparing these data with medians of LI breakdown voltage of fresh olive oil and rapeseed oil, the values determined for both oils were definitely higher. The paper [67] compared, however, mineral oil, synthetic ester and natural ester in terms of LI breakdown voltage determined in point-sphere electrode system and two gap distances 5 and 10 cm. The tests, which were performed for a fast rise time and a long tail voltage waveform (0.5/1400 μs), confirmed worse ester properties. For 5 cm gap and positive polarity, the difference was not high and reached few kV, while for 10 cm and positive polarity and both gaps at negative polarity, the differences reached even 50 kV.

The widest studies on breakdown and streamer behavior in nonuniform fields were described in [33]. The authors compared synthetic ester, natural ester, and mineral oil at the gap distances from 25 to 100 mm for positive polarity and from 15 to 75 mm for negative ones. The results indicated that only for 25 mm gap distance and positive polarity Li breakdown voltages for all three liquids tested were similar. In the case of the rest of the gap distances and both polarities, ester liquids had always lower breakdown voltages than mineral oil. They noticed also that as the electrode gap increases, the differences between esters and mineral oil becomes higher and higher. From the point of view of propagation modes, the general conclusion is that for small gap distances slow propagating streamers led to breakdown independently of the liquid kind. However, for longer gaps, acceleration voltage in mineral oil is higher than breakdown voltage, but in both types of esters acceleration voltages are equal to breakdown voltage. Hence, this may be concluded without any doubts, that ester liquids have worse properties at voltages of impulse nature. It is important to point out that these differences are more clearly visible for positive polarity, where transition of the streamers from second to third mode or directly from second to fourth mode occurs definitely in a step manner. As negative polarity is considered, it is unambiguous that indication of change in propagation mode is not as easy as in the case of positive polarity. This observation concerns both ester liquids and mineral oils.

As was mentioned above, when propagation mode changes, streamer current as well as emitted light change too as strongly connected with propagation mode. Both current and light may be an excellent indicator to reflect the general nature of streamers developing in a given liquid giving valuable data on the intensity of pre-breakdown processes (increase in pulse density over time and their frequency). The light emission or streamer current during were studied in esters and mineral oil in [18, 19, 32–39, 43–46, 59–61, 64, 68]. Most of the papers reported that evident differences exist between light and current characteristics corresponding with streamers in esters and mineral oil [32, 36, 38, 46, 61]. The intensity of emitted light was always higher when streamers developed in esters, and this finding concerns equally positive and negative polarity of LI. The light pulses registered were characterized by higher peak values and higher frequencies when observing streamers in esters. It may indicate that ionization processes might be of high intensity or more branches correspond with streamer propagation in ester liquids. Example of the results in this field is presented in Figure 6.13.

However, in some works the authors indicated that current and light characteristics are similar to each other when comparing esters with mineral oil [34, 60, 64]. The confirmation in this field may be the photographs of the streamers where their branches are mainly compared. Such photographs, taken out especially for positive polarity, do not differ from each other strongly and

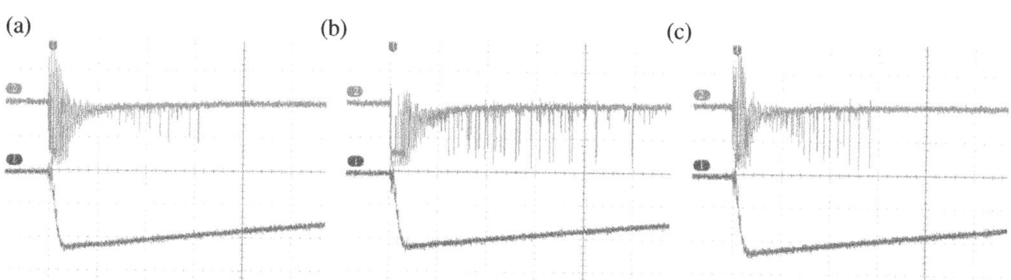

Figure 6.13 Examples of light and voltage waveforms registered at negative polarity of LI voltage, $t = 4\,\mu s/div.$, $V = 50\,kV/div.$, 1 – light, 2 – voltage: (a) mineral oil, (b) synthetic ester, (c) natural ester.

sometimes are even identical without any visible changes between these registered in ester liquids and mineral oil.

A kind of experimental technique that may be a challenge in comparative analysis of prebreakdown and breakdown phenomena in different dielectric liquids is optical spectrometry. Applying this technique has been limited practically for testing the mineral oils [47–49], and only in [50], some results related to the esters were presented. The studies described in [50] showed that the differences between esters and mineral oil exist in the spectra registered at whole discharge process finished with breakdown. The authors' analysis based on the knowledge that the characteristic wavelengths from the spectrum are associated with distinctive radiation frequencies, which represent energy radiated during the processes of atom excitation (according to the relationship $E = h\nu$, where h is Planck constant, and ν is a frequency radiation) or with energy from rotational-vibrational excitations of the molecules. Hence, the analysis made indicated that the processes differ from each other considering esters and mineral oil. For ester liquids, breakdown spectra show that the relationships between radiation from excited hydrogen atoms (mainly Hα Balmer lines) and continuum around the wavelengths of 480–520 nm are similar, so similar intensity of ionization processes occur at both levels. However, for mineral oil in typical breakdown spectrum intensity connected with Hα Balmer line is on much higher level than other wavelengths including, for example, C_2 Swan band at 516.5 nm. Thus, comparing the liquids, this may be assumed that the processes that influence emission of light in the range of lower wavelengths are more intense in the case of the ester liquids tested. However, the results obtained should be treated as of general nature. Spectrometry-based studies have to be applied for the studies focused on registration of streamer spectra in different stages of their development, including independently initiation and propagation and different propagation modes. Thus, it may consist a challenge for future studies that might explain some findings on ester liquid behavior in prebreakdown phase.

6.4.4 Influence of Nanoparticles on Prebreakdown Phenomena in Ester Liquids

Nanoparticles were preliminary applied in ester liquids in order to improve their cooling properties. It was done due to significantly higher viscosity of the esters, which is their disadvantage and causes worse heat dissipation from transformer windings to air than in the case of less-viscous mineral oils [69–72]. However, further works have shown that in addition to improving thermal properties of esters through adding the nanoparticles, they may improve also the dielectric properties of the esters. The nanoparticles that are used in the studies focused on improvement of dielectric strength of liquids are TiO_2, Fe_3O_4, SiO_2, Al_2O_3, ZnO, and less frequent CuO, ZrO_2, and C60.

As example, the work [73] was focused on the lightning impulse breakdown voltage improvement by adding Fe_3O_4 nanoparticles into vegetable oil. The authors, by performing the studies in point-to-plane electrode system, found that nanoparticles have generally a beneficial effect on the LI breakdown voltage of tested ester – positive and negative LI breakdown voltages increased in relation to fresh liquid by 37 and 12%, respectively. The latest studies described in [74, 75] showed similar effect of nanoparticles. In a quasi-uniform electric field, by adding the Fe_3O_4, Al_2O_3, and SiO_2 nanoparticles, negative LI breakdown voltage of natural and synthetic ester significantly increased. The best lightning performance was obtained for Fe_3O_4 nanoparticles in synthetic ester and for Al_2O_3 nanoparticles in natural ester. However, it is important to point out that concentration of nanoparticles and their size significantly influence the breakdown voltage results. In general, this was stated that there is an optimal concentration of given type and size of nanoparticles at which the best breakdown properties are obtained. Exemplary in [76], the best improvement in the case of synthetic ester was noticed for Fe_3O_4 (50 nm) nanoparticles – it was 48% in relation to the fresh liquid. When adding Al_2O_3, the improvement was 35 and 25% for 13 and 50 nm nanoparticles, respectively. In turn, in [77], this was reported that the breakdown properties of natural ester were improved up to a certain optimum value of Fe_3O_4 nanoparticles concentration equal to 0.008% (circa 20% with respect to the fresh liquid). However, beyond this value, breakdown voltage dropped even below the breakdown voltage corresponding to the fresh sample.

There are few approaches explaining the influence of nanoparticles on the breakdown voltage value of dielectric liquids (including esters) and on ionization processes (streamer propagation) preceding breakdown. In most cases, the explanation is directed into the process of trapping electrons by nanoparticles and of slowing the streamer propagation [78]. In details, based on the [74, 79], the authors explained the fact of improvement of breakdown voltage of dielectric-liquids-based nanofluids as due to the electron trap sites, which may be created by nanoparticles around the electrodes. Nanoparticles reduce the number of electrons and increase the inception voltage. Finally, due to reduction of electrons moving toward the opposite electrode and simultaneously, reduction of electrons accumulated at the nanoparticle-surrounding liquid interfaces (creation of the so-called double layer), the streamers are slowed down, and this delay the breakdown and thus increase breakdown voltage. However, when the double layer mentioned is saturated, it may contribute to the streamer acceleration and the breakdown voltage decrease. From this, the explanation of the noticed optimal concentration of nanoparticles may be presented. In contrast, the considerations presented in [77, 78] indicated on nanoparticles as electronic scavengers that reduce the charge density and inhibit the streamers development in nanofluids. It causes an increase in breakdown voltage. In turn, in [80], the authors pointed out that the conductive nanoparticles may capture very rapidly fast moving electrons and convert them into slow negatively charged nanoparticles. This makes the streamers slowing down and improves breakdown voltage. On this basis, the better results obtained with the use of Fe_3O_4 nanoparticles may be explained. They are solely conductive in contrast to some insulating nanoparticles.

6.5 Influence of Temperature on Prebreakdown Phenomena in Natural Ester Liquids

It is a well-established fact that moisture has an effect on the dielectric behavior of any transformer insulation [81]. The total amount of moisture is distributed among the solid and liquid insulation, the migration between both depends on temperature [82]. Apart from the generalized effects of temperature on liquid breakdown properties, it has been reported that breakdown strength of a dielectric liquid decreases with increase in stress magnitude and duration [83]. This adequate amount of high field stress gives rise to growth in zero pressure points, which favors the initiation of partial discharges in the liquid [84]. It is to be understood that oil temperature plays a significant role on the hydrostatic pressure and formation of weak pressure points. Hypothetically, during this process leading to initiation of a streamer under high field stress, the temperature of oil at this location will be increased. Thus witnessing a detectable dependence of temperature in speeding up the breakdown processes in liquids.

Hence, it is known that temperature of the liquid certainly influences the ionization process, moisture saturation, and the streamer behavior. This contributes to the possible changes in the prebreakdown and breakdown behavior at different temperatures. Thus, it is essential to investigate the prebreakdown events of alternative dielectric liquids at different temperatures. To add with the previous sentence, this section presents the results of the temperature dependence of the AC breakdown behavior of two commercial natural esters. The experiments are performed on a point-to-plane HV electrode configuration. All the tests are performed at a needle tip radius of 27.3 μm and an inter electrode gap of 10 mm. The details of the test cell and experimental setup are shown in Figure 6.14.

The test temperatures are selected based on the pourpoint of the liquids. For this study, −5 °C, 0 °C, and 20 °C have been adopted as the test temperatures. The partial discharge pre-inception voltage (PDPIV), partial discharge inception voltage (PDIV), and breakdown voltage (BDV) are reported in this section. To understand the liquid fluidity, changes in liquid viscosity at different temperatures is measured. The change in viscosities of both the liquids are shown in Figure 6.15.

The oxidative aging of insulating fluids produces large molecular weight compounds leading to the increase of oil viscosity. It is known that ester fluids have a high viscosity and hence the viscosity of synthetic esters and natural ester fluids is higher for nonaged and aged oils [85]. The changes in nonaged and aged ester liquids are reported in [85]. For the tested fluids (nonaged), an increase in the viscosity is noticed with decrease in temperature. The viscosity of NE2 is found to be higher than NE1 at all the test temperatures.

According to [44], and streamer theory proposed in [27], a pre-inception current exists below the inception voltage, called pre-inception voltage. It is also reported that pre-inception current is not witnessed with light emission; however, the same may be detected by the current measurement. The pre-inception voltage for all the measurements was plotted as a function of temperature and is presented in Figure 6.16a. It is seen that the pre-inception voltage increases with temperature. It is to be noted that at these voltages the ionization will be at the onset and hence the PMT will not capture any signal. After the pre-inception voltage, the probability of ionization will be increased and the tendency for streamer growth will be increased.

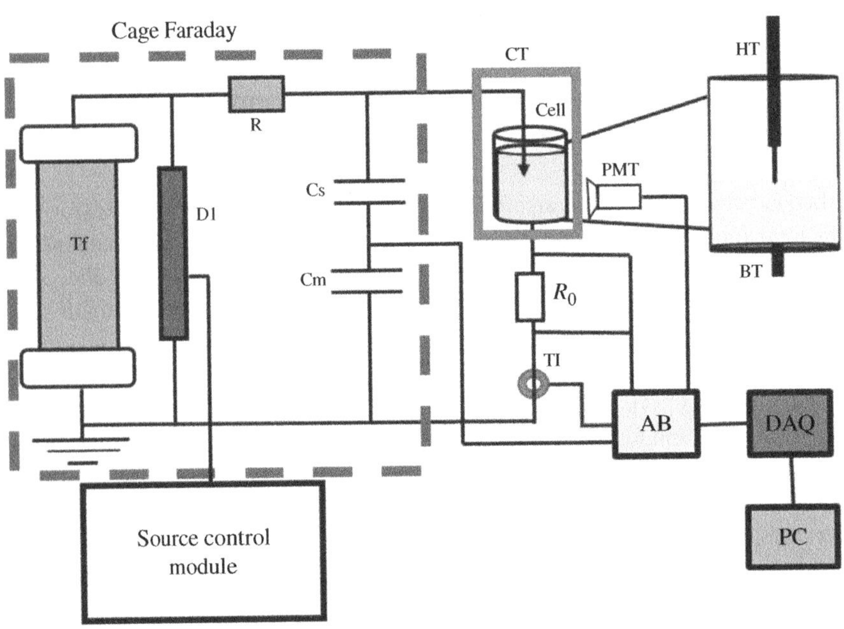

Figure 6.14 Schematic of the experimental setup. Tf: Transformer; D1: Capacitive divider 1 (for applied voltage); R: Current limiting resistor; Cs, Cm: capacitive divider 2; Electrodes: (HT: high voltage and BT: ground); CT: Climatic chamber (for temperature control); PMT: Photomultiplier Tube; R_0: Limiting resistance (leakage current); TI: Current transformer.

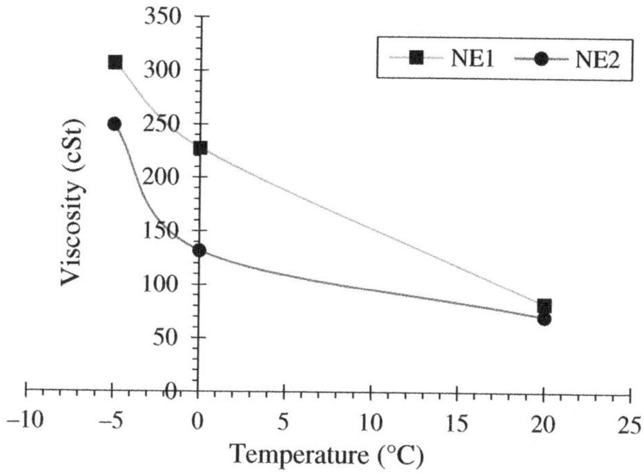

Figure 6.15 Viscosity of the liquids with change in temperature.

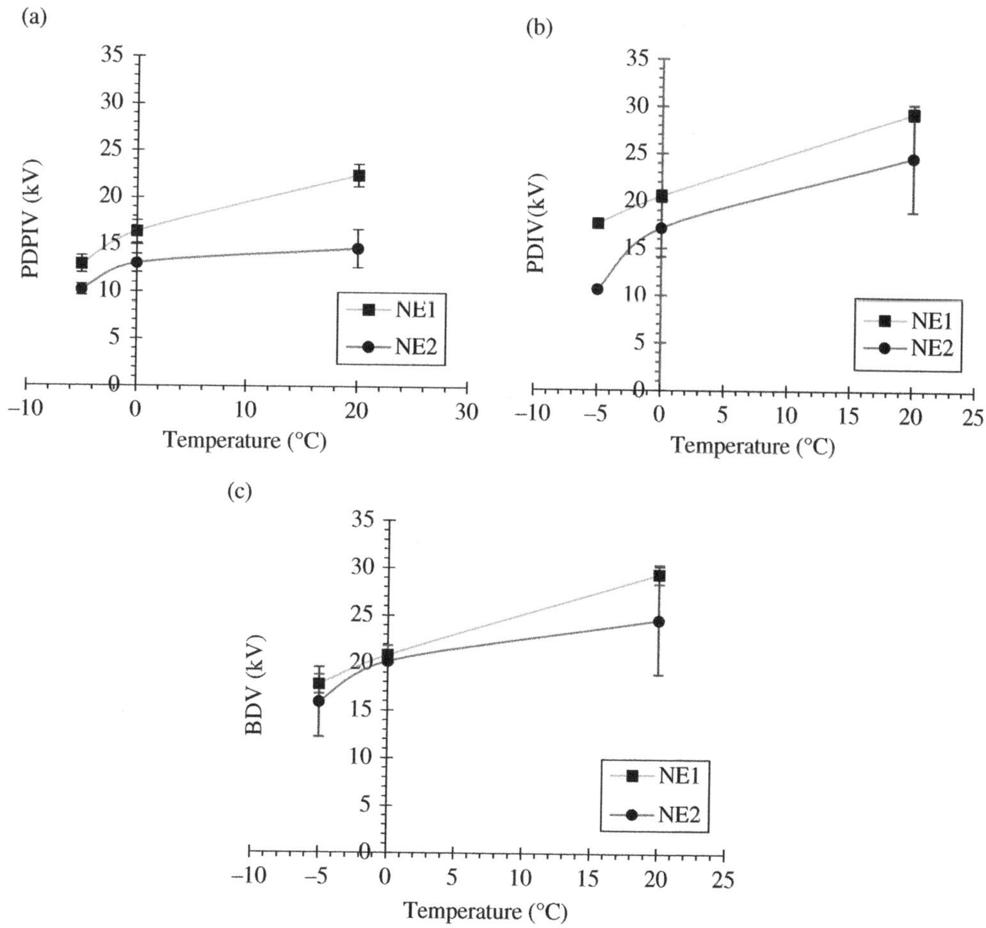

Figure 6.16 PD and BDV behavior of natural esters with change in temperature. (a) PDPIV for NE1 and NE2, (b) PDIV for NE1 and NE2, (c) BDV for NE1 and NE2.

The magnitude of the applied voltage at which the first noticeable streamer evolves is referred to the partial discharge inception voltage (PDIV). Streamer inception voltage is an important parameter to understand the dielectric performance of the insulating medium. In the present study, the level of applied voltage at which the first streamer was observed for NE1 and NE2 at different temperatures is shown in Figure 6.16b. The results seem to indicate a significant influence of temperature on the streamer inception voltage. Beyond the PDIV, the streamer propagation is accelerated by accumulation of electron avalanches in successive vapor bubbles. This process increases cumulatively with electron ionization up to the breakdown. The streamer then disappears, and the electric field becomes too small with production of a string of micro-bubbles that dissolve in the liquid [86, 87].

The transition from initiation of streamers to breakdown condition is not always rapid and the journey of streamers may be characterized by several aided factors. Therefore, temperature plays a significant role in the inception of the first electron that is responsible for ionization. It is noticed that PDIV increases with temperature in both natural esters. From Figure 6.16b, it is seen that streamer initiation and propagation increase with increase in temperature.

One recall that the viscosity of NE1 is higher than that of NE2 at all the test temperatures. Hence, it may be understood that viscosity and temperature influence the generation of streamers in NE1 and NE2. It is to be understood that hydrostatic pressure will have a significant influence on the propagation of the streamer and hence on the streamer velocity. However, the influence of viscosity and hydrostatic pressure on streamer propagation is yet to be foreseen.

Breakdown is a dynamic process that occurs when the corona streamers reach the opposite electrode. It is important to correlate the breakdown voltage to the temperature to better understand the breakdown strength of NE1 and NE2. The changes in breakdown voltage of NE1 and NE2 with temperature are presented in Figure 6.16c. It is seen that with reduction in temperature, the breakdown voltage decreases due to the fact that moisture saturation limit in the oils reduces with reduction in temperature [88]. Hence, a decrease of temperature increases the water-in-oil relative saturation and thus decreases the breakdown voltage.

The characteristic voltages of the PDs and the BDV reduce when temperature reduces from +20 °C to −5 °C. The results allow understanding the temperature dependence of natural ester oils dielectric properties for their application in transformers serving in cold regions.

6.6 Influence of Thermal Aging on Prebreakdown Phenomena in Synthetic Ester Liquids

It is known that composition of the liquid changes with aging, which introduces several aging byproducts in the bulk of the liquid. These aging byproducts deteriorate the liquid properties and its behavior. The change in liquid properties with aging may be accessed by various electrical, physical, and chemical tests. These decay particles act as local conducting particles and aid in the breakdown process of the liquid. In this section, one mineral oil (MO) and two synthetic esters (SE1 and SE2) are investigated to understand the influence of thermal aging on PD behavior. Thermal aging is performed at 120 °C in presence of cellulose (2.6% nitrogen) and copper in a closed environment. Thermal aging is performed for two weeks, four weeks, and six weeks. After every aging duration, the liquids are subjected to AC stress under point-to-plane configuration. The details of the experimental and test cell are shown in Figure 6.14. Needle tips radii of 0.93 μm (N1), 5.46 μm (N2), and 10.75 μm (N3) were used with an inter-electrode distance of 10 mm. At every aging history, partial discharge inception voltage (PDIV) and breakdown voltage (BDV) are measured for the three different tips at room temperature.

The electrode geometry (tip radius and inter-electrode gap) has a significant impact on the PDIV in liquids [39, 66, 86]. Electrode tip radius has a more significant effect on PDIV than the inter-electrode gap. The PDIV for MO, SE1, and SE2 and tip radii of 0.93 μm (N1), 5.46 μm (N2), and 10.75 μm (N3) at different aging durations are presented in Figure 6.17a.

The PDIV is increased with the electrode tip radii, and this trending is similar in MO, SE1, and SE2. This is because the electric field intensity for a given applied voltage decreases with the increase in tip radius. Further, it is noticed that the PDIV is reduced with thermal degradation. This is because aging introduces colloidal and soluble particles, which act as local conducting particles and aids in ionization. Subsequently, initiation of streamers leading the prebreakdown processes is evident for relatively lower voltages. Considering liquid degradation and electrode tip radius as influencing parameters, SE1 shows higher PDIV values when compared to that of SE2 and are comparable to MO.

Breakdown voltage is the voltage at which liquid loses its insulating properties and starts conducting. The breakdown voltage measurement is a significant way of defining the value of voltage up to which an insulating material will withstand the voltage stress. The BDVs of MO, SE1, and SE2 at different aging conditions for three different needle tip radii have been shown in Figure 6.17b. As expected, the BDVs of all the samples increase with needle electrode tip radius. Considering the degradation of liquids, the BDV values decrease with increase in liquid deterioration. For the present experimental condition, the BDVs of SE1 are better than SE2. In addition, the breakdown voltages of SE1 are comparable to MO.

6.7 Conclusions and Future Scope

The studies on the initiation and development of streamers in liquid dielectrics, although conducted for many years, still do not fully explain the phenomena that determine them. The hypotheses formulated in this area are becoming recognized, although most researchers emphasize the

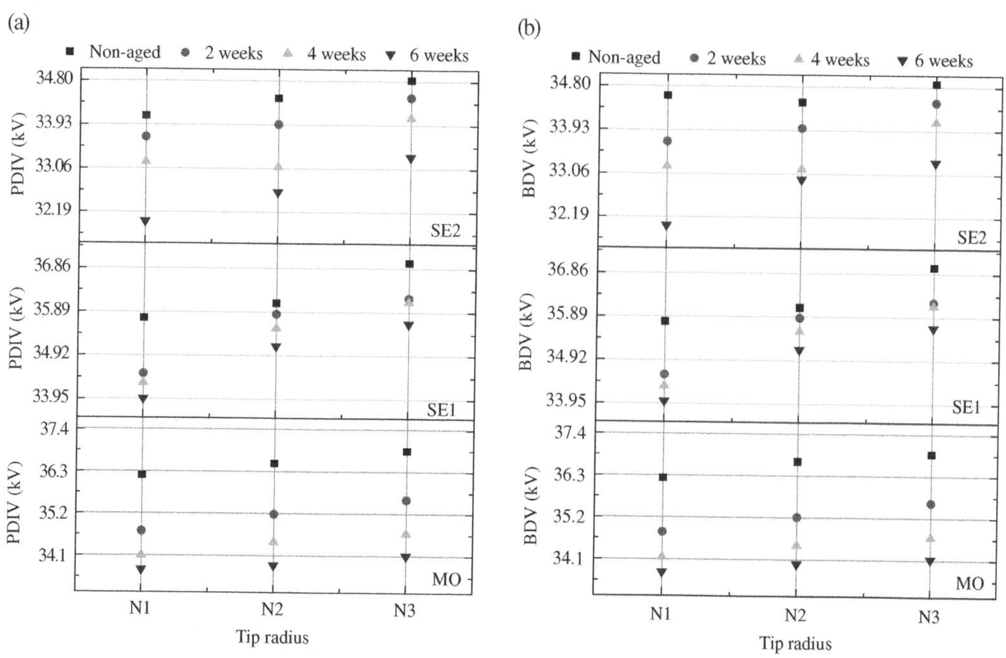

Figure 6.17 PD behavior of esters with thermal aging. (a) PDIV for different thermal degradations and tip radii, (b) BDV for different thermal degradations and tip radii.

constant need to look at the phenomena responsible for the initiation of streamer in liquids, as well as their propagation finishing with breakdown. The theories developed on the basis of studies carried out on liquid hydrocarbons and mineral oils are also applied to the biodegradable synthetic and natural esters. The comparison of individual liquids under the same testing conditions, in addition to the obvious aspect determining the ability of a given liquid to protect against a specific type of overvoltages, also made it possible to refine the already known knowledge about prebreakdown phenomena in dielectric liquids.

The presented considerations clearly indicated that bubble theories for the initiation of streamers in liquids may apply in the same way to both mineral oils and esters. The results of the experimental studies showed that the initiation is essentially independent of what kind of liquid is considered. It can therefore be assumed that the generation of the initiating gas bubble takes place in a similar way in each of the liquids, since the initiation voltages, and thus the initiation electric field stress, are practically the same for all liquid kinds.

In the case of propagation of already initiated streamers, the esters show lower resistance to the appearance of fast streamers of the third mode than the mineral oils. The acceleration voltage is at least a few tens of kilovolts lower in the case of esters. In addition, the lightning impulse breakdown voltage of esters is generally lower than corresponding breakdown voltage of mineral oil when considering the results obtained using the ASTM Standard (quasi-uniform field), as well as nonstandard point-to-plane electrode configuration. This latter is especially visible when considering long gap distances where the breakdown in esters occurs as a result of the development of the third mode streamers.

Referring the observations quoted to the presented theories of streamer propagation in liquids, it can be assumed that ester liquids are more susceptible to the appearance of the streamers, which propagate as a result of ionization in the liquid phase. These are the streamers that are more energetic than the second mode streamers, so when breakdown occurs, it is characterized by a more intense impact on the solid insulation elements.

What is the reason of the worse properties of esters in terms of resistance to lightning overvoltages is difficult to clearly define. It can be assumed that the molecular structure of esters in which, for example, the electronegative oxygen atoms exist may cause this situation. After all, in hydrocarbons, the streamers were accelerated when molecules with a low ionization potential were added to them, and this is what oxygen molecules are. Considering additionally that the theory of possible ionization of hard molecules in a liquid in a high electric field is true, the generation of the third mode streamers, which, according to this theory, are the result of this type of ionization, can occur in esters at a much lower voltage than in the case of mineral oil, because in esters it is easier to directly ionize such molecules or to vibrate them.

From the above, the studies on prebreakdown phenomena in ester liquids are still needed, especially that year by year, the new products are introduced to the transformer market, and these products have started to differ from each other by molecular structure, which may relate to the physical aspects of initiation and propagation of the streamers in these new dielectric liquids.

References

1 Crowe, R.W., Sharbaugh, A.H., and Bragg, J.K. (1954). Electric strength and molecular structure of saturated hydrocarbon liquids. *J. Appl. Phys.* 25: 1480–1484.

2 Sharbaugh, A.H., Crowe, R.W., and Cox, E.B. (1956). Influence of molecular structure upon the electric strengths of liquid hydrocarbons. *J. Appl. Phys.* 27: 806–808.

3 Lewis, T.J. (1956). Mechanism of electrical breakdown in saturated hydrocarbons liquids. *J. Appl. Phys.* 27: 645–650.

4 Ward, B.W. and Lewis, T.J. (1960). A new statistical theory for the breakdown of liquid hydrocarbons. *J. Electrochem Soc.* 107: 191–195.
5 Adamczewski, I. (1969). *Ionization, conductivity and breakdown in liquid dielectrics*. UK.
6 Sharbaugh, A.H., Devins, J.C., and Rzad, S.J. (1978). Progress in the field of electric breakdown in dielectric liquid. *IEEE Trans. Dielectr. Electr. Insul.* 13: 249–276.
7 Beroual, A. and Tobazeon, R. (1986). Pre-breakdown phenomena in liquid dielectrics. *IEEE Trans. Electr. Insul.* 21 (4): 613–627.
8 Beroual, A. (1993). Electronic and gaseous processes in prebreakdown phenomena of dielectric liquids. *J. Appl. Phys.* 73: 4528–4532.
9 Lesaint, O. and Gournay, P. (1994). Initiation and propagation thresholds of positive prebrakdown phenomena in hydrocarbon liquids. *IEEE Trans. Dielectr. Electr. Insul.* 1: 702–708.
10 Tobazeon, R. (1994). Prebreakdown phenomena in dielectric liquids. *IEEE Trans. Dielectr. Electr. Insul.* 1: 1132–1147.
11 Beroual, A. et al. (1998). Liquid dielectric committee international study group - IEEE DEIS, propagation and structure of streamers in liquid dielectrics. *IEEE Electr. Insul. Mag.* 14 (2): 6–17.
12 Lesaint, O. and Massala, G. (1998). Positive streamer propagation in large oil gaps: experimental characterization of propagation mode. *IEEE Trans. Dielectr. Electr. Insul.* 5: 360–370.
13 Denat, A. (2006). High field conduction and prebreakdown phenomena in dielectric liquids. *IEEE Trans. Dielectr. Electr. Insul.* 13: 518–525.
14 Lesaint, O. (2016). Prebreakdown phenomena in liquids: propagation modes and basic physical properties. *J. Phys. D: Appl. Phys.* 49 (49): 14401–14422.
15 Lu, W., Liu, Q., and Wang, Z.D. (2017). Pre-breakdown and breakdown mechanisms of an inhibited gas to liquid hydrocarbon transformer oil under negative lightning impulse voltage. *IEEE Trans. Dielectr. Electr. Insul.* 24: 2809–2818.
16 Linhjell, D., Lungard, L.E., Unge, M., and Hjortstam, O. (2020). Pressure-dependent propagation of streamers under step voltage in a long point-plane gap in transformer oil. *IEEE Trans. Dielectr. Electr. Insul.* 27: 1595–1603.
17 Rozga, P. (2011). The influence of paper insulation on the prebreakdown phenomena in mineral oil under lightning impulse. *IEEE Trans. Dielectr. Electr. Insul.* 11: 720–727.
18 Rozga, P., Stanek, M., and Cieslinski, D. (2013). Comparison of properties of electrical discharges developing in natural and synthetic esters at inception voltage. *Annual Report Conference Electrical Insulation Dielectrical Phenomena*, Chenzhen, China (20–23 October 2013), pp. 891–894.
19 Rozga, P. (2015). Streamer propagation in small gaps of synthetic ester and mineral oil under lightning impulse. *IEEE Trans. Dielectr. Electr. Insul.* 22: 2754–2762.
20 Hizal, E.M. and Dincer, S. (1982). Breakdown time lags and pre-breakdown phenomena in transformer oil, effects of hydrostatic pressure. *J. Electrost.* 12: 333–343.
21 Hebner, R.E. (1988). *Measurement of Electrical Breakdown in Liquids*, NATO ASI series, vol. B193. Plenum press.
22 Denat, A., Gosse, J.P., and Gosse, B. (1988). Electrical conduction of purified cyclohexane in a divergent electric field. *IEEE Trans. Electr. Insul.* 23: 545–554.
23 Kattan, R., Denat, A., and Lesaint, O. (1989). Generation, growth, and collapse of vapor bubbles in hydrocarbon liquids under a high divergent electric field. *J. Appl. Phys.* 66: 4062–4066.
24 Watson, P.K., Chadband, W.G., and Sadeghzadeh-Araghi, M. (1991). The role of electrostatic and hydrodynamic forces in the negative-point breakdown of liquid dielectrics. *IEEE Trans. Electr. Insul.* 26: 543–559.
25 Yamashita, H., Yamazawa, K., and Wang, Y.S. (1998). The effect of tip curvature on the pre-breakdown streamer structure in cyclohexane. *IEEE Trans. Dielectr. Electr. Insul.* 5: 396–401.

26 Lewis, T.J. (1998). A new model for primary process of electrical breakdown in liquids. *IEEE Trans. Dielectr. Electr. Insul.* 5: 306–315.

27 Dumitrescu, L., Lesaint, O., Bonifaci, N., and Denat, A. (2001). Study of streamer inception in cyclohexane with a sensitive charge measurement technique under impulse voltage. *J. Electrost.* 53: 135–146.

28 Lesaint, O. and Top, T.V. (2002). Streamer initiation in mineral oil, part I: electrode surface effect under impulse voltage. *IEEE Trans. Dielectr. Electr. Insul.* 9: 84–91.

29 Top, T.V. and Lesaint, O. (2002). Streamer initiation in mineral oil, part II: influence of a metallic protrusion on a flat electrode. *IEEE Trans. Dielectr. Electr. Insul.* 9: 92–96.

30 Torshin, Y.V. (2003). Prediction of breakdown voltage of transformer oil from predischarge phenomena. *IEEE Trans. Dielectr. Electr. Insul.* 10: 933–941.

31 Kolb, J.F., Joshi, R.P., Xiao, S., and Schoenbach, K.H. (2008). Streamers in water and other dielectric liquids. *J. Phys. D Appl. Phys.* 41: 234007.

32 Duy, C.T., Lesaint, O., Denat, A., and Bonifaci, N. (2009). Streamer propagation and breakdown in natural ester at high voltage. *IEEE Trans. Dielectr. Electr. Insul.* 16: 1582–1594.

33 Liu, Q. and Wang, Z.D. (2011). Streamer characteristic and breakdown in synthetic and natural ester transformer liquids under standard lightning impulse. *IEEE Trans. Dielectr. Electr. Insul.* 18: 285–294.

34 Dang, V.-H., Beroual, A., and Perrier, C. (2012). Investigations on streamers phenomenain in mineral, synthetic and natural ester oils under lightning impulse voltage. *IEEE Trans. on Dielectr. and Electr. Insu* 19 (5): 6–13.

35 Denat, A., Lesaint, O., and McCluskey, F. (2015). Breakdown of liquids in long gaps: influence of distance, impulse shape, liquid nature, and interpretation of measurements. *IEEE Trans. Dielectr. Electr. Insul.* 22 (5): 2581–2591.

36 Rozga, P., Stanek, M., and Rapp, K. (2018). Lightning properties of selected insulating synthetic esters and mineral oil in point-to-sphere electrode system. *IEEE Trans. Dielectr. Electr. Insul.* 25: 1699–1705.

37 Huang, Z. et al. (2018). Streamer characteristics of dielectric natural ester-based liquids under long gap distances. *AIP Adv.* 8 (10): 1–12.

38 Wang, K., Wang, F., Shen, Z. et al. (2020). Breakdown and streamer behavior in homogeneous synthetic trimethylolpropane triesters insulation oil. *IEEE Trans. Dielectr. Electr. Insul.* 27: 1501–1507.

39 Mohan Rao, U., Fofana, I., Beroual, A. et al. (2020). A review on pre-breakdown phenomena in ester fluids: Prepared by the international study group of IEEE DEIS liquid dielectrics technical committee. *IEEE Trans. Dielectr. Electr. Insul.* 27: 1546–1560.

40 IEC60897 (1987). Methods for the determination of the lightning impulse breakdown voltage of insulating liquids. IEC Standard.

41 ASTM-D3300 (1994). Dielectric breakdown voltage of insulating oils of petroleum origin under impulse conditions. ASTM standard.

42 Liu, Q. and Wang, Z.D. (2013). Breakdown and withstand strengths of ester transformer liquids in a quasi-uniform field under impulse voltages. *IEEE Trans. Dielectr. Electr. Insul.* 20 (2): 571–579.

43 Rozga, P. and Stanek, M. (2017). Comparative analysis of lightning breakdown voltage of natural ester liquids of different viscosities supported by light emission measurements. *IEEE Trans. Dielectr. Electr. Insul.* 24 (2): 991–999.

44 Hestad, O.L., Ingebrigtsen, S., and Lundgaard, L.E. (2005). Streamer initiation in cyclohexane, Midel 7131 and Nytro 10x. *IEEE Int. Conf. Dielectr. Liq. (ICDL)*: 123–126.

45 Liu, R., Tornkvist, C., Chandramouli, V. et al. (2009). Ester fluids as alternative for mineral oil: the difference in streamer velocity and LI breakdown voltage. *Annual Report Conference Electrical Insulation Dielectrical Phenomena (CEIDP)*, Virginia Beach, VA, USA (18–21 October 2009), pp. 543–548.

46 Rozga, P. (2016). Streamer propagation in a non-uniform electric field under lightning impulse in short gaps insulated with natural ester and mineral oil. *Bull. Polish Ac. Sci., Tech. Sci.* 64 (1): 171–179.

47 Barmann, P., Kroll, S., and Sunesson, A. (1996). Spectroscopic measurements of streamer filaments in electric breakdown in dielectric liquids. *J. Phys. D, Appl. Phys.* 29 (5): 856–863.

48 Denat, A., Bonifaci, N., and Nur, M. (1998). Spectral analysis of light emitted by streamers in hydrocarbon liquids. *IEEE Trans. Dielectr. Electr. Insul.* 5 (3): 382–387.

49 Ingebrigtsen, S., Bonifaci, N., Denat, A., and Lesaint, O. (2008). Spectral analysis of the light emitted from streamers in chlorinated alkane and alkene liquids. *IEEE Int. Conf. Dielectr. Liq. (ICDL)*: 1–4.

50 Rozga, P. and Tabaka, P. (2018). Comparative analysis of breakdown spectra registered using optical spectrometry technique in biodegradable ester liquids and mineral oil. *IET Sci. Meas. & Tech.* 12 (5): 684–690.

51 Zaengl, W.S. (1988). Electronegative gases, present state of knowledge and application, future prospects. *Nordic Symposium on Electric Insulation*, Trondheim (June 13–15, 1988), pp. I1–I39.

52 Kuffel, E., Zaengl, W.S., and Kuffel, J. (2000). *High Voltage Engineering – Fundamentals*. Butterworth-Heinemann.

53 Hauschild, W. (2015). Comments on "electrical breakdown in dielectric liquids—a short overview" by Pär Wedin. *IEEE Electr. Insul. Mag.* 31 (3): 6–7.

54 Kim, M. and Hebner, R.E. (2006). Initiation from a point anode in a dielectric liquid. *IEEE Trans. Dielectr. Electr. Insul.* 13 (6): 1254–1260.

55 Rozga, P. and Hantsz, D. (2013). Initiation of discharges in the setup of paper insulated electrodes in oil, 2013. *Annual Report Conference on Electrical Insulation and Dielectric Phenomena*, Shenzhen, China (20–23 October 2013), pp. 913–916.

56 Rozga, P. (2014). Using the 3D electrical field distribution simulation for assessment of discharge initiation process in oil. *Przegląd Elektrotechniczny* 90 (6): 216–221.

57 Gournay, P. and Lesaint, O. (1993). A study of inception of positive streamers in cyclohexane and pentane. *J Phys. D: Appl. Phys.* 26: 1966–1974.

58 Hemmer, M., Julliard, Y., Badent, R. et al. (2001). Streamer inception and propagation in rape-seed oils and mineral oils. *IEEE Annual Report Conference Electrical Insulation Dielectrical Phenomena (CEIDP)*, Kitchener, ON, Canada (14–17 October 2001), pp. 548–551.

59 Rozga, P. (2016). Streamer propagation and breakdown in a very small point-insulating plate gap in mineral oil and ester liquids at positive lightning impulse voltage. *Energies* 9 (467): 1–12.

60 Reffas, A., Moulai, H., and Beroual, A. (2018). Comparison of dielectric properties of olive oil, mineral oil, and other natural and synthetic ester liquids under AC and lightning impulse stresses. *IEEE Trans. on Dielectr. Electr. Insul.* 25 (5): 1822–1830.

61 Rozga, P. (2016). Using the light emission measurement in assessment of electrical discharge development in different liquid dielectrics under lightning impulse voltage. *Electr. Power Syst. Res.* 140: 321–326.

62 Lesaint, O. and Gournay, P. (1994). On the gaseous nature of positive filamentary streamers in hydrocarbon liquids. I: influence of the hydrostatic pressure on the propagation. *J. Phys. D: Appl. Phys.* 27 (10): 2111–2116.

63 Liu, R., Tornkvist, C., Chandramouli, V. et al. (2009). Ester fluids as alternative for mineral oil: the difference in streamer velocity and LI breakdown voltage. *IEEE Conference Electrical Insulation Dielectrical Phenomena (CEIDP)*, Virginia Beach, VA, USA (18–21 October 2009), pp. 543–548.

64 Sitorus, H.B.H., Beroual, A., Setiabudy, R., and Bismo, S. (2015). Pre-breakdown phenomena in new vegetable oil – based Jatropha Curcas seeds as substitute of mineral oils in high voltage equipment. *IEEE Trans. Dielectr. Electr. Insul.* 22: 2442–2448.

65 Thien, Y.V., Azis, N., Jasni, J. et al. (2018). Pre-breakdown streamer propagation and breakdown characteristics of refined bleached and deodorized palm oil under lightning impulse voltage. *IEEE Trans. Dielectr. Electr. Insul.* 25 (5): 1614–1620.

66 Zhou, J. 1, Li, J., He, J. et al. (2018). Study on the propagation characteristic of streamer in long oil gap of natural ester. *IEEE Int. Conf. on the Propert.Appl. Dielectr. Materials*: 872–875.

67 Ngoc, M.N., Lesaint, O., Bonifaci, N. et al. (2010). A comparison of breakdown properties of natural and synthetic esters at high voltage. *Proceedings of the 2010 Annual Report Conference on Electrical Insulation and Dielectic Phenomena*, West Lafayette, IN, USA (17–20 October 2010).

68 Wong, P.P. and Forster, E.O. (1982). The dynamice of electrical breakdown in liquid hydrocarbons. *IEEE Trans. Electr. Insul.* 17: 203–220.

69 Godson, L., Raja, B., Lal, D.M., and Wongwises, S. (2010). Enhancement of heat transfer using nanofluids—an overview. *Renew. Sustain. Energy Rev.* 14: 629–641.

70 Rafiq, M., Lv, Y., and Li, C. (2016). A review on properties, opportunities, and challenges of transformer oil-based nanofluids. *J. Nanomater.*: 1–23.

71 Huang, Z., Li, J., Yao, W. et al. (2019). Electrical and thermal properties of insulating oil-based nanofluids: a comprehensive overview. *IET Nanodielectr.* 2: 27–40.

72 Puspitasari, P., Permanasari, A.A., Shaharun, M.S., and Tsamroh, D.I. (2020). Heat transfer characteristics of NiO nanofluid in heat exchanger. *AIP Conf. Proc.* 2228: 030023.

73 Li, J., Zhang, Z., Zou, P. et al. (2012). Preparation of a vegetable oil-based nanofluid and investigation of its breakdown and dielectric properties. *IEEE Electr. Insul. Mag.* 28: 43–50.

74 Beroual, A. and Khaled, U. (2020). Statistical investigation of lightning impulse breakdown voltage of natural and synthetic ester oils-based Fe3O4, Al2O3 and SiO2 nanofluids. *IEEE Access* 8: 112615–112623.

75 Khaled, U. and Beroual, A. (2020). Lightning impulse breakdown voltage of synthetic and natural ester liquids-based Fe3O4, Al2O3 and SiO2 nanofluids. *Alex. Eng. J.* 59: 3709–3713.

76 Khaled, U. and Beroual, A. (2019). AC dielectric strength of synthetic ester-based Fe3O4, Al2O3 and SiO2 nanofluids—conformity with normal and weibull distributions. *IEEE Trans. Dielectr. Electr. Insul.* 26: 625–633.

77 Peppas, G.D., Danikas, M.G., Bakandritsos, A. et al. (2016). Statistical investigation of AC breakdown voltage of nanofluids compared with mineral and natural ester oil. *IET Sci. Meas. Technol.* 10: 644–652.

78 Du, Y., Lv, Y., Li, C. et al. (2011). Effect of electron shallow trap on breakdown performance of transformer oil-based nanofluids. *J. Appl. Phys.* 110: 104104.

79 Beroual, A., Khaled, U., and Alghamdi, A.M. (2020). DC breakdown voltage of synthetic ester liquid-based nanofluids. *IEEE Access* 8: 125797–125805.

80 Sima, W., Shi, J., Yang, Q. et al. (2015). Effects of conductivity and permittivity of nanoparticle on transformer oil insulation performance: Experiment and theory. *IEEE Trans. Dielectr. Electr. Insul.* 22: 380–390.

81 Fofana, I. (2013). 50 years in the development of insulating liquids. *IEEE Electr. Insul. Mag.* 29 (5): 13–25.

82 Saha, T.K. and Purkait, P. (2008). Investigations of temperature effects on dielectric response measurements of transformer oil-paper insulation system. *IEEE Trans. Power delivery Insul.* 23 (1): 252–260.

83 Bell, W.R. and Danikas, M. (1982). Factors affecting the breakdown strength of the transformer oil. *IEEE International Conference on Electrical Insulation*, Philadelphia, PA, USA (7–9 June 1982), pp. 264–267.

84 Krasucki, Z. (1966). Breakdown of liquid dielectrics. *Proc. Math. Phys. Eng. Sci.* 294 (1438): 393–404.

85 Mohan Rao, U., Fofana, I., Jaya, T. et al. (2019). Preliminary study of ester-based fluid for application in transformers serving in cold climatic regions. *14th CIGRE Canada Expo on Power Systems*, Montreal (September 2019).

86 Rozga, P. and Stanek, M. (2016). Characteristics of streamers developing at inception voltage in small gaps of natural ester, synthetic ester and mineral oil under lightning impulse. *IET Sci. Meas. Technol.* 10 (1): 50–57.

87 Sékongo, B., Mohan Rao, U., Fofana, I. et al. (2020). Temperature dependence of the pre-breakdown and breakdown phenomena in natural esters under AC stress. *IET Sci. Meas. Technol.* 14 (9): 762–769.

88 Przybylek, P. (2016). Water saturation limit of insulating liquids and hygroscopicity of cellulose in aspect of moisture determination in oil-paper insulation. *IEEE Trans. on Dielectr. Electr. Insul.* 23 (3): 1886–1893.

7

Miscibility and Engineering Application of a Novel Mixed Fluid

Jian Hao, Ruijin Liao, Lijun Yang, Dawei Feng, Wenyu Ye, Chenyu Gao, and Xin Chen

State Key Laboratory of Power Transmission Equipment & System Security and New Technology, Chongqing University, Shapingba, Chongqing, China

7.1 Introduction

Mineral oil-paper insulation is the main insulation form for power transformers, which gradually ages under operating conditions, resulting in decline in its insulation and mechanical properties, thus safe operation of transformers is seriously threatened. Faults caused by insulation aging account for an important part of transformer failure, thus delaying the aging of oil–paper insulation is extremely important to ensure the safe and reliable operation of transformers. With the increase of old transformers that have been in use for more than 15–20 years, scientifically and effectively extending their insulation service life and improving their reliability are major requirements for ensuring the safety of the power grid.

Taking performance improvement of insulation oil as the starting point, research on a new insulation oil that can delay chain scission rate of cellulose insulation paper and replace mineral oil in in-service transformers directly is an effective solution to prolong the insulation service life of old transformers. Although a large number of studies have shown that vegetable oils have the ability to delay the chain scission rate of cellulose, their viscosity and dielectric loss are high, and their key parameters cannot satisfy mineral oil standard requirements, thus they cannot be directly filled in power transformers in-service. In this chapter, mineral insulation oil is used as base oil, vegetable oil is used as antiaging basic element, and adjustment methods of viscosity and dielectric loss are adopted to develop a mineral oil–vegetable oil mixed antiaging transformer oil (hereinafter referred to as mixed insulation oil) to achieve in the goal that extending the insulation service life of transformers in-service.

The mixed insulation oil can complement the performance advantages of single oil product. Based on the previous research on two-element mixed insulation oil, Chongqing University successfully developed a novel three-element mixed insulation oil that meets the requirements of GB 2536-2011, and the formula is 76 vol.% mineral oil +19 vol.% soybean oil + 5 vol.% palm oil. This mixed insulation oil makes up for the high dielectric loss and kinematic viscosity of the traditional two-element mixed insulation oil. Compared with mineral oil, the three-element mixed insulation oil (hereinafter referred to as mixed insulating oil) has higher power frequency breakdown voltage and better fire resistance, and can significantly delay the thermal-aging process of oil–paper

Alternative Liquid Dielectrics for High Voltage Transformer Insulation Systems: Performance Analysis and Applications, First Edition. Edited by U. Mohan Rao, I. Fofana, and R. Sarathi.

insulation. The biodegradation rate of three-element mixed insulation oil is about 30%. At present, the three-element mixed insulation oil has been used in 10 kV distribution transformer.

7.2 Need and Research Progress of Mixed Insulating Liquids

Mixed insulation oil is one of the main directions in the development of new liquid dielectric, which could complement the advantages of single oil [1–4]. Mineral oil and natural esters or synthetic esters could be perfectly mixed together [1, 3]. Mixed insulation oil, which uses mineral oil with natural ester (synthetic ester or modified ester) in a fixed ratio, has the unique advantages of taking into account the advantages and disadvantages of both mineral oil and ester oils [1, 2, 4, 5]. In 2002, I. Fofana from University of Quebec in Canada and V. Wasserberg from University of Hannover in Germany et al. were the first to point out that the volume ratio of mineral oil to synthetic ester was 8 : 2, which was the best ratio. With the increment of the ratio of synthetic ester, the dielectric dissipation factor and kinetic viscosity of mixed insulation oil increased [6, 7]. In 2006, C. Perrier from Areva corporation and A. Beroual from University of Lyon et al. found that the breakdown voltage under power frequency and the anti-aging performance of the mixed oil which composed of 80 vol.% mineral oil +20 vol.% synthetic ester could be effectively improved [8]. From 2009 to 2012, the team of Professor Ruijin Liao from Chongqing University developed two element mixed insulation oil which is composed of 80 vol.% mineral oil +20 vol.% olive oil. The key parameters of it could meet the requirement of GB2536-90 in China and could effectively retard the thermal aging of oil–paper insulation system [9–12]. The team developed a new kind of two-element mixed insulation oil in the following years, which comprising of 80 vol.% mineral oil + 20 vol.% rapeseed oil and has been successfully applied in 10 kV distribution transformer [13]. However, the kinetic viscosity and dielectric dissipation factor of it do not meet the new requirements of GB2536-2011 in China.

In 2017, Lingfeng Kou et al. tested the breakdown voltage, flash point, and other properties of mixed insulation oil of different proportion and have been compared with natural ester and mineral oil under the same condition. It is found that the mixed insulation oil can effectively improve the flash point, ignition point, and other parameters of mineral oil. However, the dielectric loss and kinetic viscosity of the mixed insulation oil is a little bit high, which has a certain effect on the cooling performance of the transformer [14]. In the same year, Yang Xu et al. studied the breakdown voltage, acid value, viscosity, ignition point, flash point, pour point, and some other physical and chemical properties of mixed insulation oil, which is composed of 80 vol.% mineral oil + 20 vol.% FR3 ester oil and 70 vol.% mineral oil + 30 vol.% FR3 ester oil. Compared to traditional mineral oil, the power frequency breakdown voltage of the two kinds of mixed insulation oil has increased by 8.6 and 4.3%. The acid value has increased by 60 and 20% but was still lower than 0.005 mg KOH/g and could meet the requirements of IEC 60296: 2012. The flash point of 80 vol.% mineral oil + 20 vol.% FR3 ester oil and 70 vol.% mineral oil + 30 vol.% FR3 ester oil has increased by 7.0 and 11.4%, while the pour point has increased by 17.5 and 12.3%, which shows that the fire resistance of mixed insulation oil has been improved. Besides, the mixed insulation oil which is composed of 30 vol.% FR3 ester oil has an improvement in ignition point [15]. K. Takahashi et al. mixed aged mineral oil with rapeseed oil, olive oil, and soybean oil with a volume ratio of 8 : 2 and studied the negative lightning discharge performance of three kinds of mixed insulation oil and mineral oil under needle-plane electrodes. Results show that the shape of discharge in mineral oil was shrubby while that in mixed insulation oil was obvious dendritic shape under the same applied voltage. The streamer velocity of mixed insulation oil is between new mineral oil and aged mineral oil, and it is weakly correlated with the amplitude of applied voltage before lightning impulse breakdown [16]. In 2018, Abderrahmane Beroual et al. proposed a kind of mixed insulation oil which composed of 80 vol.% mineral oil and 20 vol.% jatropha oil. Compared to mineral oil, the average power frequency breakdown voltage and DC breakdown voltage of it has increased by 12.0 and 28.5%, respectively [17]. From the perspective of fire safety performance, G. Dombek et al. studied the flash point, ignition point, and other properties of mixed insulation oil which composed of mineral oil and synthetic ester. Results show that with the increment of synthetic ester, the opening flash point, closing flash point, and ignition point of the mixed insulation oil all rose gradually [18]. A. Be ldjilali et al. investigated the isothermal discharge current characteristics of mixed insulating oil and presented a kind of mixed insulation oil which composed of 15 vol.% olive oil and 85 vol.% mineral oil. Compared with mineral oil, the mixed insulation oil significantly reduced the transient discharge current and the electron mobility and has good insulation performance [19].

7.3 Preparation Method for the New Mixed Insulating Oil

7.3.1 Selection of the Base Oil

The base oil is comprised of traditional mineral oil and appropriate natural esters in accordance with the performance target of the new type of mixed insulation oil that combines the advantages of blended mineral oil and natural esters. Natural esters are derivatives of plant seeds, and their main component is fatty acid triglyceride. The difference among various natural esters in terms of physical and chemical properties is the types of acid that constitute the radical group of triglycerides [20]. Fatty acid can be divided into four types, namely, saturated, monounsaturated, di-unsaturated, and polyunsaturated, depending on the saturation degree. The oxidation stability of these acids decreases in turn [21], thus natural esters with high saturation are preferred due to their good oxidation stability. Nevertheless, high oil saturation indicates high kinematic viscosity and pour point [22], which must be avoided in practical engineering. Table 7.1 shows several widely used natural esters and their percentages of fatty acids.

Table 7.1 Characteristics of base oil samples.

	Percentage of fatty acid composition (%)			
Natural esters	**Saturated**	**Monounsaturated**	**Diunsaturated**	**Triunsaturated**
Cottonseed oil	17.8	51.8	0.2	25.8
Safflower oil	12.2	74.2	0.4	8.5
Sunflower seed oil	19.6	65.7	—	10.5
Corn oil	24.2	58	0.7	12.6
Soybean oil	22.5	51	6.8	14.2
Rapeseed oil	55.9	22.1	11.1	7.9
Palm oil	47.5	9	—	43.5

Source: Based on McShane [22].

Table 7.2 Main parameters of the three base oils.

Oil	Mineral oil	Soybean oil	Palm oil
Relative permittivity (90 °C)	2.12	2.86	2.89
tan δ (90 °C, %)	0.0536	0.5085	1.23
Kinematic viscosity (40 °C) (mm^2/s)	9.87	30.25	5.06
Acid value (mg KOH/g)	0.0056	0.01245	0.011
Resistivity (90 °C) ($\times 10^{10}\,\Omega\cdot m$)	246	3.027	2.268

Soybean oil was selected as one of the base oils due to its oxidation stability, viscosity, output, and price. Our test results showed that the dielectric dissipation factor and acid values of refined soybean oil are lower than those of many natural esters. Palm oil fatty acid ester (PFAE), a modified ester with a low viscosity [23], was also selected as a viscosity regulator to ensure that the viscosity of the new mixed insulation oil is low. Thus, the new mixed insulation oil consists of mineral oil, soybean oil, and PFAE. Table 7.2 shows the main parameters of mineral oil, soybean oil, and PFAE.

7.3.2 Determination of the Proportion

According to IEC 60296:2012, the viscosity of oil must be less than 12 mm^2/s (40 °C), and tan δ should be less than 0.5% (90 °C). The kinematic viscosity and dielectric dissipation of previous mixed oils (80% mineral oil + 20% esters) [11, 24] do not satisfy the IEC 60296:2012 requirement. Therefore, these two parameters were adopted as the criteria in determining the oil proportion in the current study. The viscosity and tan δ parameters of the mixed insulation oil with different volume proportions is shown in Table 7.3 under the condition that the 20% volume ratio of natural ester is constant (the moisture content of each sample is approximately 25 ppm). The results indicated that the viscosity of the mixed oil gradually decreased with the addition of PFAE. The existence of PFAE helped reduce viscosity. However, the tan δ of the mixed oil increased. Table 7.3 shows that the appropriate appending proportion of PFAE is approximately 6%.

The proportion of the three base oils was further adjusted to reduce viscosity and tan δ. The results in Table 7.4 show that the viscosity and tan δ of the mixed oil decreased with a decrease in the proportion of soybean oil. However, although soybean oil has high viscosity, dissipation factor, and acidity, its proportion must not be reduced to a large extent to ensure excellent three-element mixed insulation advantages. In addition, when the proportion of total natural esters was more

Table 7.3 Viscosity and tanδ of mixed oil with different proportions of palm oil.

Oil proportion (volume ratio)				
Soybean oil (%)	Mineral oil (%)	Palm oil (%)	Viscosity (40 °C, mm^2/s)	tan δ (%, 90 °C)
20	80	0	12.6	0.26
20	77	3	12.4	0.32
20	74	6	11.7	0.44
20	71	9	11.4	0.501

Table 7.4 Results of further adjustment on the proportion.

Oils proportion (volume ratio)					
Soybean oil (%)	**Mineral oil (%)**	**Palm oil (%)**	**Viscosity (40 °C, mm²/s)**	**tan δ (%, 90 °C)**	**Acidity (mg KOH/g)**
20	74	6	11.716	0.4415	0.01
19	75	6	11.625	0.4184	0.008
18	76	6	11.563	0.3627	0.007
20	75	5	11.868	0.4121	0.009
19	76	5	11.655	0.3610	0.007

than 25%, the acid value exceeded the requirement of 0.01 mgKOH/g in IEC 60296:2012. Therefore, 76% mineral oil + 19% soybean oil + 5% PFAE was determined to be the ideal oil proportion.

7.3.3 Improvement of Oxidation Stability

Antioxidants were added to improve the oxidation stability of the three-element mixed insulation oil (76% mineral oil + 19% soybean oil + 5% PFAE oil). Investigation of the compound antioxidants indicated that the performance of the antioxidants was depended on the base oils [25, 26]. To determine the ideal antioxidants, we added various antioxidants to the mixed oil at a proportion of 0.2 wt% because the IEC standard stipulates that the antioxidant content must not exceed 0.4 wt%. The added antioxidants included tea polyphenols, BASF 1076, t-butyl hydroquinone (TBHQ), di-t-butyl p-cresol (T501), 2,6-butyl liquid diphenylamine antioxidant (L57), and high-purity alkylated phenyl-α-naphthylamine (L06). Pressure differential scanning calorimetry (PDSC) is suitable for measuring oxidation stability [27]. The initial oxidation temperature (IOT) for each oil sample was measured with PDSC to determine the suitable antioxidants for the three-element mixed insulation oil. Table 7.5 shows the IOT test results. The mixed oil added with T501, T511, L57, and L06 had a relatively high IOT. Thus, this oil was selected for further research. T501 and T511 are phenolic antioxidants, and L57 and L06 are amine antioxidants.

Any two among T501, T511, L57, and L06 were combined to constitute a different compound antioxidant. Each pair of the compound antioxidant was added to the three-element mixed insulation oil. Table 7.6 presents the IOT test results. The IOT of the mixed oil with compound additives was more efficient than that with single additives. The maximum IOT was the combination of antioxidants T511 and L06. The OIT of the mixed oil with 0.2 wt% T511 + 0.2 wt% L06 was higher (17.3 °C) than that of the selected mineral oil and higher (13.2 °C) than that of the mixed insulation oil without antioxidants. Therefore, the ideal compound antioxidant for the three-element mixed insulation oil is 0.2 wt% T511 + 0.2 wt% L06.

7.3.4 Stability Overall Performance

Table 7.7 presents the main parameters of the novel three-element mixed insulation oil (76 vol.% mineral oil + 19 vol.% soybean oil + 5 vol.% PFAE oil + 0.2 wt% T511 + 0.2 wt% L06). The key parameters of this insulation oil satisfy the IEC 60296:2012 requirement for mineral oil. Thus, this novel insulation oil can be used in power transformers as a substitute for mineral oil without changing the structure of transformers.

Table 7.5 IOT of mixed oil with different antioxidants.

Antioxidant	Temperature (°C)
Selected mineral oil	197.1
Selected soybean oil	195
PFAE oil	226.5
Three-element mixed insulation oil without antioxidants	201.2
T501	205.1
TBHQ	202.2
L57	207.1
L06	209.6
BASF 1076	204.3
T511	208
Tea polyphenols	202.3

Table 7.6 IOT of mixed oil with different antioxidants.

Temp (°C)	0.2% T501	0.2% T511	0.2% L57	0.2% L06
0.2%T501	207.5	—	—	—
0.2%T511	208.8	211.3	—	—
0.2%L57	211.2	213.2	209	—
0.2%L06	212.1	214.4	211.3	213.5

Table 7.7 Parameters of the novel three-element mixed insulation oil.

Parameter	Mixed oil	IEC 60296:2012
Kinematic viscosity (40 °C, mm^2/s)	11.65	≤12
Moisture content (ppm)	14	≤30
Breakdown voltage (2.5 mm, kV)	78.6	≥70
Density (20 °C, kg/m^3)	886.8	≤895
tan δ (90 °C, %)	0.0047	≤0.005
Appearance	Transparent, sediment-free, and suspended	Transparent, sediment-free, and suspended
Acid value (mg KOH/g)	0.007	≤0.01
Total sulfur content (mass fraction, %)	0.0265	≤0.15
Corrosive sulfur	Non-corrosive	Non-corrosive
Furfural content (mg/kg)	Not detectable (≤0.01)	≤0.05
Flash point (°C)	147	≥135
Polycyclic aromatic hydrocarbon (PCA) content (%)	1.6	≤3
Polychlorinated biphenyl (PCB) content (%)	Not detectable	Not detectable
Relative permittivity (90 °C)	2.32	—

7.3.5 Performance of Novel Three-Element Mixed Insulating Oil

In oil-immersed transformers, cooling is a key function of insulation oil and is mainly achieved by oil circulation. Heat dissipation and cooling performance are closely related to the kinematic viscosity of the insulation oil. Insulation oil with a low kinematic viscosity is favorable for transformers. The kinematic viscosity of the novel three-element mixed insulation oil and that of mineral oil were measured from 0 to 100 °C according to ISO 3104 (Figure 7.1). Temperature exerted a significant effect on kinematic viscosity. The kinematic viscosities of the novel three-element mixed insulation oil and mineral oil decreased through the exponential decay law. Kinematic viscosity rapidly decreased with the increase in temperature. Afterward, it gradually decreased. The kinematic viscosity of the novel three-element mixed insulation oil was only slightly higher than that of mineral insulation oil at all tested temperatures. However, the difference in kinematic viscosity between the two oils became increasingly small with the increase in temperature from 40 to 100 °C. This result means that the fluidity of the novel three-element mixed insulation oil was equivalent to that of mineral oil in the field operation transformer.

Thermodynamic properties, such as specific heat capacity and thermal conductivity, are also related to the heat dissipation and cooling performance of insulation oil. Thus, the thermodynamic

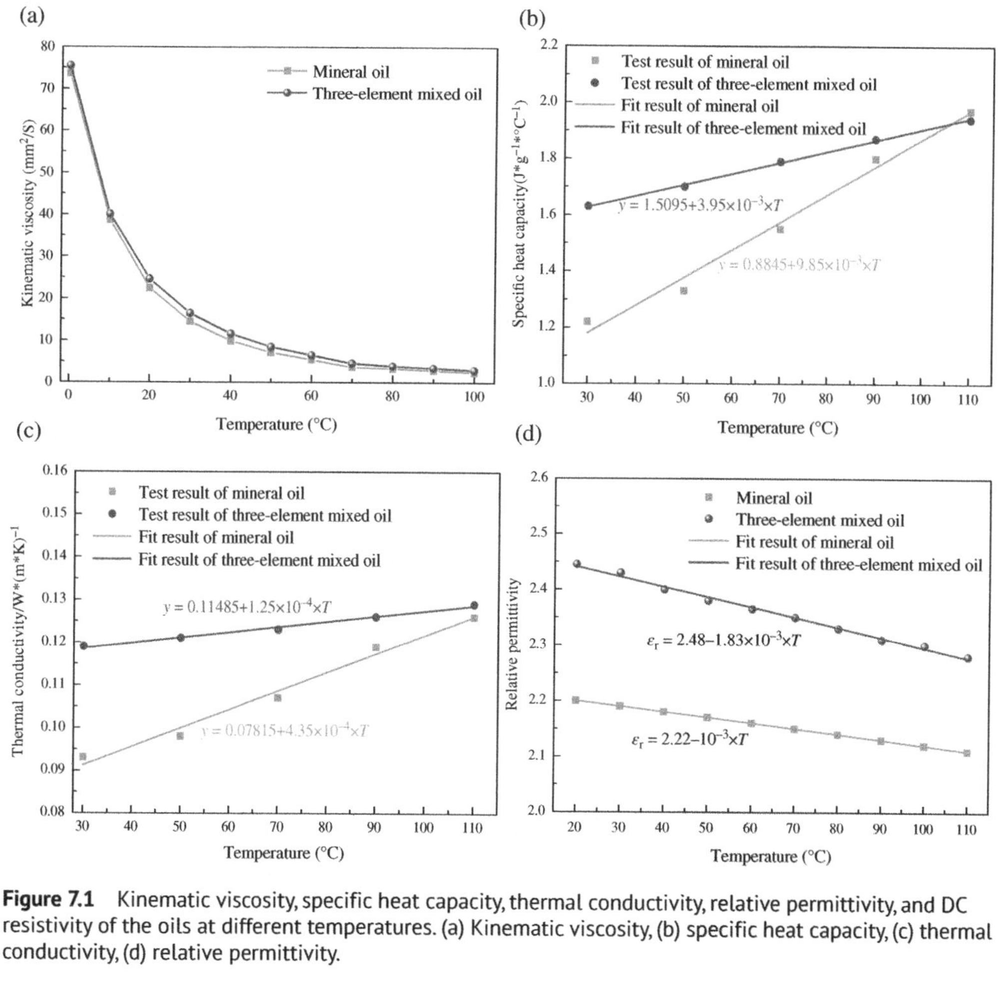

Figure 7.1 Kinematic viscosity, specific heat capacity, thermal conductivity, relative permittivity, and DC resistivity of the oils at different temperatures. (a) Kinematic viscosity, (b) specific heat capacity, (c) thermal conductivity, (d) relative permittivity.

properties of the two oils were measured at different temperatures. Figure 7.1a and b shows the results. The specific heat capacity of both insulation oils gradually increased with the linear increase in temperature. The novel three-element mixed insulation oil had a higher specific heat capacity than mineral oil between 30 and 100 °C. However, the specific heat capacity of mineral oil presented a higher growth rate, as indicated by the slope of the linear fitting curve shown in Figure 7.1b. Figure 7.1c shows that the thermal conductivity of the novel three-element mixed insulation oil was higher than that of mineral oil between 30 and 110 °C. However, the thermal conductivity of mineral oil had a higher growth rate, as indicated by the value of the slope of the linear fitting curve shown in Figure 7.1c.

The average operation temperature of transformer oil is usually between 30 and 75 °C [28]. The large specific heat capacity indicated that the increase in the temperature of the novel three-element mixed insulation oil was smaller than that of mineral oil when the oils absorbed the same amount of heat. The large thermal conductivity of the novel three-element mixed insulation oil meant that its transfer heat capability was better than that of the mineral oil.

Figure 7.1d shows a comparison of the relative permittivity of the novel three-element mixed insulation oil and mineral oil. The relative permittivity of both oils decreases with a linear increase in temperature. The increase in temperature led to volumetric expansion. Thus, the number of polarized particles decreased within the unit volume, which led to a decrease in relative permittivity. Meanwhile, the increment in temperature aggravated the heat movement of the polar molecules, which hindered the orientation of the dipoles in the electric field and reduced relative permittivity. For mineral oil, the decrease in relative permittivity was mainly due to its weak polar dielectric. Natural esters are triglycerides and polar molecules. Therefore, the polarity of the mixed oil was stronger than that of mineral oil. In addition to the first reason, the decrease in relative permittivity of the mixed oil was also due to the second reason above. Hence, the decline slope of the relative permittivity of the mixed oil was larger than that of mineral oil. In addition, the relative permittivity of the mixed oil was higher than that of mineral oil mainly due to the stronger polarity of the former than the latter.

Vegetable oils delay the aging of insulation paper mainly due to their very high saturation water content [11]. Saturation water content is also important for the novel three-element mixed insulation oil. The water absorption rate of the novel three-element mixed insulation oil and mineral oil was compared from 15 to 90 °C at a 15 °C interval. The moisture content in the oils was measured with a Karl Fisher coulometer according to IEC 60814. The fitting equation is shown in Eq. (7.1) and Figure 7.2a.

$$M = y_0 + A_1 * e^{-t/b} \tag{7.1}$$

where M is the absolute moisture content in oil in ppm; t is the water absorption time in hour(s); and y_0, A_1, and b are constants.

According to the fitting equation, the increment trend of the absolute moisture content in the two oils was different. The absolute moisture content of the mixed oil was 180.6 ppm at 30 °C, which is 3.96 times as large as that of mineral oil. This condition mainly depends on the structure and group of oil molecules. The mixed oil possessed larger solubility of water than mineral oil due to its proportion of natural esters, whose molecules display strong polarity and can thus absorb around 20–30 times more moisture than mineral oil [29]. Furthermore, the saturation water content of the novel three-element mixed insulation oil and mineral oil was compared at different temperatures. Figure 7.2b shows that the saturation water content increased with an exponential increase in temperature. The increase rate of the saturation water content of the novel three-element mixed insulation oil was higher than that of mineral oil. The saturated water content of the novel three-element mixed insulation oil was approximately 3–4 times as large as that of miner.

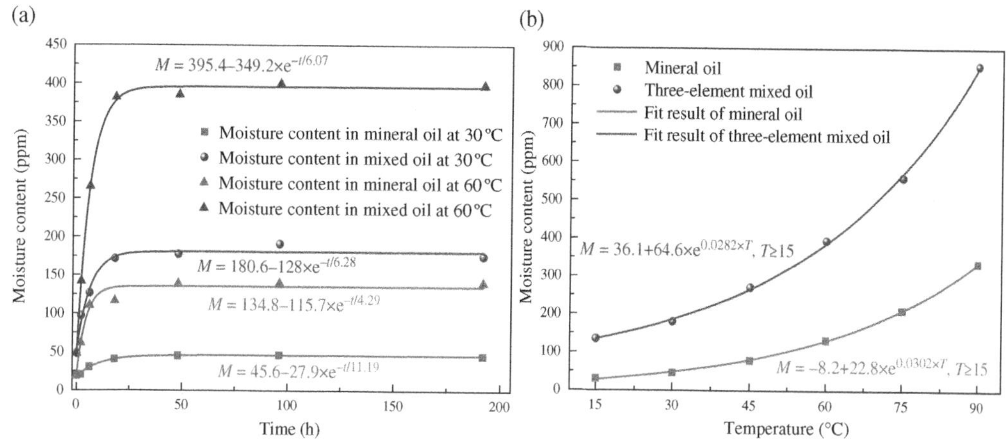

Figure 7.2 Variation in the moisture content and saturated water content of the two oils at 30 and 60 °C. (a) Variation in the moisture content. (b) Saturated water content.

7.4 Thermal Aging Characteristics of the New Mixed Insulation Oil–Paper Insulation and Its Delaying Thermal Aging Mechanism

7.4.1 Introduction

The cellulose paper and pressboard utilized in this experiment are listed in Table 7.8, and some copper bars were added in the oil–paper system to simulate the transformer environment. All these solid materials were provided by Yadongya Transformer Co., Chongqing, China. The main object for analysis in the current study is cellulose paper, which consists of 90% cellulose polymers, ~6–7% hemicellulose, and ~3–4% lignin [30]. The two insulation oils used were the three-element mixed insulation oil and Karamay 25# naphthenic mineral oil, and their main parameters are listed in Table 7.9.

Both insulation oils were dried at 90 °C/50 Pa for 48 hours to remove the dissolved gas and most of the moisture, and the moisture contents of the mixed oil and mineral oil after drying were approximately 14 and 10 ppm, respectively. The cellulose paper and pressboard were also dried at 90 °C/50 Pa for 48 hours to maintain their moisture contents below 1%. A total of 36 glass bottles were prepared for this thermal-aging experiment, and these bottles were divided into four groups, as shown in Table 7.10. Initially, 11 bottles were filled with 1 l of the dried mixed oil (Group B), and the remaining bottles were filled with 1 l of the dried mineral oil. The mass ratio of the insulation oil, cellulose paper, and cellulose pressboard was controlled to be 20 : 1 : 1. Thus, the corresponding mass of the dried cellulose paper and pressboard was added in each bottle. A copper bar equal to the mass of the cellulose paper was added to each bottle to simulate the transformer's condition. Thereafter, these bottles were placed in 60 °C/50 Pa for 24 hours to complete the oil-impregnation process. Then, the bottles were sealed under a nitrogen atmosphere and placed in a 130 °C thermostat.

Four hours later, two bottles, one from each group (A and B), were separately taken out as initial samples. Thereafter, samples in Groups A and B were periodically obtained based on previous experience. On the 20th day of aging, the DP values of cellulose paper dropped to approximately 600–700. In order to verify that the three-element mixed insulation oil has the ability to delay the aging transformer oil–paper insulation system, an aging oil change test was carried out. The aged mineral oil in Groups C and D was then replaced with unused mineral and mixed oil, respectively.

Table 7.8 Parameters of cellulose polymers.

Material	Density (g/cm^3)	Thickness (mm)	Provider
Cellulose paper	0.933	0.108	Yadongya Transformer Co., Chongqing, China
Cellulose pressboard	1.20	0.45	

Table 7.9 Parameters of the three-element mixed insulation oil and 25# mineral oil.

Property	Three-element mixed insulation oil	Mineral oil
Appearance	Transparent, free water or suspended matter, yellow	Transparent, free water or suspended matter, colorless
Kinematic viscosity (40 °C, mm^2/s)	11.65	9.2
Moisture content (ppm)	14	11
Breakdown voltage (2.5 mm, kV)	78.6	72
Density (20 °C, kg/m^3)	886.8	890
tan δ (90 °C)	0.0047	0.001
Acid value (mg KOH/g)	0.007	0.004
Relative permittivity (90 °C)	2.32	2.12

Table 7.10 Sample grouping and oil treatment.

Group	A	B	C	D
Bottle number	11	11	7	7
Initial oil	Mineral oil	Mixed oil	Mineral oil	Mineral oil
Oil replacement	No	No	Yes	Yes
Time for oil replacement	—	—	20th day	20th day
Oil replaced with	—	—	Mineral oil	Mixed oil

Subsequently, all samples in these four groups were placed at 60 °C/50 Pa for 24 hours. These samples were then sealed in a nitrogen atmosphere. Finally, the four group samples were returned to the 130 °C thermostat for continued aging and periodically sampled for parameter measurement.

The main parameters, such as DP values, breakdown voltage, and furfural content in oil, were measured during the aging process to analyze the aging characteristics of cellulose paper. Previous works indicate that there is a direct correlation between the DP of cellulose paper and its mechanical strength [31], and the mechanical strength of cellulose paper is very important in the power industry, thus the DP values are regarded as the most direct parameter to mirror aging extent of cellulose paper [32, 33]. The tested parameters and their standards are listed in Table 7.11. For each sample, breakdown voltage was measured nine times, the other parameters were tested three times, and the average value was obtained as the final result.

Table 7.11 Test method and standard of parameter measurement.

Parameter	Method	Standard	Equipment
DP value of cellulose paper	Pulp viscosity	IEC 60450	NCY viscosity meter
Acidity of oil	KOH solution titration	IEC 62021	—
Infrared spectrometry	Attenuated total reflectance	—	Nicolet iS5

The DP value of cellulose paper is the most important parameter in the current study. Thus, the measurement process is outlined in detail. According to IEC 60450-aml, the specific steps of the DP value test can be described as follows: first, the cellulose-paper sample is degreased with normal hexane; then, the cellulose paper is cut into pieces, and a certain mass of cellulose paper is weighted in accordance with the standard; next, the cellulose-paper fragments are placed into the cupriethy lenediamine solution to dissolve, and a vibration gauge is used to promote dissolution; finally, the viscosity of the mixed solution is measured by viscometer, and the DP values are calculated using the measurement results [34]. It should be noted that the DP value tested according to the standard is the viscosity average molecular weight of cellulose paper, which is mainly determined by macromolecules in cellulose paper and able to accurately mirror the mechanical properties of cellulose paper.

7.4.2 DP Values of Cellulose Paper

Cellulose paper mainly comprises cellulose, and the DP value of cellulose is the average amount of glucose for the cellulose chain. Existing works have shown that DP values are closely related to the condition of the cellulose paper, such as mechanical strength and tenacity [31]. Thus, DP values are used to mirror the aging extent of cellulose paper, and the DP values of the cellulose paper in Groups A–D were tested and are shown in Figure 7.3.

The DP values of the cellulose paper rapidly decline in the initial aging period and then decrease at a low rate due to the structure of cellulose paper. The cellulose paper can be divided into crystalline and amorphous zones in a microcosm, and the two types of zones are alternately joined to form cellulose paper. The molecular arrangement in amorphous areas is loose and disorderly compared with that in crystalline zones. Thus, amorphous areas are more likely to be broken due to the effects of heat, oxygen, and moisture [35]. Given that the crystalline and the amorphous areas generally alternate, the destruction of the amorphous areas leads to a rapid decrease in the DP values of the cellulose paper. However, the performance of the crystallization areas is relatively stable. Thus, the decreased rate of the DP values slows down in the later aging period.

The comparison between Groups A and B shows that the decreasing rate of DP values in Group B is apparently slower than that in Group A. This phenomenon confirms that the three-element mixed insulation oil can retard the aging rate of cellulose paper. The retardation effect was evident from the beginning of the aging experiment. On the 90th day, the DP values of cellulose paper impregnated by mineral oil dropped to 228, whereas those of cellulose paper impregnated by three-element mixed insulation oil remained above 400. Groups C and D had oil replacement, and the cellulose paper exhibited different aging trends due to the different oils used. The aging rate of the cellulose paper was almost unchanged after oil replacement with unused mineral oil, and the DP values of cellulose paper in Group C almost coincided with those in Group A. By contrast, the aging rate of cellulose paper was apparently reduced after oil replacement with the three-element mixed insulation oil, and the DP values of the cellulose paper in Group D were considerably higher

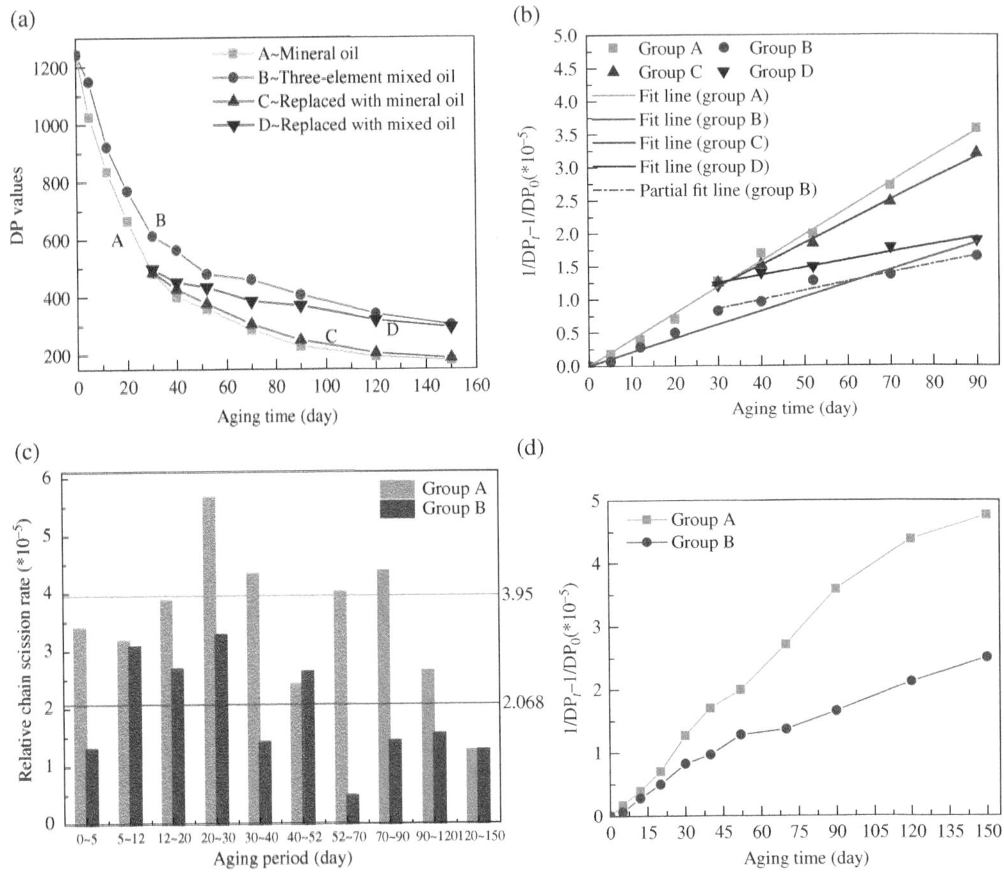

Figure 7.3 Test results of degree of polymerization (DP) values during the aging process and relative chain-scission rate of cellulose in each aging period. (a) DP values, (b) zero-order kinetic fitting results, (c) calculated $k_{t1\sim t2}$, (d) slope of the family of two points for Groups A and B. *Source:* Based on Zhang and MacAlpine [31].

than those in Groups A and C. When the DP values of the cellulose paper in Group A dropped to 228 on the 90th day, those of the cellulose paper in Group D were 370. The comparison of aging introduction shows that mixed oil can not only delay the aging of new cellulose but also further reduce the aging rate of aging cellulose.

The mixed oil delayed the aging rate of the cellulose paper due to the vegetable oils in the three-element mixed insulation oil. Vegetable oil comprises polar triglycerides; thus, mixed oil can hold additional water, and saturated-water content is 3–4 times that of mineral oil [2]. Based on the moisture equilibrium between oil and cellulose paper [36], some moisture in paper diffuses into insulation oil, reducing the moisture content in cellulose paper. Simultaneously, ester bonds of vegetable oil may hydrolyze [37], thereby consuming water and decreasing moisture content. Although this process generates macromolecular acids, their effect on the aging rate of cellulose paper is not significant [38]. Moisture is the number-one enemy of cellulose polymer aging [3], and the retardation of the aging rate of cellulose paper by mixed oil can be attributed to the reduction in moisture content in the cellulose polymer.

A zero-order kinetic equation is usually adapted to fit the variation of DP values during the aging process of cellulose paper to analyze the aging rate of cellulose paper, as shown in Eq. (7.2) [39].

$$1/\mathrm{DP}_t - 1/\mathrm{DP}_0 = k_0 \cdot t \tag{7.2}$$

where DP_0 denotes the initial DP value of cellulose paper; t is aging time (days), DP_t is DP value at time t, k_0 is a constant, which indicates the relative aging rate of cellulose paper in Eq. (7.1).

Here is a simple algorithm to understand Eq. (7.2). Cellulose paper is mainly made of cellulose, whose molecular formula is $(C_6H_{10}O_5)_n$. Understandably, the number of chain scissions is approximately equal to the increased number of cellulose chains because each breakage of a cellulose chain means that the number of molecules increases by one. Supposing m is the mass of the used cellulose paper, the number of chain scissions (N) during aging time $0 \sim t$ can be calculated, as shown in Eq. (7.2), in which 162 is the molar mass of the cellulose monosaccharide.

Equation (7.3) is a variant of Eq. (7.2), which suggests that $1/DP_t - 1/DP_0$ is a multiple of the number of chain scissions.

$$N = m/(162 \times DP_t) - m/(162 \times DP_0) \tag{7.3}$$

where N is the number of chain scissions, in mol; m is the mass of the cellulose paper, in g; and 162 is the molar mass of the cellulose monosaccharide, in g/mol. Comparing Eqs. (7.1) and (7.3), k_0 represents the relative chain-scission rate of cellulose over time t, as shown in Eq. (7.5), in which N/t is the real chain-scission rate. The chain scission of cellulosic polymers means the aging of the polymer; thus, k_0 can also be considered as a relative aging rate.

$$1/DP_t - 1/DP_0 = \frac{162}{m} \times N \tag{7.4}$$

$$k_0 = \frac{162}{m} \times \frac{N}{t} \tag{7.5}$$

Here, $1/DP_t - 1/DP_0$ was calculated according to the data in Figure 7.3. The DP values of cellulose paper impregnated with mineral oil (Groups A and C) had fallen below 250 on the 90th day, so results before 90 days were fitted with a zero-order kinetic equation. For comparison, the DP values in Groups B and D conducted the same fitting, although there were still two points (120th and 150th day) not included; the fitting results are shown in Figure 7.3b and Table 7.12.

Group A showed good fitting effect with a zero-order kinetic equation before the 90th day, which was mainly because the relative chain-scission rate calculated at each aging period was around the average values (k_0) during 0–90 days. Based on the above analysis, the relative chain-scission rate of cellulose in each aging period could be calculated with Eq. (7.6), which evolved from Eq. (7.2). The difference is that Eq. (7.2) is for the entire aging process, while Eq. (7.6) is for each aging period. The calculation results with Eq. (7.6) for Groups A and B are presented in Figure 7.3c. In fact, the calculated results in Figure 7.3c are exactly the slope of the adjacent two points for Groups A and B, as shown in Figure 7.3d. The relative chain-scission rates of different aging periods in Group A fluctuated around the average value of 3.95 before 90 days, suggesting that the

Table 7.12 Fitting results of DP values with kinetic equation.

Group	k_0 ($\times 10^{-5}$)	R-Square
A	3.95	0.998
B	2.068	0.978
C	3.268	0.994
D	1.157	0.951
Partial fit line (group B)	1.347	0.942

chain-scission rate was nearly the same. However, the chain-scission rate of cellulose paper impregnated with mineral oil decreased significantly after 90 days. [40]. This indicates the chain-scission rate is very slow when DP value decreases to a low value, which is around 200 for cellulose paper impregnated by mineral oil [40]. For Group B, however, not only was the chain-scission rate lower than that in Group A in each aging period, but the average chain-scission rate was also significantly decreased in the late aging period, especially for days 52–150, during which DP values were below 500. This result suggests that the mixed oil retarded the aging of the cellulose paper mainly at the late aging period.

$$1/\mathrm{DP}_{t2} - 1/\mathrm{DP}_{t1} = k_{t1\sim t2} \cdot (t2 - t1) \tag{7.6}$$

Insulation oil was replaced in Groups C and D, and data in the late aging period (30–90 days) were fitted with Eq. (7.1), and fitting results were satisfactory. Similarly, results in Group D showed that the mixed oil can significantly delay the aging rate of cellulose paper in the late period.

Fitting results in Table 7.12 show that the aging rate of cellulose paper in Group B is apparently slower than that in Group A. This phenomenon confirms that the three-element mixed-insulation oil strongly retards the aging rate of cellulose paper. The oil in Group C was replaced with unused mineral oil, thereby decreasing the aging rate of cellulose paper to 0.83 times of that in Group A. The mineral oil in Group A underwent a long aging time and produced additional aging by products such as acid, which further accelerate the aging of the cellulose paper [38]. The cellulose paper in Group C slowly slightly degraded. The oil in Group D was replaced with three-element mixed-insulation oil, and the aging rate of cellulose paper decreased to approximately 0.29 times the aging rate of that in Group A, which shows the delaying effect of the mixed oil on the further aging rate of aged cellulose paper.

Cellulose paper loses most of its mechanical strength when DP values decrease below 250. Assuming that DP values drop to 250 as the life endpoint of cellulose paper, then the thermal-aging life of cellulose paper in Groups A and B could be calculated with Eq. (7.7), which is a variation of Eq. (7.2), and the results are listed in Table 7.6.

$$t = \left(1/\mathrm{DP}_t - 1/\mathrm{DP}_0\right)/k_0 \tag{7.7}$$

Considering the change in aging rate in Groups C and D, the lifetime is divided into two parts. Thermal-aging life in Group C after the 30th day was calculated with the fitting results for Group C in Table 7.12, and then total aging life can be calculated, as shown in Table 7.6. For Group D, DP value only reduced to 294.4 after 150 days of aging; thus, its accurate aging life could not be obtained based on the experiment results. Similar to Group C, aging life after 30 days was estimated based on the fitting results in Table 7.5, and the result was 171.9 days; thus, the total lifespan of cellulose paper in Group D could be recorded as 201.9 days.

Comparing the results in Groups A and B, the aging life of cellulose paper was significantly improved by the mixed oil, which increased by 86.8% when compared with the cellulose paper in Group A. Supposing that mineral oil is replaced with mixed oil when the DP is approximately 600 (Group D), the aging life of cellulose paper is extended to 2.5 times. The retarding-aging effect of mixed oil is mainly reflected in the period of low DP values, especially those below 500, during which aging rate was considerably reduced. Therefore, oil replacement with mixed oil in the later aging stage could also delay the further aging rate of the aged cellulose paper (Table 7.13).

An interesting phenomenon is that cellulose paper life in Group B was significantly lower than that in Group D, which was mainly due to the different methods used in the fitting. In fact, the aging rate of cellulose paper in Group B should have been near that in Group D in the late aging

Table 7.13 Expected life of cellulose paper in thermal aging test (130 °C).

Group	Life after 30th day	Total life (*d*)
A	—	80.9
B	—	151.1
C	59.5	89.5
D	171.9	201.9

period due to the consistent action of the mixed oil, which was proved by that the slope of the partial linear fitting of Group B was close to that in Group D in the late aging period in Figure 7.3b. Supposing the aging life of cellulose paper after the 90th day in Group B is roughly estimated using the slope of partial linear fitting result, the remaining life after 90 days would approximately be 114.4 days. This suggests that the total life of cellulose paper in Group B was 204.4 days, which is closed to the estimated lifespan in Group D. When oil replacement was conducted, the DP values of cellulose paper in Group B were higher than those in Group D. However, the oil in Group B underwent longer aging time, resulting in more byproducts that accelerate the aging rate of cellulose paper [38], so the final estimated life was nearly the same as that in Group D.

Someone may question the above estimation methods of Group B and Group D. In fact, after 90 days, there are two more points that were tested during the experiment, which could be used to verify the correctness of the estimation. The k_0 used in the estimation in Groups B and D was 1.347×10^{-5} and 1.157×10^{-5}, respectively. Assume k_0 remains unchanged after 90 days, the DP values of cellulose paper in Groups B and D at 120 and 150 days could be calculated with Eq. (7.8). Take the 120th day in Group B as an example: if $t1$ equals 120 and $t2$ equals 90, then Eq. (7.9) can be obtained. The measured DP_{90} is 406.7, so the DP value on the 120th day can be calculated with k_0 ($1.347 * 10^{-5}$). All calculated results are shown in Table 7.14. The results show that the calculated results are very close to the measured DP values, which proves that partial fitting in Groups B and D can be used to predict the subsequent development trend of DP values.

$$\left.\begin{aligned}1/\mathrm{DP}_{t1}-1/\mathrm{DP}_0=k_0\cdot t1\\1/\mathrm{DP}_{t2}-1/\mathrm{DP}_0=k_0\cdot t2\end{aligned}\right\}\Rightarrow 1/\mathrm{DP}_{t1}-1/\mathrm{DP}_{t2}=k_0\cdot\left(t1-t2\right) \tag{7.8}$$

$$1/\mathrm{DP}_{120}-1/\mathrm{DP}_{90}=k_0\cdot\left(120-90\right) \tag{7.9}$$

where $t1$, $t2$, 90, and 120 denote aging time, in days.

Although lifespan estimation in Groups B and D is not absolutely precise, they showed apparently longer lifespan than that in Groups A and C, because the DP values of cellulose paper immersed in the mixed oil (Groups B and D) stayed around 300 at 150th day, which far exceeded the 81-day life of cellulose paper immersed in mineral oil (Group A). The results demonstrate that the three-element mixed insulation oil can retard the aging rate of cellulose paper, which is hopeful to replace aged mineral oil in old transformers to extend the aging life of cellulose paper in field transformers [41].

7.4.3 Mechanism of Delaying Thermal Aging

Oxidation stability is an important parameter for the stability of insulating oils. The initial oxidation temperatures of the new three-element mixed insulation oil and new mineral oil were measured by

Table 7.14 Calculated and measured DP values after 90 days.

	Calculated DP value		Measured DP value		
Group	**120th day**	**150th day**	**90th day**	**120th day**	**150th day**
B	349.3	306.1	406.7	341.8	302.3
D	322.0	284.9	370.1	319.5	294.4

pressure differential scanning calorimetry (PDSC), the result is shown in Figure 7.4a. According to the PDSC curve of four types of insulation oils, the initial oxidation temperature (IOT) of mineral oil and the new three-element mixed insulation oil is 197.1, 214.4. The IOT of the new three-element mixed insulation oil is 17.3 °C higher than the new mineral oil. The new mixed insulation oil has better oxidation stability than new mineral oil, which is one of the reasons for the mixed oil to delay the thermal aging of cardboard. Besides, thermogravimetric decomposition characteristics of different insulating oils are shown in Figure 7.4b and c. The decomposition rate of mineral oil reached the maximum 1.71%/°C at 214.19 °C. The initial decomposition temperature was 168.27 °C, and mineral oil was completely decomposed before the temperature reached 250 °C. Compared with

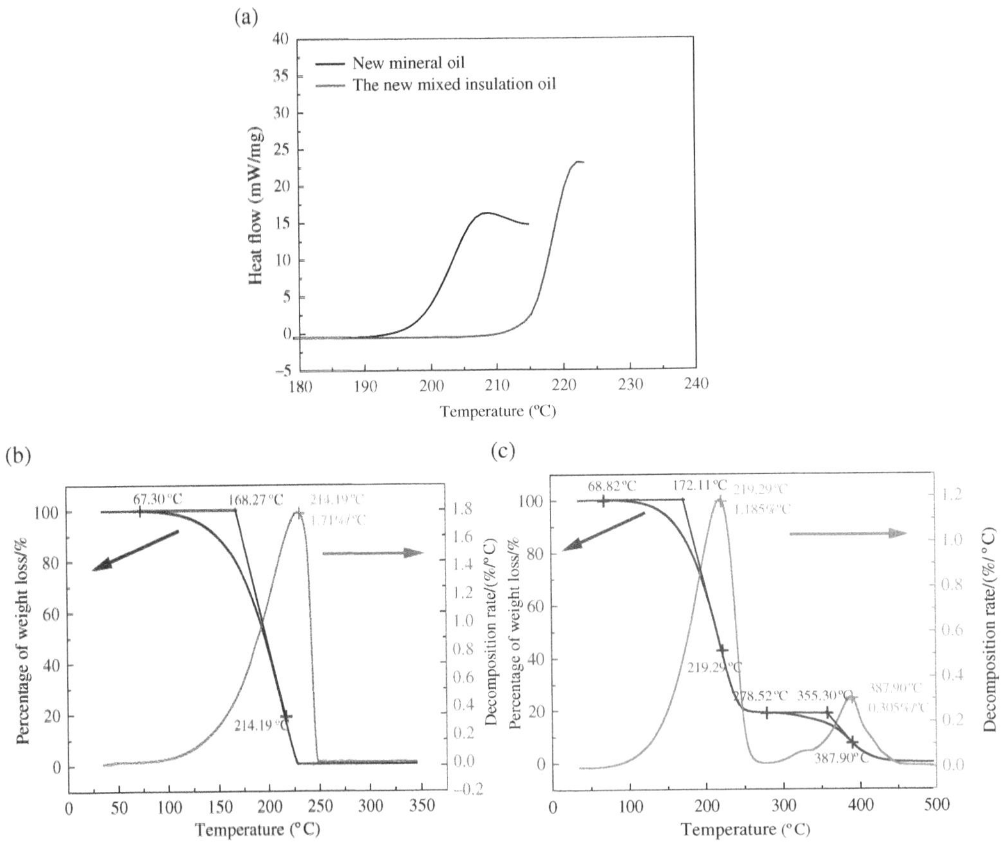

Figure 7.4 PDSC and thermogravimetric decomposition characteristics of different insulating oils. (a) PDSC results, (b) thermogravimetric decomposition characteristics of mineral oil, (c) thermogravimetric decomposition characteristics of three-element mixed insulation oil.

mineral oil, three-element mixed insulation oil has two peaks of decomposition rate, 1.185%/°C at 219.29 °C and 0.305%/°C at 387.90 °C, respectively. Because of low decomposition rate and two stages of thermal decomposition, it was not until 450 °C that three-element mixed insulation oil was fully thermally decomposed. It can be concluded that three-element mixed insulation oil is more stable than mineral oil at high temperature and thus has better protective effect on cellulose.

On the other hand, the main component of insulating paper is cellulose, which is polymerized by D-glucose via β-1,4 glycosidic bonds. The oil–paper insulation system cannot be absolutely dry, and there will be a certain amount of water dispersed in the insulation paper and the insulation oil. The acid values of the new mineral oil and the three-element mixed insulation oil are 0.004 and 0.007 mg KOH/g, respectively. The pH values are both about 6.2, indicating that there are traces of acidic substances in the insulating oil. During the aging process, insulating oil and insulating paper will also react with the trace oxygen that exists to produce acidic substances. Therefore, as the aging time increases, the acidic substances in the oil–paper system increase. The measured results of the insulating oil acid value are shown in Figure 7.5a. During the entire aging process, both moisture and acid exist in the oil–paper system. In an environment where water and acid coexist, water can ionize molecular acid into H^+, but the ionized H^+ is difficult to exist alone, and it is easy to form H_3O^+ with H_2O. H_3O^+ easily reacts with the oxygen atoms in the cellulose unit, including the oxygen atoms on the pyran ring and the glycoside oxygen atoms. The acid catalyzed cellulose hydrolysis process is shown in Figure 7.5b.

In this case, a composite system of three-element mixed insulation oil and mineral oil with paper were constructed using molecular simulation technology. Add water and hydronium ions to the two composite systems, respectively. The content of water and hydronium ions is shown in the Table 7.15. First, the AC (Amorphous Cell) module of Materials Studio is used to construct a cellulose model. The degree of polymerization of cellulose is 5, with a total of 10 molecular chains. When constructing the oil–paper composite model, the cellulose model was used and the oil–paper mass ratio was about 10 : 1. Then, the Build Layer Tool was used to compose the oil model and the cellulose model into a composite model to obtain the three-element mixed insulation

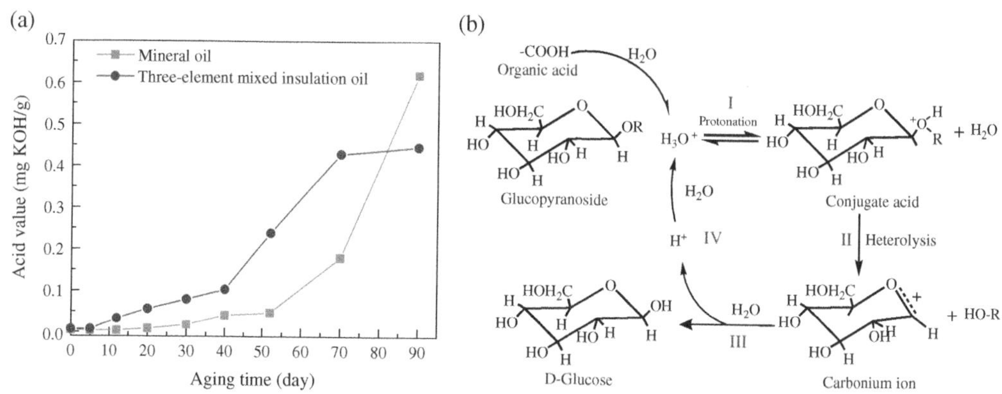

Figure 7.5 Acid values of insulation oils and acid catalyzes the hydrolysis of cellulose during the aging process. (a) Acid values, (b) acid catalyzes the hydrolysis process.

Table 7.15 Addition mass ratio of H_2O and H_3O^+ in two insulation oils (%).

Insulation oil	Mineral oil			Three-element mixed insulation oil		
H_2O	1			1		
H_3O^+	1	3	5	1	3	5

oil–cellulose and mineral oil–cellulose models. At this time, the three-element mixed insulation oil was separated from the insulating paper, which was equivalent to the state of the oil–paper system before being immersed in oil. In the next step, the Forceite module was used to perform geometric optimization, annealing, and balance processing on the two composite models. Among them, the annealing temperature range was 300–500 K, and a total of five annealing cycles were performed. The constant-pressure and constant-temperature ensemble (NPT) was used for 100 ps balance. The temperature was set to 343 K, and the pressure was set to a standard atmospheric pressure of 1.013×10^{-4} Gpa. The force field selected the polymer consistent force field (PCFF). Finally, the canonical ensemble (NVT) was used to perform molecular dynamics simulation at 100 ps, with a step size of 1 fs. The trajectory was used to calculate interaction energy, hydrogen bond, and other parameters. The other simulation parameters is shown in Table 7.16.

Moisture is a strong polar substance, and the polarity of cellulose is obviously stronger than that of insulating oil. During the balance process of oil–paper system, moisture can diffuse into insulating paper, which is consistent with the moisture distribution law measured in practice. The molecular dynamics simulation results in this paper show that the water molecules and H_3O^+ added to the insulating oil diffuse into the insulating paper, and in the area near the coexistence of the insulating oil and the insulating paper, a large amount of H_3O^+ forms hydrogen bonds with the carbonyl group in the three-element mixed insulation oil.

In order to analyze the state of water and H_3O^+ in two kinds of insulating oil–paper systems, the interaction energy and hydrogen bond between insulating oil, insulating paper, water and H_3O^+ are calculated, as shown in Tables 7.17 and 7.18. In terms of the number of hydrogen bonds, a large number of H_3O^+ and three-element mixed insulation oil molecules form hydrogen bonds, while mineral oil cannot form hydrogen bonds with H_3O^+, and the number of hydrogen bonds between mineral oil and H_3O^+ is 0. The three-element mixed insulation oil can bind H_3O^+ through hydrogen bonds, making H_3O^+ unable to participate in the process of accelerating cellulose hydrolysis. In the three-element mixed insulation oil–paper system containing 1% H_3O^+, the total number of

Table 7.16 Simulation parameters.

Conditions	Summation methods	Conditions	Methods
Force field	PCFF	Thermostat	Nose
Van Der Waals	Atoms based	Barostat	Berendsen
Electrostatics	Eward	Boundary structure	Periodic boundary structure

Table 7.17 Hydrogen bonds and interaction energy in three-element mixed insulation oil–paper system.

Parameters	Hydrogen bond (pcs)			Interaction energy (kcal/mol)		
H_3O^+ content in the system	1%	3%	5%	1%	3%	5%
Cellulose and H_3O^+	28	59	87	−710.0	−1220.2	−1648.7
Cellulose and H_2O	2	1	0	−1.23	0.48	−0.3
Cellulose and insulation oil	0	0	0	−31.4	−15.1	−313.7
Insulation oil and H_3O^+	33	43	44	−664.5	−1004.5	−1185.6
Insulation oil andH_2O	8	2	1	−52.8	−28.1	−56.3
H_2O and H_3O^+	24	26	27	−753.5	−984.3	−979.2

Table 7.18 Hydrogen bonds and interaction energy in mineral oil–paper system.

Parameters	Hydrogen bond (pcs)			Interaction energy (kcal/mol)		
H_3O^+ content in the system	1%	3%	5%	1%	3%	5%
Cellulose and H_3O^+	29	61	106	−583.3	−1182.4	−1935.7
Cellulose and H_2O	3	5	2	−1.8	−4.0	−1.1
Cellulose and insulation oil	0	0	0	−446.0	−493.0	−640.0
Insulation oil and H_3O^+	0	0	0	61.89	137.3	260.7
Insulation oil and H_2O	0	0	0	−66.3	−67.6	−87.2
H_2O and H_3O^+	27	27	28	−879.6	−1008.1	−1036.5

hydrogen bonds formed is 71, which is 2.2 times of the total number of hydrogen bonds in the mineral oil–paper system containing 1% H_3O^+, indicating that the three-element mixed insulation oil–paper system can effectively bind H_3O^+. Therefore, three-element mixed insulation oil can slow down the aging of pressboard well.

7.5 Mechanism of Property Enhancement of the New Mixed Insulation Oil on Power Frequency Breakdown of Oil–Paper Insulation

7.5.1 Introduction

The three-element mixed insulation oil, mineral oil, and cellulose pressboard were used in this experiment. The parameters of insulation oils and pressboard are consistent with Section 7.4.1. Oil samples with varying moisture contents were obtained by moisture absorption. All the oil was first dried at 90 °C/50 Pa for 48 hours to remove most of the moisture and dissolved gas. Thereafter, some oil was taken out for moisture absorption in a sealed box with 100% relative humidity. The moisture content in oil is controlled by the length of time for moisture absorption. The moisture content for unused mineral oil specified in IEC 60296 should be less than 30 ppm, so the maximum moisture content of mineral oil in this experiment is chosen to be around 30 ppm. The moisture content of the mixed oil is based on its saturated water content. Next, all the oil samples were degassed and placed in a sealed bottle for two days to allow the water to be distributed uniformly in the oil. Finally, the AC breakdown voltage of the oil samples was tested according to IEC 60156, and the moisture contents of the oil before and after the AC breakdown measurement were measured with a Karl Fisher coulometer. The standard oil cup with a plate–plate electrode was bathed in oil to obtain different temperatures (15, 30, 45, 60, 75, and 90 °C). Each sample was tested nine times, and the results were fitted with the two-parameter Weibull distribution function, whose cumulative density function is shown in Eq. (7.10). Where t is the measured variable, denotes breakdown voltage in this paper, $F(t)$ is the probability of failure at a voltage less than or equal to t. α is the scale parameter and β is the shape parameter.

$$F(t;\alpha,\beta)=1-\exp\left\{-\left(\frac{t}{\alpha}\right)^{\beta}\right\} \tag{7.10}$$

The insulation pressboards were cut into 6 cm × 6 cm squares. The pressboards were dried at 90 °C/50 Pa for different time, and they were taken out in turn and immersed in mineral oil (11 ppm) and three-element mixed insulation oil (14 ppm) immediately. Subsequently, all samples were severally placed at 60 °C/50 Pa for 24 hours to complete oil immersion process. Thereafter, oil–paper (pressboard) samples were separately sealed in glass bottles and placed in room temperature (25 °C) for 20 days. Finally, the moisture content was measured using a Karl Fisher coulometer in accordance with IEC 60814. The moisture contents of insulation paper and pressboard, which were dried for different time, were 0.9% (0.85 ~ 0.95%), 1.6% (1.5 ~ 1.7%) and 4.3% (4.2 ~ 4.4%), respectively.

Different test temperatures in this experiment (25–90 °C) were controlled with oil bath. Two oil bath boilers were used to control the test temperature. The one kept the samples at the temperature to be measured, and the other one was used to keep the test cell at the same temperature. Silicone oil was used in both oil bath boilers, in which a temperature sensor existed to monitor the temperature of the silicone oil. The test cell is made of stainless steel and has good thermal conductivity, thus mineral oil or mixed oil in the test cell, depending on the oil to be tested, can be heated by oil bath. The temperature of the insulation oil in the test cell was measured using a thermometer with ± 0.5 °C error range. Measurement on the temperature in test cell was conducted every three breakdown tests to make sure the temperature was constant. Due to the heating of the oil bath, the temperature in the test cell kept unchanged throughout the test. The AC breakdown voltage of the oil-immersed insulation paper (pressboard) was tested using an equal diameter electrode in accordance with IEC 60243 [42]. The rise rate of voltage chose 1 kV/s. three-minute interval was set between two breakdown tests.

7.5.2 Oils Breakdown Voltage with Different Moisture Contents

Figure 7.6 shows the measurement of AC breakdown voltage for the three-element mixed insulation oil and mineral oil at 45 °C. The moisture contents of the oils before and after the AC breakdown test are also presented. Take the black line in Figure 7.6a as an example. The moisture contents before and after the test are 9.3 and 9.8 ppm, respectively. The moisture measurement indicates that moisture content undergoes minimal changes in the test process. Figure 7.6b uses the same expression method and shows same trends. Therefore, the moisture content before the test was adopted in this work to conveniently represent moisture content. As moisture content in oil increases, the breakdown voltages of the mineral oil and the three-element mixed insulation oil decrease. Specifically speaking, as the moisture content in oil increases, the breakdown voltage of the insulation oil initially drops slowly, drops drastically when the moisture content exceeds a certain value, and finally it drops slowly as breakdown voltage is already low. In summary, almost all insulation oils conform to this law [43].

Scale parameter (the value at 63.2%) of the Weibull distribution can be regarded as the average value of insulation failure, and it is redrawn in Figure 7.6c to compare the average breakdown voltages of these two oils. The AC breakdown voltage of the novel three-element mixed insulation oil is substantially higher than that of the mineral oil at the same absolute water content. Take the result at 60 °C as an example. The breakdown voltage of mineral oil with 43 ppm is about 35 kV, whereas that of the mixed insulation oil is about 65 kV at the same moisture content. The moisture content in the three-element mixed insulation oil is about 125 ppm when breakdown voltage decreases to 35 kV.

For water in oil, it can be divided into four states, dissolved water, suspended fine droplets, thin water film, and water drops [44]. Breakdown voltage discussed here is based on short and uniform oil gaps, which is mainly affected by dissolved water [44]. Water molecules are polar molecules, thus the increase of moisture content in oil, especially for isolated water molecules, will lead to an increase in dipole moment [45], resulting in a significant increase in the ionic conductivity of the insulation oil, thereby decreasing the breakdown voltage [45, 46]. As a result, the breakdown voltage of both the oils decrease with increasing moisture content.

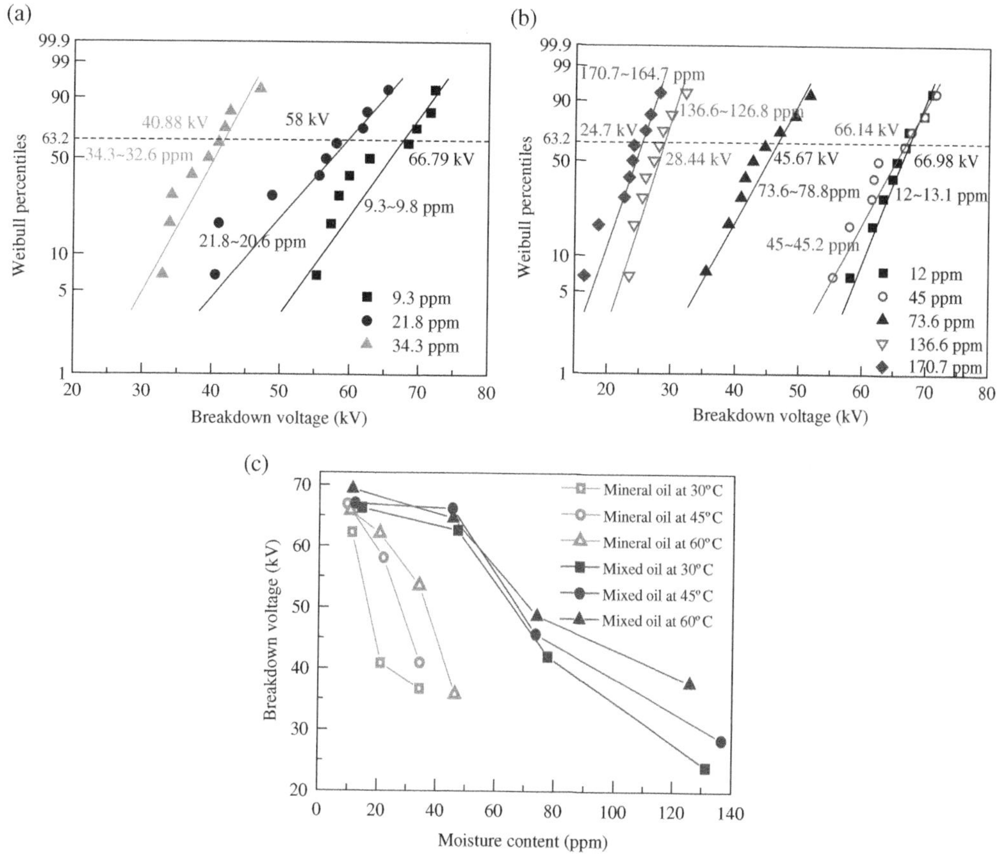

Figure 7.6 Weibull distribution of AC breakdown voltage of mineral oil and the three-element mixed insulation oil at 45 °C. (a) Mineral oil, (b) three-element mixed insulation oil, (c) breakdown voltage comparison.

7.5.3 Oils Breakdown Voltage with Different Temperatures

For comparison of the results at different temperatures, the breakdown voltages at 63.2% in the Weibull distribution of the parts of the oil samples are redrawn in Figure 7.7. The results show that increasing temperature increases the breakdown voltages of both oils, but the increasing trend is slightly different for insulation oil with different moisture contents. For the oil samples with low moisture content (<15 ppm), the breakdown voltage slightly increases as temperature rises, and the breakdown voltage of mixed insulation oil is slightly higher than that of mineral oil at each temperature point. A low moisture content (<15 ppm) keeps the breakdown voltages of both oils relatively high even at 15 °C; thus, not much room is available to augment the breakdown voltage because there is an upper limit to the breakdown voltage for a dielectric liquid. Note that breakdown voltage slightly decreases when the temperature exceeds 75 °C. However, for the oil samples with high moisture content (mineral oil 33 ppm and the mixed oil 77 ppm), the breakdown voltage is relatively low at 15 °C. With the increase of temperature, breakdown voltages of both oils continuously increase, and they do not show any decline when the temperature reaches 90 °C.

To date, it is currently widely accepted that the breakdown voltage of insulation oil is determined by the relative moisture content rather than absolute water content [6, 47]. Absolute moisture

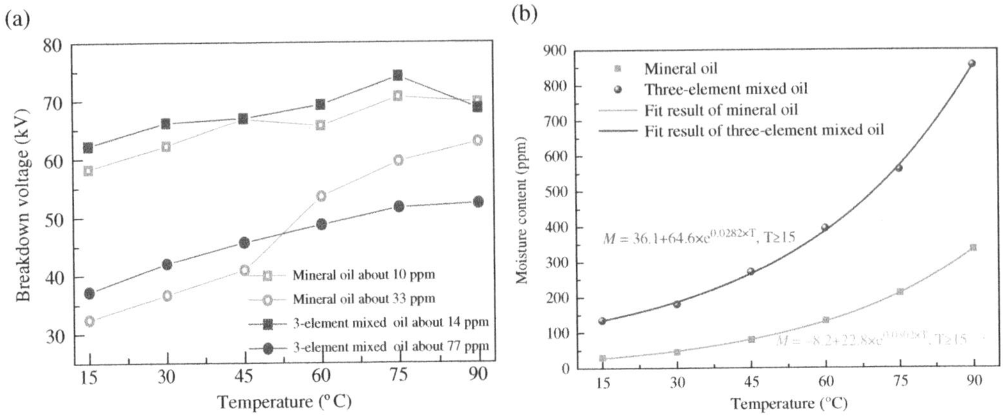

Figure 7.7 Comparison of Weibull breakdown voltages and saturated water content of mineral oil and three-element mixed insulation oil at different temperatures. (a) Comparison of Weibull breakdown voltages, (b) saturated water. *Source:* Based on Liao et al. [2].

content is the actual moisture content present in oil, and it can be measured with a Karl Fisher coulometer in accordance with IEC 60814. The saturated moisture content is the maximum dissolved moisture content in a certain temperature and pressure. Relative moisture content is calculated from the absolute moisture content, which is the result of the absolute moisture content divided by the saturated moisture content [6]. With the increase of temperature, saturated water content of the both insulation oils increase, which are presented in Figure 7.7 [2]. For the insulation oil with a certain absolute water content, relative water content decreases when the temperature rises, increasing the breakdown voltage of the insulation oil. This is the reason for the phenomenon in Figure 7.7. The breakdown voltage of insulation oil with low moisture content slightly decreases when the temperature exceeds 75 °C in Figure 7.7. With the increase of temperature, insulation oil is more likely to generate liquid vapor, resulting in low-density areas and bubbles between the electrodes, which facilitates the formation of streamer and finally lead to breakdown [44]. Comparatively speaking, temperature has minimal influence on breakdown voltage as liquid vapor cannot easily form and affect breakdown voltage as directly as moisture in oil could. Therefore, the breakdown voltage continuously increases as the temperature rises when moisture content in oil is relatively high.

7.5.4 Oil Breakdown Voltage Under Combined Effects of Moisture and Temperature

Figure 7.8a and b show the combined effects of moisture and temperature on the breakdown characteristics of mineral oil and three-element mixed oil, respectively. The stereogram demonstrates that the three-element mixed insulation oil is able to maintain breakdown strength with little decline when the moisture content is below 50 ppm. When the moisture content continues to increase, this invariability is disrupted, and the breakdown voltage rapidly declines with the increase in moisture content. Therefore, moisture content in the three-element mixed insulation oil is best controlled below 50 ppm by oil filters to maintain excellent performance in actual engineering. This interesting phenomenon cannot be observed in mineral oil. The difference between mixed insulation oil and mineral oil is that the former contains vegetable oil, which is composed of triglycerides and is different from the molecule structure in mineral oil. In other words, this

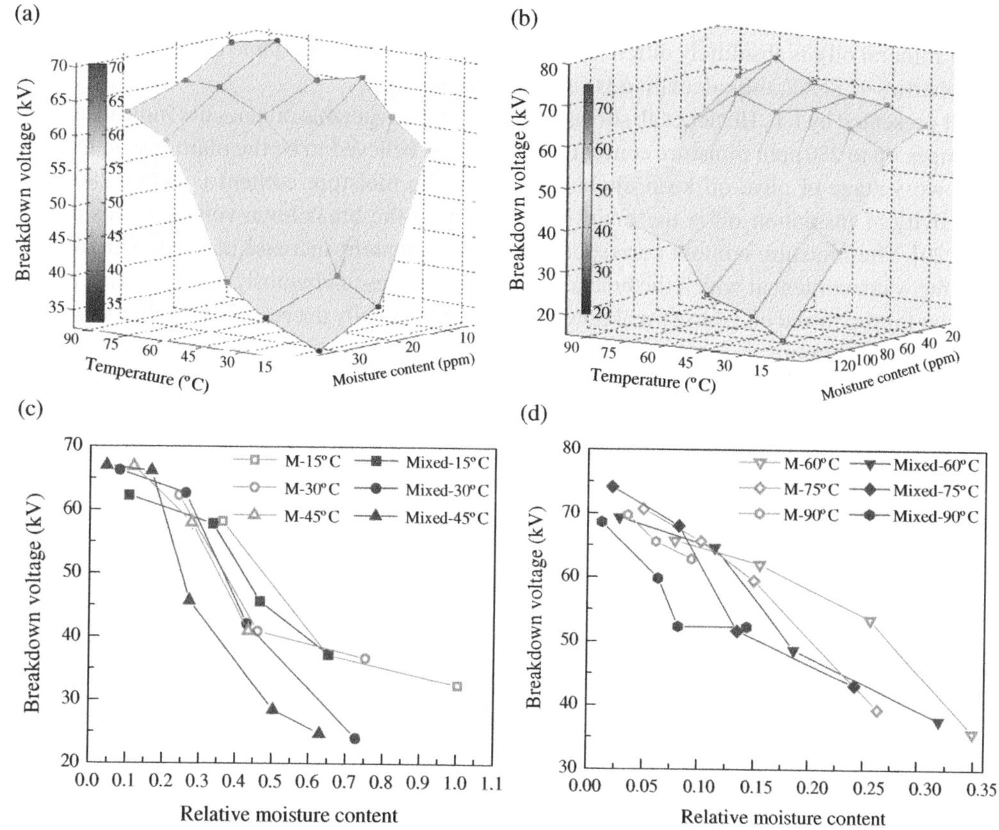

Figure 7.8 AC breakdown voltage of insulation oil with the combined effects of moisture and temperature and relative moisture content. (a) Mineral oil, (b) three-element mixed insulation oil, (c) mineral oil, (d) three-element mixed insulation oil.

property that keeps breakdown voltage hardly decrease within a certain moisture content is the characteristic of vegetable oils. Since the three-element mixed oil contains vegetable oil, the three-element mixed oil also has the above characteristics. This is also the main reason that three-element mixed oil has apparently higher breakdown voltage than mineral oil.

Therefore, the initial moisture content at which the breakdown voltage begins to drastically decrease is related to the amount of vegetable oil in the mixed insulation oil. Changing the proportion of vegetable oil in mixed insulation oil alters the initial moisture content. The high proportion of vegetable oil in mixed insulation oil equates to a high initial moisture content. However, the maximum proportion of vegetable oil in mixed insulation oil is close to 100%. The three-element mixed insulation oil contains 19% soybean oil, 5% PFAE, and 76% mineral oil. The results in Figure 7.8 show that the initial moisture content for the three-element mixed insulation oil is about 50 ppm. If the proportion of vegetable oil is increased to 100%, that is, the proportion of vegetable oil is expanded to 4–5 times of that in the three-element mixed insulation oil, then the initial moisture content should also increase by 4–5 times, which is about 200–250 ppm. In other words, the initial moisture content of vegetable oil at which the breakdown voltage starts to drop drastically should be in the range of 200–250 ppm. This inference is consistent with the experimental results of I. L. Hosier et al., which show that initial moisture content in olive oil at which the breakdown voltage begins to drop drastically is about 280 ppm [22]. Note that PFAE is a modified ester whose performance, such as low viscosity and low thermal decomposition temperature (near to those of mineral oil), is absolutely different from that of traditional vegetable oils. Given the unusual properties of PFAE, the calculation is vague.

Result presented by I. L. Hosier et al. shows conductivity of vegetable oil does not indicate apparent changes up to 280 ppm moisture content [22], which is believed to be the main reason for that breakdown voltage of olive oil keep unchanged when the moisture content is below 280 ppm. Conductivity of insulation oil is the main factor affecting the breakdown voltage of insulating oil [19, 20]. The moisture content corresponding to the apparent increase of oil conductivity is exactly the water content at which the breakdown voltage begins to obviously decrease. This is also the reason for the mixed oil whose breakdown voltage dramatically decreases only when moisture content exceeds 50 ppm.

For transformers (voltage class: 35 kV and below) immersed in mineral oil, operation regulations in China generally stipulate that breakdown voltage should not be lower than 35 kV during the use of mineral oil, and the corresponding moisture content should be 30 ppm. In the three-element mixed insulation oil, the moisture content should be controlled within roughly 80 ppm under the same requirements. Such setting greatly increases the margin of moisture content in the insulation oil. According to the results in Figure 7.8a and b, the relative moisture contents of insulation oil can be calculated. Figure 7.8c and d shows the breakdown voltage of the two insulation oils varying with relative moisture contents at different temperatures. Assume that the breakdown voltage of insulating oil is thoroughly determined by relative moisture content, the breakdown voltages of oil at the same relative moisture content should be close. Figure 7.8 demonstrates that, at the same temperature, breakdown voltages of the two oils are nearly the same at same relative moisture content, especially at low temperatures. However, when the temperature increases, the breakdown voltage of mixed oil is lower than that of mineral oil at high relative moisture contents, particularly 90 °C at which the breakdown voltage of mixed oil is lower than that of mineral oil all the time.

For the same relative moisture content, the absolute moisture content in mineral oil being apparently lower than that in mixed insulation oil is mainly a result of the low saturated moisture content in mineral oil. When the relative moisture content is low, the two oils have similar breakdown strengths because the mixed insulation oil possesses the ability to maintain its breakdown strength within a certain range of absolute moisture content. However, the ability to maintain breakdown strength is limited for mixed oil. When it comes to high relative moisture content, the mixed oil has a significantly larger ionic conductivity because of high absolute moisture content [45], leading to faster breakdown. Thus, it can be concluded that the breakdown voltage of insulating oil is roughly related to relatively moisture content at the same temperature, but other factors also affect, such as absolute moisture content and property of insulation oil.

7.5.5 Comparison of AC Breakdown Characteristics of Composite Insulation with Different Temperatures and Moisture Contents

With the change in temperature and moisture content, the breakdown voltages of three-element mixed insulation oil and mineral oil immersed papers (pressboard) present the same change trend, as shown in Figure 7.9a. That is, for low moisture content, increasing the temperature slightly decreases the breakdown voltage of oil-immersed paper (pressboard). Only with the apparent simultaneous increase in temperature and moisture content, the breakdown voltage of the oil-immersed paper (pressboard) decreases significantly. Therefore, taking mixed oil-immersed pressboard as an example, the effects of temperature and moisture content on the breakdown voltage are analyzed, especially for the comparison of 0.9 and 4.3% moisture content.

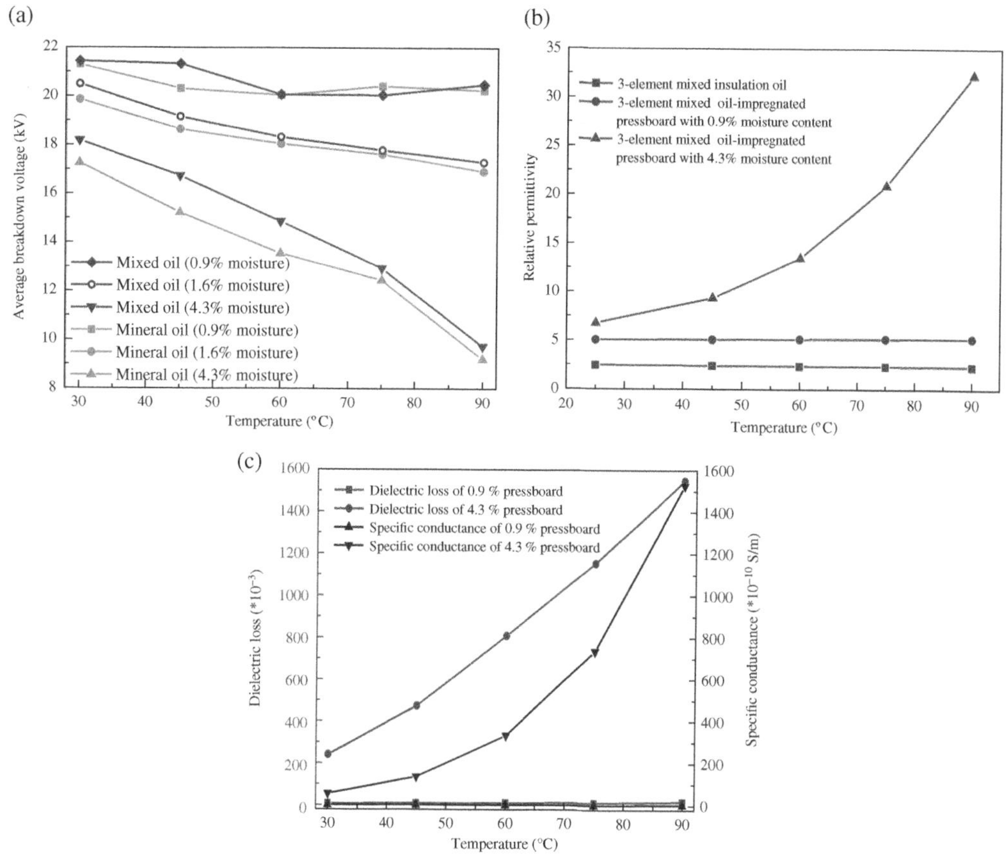

Figure 7.9 Breakdown voltage, relative permittivity, dielectric loss, and specific conductance of pressboard varying with relative moisture content at different temperatures. (a) Breakdown voltage, (b) relative permittivity, (c) dielectric loss and specific conductance.

The breakdown of oil-immersed pressboard is a complex process that is affected by numerous factors. On one hand, relative permittivity is one of the main factors that affect the breakdown voltage, as shown in Figure 7.9b. The relative permittivity of the oil slightly changes with the temperature, mainly due to the increasing thermal motion decreases the orientational polarization of oil molecules [2]. Relative permittivity slightly increases for oil-impregnated pressboard with low moisture content (0.9%). The relative permittivities of solid dielectrics change differently with the liquid dielectrics as temperature increases, mainly due to solid molecules are bound. Increase in temperature enhances motion of solid molecules, allowing these molecules in solid to undergo orientational polarization, thus the relative permittivity increases [48]. Water molecules are strong polar molecules, and increasing moisture content increases the relative permittivity of pressboard, especially for high temperatures, at which the relative permittivity of oil-impregnated pressboard (4.3%) apparently increases.

For oil–paper composite insulation, the relative permittivity of insulation paper is higher than that of insulation oil, thus the insulation oil undergoes higher field strength when subjected to a certain voltage. However, the breakdown strength of liquid dielectric is generally lower than that of solid dielectric [49], thus the insulation oil becomes a weak point in oil–paper composite insulation, which may cause partial discharge or breakdown, and then the breakdown of the oil–paper composite insulation. For pressboards with 0.9% moisture content, the difference in relative permittivity between pressboard and insulation oil increases slightly with increasing temperature, and electric field becomes more uneven, thus breakdown occurs more easily and breakdown voltage slightly decreases. For 4.3% pressboards, the difference in relative permittivity increases dramatically with increasing temperature, resulting in an apparent uneven electric field distribution, thus the breakdown voltage is significantly reduced. The variation of the relative permittivity is consistent with the variation of the breakdown voltage of oil-immersed pressboard, that is, it changes apparently only when the temperature and moisture content significant increase simultaneously.

On the other hand, the increase in conductance loss and dielectric loss is the other major reason for pressboard breakdown. The dielectric loss and specific conductance of pressboard were measured with broadband dielectric spectrum, and the results at 50 Hz were shown in Figure 7.9c. Thermal breakdown is a major breakdown form of solid insulation, and A. Küchler proposed that long durations (more than one second) with high voltage that causes breakdown is considered to be thermal breakdown, during which the heat generated in the dielectric can lead to thermal destruction of the insulation material [49]. The rise rate of voltage is 1 kV/s for 0.45 mm pressboard, which suggests the voltage acted on the pressboard is high during the several seconds before breakdown. Heat is continuously generated by polarization and conduction losses in an alternating electric field. The pressboard experiences difficulty in dissipating heat due to high test temperature, leading to heat accumulation and a further increase in temperature of the pressboard. Heat accumulation leads to carbonization, damage of the pressboard, and then thermal breakdown occurs easily [50]. The increase of the moisture content in the pressboard causes a sharp increase in dielectric and conductance losses. During the breakdown process, the heat in the pressboard will be more and cause breakdown. When moisture content and temperature increase at the same time, a large amount of heat is generated and is difficult to dissipate, which is more likely to cause carbonization, damage of the pressboard, and then thermal breakdown becomes extremely easy, thus the breakdown voltage drops dramatically [51].

A COMSOL electric field simulation of the test electrode is shown in Figure 7.10, in which Figure 7.10a is the main view of the test electrode and Figure 7.10b is the simulation result of electric field at the chamfering. The transient result shows that the electric field in the oil at the chamfering of the electrode is high. In the plane where the upper surface of the pressboard is located, the electric field distribution in the radial direction from the electrode center is shown in Figure 7.10c. The relative permittivity of insulation oil is lower than that of a pressboard; thus, oil withstands a high electric field at the chamfering of the electrode. When the same voltage is applied, the electric field of the three-element mixed insulation oil at the chamfering of the electrode is relatively small due to high relative permittivity. The results in Figure 7.10c show that maximum electric field of the three-element mixed insulation oil (1.58×10^4 V/mm) is slightly lower than that of the mineral oil (1.67×10^4 V/mm) at the same time. The high electric field in the oil causes a partial discharge on the upper and lower surfaces of the pressboard and may thus cause damage to the insulation pressboard and lead to breakdown [50]. The breakdown strength of the three-element mixed insulation oil is considerably higher than that of the mineral oil, and the electric field around the chamfering is lower than that of the mineral oil. Thus, partial discharge and breakdown are incredibly difficult to produce the mixed insulation oil at the chamfering, which increases the breakdown voltage of the mixed insulation oil-impregnated pressboard.

The dielectric loss of the pressboard impregnated with the three-element mixed insulation oil is measured, and the results are shown in Table 7.19. The dielectric loss substantially increases as moisture content and temperature increase, which may result in a large amount of heat generated by the pressboard in an alternating electric field. Heat accumulation leads to carbonization,

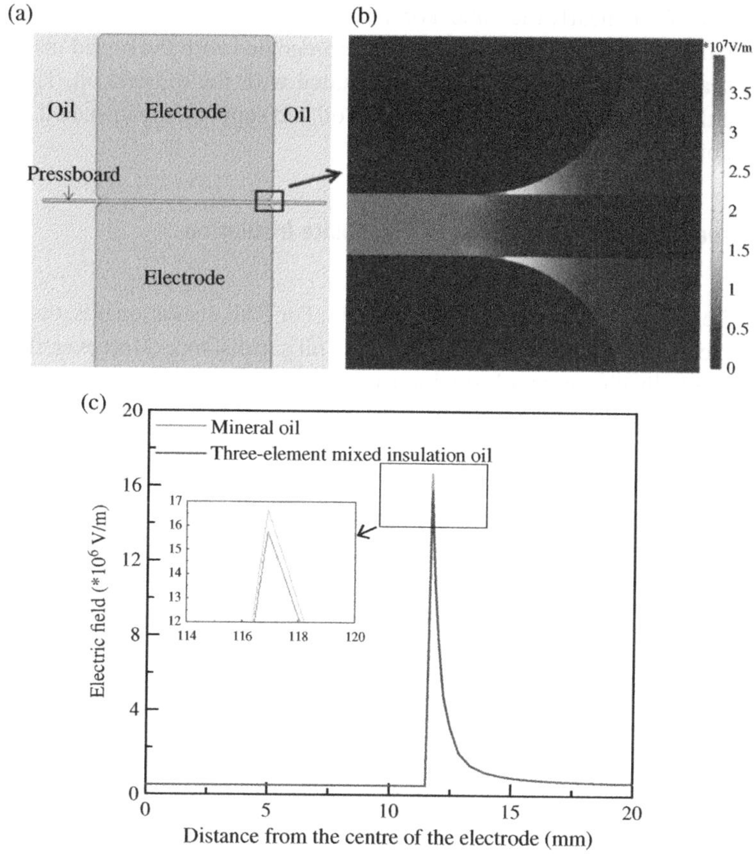

Figure 7.10 Electrode structure and the transient simulation result of electric field distribution at the chamfering of the electrode and electric field distribution in the radial direction from the electrode center for the three-element mixed insulation oil. (a) Electrode structure, (b) electric field at the chamfering, (c) electric field distribution.

Table 7.19 Dielectric loss of the new three-element mixed insulation oil-impregnated pressboard at different temperatures.

Temperature (°C)	0.9% moisture content	4.3% moisture content
30	7.77×10^{-3}	240.1×10^{-3}
45	8.75×10^{-3}	472.4×10^{-3}
60	10.3×10^{-3}	807.8×10^{-3}
75	12.5×10^{-3}	1154×10^{-3}
90	18.4×10^{-3}	1549.8×10^{-3}

damage of the pressboard, and thermal breakdown. Increased temperature causes difficulty in heat dissipation; thus, the heat easily accumulates in the pressboard, and thermal breakdown occurs. We can infer that when the moisture content is less than 0.9%, which reflects the situation in field transformers, the breakdown characteristics of the pressboards impregnated with the mixed insulation oil and mineral oil are nearly the same. For the pressboard with moisture content between 0.9 and 4.3%, the breakdown voltage of the pressboard impregnated with the mixed insulation oil is slightly higher than that of the pressboard impregnated with the mineral oil. This result shows good property of the three-element mixed insulation oil and supports the mixed oil as a direct substitute of mineral oil in oil transformers.

7.5.6 Comparison of AC Breakdown Characteristics of Composite Insulation with Oil Gap

The Weibull distribution of the test results are shown in Figure 7.11. For both insulation oils, the α increases slowly and then increases apparently with the increase in oil gap distance. Moreover, the slope of the Weibull fitting gradually changes with the increase in oil gap distance. The slope in Weibull fitting is related to dielectric materials because different materials have various breakdown error ranges. The breakdown voltage range of the 0.108 mm insulation paper is approximately 1.5 kV according to the measurement result in Figure 7.11, whereas that of insulation oil may reach 20 kV or even 30 kV. Small breakdown voltage range indicates large slope of the fit line in the

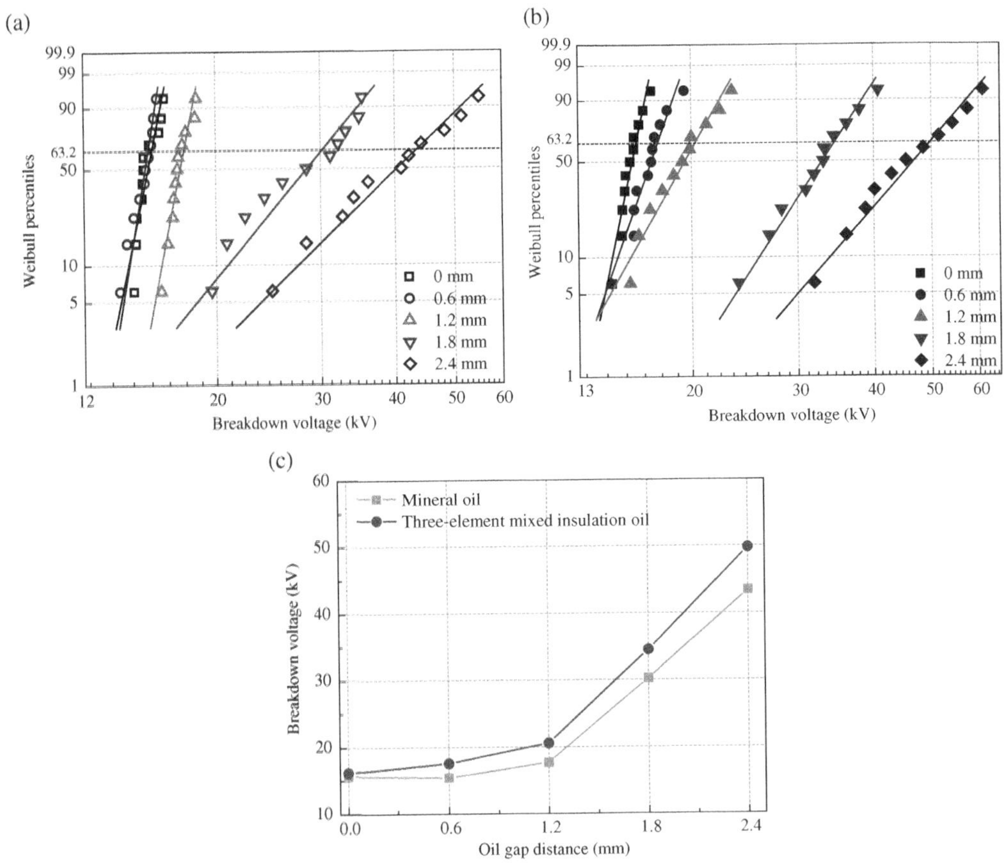

Figure 7.11 Breakdown voltages of oil-immersed pressboards with different oil gap distances. (a) Pressboard immersed with mineral oil, (b) pressboard immersed with three-element mixed insulation oil. (c) Breakdown voltage of pressboard under different oil gap distance.

Weibull distribution. Therefore, the change of the fit line slope in Figure 7.11 suggests that the breakdown voltage is initially decided by the oil-immersed pressboard, followed by the combined structure of pressboard and oil gap, and eventually by the insulation oil.

For comparison, Figure 7.11c shows the α obtained from Figure 7.11a and b. The breakdown voltage of the mixed oil-immersed pressboard with oil gap (for the same oil gap) is higher than that of the mineral oil-immersed pressboard with oil gap within the scope of the experiment. This result shows the superiority of the breakdown performance of the pressboard immersed by the three-element mixed insulation oil.

To clarify this phenomenon, a COMSOL simulation about electric field around the electrode is performed. The relative permittivity of insulation materials used in the simulation is measurement result at 25 °C, which are shown in Table 7.20. The applied voltage is a 50 Hz sinusoidal voltage. Figure 7.12a–d shows the electric field distribution in the insulation oil gap and oil-immersed pressboard along the radius of the electrode at the same simulation time, at which the applied voltage amplitude is approximately 3.536 kV. As the oil gap distance increases, the field strength in the insulation oil and in the pressboard decreases at the same electrode voltage. The electric fields at the middle of the oil gap and the pressboard are counted, as shown in Table 7.21.

Table 7.20 Relative permittivity of insulation materials.

Materials	Relative permittivity
Mineral oil	2.195
Mixed oil	2.44
Mineral oil-immersed pressboard	5
Mixed oil-immersed pressboard	5.05

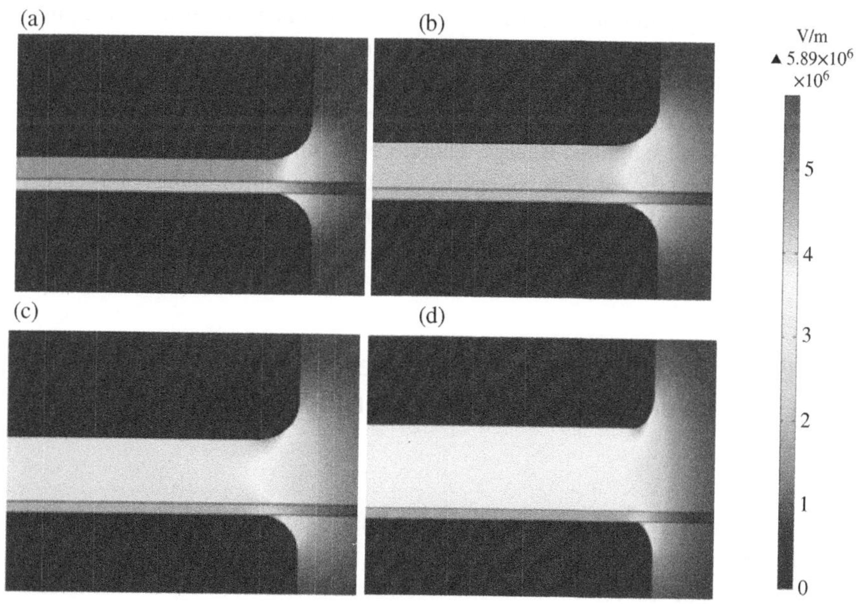

Figure 7.12 Transient simulation result of electric field distribution at the chamfering of the electrode for the three-element mixed insulation oil. (a) 0.6 mm, (b) 1.2 mm, (c) 1.8 mm, (d) 2.4 mm.

Table 7.21 Electric fields (kV/mm) at the middle of the oil gap and the oil-immersed pressboard.

Distance of oil gap (mm)	0.6	1.2	1.8	2.4
Mineral oil	4.833	2.655	1.83	1.397
Mixed oil	4.747	2.629	1.818	1.389
Pressboard immersed by mineral oil	2.121	1.166	0.804	0.613
Pressboard immersed by mixed oil	2.293	1.27	0.878	0.671

The calculation results in Table 7.21 indicate that in addition to the results presented in Figure 7.12, the electric field strength ratio of the pressboard to the insulation oil is the same, which is only determined by the relative dielectric constant of the two dielectrics instead of the oil gap distance. The electric field strength of the insulation oil is evidently higher than that of the pressboard due to higher relative permittivity of the pressboard; however, the breakdown field strength of the liquid dielectric (insulation oil) is always lower than that of the solid dielectric (pressboard) [49]. Therefore, the breakdown initially occurs in the insulation oil in this structure. Comparison in Table 7.21 shows that the electric field of the mixed insulation oil is smaller than that of the mineral oil due to the larger relative permittivity of the mixed oil, and the breakdown strength of the mixed oil is larger than that of the mineral oil [2]. Because mineral oil is easier to breakdown than three-element mixed oil, its structure of oil gap and pressboard is more easy to breakdown than that of three-element mixed oil. In comparison, the electric field is relatively evenly distributed between pressboard and the mixed oil, thus the mixed oil has a higher breakdown voltage in this structure. The three-element mixed insulation oil can match the insulation paper for improved insulation effect due to its high relative dielectric constant.

The preceding electric field distribution in insulation oil and pressboard can also address the slope of the fit line in Figure 7.11. When the oil gap is small, the breakdown voltage of the oil gap is also small. Subsequently, the oil gap initially breaks down, and all the voltages then act on the pressboard. However, such voltage cannot result in the breakdown of the pressboard, and breakdown occurs when the voltage further rises. Therefore, the error range is determined by the breakdown of the pressboard when the oil gap is small. By contrast, although the oil gap also breaks down initially when the oil gap is large, the voltage applied is large. Therefore, the entire voltage after the break-down of the oil gap acts on the pressboard. This voltage exceeds the breakdown voltage of the pressboard, which causes the pressboard to breakdown instantaneously. Therefore, the error range is determined by oil when the oil gap is large, which changes the slope of the fitted line (Figure 7.11).

7.6 Enhancing Effect and Mechanism of the New Mixed Insulation Oil on Flashover Voltage of Oil–Paper Insulation

7.6.1 Introduction

The three-element mixed insulation oil, mineral oil, and cellulose pressboard were used in this experiment [52–57]. The parameters and preprocessing step of insulation oils and pressboard are consistent with Section 7.4.1. The test cell contains two electrode structures, which are the needle-plate and finger-finger electrodes made by copper. In the internal structure of the transformer, slightly non-uniform electric field exists between adjacent windings, the low-voltage winding, and the core shield, while the extremely non-uniform electric field is the most common electric field in transformer. Therefore, needle-plate electrode and finger-finger electrode are used in this experiment to simulate extremely non-uniform electric field and slightly non-uniform electric field, respectively, in order to study the surface flashover characteristics of three-element mixed insulation oil-insulation pressboard (3EMO-IP) and mineral oil-insulation pressboard (MO-IP). The experimental temperature was 25 °C, and the internal fluid volume of test cell controlled at 1.5 l to completely impregnate the electrode. After every 20 times of flashover, new electrodes will be used to reduce the influence of shape changes on the experiment.

The oil-impregnated pressboard was removed from oil and placed on the epoxy block. The electrodes were placed with axis coincidence, and the distance between the electrodes was set at planned value. After the dried insulating oil was filled into the plexiglass test cell, the bolts above the electrode were adjusted to closely fit the electrode and pressboard. First, surface flashover voltage of MO-IP and 3EMO-IP were obtained under needle-plate and finger-finger electrode, respectively. The distances of oil gap between positive and negative electrode varied from 5 mm up to 20 mm with 5 mm increment. The rise rate of AC voltage was controlled at 3 kV/s, and each sample was tested 15–18 times. Second, after 30 times flashover at 10 mm oil gap under needle-plate and finger-finger electrode, the surface damage of cellulose insulation pressboard was studied by microscope. The observation position focused on the pressboard surface at the tip of needle and finger electrode, and the microscope was OLYMPUS-BX53 at the magnification of 10x.

7.6.2 Surface Flashover Voltage of Oil-Cellulose Insulation Pressboard

The surface flashover voltage under needle-plate electrode and finger-finger electrode at different electrode distance were fitted with the two-parameter. The Weibull distribution of surface flashover voltage for the 3EMO-IP and MO-IP under needle-plate electrode at different oil gap distance is presented in Figure 7.13. The R-Square (R^2) of each Weibull distribution obtained by linear fitting is also shown in Table 7.22. As the oil gap between needle electrode and plate electrode increases, the Weibull distribution line moves from left to right direction. This means that the surface flashover voltages of the two kinds of oil–cellulose insulation pressboard show upward trend. The values of R^2 were basically above 0.85, which demonstrate the surface flashover voltage was fitted with the two-parameter Weibull distribution, and the equivalent surface flashover voltage values can be obtained at any certain failure probability. In order to compare the surface flashover voltage of pressboard immersed in 3EMO and MO clearly, the equivalent surface flashover voltages at 5% probability in the Weibull distribution and the average values are shown in Table 7.22 and Figure 7.13s. Compared with MO-IP, the surface flashover of 3EMO-IP occurs at higher voltage. Take the result at oil gap 10 mm as an example, the flashover voltage at 5% probability and average values of 3EMO-IP is 15.90 and 8.36% higher than those of MO-IP, respectively.

The Weibull distribution of surface flashover voltage of 3EMO-IP and MO-IP under finger-finger electrode is presented in Figure 7.14. Similar to the needle-plate electrode, the surface flashover voltage of oil–cellulose insulation pressboard has a significant increase with the distance of oil gap. Figure 7.14c and Table 7.23 show the equivalent surface flashover voltages at 5% probability and the average values. Compared with MO-IP, 3EMO-IP has higher flashover voltage at the same oil gap. For instance, for 3EMO-IP, its equivalent flashover voltages at 5% probability are 1.05 and 1.07 times higher than those of MO-IP at 10 and 20 mm, as shown in Figure 7.14. Besides, the difference in surface flashover voltage between MO-IP and 3EMO-IP is more significant under needle-plate electrode. According to the average of the differences between MO-IP and 3EMO-IP at the four gap distances, the equivalent flashover voltage of 3EMO-IP is 12.02% higher than that of MO-IP under needle-plate electrode, while this number decrease to 4.91% under finger-finger electrode.

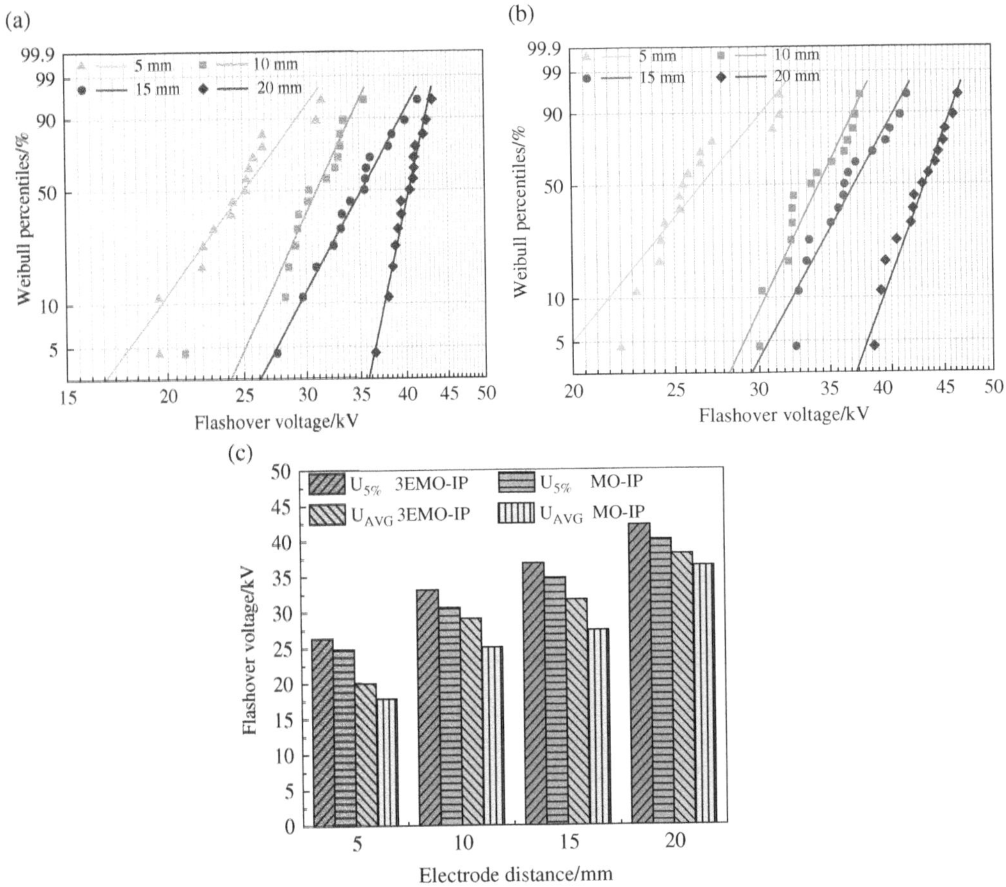

Figure 7.13 Weibull distributions of surface flashover voltages and comparison of surface flashover voltages under needle-plate electrode. (a) Surface flashover voltages of MO-IP, (b) surface flashover voltages of 3EMO-IP, (c) comparison of surface flashover voltages.

Table 7.22 The equivalent flashover voltages at 5% probability and average values of 3EMO-IP and MO-IP under needle-plate electrode.

	3EMO-IP			MO-IP		
Gap/mm	U_{AVG}/kV	$U_{5\%}$/kV	R^2	U_{AVG}/kV	$U_{5\%}$/kV	R^2
5	26.29	20.05	0.86	24.82	17.88	0.89
10	33.17	29.08	0.96	30.61	25.09	0.79
15	36.81	31.68	0.97	34.81	27.44	0.96
20	42.22	38.08	0.98	40.15	36.41	0.96

The streamer develops in insulating oil, along the oil–pressboard interface. So the breakdown property of insulating oil is also an essential factor for surface flashover process. The Weibull distribution of breakdown voltage for 3EMO and MO under needle-plate electrode and finger-finger electrode are presented in Figure 7.15, and the oil gap between positive and negative electrodes

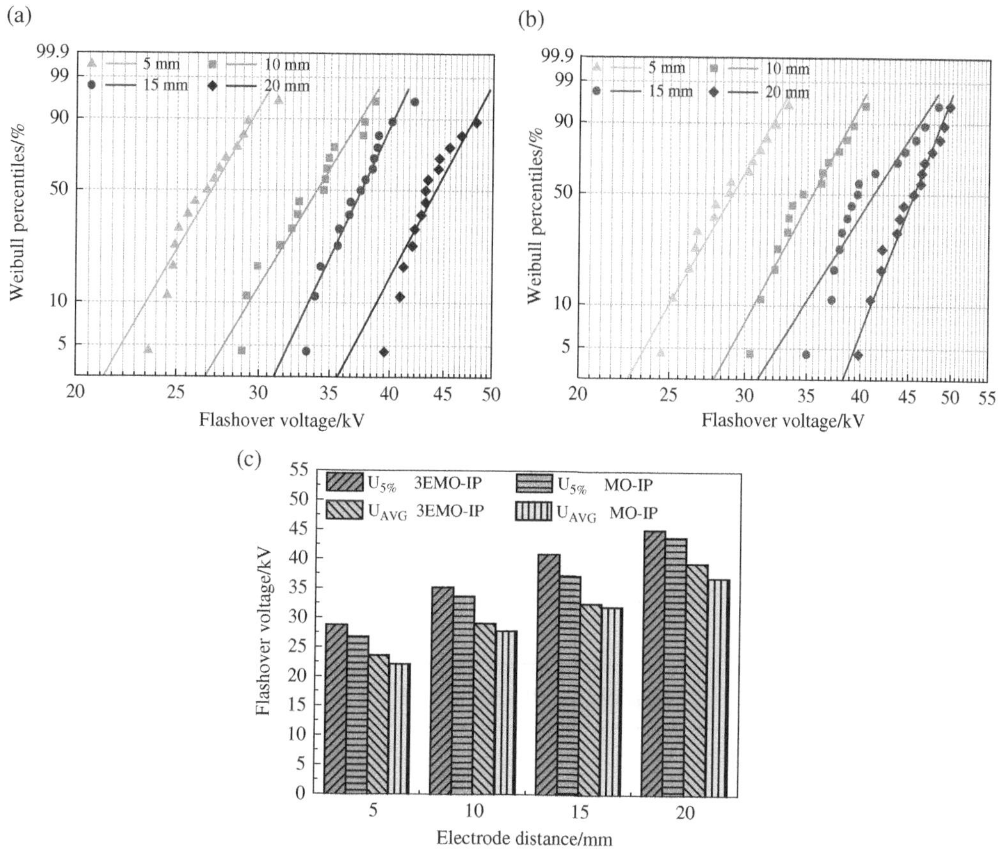

Figure 7.14 Weibull distributions of surface flashover voltages and comparison of surface flashover voltages under finger-finger electrode. (a) MO-IP, (b) 3EMO-IP. (c) Comparison of surface flashover voltages.

Table 7.23 The equivalent flashover voltages at 5% probability and average values of 3EMO-IP and MO-IP under finger-finger electrode.

	3EMO-IP			MO-IP		
Gap/mm	U_{AVG}/kV	$U_{5\%}$/kV	R^2	U_{AVG}/kV	$U_{5\%}$/kV	R^2
5	28.77	23.57	0.99	26.75	22.14	0.96
10	35.11	29.05	0.98	33.65	27.75	0.97
15	40.85	32.52	0.84	37.19	31.96	0.96
20	45.13	39.39	0.98	43.89	36.90	0.88

was set at 10 and 20 mm, respectively. The equivalent surface flashover voltages at 5% probability and the average values are shown in Table 7.24. Due to the greater distortion of the electric field, 3EMO and MO have lower breakdown voltages under needle-plate electrode than those under finger-finger electrode. Compared with MO, the breakdown voltage of 3EMO is higher, which is an important reason for higher surface flashover voltage of 3EMO-IP.

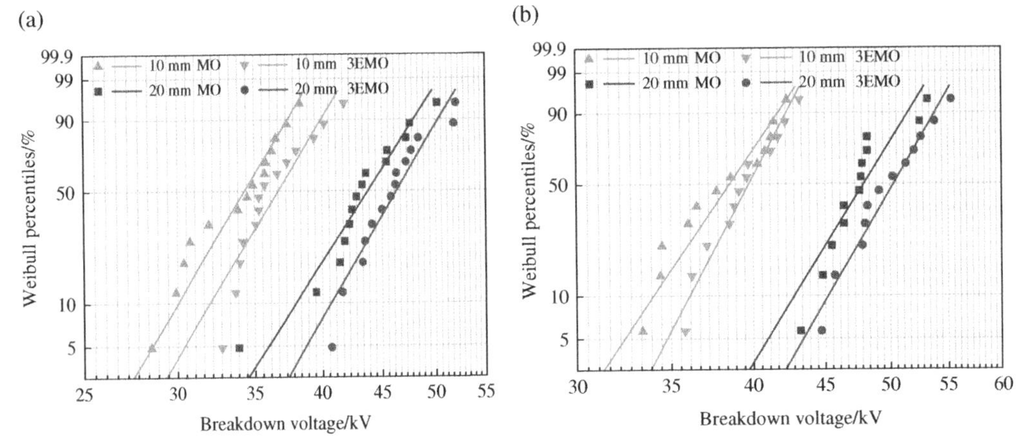

Figure 7.15 Weibull distributions of breakdown voltages under needle-plate and finger-finger electrode. (a) Needle-plate electrode, (b) finger-finger electrode.

Table 7.24 The equivalent breakdown voltages at 5% probability and average values of 3EMO and MO.

		3EMO			MO		
Electrode	**Gap/mm**	U_{AVG}/kV	$U_{5\%}$/kV	R^2	U_{AVG}/kV	$U_{5\%}$/kV	R^2
Needle-plate	10	36.39	30.45	0.96	33.91	28.50	0.79
Electrode	20	45.97	38.75	0.98	43.34	35.93	0.96
Finger-finger	10	39.43	34.77	0.98	38.09	32.36	0.97
Electrode	20	49.89	43.42	0.98	47.63	40.90	0.88

7.6.3 Surface Flashover Difference Analysis

The distribution of electric field strength between insulating oil and insulating paperboard can be calculated in Eq. (7.11) [58].

$$E_{Oil} / E_p = \varepsilon_{rp} / \varepsilon_{roil} \tag{7.11}$$

where ε_{roil} and ε_{rp} are the relative permittivity of insulating oil and oil-impregnated pressboard. E_{oil} and E_p are the electric field intensities of oil and insulating pressboard. Because the ratio of oil to pressboard in relative permittivity is smaller, the distortion at oil–pressboard interface in 3EMO-IP is less than that in MO-IP.

A COMSOL electric field simulation of the two kind of test electrodes for the two kinds of oil-cellulose insulation pressboard system was studied by using electrostatic module, and the simulation results at 0.005 seconds of 3EMO-IP are shown in Figure 7.20. The applied voltages were the AC voltages, and the amplitudes were 40 kV under needle-plate electrode and 45 kV under finger-finger electrode, respectively, according to the experiment result. The two electrodes were set to AC voltage and zero potential, respectively. Under needle-plate electrode, the needle electrode was connected to the high voltage, and the plate electrode was grounded. The gap of positive and negative electrode was set at 20 mm. In the three-dimensional simulation structure, the x direction and the y direction are, respectively, parallel and perpendicular to the axis of the electrodes, and the z direction is perpendicular to the surface of the pressboard. The result shows that the electric field intensity at the tip of needle electrode is significantly higher than that of finger-finger electrode, as shown in Figure 7.16.

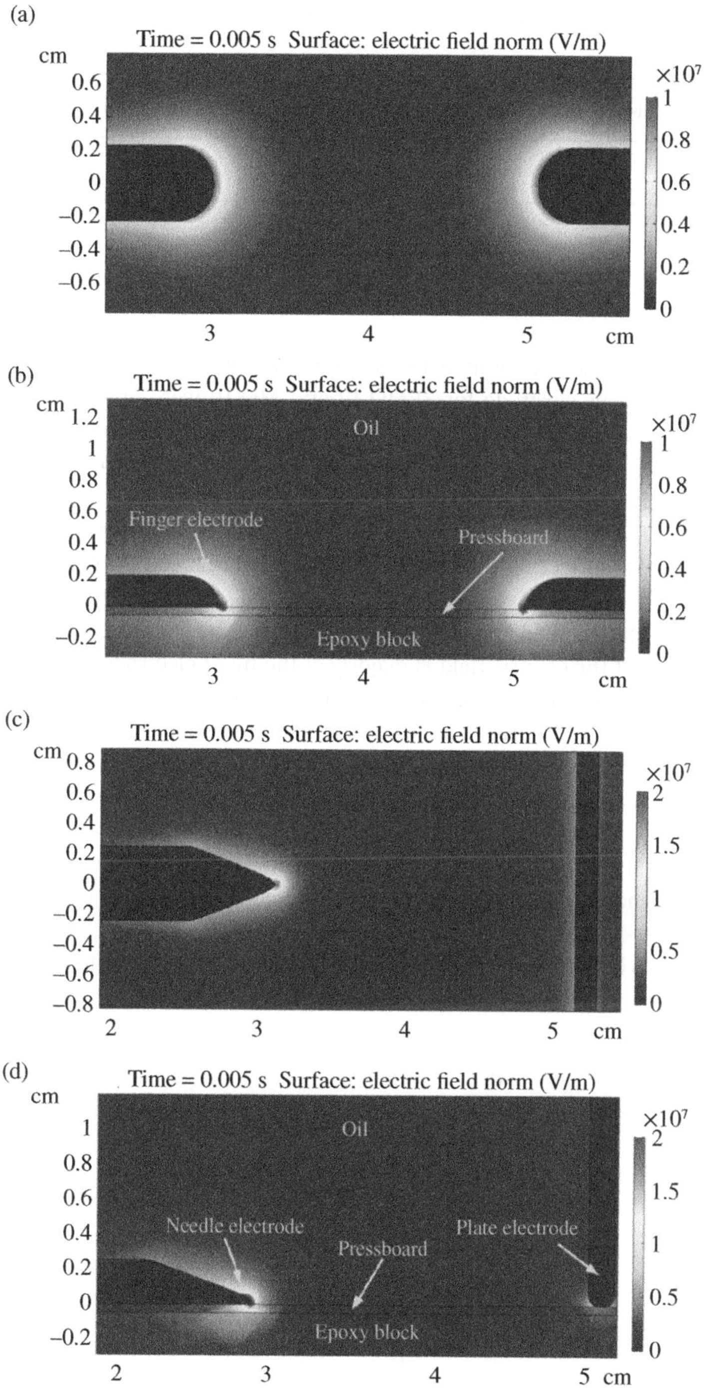

Figure 7.16 Electric field distribution simulation result of the 3EMO-IP under needle-plate electrode and finger-finger electrode. (a) Finger-finger electrode at distance 20 mm (x–y plane, z = 0 mm), (b) finger-finger electrode at distance 20 mm (x–z plane, y = 0 mm), (c) needle-plate electrode at distance 20 mm (x–y Plane, z = 0 mm), (d) needle-plate electrode at distance 20 mm (x–z Plane, y = 0 mm).

Table 7.25 Magnitude of electric field intensity (E_m), the components on x-axis (E_x) and z-axis (E_z) along oil-impregnated pressboard.

	3EMO-IP ($\times 10^7$ V/m)			MO-IP($\times 10^7$ V/m)		
Electrode	E_x	E_z	E_m	E_x	E_z	E_m
Needle-plate electrode	3.8300	1.2367	4.0247	3.9126	1.3915	4.1527
Finger-finger electrode	2.2158	0.6539	2.3128	2.2583	0.6548	2.3531

The magnitude of electric field intensity (E_m), the components of electric field intensity on x-axis (E_x) and z-axis (E_z) at the tip of needle/finger electrode for 3EMO-IP and MO-IP were presented in Table 7.25. Whether at the tip of the needle electrode or the finger electrode, it can be concluded that E_m, E_x, and E_z in 3EMO-IP are slightly lower than those in MO-IP. The lower component of electric field intensity on x-axis (E_x) means that the electric stress provided by the external electric field is relatively small during the process of streamer development, and the lower z-axis component (E_z) also reduces the damage to the cellulose surface by weakening the interaction between the streamer and oil-impregnated pressboard.

Figure 7.17 shows the variation of electric field intensities along the axis of the two electrodes for 3EMO-IP and MO-IP. The distortion of the electric field is majorly at the tip of electrode, and the magnitude of electric field of MO-IP is slightly higher than that of 3EMO-IP, as shown in Figure 7.17. In order to further compare the difference in electric field intensity between 3EMO-IP and MO-IP, a position in the oil and paper at a distance of 0.1 mm from the oil–paper interface was chosen to compare the electric field strength, as shown in Table 7.26. Because of lower relative permittivity for insulating oil, the electric field intensity in oil (E_o) is higher than that in pressboard (E_p). Compared with MO-IP, 3EMO-IP not only has lower electric field intensities in oil and pressboard but also has the lower difference between E_o and E_p, which reflect the relatively uniform distribution of electric field in 3EMO-IP.

The difference of flashover voltages between 3EMO-IP and MO-IP can also be analyzed from the interface charge. The interface charges are divided into free surface charge from the space charge

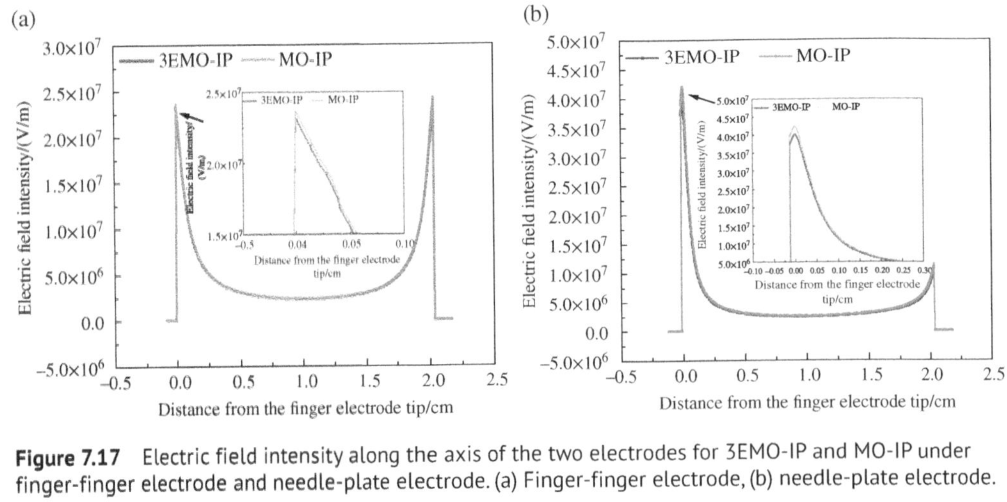

Figure 7.17 Electric field intensity along the axis of the two electrodes for 3EMO-IP and MO-IP under finger-finger electrode and needle-plate electrode. (a) Finger-finger electrode, (b) needle-plate electrode.

Table 7.26 Magnitudes of electric field intensity in oil and pressboard, 0.1 mm away from the oil-pressboard interface.

Electrode	3EMO/($\times 10^7$ V/m)			MO/($\times 10^7$ V/m)		
	E_o	E_p	ΔE	E_o	E_p	ΔE
Needle electrode	3.1264	2.4497	0.6767	3.2369	2.1581	1.0788
Finger electrode	1.3827	1.2910	0.0917	1.4023	1.2059	0.1964

and bound surface charge from the polarization charge. Jin Hyun Choi et al. [59] studied the distribution of the charge at oil–pressboard interface. They proposed that the dominant space charge density is due to the positive ions in the streamer head, while the negative bound surface charge is dominant around the streamer head region, and the positive free surface charge remains around the back of the streamer head. Because of the attraction force between space charge and bound surface charge around the streamer head, the creeping discharge developed continuously. The repulsion force around the back of the streamer head results in the force pushing forward.

Owing to these forces, faster surface discharge will be achieved with the higher electric field intensity. Compared with the 3EMO-IP, there is more permittivity difference between MO and its impregnated pressboard, which means larger electric field concentrated at the interface of oil–pressboard. Therefore, the greater the field concentration at the interface, the greater the space charge generated by ionization, while the negative surface charge density at the oil–pressboard interface and the attractive force between the space charge and the surface charge increase. Besides, the higher AC breakdown voltage of 3EMO under needle-plate electrode and finger-finger electrode is also a reason for advantage of 3EMO-IP in flashover voltage. The 3EMO-IP also has smaller surface resistivity, which is not conducive to accumulation of surface charge during the flashover process.

7.7 Application of the New Mixed Insulation Oil: Service Experiences

7.7.1 Introduction

According to the previous study, the performance parameters of antiaging oil can meet the standard requirements of mineral transformer oil, and it has advantages over mineral oil in terms of specific heat capacity, thermal conductivity, and saturated water content [60–65]. In this section, the three-element mixed insulation oil and mineral oil were filled in mineral oil distribution transformers with different operating years, respectively. First, the oxidation stability, heat capacity and thermal conductivity, kinematic viscosities, the saturated water contents, and AC dielectric strength of the novel three-element mixed insulation oil and mineral oil were measured. The oxidation stability was tested by pressure differential scanning calorimetry (PDSC) method. Then, the parameters of transformers were tested to investigate the application of novel three-element mixed insulation oil in 10 kV distribution transformer. Two kinds of distribution transformers were used in this experiment, as shown in Table 7.27. Factory tests were carried out for the distribution transformers using the new three-element mixed insulation oil and new mineral oil, respectively.

Table 7.27 Comparison the parameters of distribution transformer filled with new mixed oil and mineral oil.

Model	Operating time	Filling insulation oil
S11-M-400/10-0.4	New	New three-element mixed insulation oil
S11-M-400/10-0.4	New	New mineral oil
SBH15-M-400/10-0.4	Seven years	New three-element mixed insulation oil
SBH15-M-400/10-0.4	Seven years	New mineral oil

7.7.2 Using the New Three-Element Mixed Insulation Oil in 10 kV Transformer

Without changing the structural design of the original transformer, the type of S11-M-400/10-0.4 ((Dyn11) new distribution transformer was filled by the new three-element mixed insulation oil and new mineral oil, respectively. Besides, the type of SBH15-M-400/10-0.4 (Dyn11) aged distribution transformer (operation seven years) was again filled by the new three-element mixed insulation oil and new mineral oil, respectively. The S11-M-400/10-0.4 new distribution transformer and SBH15-M-400/10-0.4 aged distribution transformer (operation seven years) filled by the new three-element mixed insulation oil are shown in Figure 7.18. The measurement results of the performance parameters for the new and aged distribution transformer filled by the new three-element mixed insulation oil and new mineral oil are shown in Table 7.28. According to the measurement result, all the parameters of the both distribution transformer above meet the requirement of the standard. Compared with the same distribution transformer filled by mew mineral oil, the distribution transformer filled by the novel three-element mixed insulation oil has good insulation resistance, dielectric loss, no-load and load loss, etc. The good performance of the novel three-element mixed insulation oil could ensure the safe operation of the old and new 10 kV distribution transformer.

7.7.3 Overheating and Discharge Fault Identification for Novel Three-Element Mixed Oil-Paper Insulation System

Figure 7.19 shows the concentration of dissolved gases (CDG) in TEMO and MO under different overheating temperatures [66]. Result shows that the main gas generated from two kinds of insulation oil during low temperature thermal fault and medium temperature thermal fault was CO_2. The content of H_2 and hydrocarbon gases that produced at 150 and 300 °C was little, which accounts for less than 2% of the total gases. Especially, there was no C_2H_2 produced under low temperature and medium temperature (150, 300 and 450 °C) for mineral oil and TEMO. When under high temperature (750 °C) thermal fault, the concentration of C_2H_2 that produced by mineral oil and TEMO was about 4.7 and 11 ml/l, respectively. Since the overheating time was short, the content of dissolved gases was little when compared to that of CO_2. With the increment of overheating temperature, the production of all kinds of gases increases obviously. The content of hydrocarbon gases increases fast, which leads to the rapid drop of the percentage of CO_2.

The concentration of dissolved gases in oil–paper insulation systems under different overheating temperatures is shown in Figure 7.20. The main gas which generated during overheating under different temperatures in oil–paper insulation system was CO_2. The content of CO_2 generated from oil–paper system increased much more evidently than that in pure insulating oil, which was 3–5

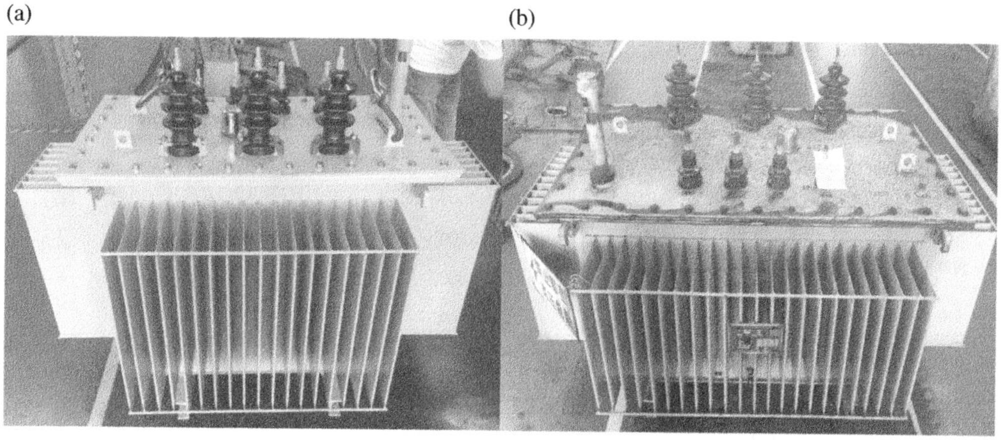

Figure 7.18 Distribution transformer filled by the new three-element mixed insulation oil. (a) S11-M-400/10-0.4 new distribution transformer, (b) SBH15-M-400/10-0.4 aged distribution transformer.

Table 7.28 The measurement results of the performance parameters for the new and aged distribution transformer filled by the new three-element mixed insulation oil and new mineral oil.

Model		**S11-M-400/10-0.4**		**SBH15-M-400/10-0.4**	
Type of insulation oil		The mixed insulation oil	Mineral oil	The mixed insulation oil	Mineral oil
No-load loss/W		561	541	152	153
Load loss/75 °C, W		4400	4428	4102	4260
Insulation resistance/MΩ	HV-LV	≥1000	≥1000	≥1000	≥1000
	HV/LV-E	≥1000	≥1000	≥1000	≥1000
	LV/HV-E	≥1000	≥1000	≥1000	≥1000
Capacitance/pF	HV/LV-E	2961	2667	6748	6806
	LV/HV-E	5362	4979	11210	9889
tan δ	HV/LV-E	0.667%	0.349%	0.496%	0.394%
	LV/HV-E	0.421%	0.312%	0.484%	0.408%
Short-circuit impedance (75 °C)/%		3.82%	3.85%	3.86%	3.74%
Transformer top oil temperature rise (°C)		51.5	52.3	/	/

times of that generated during pure oil overheating. Besides, the content of CO that generated from oil–paper insulation system was 2–3 times of that generated from pure insulation oil. The total amount of gas which generated during overheating of oil–paper insulation system was obviously more than that from overheating of pure oil. The extra gases were mainly carbon oxide. When the oil–paper insulation system was overheating at a low temperature, the content of total carbon oxide would account for more than 98% of the total gases. When oil–paper insulation system was overheating under a higher temperature, the production of hydrocarbon gases increased greatly, while the total carbon oxide still accounts for more than 70% of the total gases, which was much higher than the percentage of carbon oxide in pure oil overheating. Compared with the

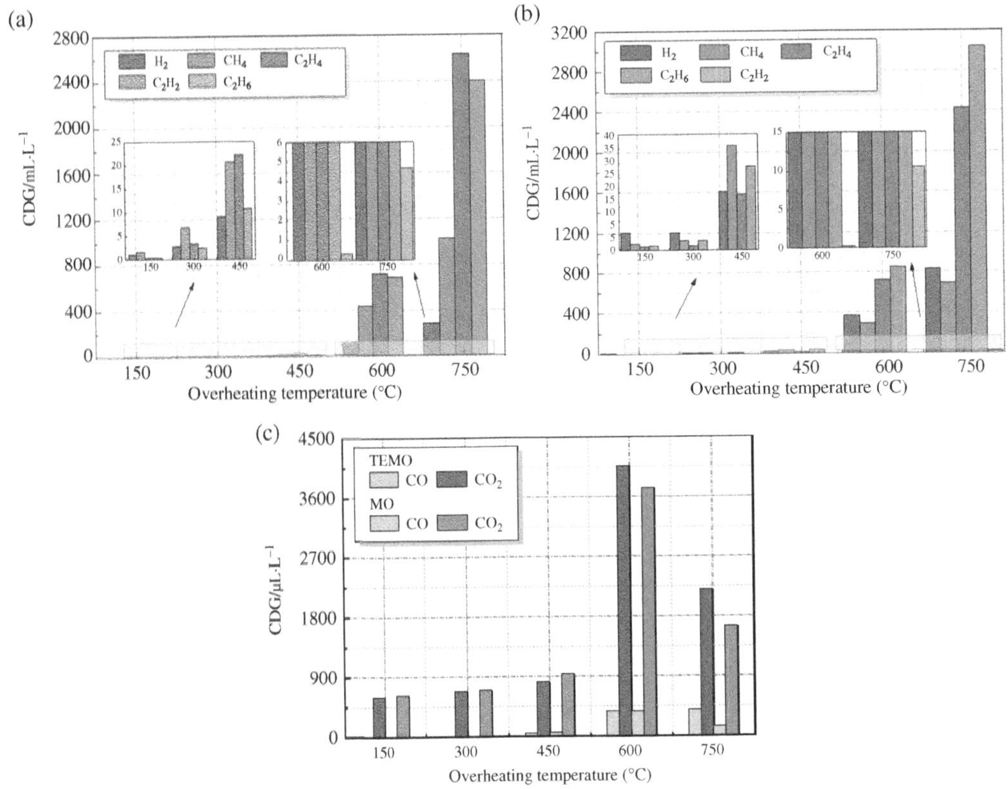

Figure 7.19 Content of dissolved gases in these five gases in MO and TEMO after overheating under different temperatures l. (a) H_2 and hydrocarbon gases in MO, (b) H_2 and hydrocarbon gases in TEMO, (c) CO and CO_2 in MO and TEMO.

results of pure oil overheating, there was still no C_2H_2 produced under low temperature (150 and 300 °C) and only a little bit C_2H_2 was produced under medium temperature (450 °C) for mineral oil and TEMO. When under high temperature (750 °C) thermal fault, the concentration of C_2H_2 that produced by mineral oil and TEMO was about 27.6 and 33.2 ml/l, respectively. It can be seen that CO and CO_2 are the main products of oil-impregnated paper overheating, while H_2 and hydrocarbon gases are less.

Figure 7.21 shows the percentage of H_2, CH_4, C_2H_4, C_2H_6, and C_2H_2 in these five gases during corona discharge and air-gap discharge. It can be seen that the proportion of H_2 in these five gases in TEMO reached 73.09 and 61.95%, respectively, after corona discharge and air gap discharge and 1.3 times of that after corona discharge. Besides, there is little difference between two kinds of oil–paper insulating system. The content of H_2 and CH_4 dissolved in oil changed distinctly after partial discharge, which means that they can be used as the characteristic gases of partial discharge fault for novel three-element mixed oil–paper insulation system.

Tables 7.29 and 7.30 show the content of dissolved gases in oil after multiple times of surface flashover of MO and TEMO under needle-plane electrodes and finger-finger electrodes. With the increase of flashover times, the content of H_2 and hydrocarbon gases that dissolved in oil increased significantly.

Figure 7.22 shows the percentage of H_2, CH_4, C_2H_4, C_2H_6, and C_2H_2 in these five gases during surface flashover with needle-plane electrode and finger-finger electrode. It can be seen that the

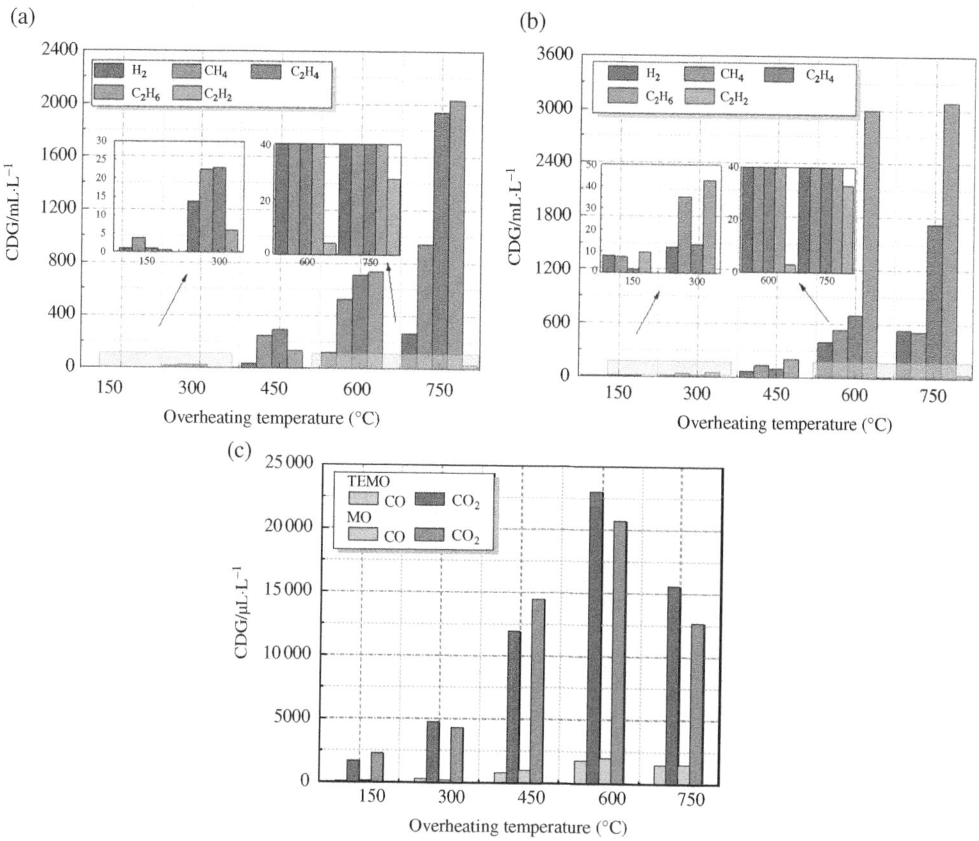

Figure 7.20 Content of dissolved gases in these five gases in MO–paper and TEMO–paper after overheating under different temperatures L. (a) H_2 and hydrocarbon gases in MO, (b) H_2 and hydrocarbon gases in TEMO, (c) CO and CO_2 in MO and TEMO.

proportion of H_2 and C_2H_2 in these five gases which dissolved in TEMO after 20 times of surface flashover is 1.55 and 1.93 times of that after 3 times of surface flashover. Under finger-finger electrodes, the content of these five gases which dissolved in oil is higher than that of needle-plane electrodes after same times of surface flashover, as shown in Tables 7.30 and 7.31. Concentration of C_2H_2 and total hydrocarbon gases that dissolved in TEMO is smaller than that in MO, which becomes much more significant with the increase of flashover times. Since the content of H_2 and C_2H_2 changes greatly after surface flashover compared to other gases, they can be treated as the characteristic gases for surface flashover breakdown fault of novel three-element mixed oil–paper insulating system.

Tables 7.31 and 7.32 show the contents of dissolved gases in MO-paper and TEMO-paper insulating system under AC breakdown fault. With the increase of breakdown times, the content of each characteristic gas dissolved in oil increased at the same time. After the breakdown of oil-impregnated pressboard, the content of H_2 and hydrocarbon gases in oil is higher when compared with the breakdown in oils. Besides, the difference between the content of H_2 and C_2H_2 is slighter than that of pure oil. After 60 times of AC breakdown of the oil-impregnated pressboard under power frequency, the ratio of the content of H_2 and C_2H_2 is 0.78, while the ratio of that in pure TEMO after same times of breakdown could reach 2.54. Concentration of H_2 and total

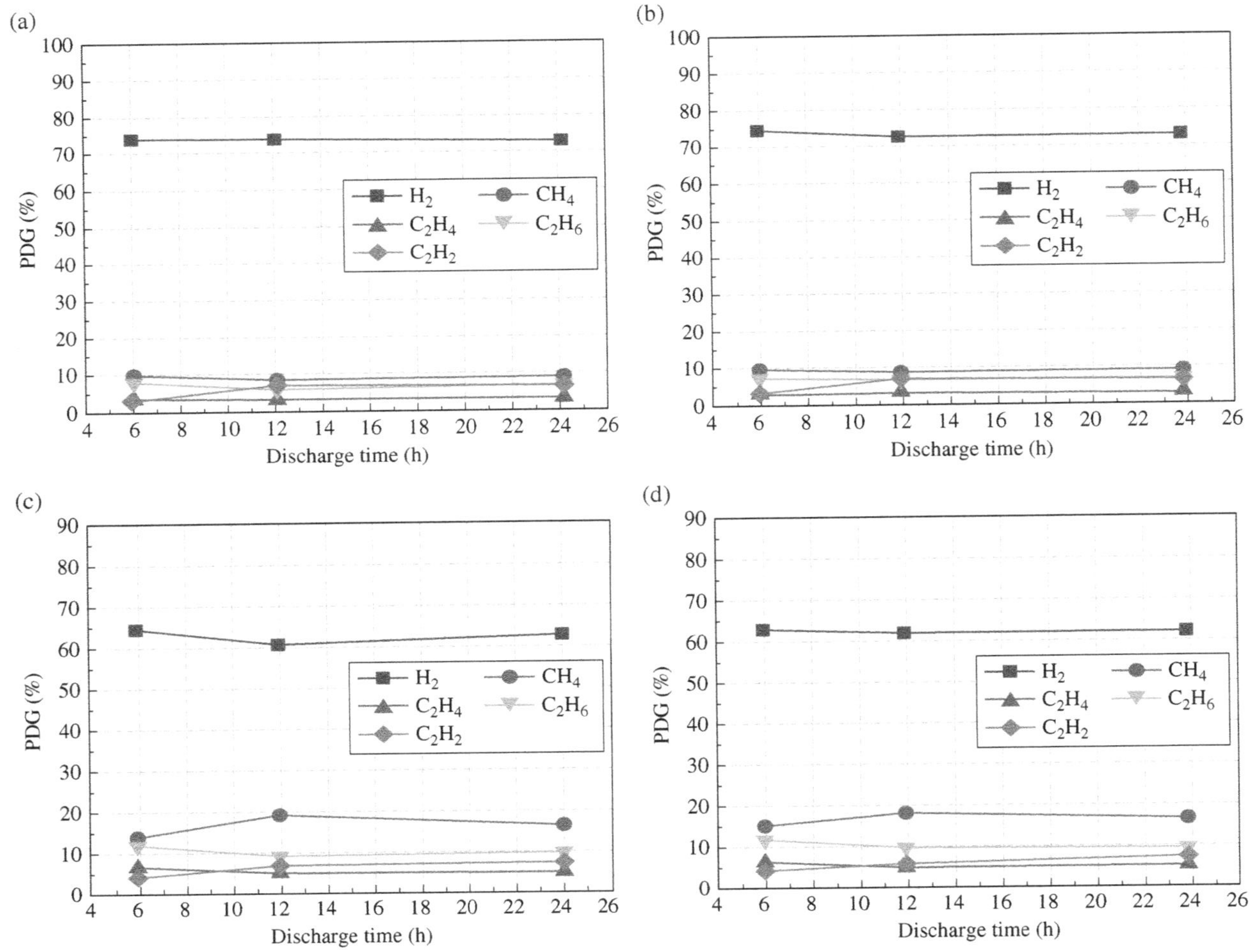

Figure 7.21 Percentage of H_2, CH_4, C_2H_4, C_2H_6 and C_2H_2 in these five gases of MO–paper and TEMO–paper after partial discharge. (a) MO-corona discharge, (b) TEMO-corona discharge, (c) MO-air-gap discharge, (d) TEMO-air-gap discharge.

Table 7.29 Content of dissolved gases in oil after multiple times of surface flashover of MO and TEMO under needle-plane electrodes.

		Surface flashover times				
Gas content in oil (μl/l)	**Oil**	**3**	**5**	**10**	**15**	**20**
H_2	MO	37	53	103	127	168
	TEMO	46	63	69	120	137
CH_4	MO	7.5	11.2	22.5	25.3	28.6
	TEMO	8.1	10.9	10.5	17.7	19.9
C_2H_4	MO	17.5	28.1	52.2	56	69.6
	TEMO	16.1	22.7	29.4	51.2	59.1
C_2H_6	MO	1.3	2	3.8	4	4.5
	TEMO	1.5	2	2.2	3.7	4.2
C_2H_2	MO	157	239	363	390	643
	TEMO	114	158	216	364	425

Table 7.30 Content of dissolved gases in oil after multiple times of surface flashover of MO and TEMO under finger-finger electrodes.

		Surface flashover times				
Gas content in oil (μl/l)	**Oil**	**3**	**5**	**10**	**15**	**20**
H_2	MO	36	44	145	140	185
	TEMO	58	50	102	114	221
CH_4	MO	6.9	9.8	27.4	27.6	31.1
	TEMO	9.7	10.3	21.2	17.5	37.7
C_2H_4	MO	15.5	19.9	61.6	63.1	81.1
	TEMO	19.5	28.9	48.2	50.5	85.4
C_2H_6	MO	1.3	1.9	4.2	4.8	5.5
	TEMO	1.9	2.7	3.7	3.9	6.2
C_2H_2	MO	146	143	446	450	761
	TEMO	124	186	313	342	554

hydrocarbon gases in novel three-element mixed oil-paper insulating system is smaller than that in MO-paper insulating system after same times of AC breakdown.

Figure 7.23. shows the percentage of H_2, CH_4, C_2H_4, C_2H_6, and C_2H_2 in these five gases after AC breakdown in insulation oil and oil–paper system. It can be seen that the proportion of dissolved H_2 in the oil is around 11 and 40%, while the proportion of dissolved C_2H_2 is around 40 and 45%, respectively, after AC breakdown of TEMO and its oil-impregnated pressboard under power frequency. Table 7.33 shows the contents of dissolved gases and Figure 7.24 shows the percentage of

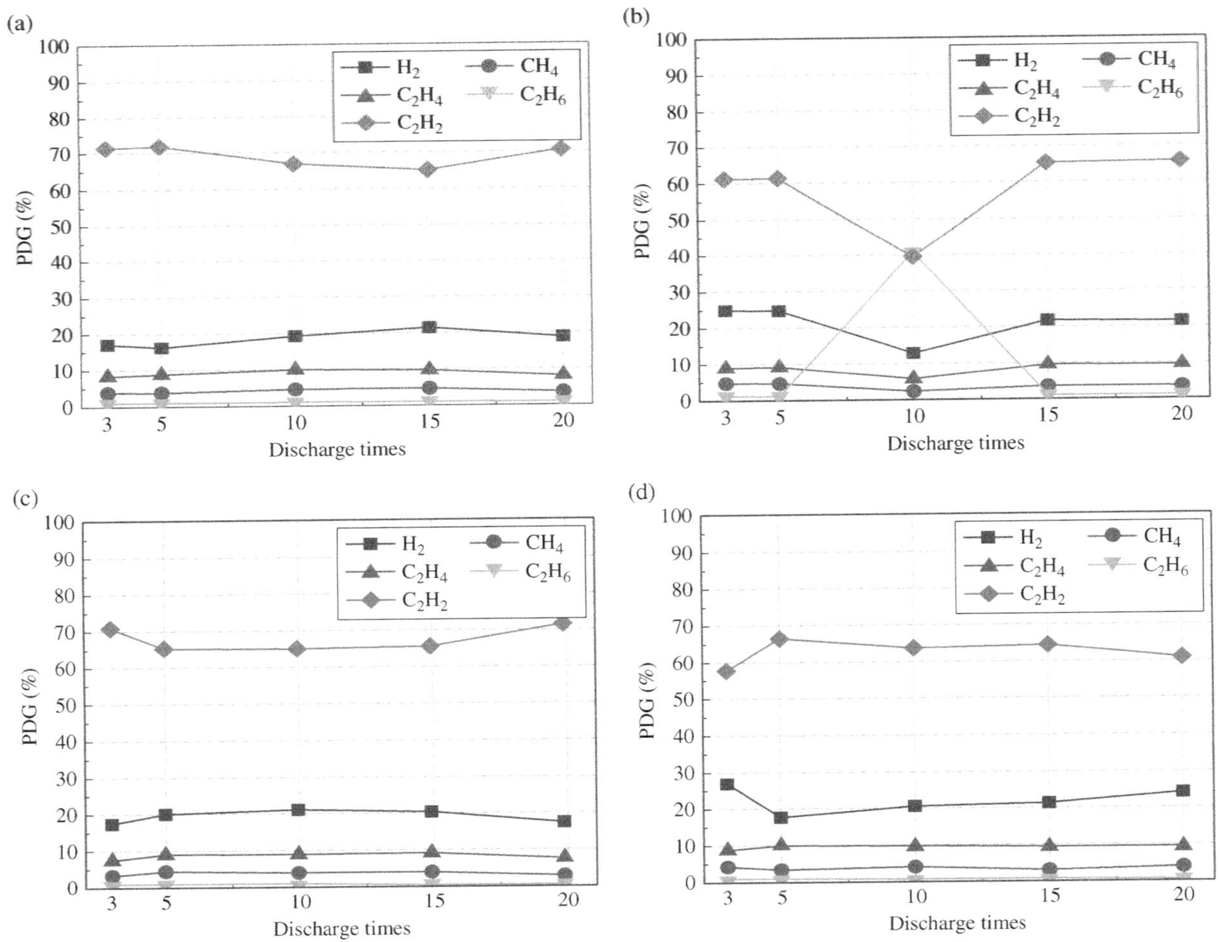

Figure 7.22 Percentage of H_2, CH_4, C_2H_4, C_2H_6 and C_2H_2 in these five gases during surface flashover of oil–paper insulating system (a) MO-needle-plane electrode, (b) TEMO-needle-plane electrode, (c) MO-finger-finger electrode, (d) TEMO-finger-finger electrode. *Source:* Based on Kuchler [49].

Table 7.31 Content of dissolved gases in oil after multiply AC breakdown of MO and TEMO.

Gas content in oil (μl/l)	Oil	Breakdown times 10	20	40	60
H_2	MO	19	34	41	34
	TEMO	1	15.1	34.6	57.4
CH_4	MO	4.7	6.4	6.9	10.1
	TEMO	1.3	3.7	10.9	13.6
C_2H_4	MO	6.8	13	13	22.5
	TEMO	3	7.9	17.9	26.3
C_2H_6	MO	0.7	1.4	1.7	2.5
	TEMO	0	0.3	1.4	2.3
C_2H_2	MO	52.1	105.1	127.4	167
	TEMO	23	71.4	133.4	204

Table 7.32 Content of dissolved gases in oil after multiply AC breakdown of MO and TEMO.

Gas content in oil (μl/l)	Oil	Breakdown times 20	40	60
H_2	MO	33	174	263
	TEMO	20.3	94	219
CH_4	MO	5.1	29.6	45.6
	TEMO	2.6	10.6	34.1
C_2H_4	MO	5.4	29.9	39.9
	TEMO	2.4	12.4	40.9
C_2H_6	MO	1.9	9.7	12.9
	TEMO	6.8	9.3	17.8
C_2H_2	MO	36.8	128.8	248.8
	TEMO	16.7	85	180

H_2, CH_4, C_2H_4, C_2H_6, and C_2H_2 in these five gases in MO and TEMO under positive polarity lightening impulse breakdown fault. With the increase of breakdown times, the concentration of each characteristic gas in the oil showed an upward trend and the content of H_2 and C_2H_2 increased significantly. After 60 times of lightning impulse breakdown in three-element mixed insulation oil under positive polarity, the concentration of H_2 in the oil increased by 120.27% compared with that after 20 times of breakdown, while the concentration of C_2H_2 in the oil increased by 127.53%. The content of each characteristic gas in TEMO is generally lower than that in MO, except for CO. After 60 times of breakdown under positive polarity, the content of H_2 and C_2H_2 in TEMO is 0.65 and 0.54 times of MO, respectively. It can be seen that the proportion of H_2 in these five gases in TEMO is slightly larger than that of MO, while the proportion of C_2H_2 is slightly smaller than that of MO

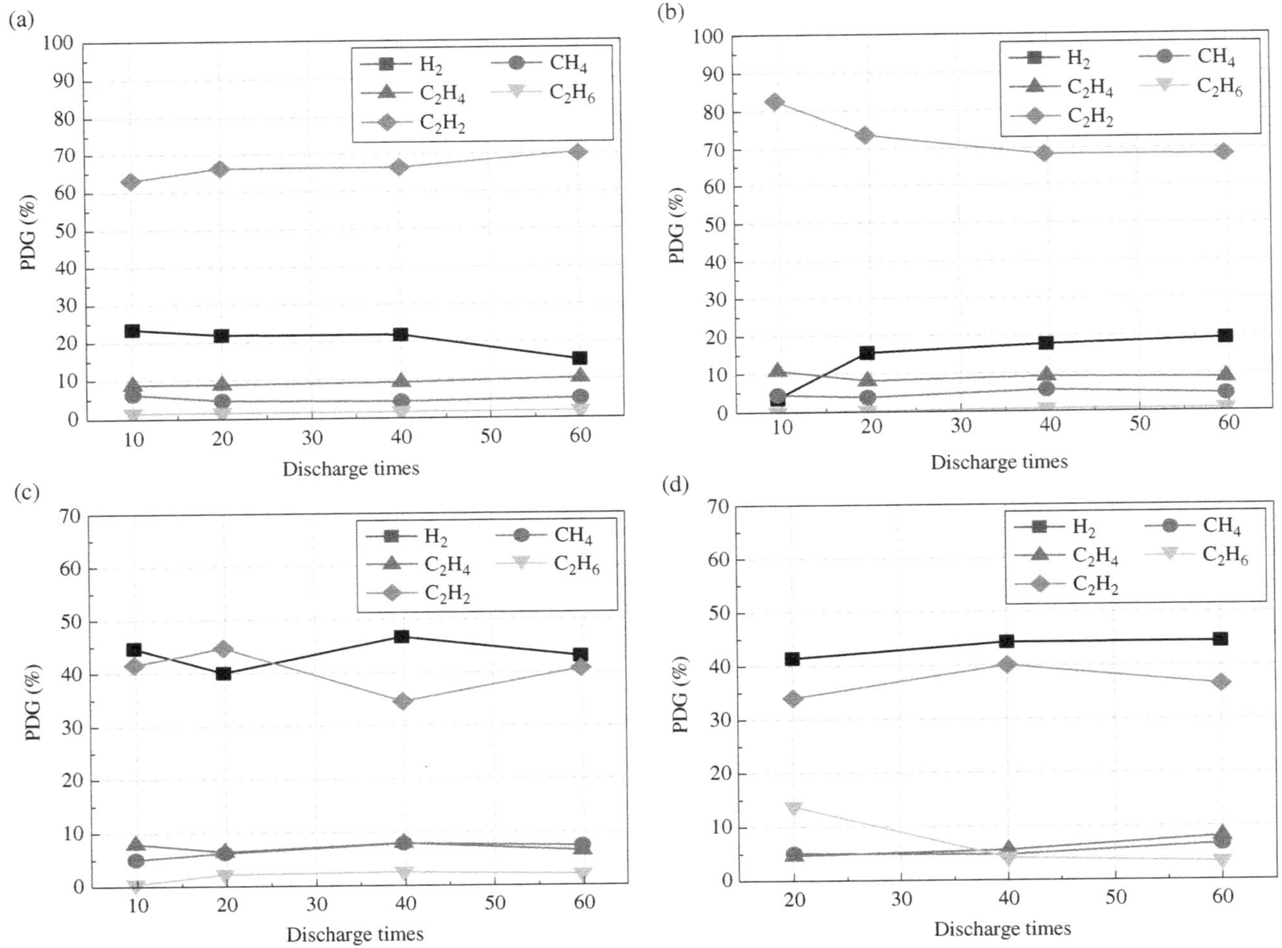

Figure 7.23 Percentage of H_2, CH_4, C_2H_4, C_2H_6, and C_2H_2 in these five gases during surface flashover of oil-paper insulating system (a) MO, (b) TEMO, (c) MO impregnated pressboard, (d) TEMO impregnated pressboard.

Table 7.33 Dissolved gases in oil after multiply positive polarity lightning impulse breakdown of MO and TEMO.

		Breakdown times		
Gas content in oil (μl/l)	**Oil**	**20**	**40**	**60**
H_2	MO	203	240	502
	TEMO	148	184	326
CH_4	MO	8.3	14.3	23.7
	TEMO	6.1	7.3	12.6
C_2H_4	MO	9.2	18	30
	TEMO	7	9.7	18.2
C_2H_6	MO	0.8	1.4	2.3
	TEMO	0.8	0.9	1.4
C_2H_2	MO	310.6	416.9	848.5
	TEMO	199.8	247.9	454.6

after 60 times of breakdown. After 60 times of lightning impulse breakdown, the proportion of total content of H_2 and C_2H_2 is as high as 72.84 and 66.71%, which can be used as the characteristic gas for lightning impulse breakdown fault of novel three-element mixed oil–paper insulating system.

Table 7.34 shows the contents of dissolved gases, and Figure 7.24 shows the percentage of H_2, CH_4, C_2H_4, C_2H_6, and C_2H_2 in these five gases in MO–paper and TEMO–paper insulating system under positive polarity lightening impulse breakdown fault. Compared with the pure insulation oil, the content of H_2 and various hydrocarbon gases dissolved in oil after multiply lightning impulse breakdown of oil-impregnated pressboard was higher, and the content of CH_4, C_2H_4, and C_2H_6 dissolved in oil was significantly increased. With the increase of breakdown times, the content of H_2 and C_2H_2 after 18 times of lightning impulse breakdown of oil-impregnated pressboard increased by 2.42 and 2.37 times, respectively, when compared with that after six times of breakdown. The characteristic gas dissolved in the TEMO is slightly less than that of MO after lightning impulse breakdown fault. After 18 times of lightning impulse breakdown, the content of H_2 in TEMO accounted for 31.94%, while the content of C_2H_2 accounted for up to 31.00%, the content of C_2H_4 accounted for up to 15.19%.

7.7.4 Fault-Type Identification Model Based on Hydrogen, Ethane, and Acetylene

The novel three-element mixed oil–paper insulating system has different dissolved characteristic gases under different overheating and discharge faults. C_2H_6, C_2H_4, and H_2 are the key gases under overheating faults for novel three-element mixed insulation oil. However, the novel three-element mixed insulation system would also generate C_2H_6 due to the peroxidation of unsaturated fatty acids during the normal thermal aging process, which could not be used as the characteristic gas. H_2 and C_2H_2 account for the highest proportion among the characteristic gases under AC breakdown fault, surface flashover breakdown fault, and lightning impulse breakdown fault, while C_2H_4 also accounts for a high proportion. It is H_2 and CH_4 that the key gases after partial discharge fault, which made it

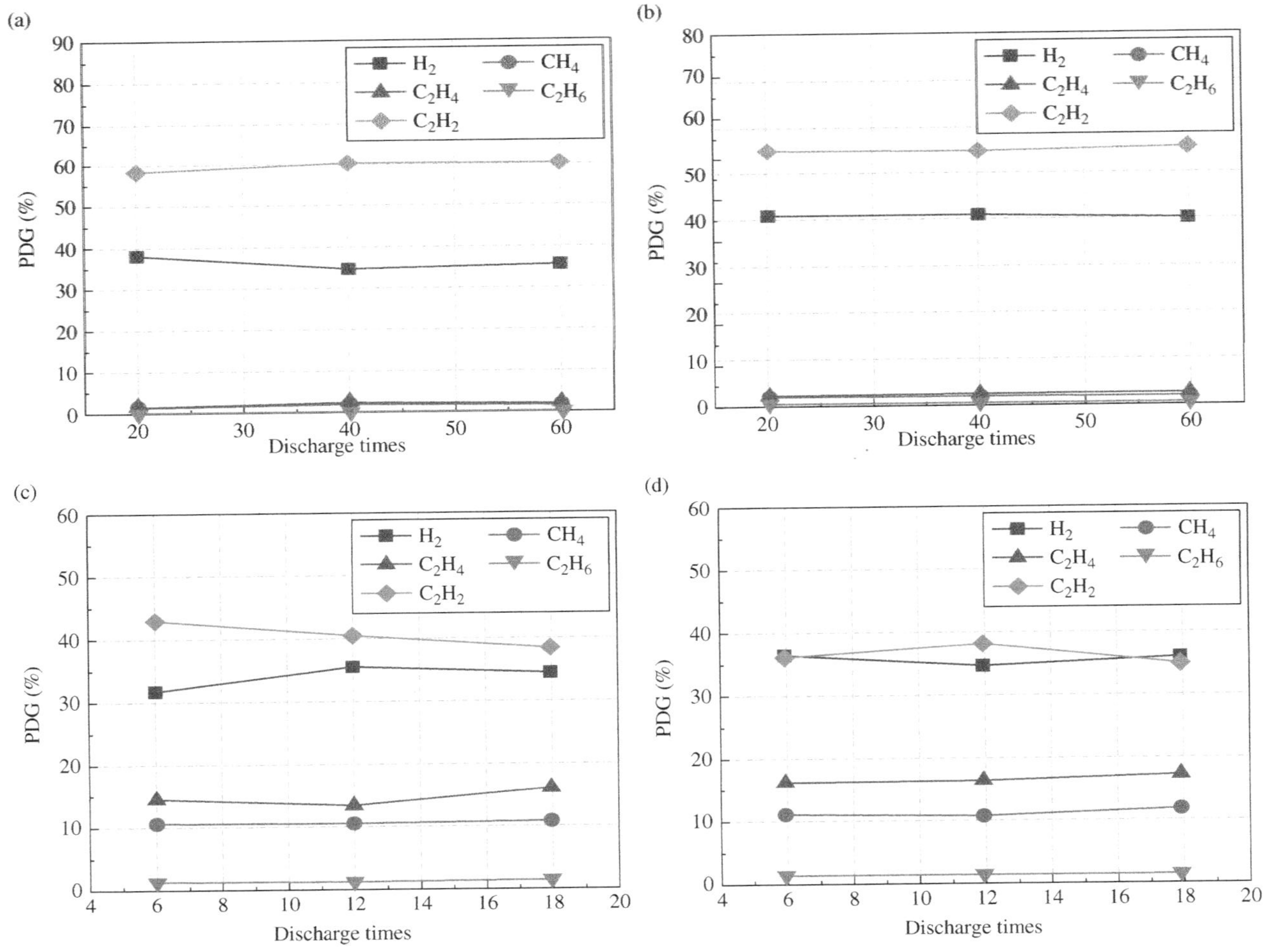

Figure 7.24 Percentage of H_2, CH_4, C_2H_4, C_2H_6 and C_2H_2 in these five gases after multiplying positive polarity lightning impulse breakdown of oil–paper insulating system (a) MO, (b) TEMO, (c) MO impregnated pressboard, (d) TEMO impregnated pressboard.

Table 7.34 Dissolved gases in oil after multiply positive polarity lightning impulse breakdown of MO- and TEMO-impregnated pressboard.

		Breakdown times		
Gas content in oil (μl/l)	**Oil**	**6**	**12**	**18**
H_2	MO	686	1118	1663.
	TEMO	513	1299	1279
CH_4	MO	204.5	337.9	535.3
	TEMO	167.9	375.4	390.6
C_2H_4	MO	302.2	522.3	790.7
	TEMO	231.7	481.1	588.9
C_2H_6	MO	19.5	33.8	49.9
	TEMO	15.6	31.9	39.5
C_2H_2	MO	679.9	1230.3	1614
	TEMO	697.1	1481.4	1426.2

different from other discharge faults. Nevertheless, it is worth noting that the content of C_2H_2 also increases with the discharge time. Since that C_2H_2, H_2, and C_2H_4 have been chosen as the characteristic gases, and the overheating and discharge faults would be diagnosed by triangle diagram diagnosis.

The C_2H_2, H_2, and C_2H_4 were selected as the x, y, and z axes, respectively. The coordinate calculation method in the triangular coordinate system corresponding to each gas sample is shown in Eq. (7.12). In the Eq. (7.12), C_x, C_y, and C_z are the coordinates of the three characteristic gases of C_2H_2, H_2, C_2H_4 in the triangular coordinate system, and $c(C_2H_2)$, $c(H_2)$, and $c(C_2H_4)$ are the contents of these three gases (μl/l). Taking the experimental data into Eq. (7.12) and taking the percentage data of three characteristic gases into the triangle coordinate system could obtain the specific positions of points, as shown in Figure 7.25, and different faults are divided by the dotted line. The boundary settings are shown in Table 7.35. The content of C_2H_4 could be used to divide different kinds of thermal fault, the content of C_2H_2 could be used to divide the partial discharge fault, while the content of H_2 could be used to divide different kinds of breakdown fault under power frequency.

$$\begin{aligned} C_x &= \frac{100c(C_2H_2)}{c(C_2H_2)+c(H_2)+c(C_2H_4)} \\ C_y &= \frac{100c(H_2)}{c(C_2H_2)+c(H_2)+c(C_2H_4)} \\ C_z &= \frac{100c(C_2H_4)}{c(C_2H_2)+c(H_2)+c(C_2H_4)} \end{aligned} \tag{7.12}$$

7.8 Conclusions and Future Scope

The mixed oil has higher oxidation stability than mineral oil and lower viscosity, dielectric loss, and acidity than natural esters. The novel three-element mixed insulation oil has a similar kinematic viscosity as mineral oil. The thermodynamic properties of the novel three-element mixed

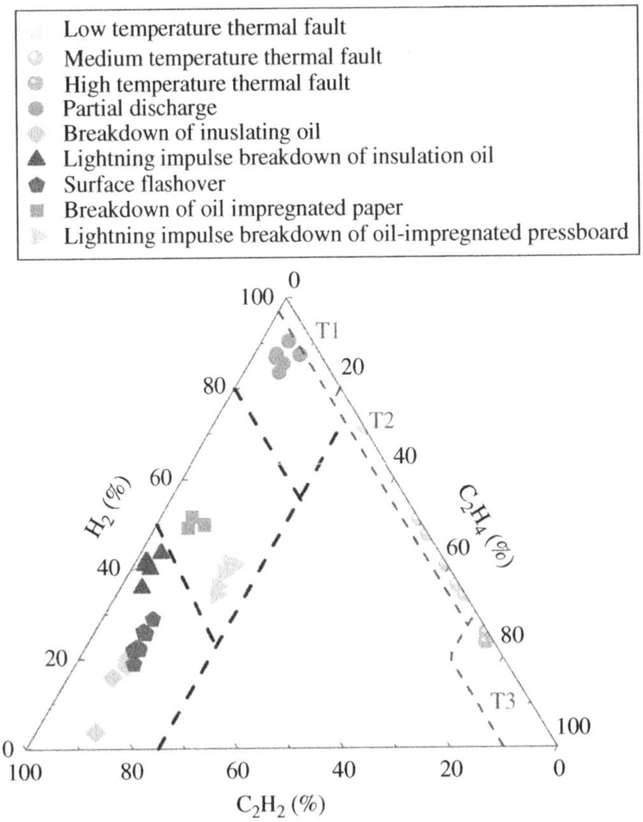

Figure 7.25 C_2H_2, H_2, C_2H_4 triangle for diagnosing thermal and electrical fault of novel three-element mixed insulating system.

Table 7.35 Boundary settings of C_2H_2, H_2, C_2H_4 triangle.

Fault types	Boundary settings
Low temperature thermal fault	$C_2H_4 = 20\%$, $C_2H_2 = 3\%$
Medium temperature thermal fault	$C_2H_4 = 20\%$, $C_2H_2 = 3\%$, $C_2H_4 = 70\%$
High temperature thermal fault	$C_2H_2 = 10\%$, $C_2H_4 = 70\%$
Breakdown discharge of insulating oil	$C_2H_4 = 25\%$, $C_2H_2 = 50\%$
Partial discharge	$C_2H_4 = 25\%$, $C_2H_2 = 20\%$, $C_2H_2 = 3\%$
Breakdown discharge of oil-impregnated paper	$C_2H_2 = 20\%$, $C_2H_4 = 25\%$, $C_2H_2 = 50\%$

insulation oil are superior to those of mineral oil. Its specific heat capacity and thermal conductivity values are higher than those of mineral oil between 30 and 100 °C. The novel three-element mixed insulation oil also has higher relative permittivity than mineral oil. However, its DC resistivity is lower than that of mineral oil. The DC resistivity difference between the two oils is small above 60 °C. The saturated water content of the novel three-element mixed insulation oil is approximately 3–4 times larger than that of mineral oil above 45 °C. The mixed insulation oil has a higher AC breakdown voltage than mineral oil under the same absolute water content.

The three-element mixed oil could delay the aging rate of cellulose paper, and this retardation was evident from the beginning. Supposing DP values drop to 250 at the end of service life, the estimated life of cellulose paper immersed with mixed oil can reach 2.5 times that of cellulose paper impregnated with mineral oil. For a mineral oil–paper system, oil replacement with three-element mixed insulation oil in the middle aging period can delay the further aging rate of the cellulose paper, and aging life may significantly increase. This result may be applied to old field transformers to increase their insulation life. Some other parameters, such as the light color of the aged paper impregnated with the mixed oil and the low furfural content in the mixed oil, also reflect that mixed oil retards the aging of the cellulose paper. In addition, mixed oil maintains its good performance throughout the aging process, which is useful for applications of mixed oil in the power industry.

The AC breakdown characteristics of the three-element mixed insulation oil are absolutely superior to those of mineral oil at the same temperature and absolute moisture content. The properties of insulation oil are important to the breakdown characteristics, and the three-element mixed insulation oil is able to maintain its breakdown strength over a range of moisture contents. At the same temperature, breakdown voltage of insulating oil is roughly related to relatively moisture content, but the absolute moisture content also affects. Insulating liquid with large saturated water content always has a relatively low breakdown voltage at the same relative moisture content, for absolute water content of the oil is high, which results in breakdown easily. With the increase in temperature for dried insulation paper (pressboard), the breakdown voltage does not decrease significantly. At 25 °C, the increase in the moisture content of the insulation paper (pressboard) cannot seemingly reduce the breakdown voltage. Only at elevated temperatures with high moisture content the insulation strength will decrease significantly, which presents a combined effect of temperature and moisture content. For the oil gap and pressboard structure, the insulation characteristics of the three-element mixed insulation oil are superior to those of the mineral oil at any oil gap distance, which benefits from its larger relative dielectric constant.

The surface flashover voltage increases with the oil gap distance for both 3EMO-IP and MO-IP under needle-plate and finger-finger electrode, respectively. The surface flashover voltage of 3EMO-IP is higher than that of MO-IP. Higher breakdown voltage of 3EMO is a reason for this advantage. Lower electric field intensities in 3EMO-IP and more uniform distribution of electric field between oil and pressboard also result in the higher flashover voltage of 3EMO-IP. Besides, the lower surface resistivity of 3EMO-IP is not conducive to the accumulation of interface charge to improve flashover voltage. After 30 times flashover at the interface of oil–cellulose pressboard, the carbonization of surface fibers on 3EMO-IP is slightly less than that on MO-IP. The better thermal stability of 3EMO can protect the fiber on pressboard surface at high temperature. The new mixed insulation oil has better oxidation stability than new mineral oil and soybean oil. The new mixed insulation oil has better thermodynamic properties than mineral oil. The fluidity of the novel three-element mixed insulation oil was equivalent to that of mineral oil in the field operation transformer. The new three-element mixed insulation oil has a higher saturated water content than new mineral oil, which is beneficial for its excellent AC breakdown behavior. The good performance of the novel three-element mixed insulation oil could ensure the safe operation of the old and new 10 kV distribution transformer.

The diagnosis method of discharge faults for novel three-element mixed oil–paper insulating system was proposed, while H_2, C_2H_2, and C_2H_4 have been chosen as the characteristic gases, and the boundary settings were determined. The content of C_2H_4 could be used to divide different kinds of overheating fault, the content of C_2H_2 could be used to divide the partial discharge fault, while the content of H_2 could be used to divide different kinds of breakdown fault under power frequency. Besides, the accuracy of the overheating and discharge diagnosis method for novel three-element mixed oil–paper insulation system has been verified.

In the future, two major tasks need to be carried out. Some aged mineral oil will remain in the process of oil replacement. Due to the residual of some aged mineral oil, the initial value of dissolved gas in the oil is different from that of the new three-element mixed oil, so the diagnostic criteria of the above diagnosis model have not been completely unified. Further research is needed to unify the diagnostic criteria. The three-element mixed oil has been applied in 10 and 35 kV transformer. It is necessary to further study the surface flashover characteristics and lightning impulse breakdown characteristics under long oil gap condition and carry out multi-physical field simulation analysis of transformer using the new three-element mixed oil, so as to promote its safety application in 110 kV and other higher voltage-level transformer.

References

1 Mohan Rao, U., Fofana, I., Jaya, T. et al. (2019). Alternative dielectric fluids for transformer insulation system: progress, challenges, and future prospects. *IEEE Access* 7: 184552–184571.

2 Liao, R., Feng, D., Hao, J. et al. (2019). Thermal and electrical properties of a novel 3-element mixed insulation oil for power transformers. *IEEE Transactions on Dielectrics and Electrical Insulation* 26 (2): 610–617.

3 Beroual, A., Sitorus, H.B.H., Setiabudy, R., and Bismo, S. (2018). Comparative study of AC and DC breakdown voltages in Jatropha methyl ester oil, mineral oil, and their mixtures. *IEEE Transactions on Dielectrics and Electrical Insulation* 25 (5): 1831–1836.

4 Fofana, I. (2013). 50 years in the development of insulating liquids. *IEEE Electrical Insulation Magazine* 29 (5): 13–25.

5 Hamdi, A., Fofana, I., and Djillali, M. (2017). Stability of mineral oil and oil–ester mixtures under thermal ageing and electrical discharges. *IET Generation, Transmission & Distribution* 11 (9): 2384–2392.

6 Fofana, I., Wasserberg, V., Borsi, H., and Gockenbach, E. (2002). Challenge of mixed insulating liquids for use in high-voltage transformers.1. Investigation of mixed liquids. *IEEE Electrical Insulation Magazine* 18 (3): 18–31.

7 Fofana, I., Wasserberg, V., Borsi, H., and Gockenbach, E. (2002). Challenge of mixed insulating liquids for use in high-voltage transformers. II. Investigations of mixed liquid impregnated paper insulation. *IEEE Electrical Insulation Magazine* 18 (4): 5–16.

8 Perrier, C., Beroual, A., and Bessede, J. (2006). Improvement of power transformers by using mixtures of mineral oil with synthetic esters. *IEEE Transactions on Dielectrics and Electrical Insulation* 13 (3): 556–564.

9 Liao, R., Liang, S., Li, J. et al. (2009). Study on the physics and chemistry characteristics and breakdown voltage of a mixed insulation oil composed of mineral oil and natural easter [J]. *Proceedings of the CSEE* 29 (13): 117–123.

10 Hao, J., Yang, L., Liao, R. et al. (2010). Mechanism investigation on delaying of the thermal aging rate of oil-paper insulation with mixture insulating oil [J]. *Proceedings of the CSEE* 30 (19): 120–126.

11 Liao, R., Liang, S., Yang, L. et al. (2012). Comparison of ageing results for transformer oil-paper insulation subjected to thermal ageing in mineral oil and ageing in retardant oil. *IEEE Transactions on Dielectrics and Electrical Insulation* 19 (1): 225–232.

12 Liang, S. (2009). *Study on Antioxidation Transformer Insulation Oil and Its Effects on Oil-immersed Paper Thermal Aging [D]*. Chongqing University.

13 Guo, P. (2014). *The Therming Aging Properties of the New Mixture Oil-pressboard Insulation and Its Application in the Distribution Transformer [D]*. Chongqing University.

14 Kou, L., Wang, J., and Song, Q. (2017). Experimental study on properties of different mixed insulating oils [J]. *Insulating Materials* 50 (2): 76–80.

15 Yu, H., Chen, R., Hu, X. et al. (2017). Dielectric and physicochemical properties of mineral and vegetable oils mixtures[C]. *19th IEEE International Conference on Dielectric Liquids (ICDL)*, Manchester, United Kingdom (25–29 June 2017).

16 Takahashi, K., Miyagi, K., and Hanaoka, R. (2017). Negative discharge response of blend oil (esterbased insulating oil with aging mineral oil) in a non-uniform field[C]. *19th IEEE International Conference on Dielectric Liquids*, Manchester, UK (25–29 June 2017).

17 Beroual, A., Sitorus, H.B.H., Setiabudy, R. et al. (2018). Comparative study of AC and DC breakdown voltages in Jatropha methyl ester oil, mineral oil, and their mixtures [J]. *IEEE Transactions on Dielectrics and Electrical Insulation* 25 (5): 1831–1836.

18 Dombek, G. and Gielniak, J. (2018). Fire safety and electrical properties of mixtures of synthetic ester/mineral oil and synthetic ester/natural ester[J]. *IEEE Transactions on Dielectrics and Electrical Insulation* 25 (5): 1846–1852.

19 Beldjilali, A., Idir, O., Saidi-Amroun, N. et al. (2018). Electrical and physicochemical properties and transient charging currents in mineral and vegetable oils mixtures[J]. *IEEE Transactions on Dielectrics and Electrical Insulation* 25 (5): 1739–1748.

20 U.S. Environmental Protection Agency (2002). BIOTEMP® Vegetable Oil-Based Insulating Dielectric Fluids. Environmental Technology Verification Report, Sacramento, California, USA, Chapter 4, EPA 600/R-02/043, pp. 25–26.

21 McShane, C.P., Luksich, J., and Rapp, K.J. (2012). Retrofilling aging transformer with ester based dielectric coolant for safety and life extension. *Cement Industry Technical Conference,* Dallas, TX, USA (4–9 May 2003).

22 McShane, C. (2000). New safety dielectric coolants for distribution and power transformers. *IEEE Industry Applications Magazine* 6: 24–32.

23 Murdiya, F., Hanaoka, R., Akiyama, H., and Miyagi, K. (2014). Creeping discharge developing on vegetable-based oil/pressboard interface under AC voltage. *IEEE Transactions on Dielectrics and Electrical Insulation* 21: 2102–2110.

24 Liao, R.J., Guo, P., Zhou, N.R. et al. (2015). The thermal aging characteristics of the new anti-aging mixed oil-pressboard insulation. *Transactions of China Electrotechnical Society* 30: 222–230.

25 Knothe, G. and Dunn, R.O. (2003). Dependence of oil stability index of fatty compounds on their structure and concentration and presence of metals. *Journal of the American Oil Chemists Society* 80: 1021–1026.

26 Mittelbach, M. and Schober, S. (2003). The influence of antioxidants on the oxidation stability of biodiesel. *Journal of the American Oil Chemists Society* 80: 817–823.

27 Kowalski, B., Gruczynska, E., and Maciaszek, K. (2000). Kinetics of rapeseed oil oxidation by pressure differential scanning calorimetry measurements. *European Journal of Lipid Science and Technology* 102: 337–341.

28 Shanker, T.B., Narasimhaiah, N.H., Antony, D., and Punekar, G.S. (2018). Effects of Transformer-oil Temperature on Amplitude and Peak Frequency of Partial Discharge Acoustic Signals. *IEEE Transactions on Power Delivery* 14: 1.

29 Darwin, A., Perrier, C., and Foliot, P. (2007). The use of natural ester fluids in transformers. *Proceedings of MATPOST Conference*, paper 0036, Lyon, France (15–16 November 2007).

30 Lin, Y., Yang, L., Liao, R. et al. (2015). Effect of oil replacement on furfural analysis and aging assessment of power transformers. *IEEE Transactions on Dielectrics and Electrical Insulation* 22: 2611–2619.

31 Zhang, C.H. and MacAlpine, J.M.K. (2006). Furfural concentration in transformer oil as an indicator of paper ageing, part 1: A review. *Proceedings of the 2016 IEEE/PES Transmission Distribution Conference and Exposition*, Latin, USA (29 October–1 November 2006), pp. 1–6.

32 Matharage, S.Y., Liu, Q., and Wang, Z.D. (2016). Aging assessment of kraft paper insulation through methanol in oil measurement. *IEEE Transactions on Dielectrics and Electrical Insulation* 23: 1589–1596.

33 Okabe, S., Ueta, G., and Tsuboi, T. (2013). Investigation of aging degradation status of insulating elements in oil-immersed transformer and its diagnostic method based on field measurement data. *IEEE Transactions on Dielectrics and Electrical Insulation* 20: 346–355.

34 Feng, D., Yang, L., Liao, R., and Zhou, L. (2018). Effect of moisture content on the production and partitioning of furfural in oil–paper insulation. *IEEE Transactions on Dielectrics and Electrical Insulation* 25: 2389–2397.

35 Yang, L., Liao, R., Sun, C., and Zhu, M. (2011). Influence of vegetable oil on the thermal aging of transformer paper and its mechanism. *IEEE Transactions on Dielectrics and Electrical Insulation* 18: 692–700.

36 Du, Y., Zahn, M., Lesieutre, B.C. et al. (1999). Moisture equilibrium in transformer paper-oil system. *IEEE Electrical Insulation Magazine* 15: 11–20.

37 Matharage, S.Y., Liu, Q., Wang, Z.D. et al. (2018). Aging assessment of synthetic ester impregnated thermally non-upgraded kraft paper through chemical markers in oil. *IEEE Transactions on Dielectrics and Electrical Insulation* 25: 507–515.

38 Lundgaard, L.E., Hansen, W., and Ingebrigtsen, S. (2008). Ageing of mineral oil impregnated cellulose by acid catalysis. *IEEE Transactions on Dielectrics and Electrical Insulation* 15: 540–546.

39 Ding, H.Z. and Wang, Z.D. (2008). On the degradation evolution equations of cellulose. *Cellulose* 15: 205–224.

40 Calvini, P. (2005). The influence of levelling-off degree of polymerisation on the kinetics of cellulose degradation. *Cellulose* 12: 445–447.

41 Feng, D., Hao, J., Liao, R. et al. (2019). Comparative study on the thermal-aging characteristics of cellulose insulation polymer immersed in new three-element mixed oil and mineral oil. *Polymers* 8: 1292.

42 IEC 60243-1 (2013). Electric strength of insulating materials – Test methods – Part 1: Tests at power frequencies.

43 Suwarno and Darmawan, M.A. (2016). Effect of water content on dielectric properties and partial discharge characteristics of silicone oil. *IEEE Conference on Power Engineering and Renewable Energy (ICPERE)* (29–30 November 2016), pp. 125–129.

44 Li, C., Cheng, T., Wang, F., and Worland, E.M. (1992). The effect of moisture on the breakdown of aged oil in underground load interrupter switches. *IEEE Transactions on Dielectrics and Electrical Insulation* 27 (5): 1054–1057.

45 Feng, D., Hao, J., Liao, R. et al. (2019). Power frequency breakdown characteristics of new 3-element mixed insulation oil and oil-impregnated pressboard. *IEEE Transactions on Dielectrics and Electrical Insulation* 26 (6): 1885–1893.

46 Suleiman, A.A., Muhamad, N.A., Bashir, N. et al. (2014). Effect of moisture on breakdown voltage and structure of palm based insulation oils. *IEEE Transactions on Dielectrics and Electrical Insulation* 21 (5): 2119–2126.

47 Liao, R., Hao, J., Chen, G. et al. (2011). A comparative study of physicochemical, dielectric and thermal properties of pressboard insulation impregnated with natural ester and mineral oil. *IEEE Transactions on Dielectrics and Electrical Insulation* 18 (5): 1626–1637.

48 Wu, G., Zhong, X., and Bao, J. (2015). Effect of temperature on dielectric spectrum characteristic of off-paper insulation. *High Voltage* 41 (12): 4081–4088.

49 Kuchler, A. (2018). *High Voltage Engineering–Fundamentals, Technology, Applications*, 215–216. Germany: Springer Vieweg.

50 Umemoto, T., Kainaga, S., Muto, H. et al. (2017). Characteristics of partial discharge in the oil gap separated by pressboard barrier in oil/pressboard composite insulation system. *IEEE International Conference on Dielectric Liquids*, Manchester, UK (29 June–3 July 2014).

51 Feng, D., Hao, J., LijunYang, R.L. et al. (2020). Comparison of AC breakdown characteristics on insulation paper (pressboard) immersed by three-element mixed insulation oil and mineral oil. *High Voltage* 5 (3): 298–305.

52 Hao, J., Gao, C., Chen, X. et al. Creeping discharge characteristics of the liquid/solid system composited of three-element mixed insulation oil and insulation pressboard. *High Voltage Engineering* https://doi.org/10.13336/j.1003-6520.hve.20200227005.

53 Chen, X., Hao, J., Feng, D. et al. (2020). Comparative study on lightning impulse breakdown and gas production characteristics of three-element mixed insulation oil and mineral oil. *Transactions of China Electrotechnical Society* 35 (4): 906–918.

54 Yang, T., Hao, J., Chen, X. et al. (2020). Study on moisture absorption, power frequency breakdown and gas production characteristics of three-element mixed insulation oil. *High Voltage Engineering* 46 (11): 4014–4022.

55 Zhu, M., Zhu, W., Chen, X. et al. (2020). Dielectric, AC breakdown, and gas generation characteristics of three-element mixed insulating oil-impregnated pressboard under different moisture contents. *Insulating Materials* 53 (11): 89–95.

56 Hao, J., Feng, D., Liao, R. et al. (2019). A novel three-element mixed insulation oil and its application in 10 kV distribution transformer. *2019 2nd International Conference on Electrical Materials and Power Equipment (ICEMPE)*, Guangzhou, China (7–10 April 2019), pp. 74–77.

57 Hao, J., Chen, X., Feng, D. et al. (2019). Gas production characteristics and AC breakdown of a new three-element mixed insulation oil. *2019 IEEE 20th International Conference on Dielectric Liquids (ICDL)*, Roma, Italy (23–27 June 2019), pp. 1–4.

58 Beroual, A., Viet-Hung Dang, M.-L., and Coulibaly, C.P. (2013). Investigation on creeping discharges propagating over pressboard immersed in mineral and vegetable oils under AC, DC and lightning impulse voltages. *IEEE Transactions on Dielectrics and Electrical Insulation* 20 (5): 1635–1640.

59 Liao, R., Hao, J., Yang, L. et al. (2010). Simulation and experimental study on frequency-domain dielectric spectroscopy of oil-paper insulation for transformers. *Proceedings of the CSEE* 30 (22): 113–119.

60 Xin, C., Jian, H., Lijun, Y. et al. (2019). Investigation on lightning impulse breakdown and dissolved gas characteristics of the novel three-element mixed insulation oil. *2019 2nd International Conference on Electrical Materials and Power Equipment (ICEMPE)*, Guangzhou, China (7–10 April 2019), pp. 526–529.

61 Feng, D., Hao, J., Liao, R. et al. (2019). Analysis on extending service life of aged transformers by oil replacement with novel 3-element mixed insulation oil. *2019 IEEE 20th International Conference on Dielectric Liquids (ICDL)*, Roma, Italy (23–27 June 2019), pp. 1–4.

62 Feng, D., Hao, J., Wang, Q. et al. (2018). AC breakdown characteristics of a novel three-element mixed insulation oil for power transformer. *2018 IEEE International Conference on High Voltage Engineering and Application (ICHVE)*, ATHENS, Greece (10–13 September 2018), pp. 1–4.

63 Chen, X., Hao, J., Liao, R. et al. AC surface flashover and gas generation difference of the cellulose insulation pressboard immersed in novel 3-element mixed oil and mineral oil. *IEEE Access* 7: 147048–147059.

64 Hao, J., Feng, D., Liu, X. et al. (2019). Excellent electrical and anti-aging performance of a novel 3-element mixed insulation oil. *21st International Symposium on High Voltage Engineering*, Budapest, Hungary (26–30 August 2019).
65 Chen, X., Hao, J., Liu, X. et al. (2019). Researches on the discharge characteristics of the novel 3-element mixed oil-pressboard insulation system under severe non-uniform electrical field. *21st International Symposium on High Voltage Engineering*, Budapest, Hungary (26–30 August 2019).
66 Jian, H., Gao, C., Xin, C. et al. (2021). Overheating and discharge fault identification for novel three-element mixed oil–paper insulation system based on new characteristic gas combination parameters. *Electrical Engineering* 103 (4): 1–16.

8

Natural Ester Nanosfluids as Alternate Insulating Oils for Transformers

Joyce Jacob, Preetha Prabhu, and Sindhu Thiruthi Krishnan

Department of Electrical Engineering, National Institute of Technology Calicut, Kerala, India

8.1 Introduction

Transformers are integral part of the power system, the efficient and reliable operation of which is dependent on their oil–paper insulation system. For over a century, conventional mineral oil insulation has served as a good dielectric and cooling agent in transformers. However, recent environmental regulations regarding the use and safe disposal of mineral oil have raised concerns about its nonbiodegradable nature. Natural esters or green oils gained popularity in the 1990s when their potential as a fully biodegradable oil with good dielectric and thermal properties were identified [1–4]. The results of tests conducted on ester oils were indicative of a possible substitute for mineral insulating oils. Further research in this direction led to the development of nanofluids that possessed better insulation and heat dissipation properties.

8.1.1 Importance of Nanofluids

The growing power demand had led to the development of the HVAC system in the 1880s followed by HVDC in the 1950s and later ultrahigh voltage generation and transmission. The key concern with the increased voltage rating is the design of an insulation system with improved dielectric and thermal properties. This triggered the research toward the development of materials with enhanced insulation properties. Meanwhile, biodegradability concerns over mineral oil have paved the way for natural ester oils. Natural esters are fully biodegradable oils that belong to the K2 class of high fire safety oils. Organization for Economic Co-operation and Development (OECD) has published certain testing guidelines regarding the toxicity of materials to prevent any harm to living organisms. Unlike mineral oil, natural ester has no toxic effect on aquatic and terrestrial organisms as per OECD 301B, OECD 201, OECD 202, and OECD 203 for biodegradability, the effect on the growth of microalgae, the effect toward daphnids, and mortality measure on fish, respectively [5].

The introduction of nanotechnology in the area of dielectrics began in 1994. T.J Lewis explained the major role of interfaces in property enhancement when particles of nanometric dimension were dispersed in a base medium [6, 7]. The development and characterization of solid nanocomposites have resulted in materials with improved dielectric, thermal, and aging properties. With the introduction of nanofluids by Choi et al. as liquids with enhanced thermal properties,

Alternative Liquid Dielectrics for High Voltage Transformer Insulation Systems: Performance Analysis and Applications, First Edition. Edited by U. Mohan Rao, I. Fofana, and R. Sarathi.

researchers considered it an opportunity to widen the possibilities of nanotechnology in dielectric liquids, thus introducing liquid nano dielectrics [8]. The development of an efficient nanofluid–paper insulation system with enhanced dielectric and thermal properties can result in high-voltage transformers with reduced volume.

Natural esters are a probable substitute for nonbiodegradable mineral oil and can satisfy the demands of insulating oils with improved dielectric properties. Though the dielectric properties of natural esters are promising, there are certain limitations to the use of ester oil. The high oxidation-prone nature of ester oils alters their properties and is a major drawback that hinders their actual implementation in transformers. The oxidative nature of natural esters can be inhibited by the addition of suitable reagents. The need for improved insulating oil requires the property enhancement of the ester oil that can be achieved by the use of suitable nanoparticles. A few other details on this discussion are also presented in the next chapter.

8.1.2 Improvement of Natural Esters

Studies conducted on different types of natural ester show that these green oils have good dielectric and thermal properties that make them suitable for use as insulating oils. However, a major concern over the use of green insulating oils is their oxidation-prone nature which makes them unstable. The addition of chemicals often termed as antioxidants is an effective way to reduce the oxidation of natural esters. Literature studies related to the nanoparticle addition in ester oils establish the enhancement in the base oil properties. This section deals with the methods to improve the oxidation stability of ester oils by the use of antioxidants and enhancement of the ester oil properties by the addition of nanoparticles.

8.1.2.1 Additives for Chemical Structure Modification

The basic constituents of natural esters are triglyceride molecules, formed by the esterification reaction between glycerol and acid groups. The structure of glycerol, fatty acid, and triglyceride molecule, and the esterification reaction are shown in Table 8.1 [9].

The triglyceride molecules may be saturated or unsaturated. Saturated triglyceride molecules as present in animals have only carbon–carbon single bonds and are in a tightly packed solid form at room temperature. Unsaturated fats, as found in ester oils (derived from plants), are loosely packed liquids with one or more double bonds in the carbon chain. The unsaturated fatty acid groups are

Table 8.1 Structure of glycerol, fatty acid, and triglyceride molecules [9].

Molecule type	Glycerol	Fatty acid	Triglyceride
Structure	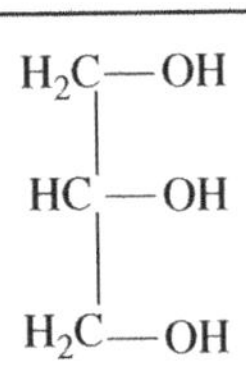	3R—COOH	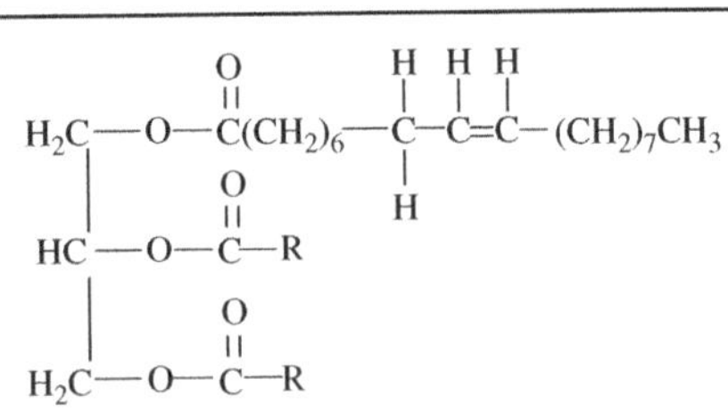
Esterification reaction	Glycerol + Fatty acid → Triglyceride		

susceptible to oxidation by the formation of free radicals. Hence, if unsaturated fatty acids are more in an ester, its oxidation tendency will be more.

The fat molecule (RH) contains an α-methylenic carbon, which is the oxidation site of the triglyceride molecule from where the oxidation process initiates. A hydrogen atom leaves the α-methylenic carbon and an alkyl free radical R• is formed. This alkyl free radical is highly susceptible to the attack of oxygen. The oxidation reaction of the alkyl free radical results in the formation of a peroxy free radical (ROO•), which again is unstable and is subjected to further oxidation [9, 10]. The final product of oxidation is hydroperoxide, the decomposition of which produces alcohols, aldehydes, ketones, and fatty acids. These free radicals initiate a chain of autocatalytic oxidation reactions. The oxidation reaction and formation of free radicals are as shown in Figure 8.1. Antioxidants are chemicals used to inhibit these oxidative reactions.

Antioxidants are of three types – natural, synthetic, and synergists. L-ascorbic acid, α-tocopherol ($\alpha - T$), and citric acid in the anhydrous form are natural antioxidants; butylated hydroxyanisole (BHA), butylated hydroxytoluene (BHT), propyl gallate (PG), pyrogallol, tert butyl hydroquinone, lauryl tert butyl hydro quinone, etc., are synthetic antioxidants; ascorbic acid, carnosic acid, citric acid, phosphoric acid, ethylene diamine tetra acetic acid, carotenes, oryzanol, rosemary extracts, etc., form synergist antioxidants. Antioxidants, which are free radical scavengers, react with the alkyl radical and peroxy radical of fatty acid in natural esters. This reaction produces fatty acids RH, ROOH, respectively, and an antioxidant radical A•. The antioxidant radical further reacts with the free radicals forming non-radical products like RA and ROOA. Also, the antioxidants get oxidized upon reaction with oxygen molecules. The concerned reactions are shown in Eqs. (8.1)–(8.5). A detailed study on the improvement in the natural ester properties by the addition of antioxidants is given by S. S. Kumar et al. in [11].

$$R\bullet + AH \rightarrow RH + A \tag{8.1}$$

$$ROO\bullet + AH \rightarrow ROOH + A\bullet \tag{8.2}$$

$$R\bullet + A\bullet \rightarrow RA \tag{8.3}$$

$$ROO\bullet + A\bullet \rightarrow ROOA \tag{8.4}$$

$$\text{Antioxidant} + O_2 \rightarrow \text{Oxidised antioxidant} \tag{8.5}$$

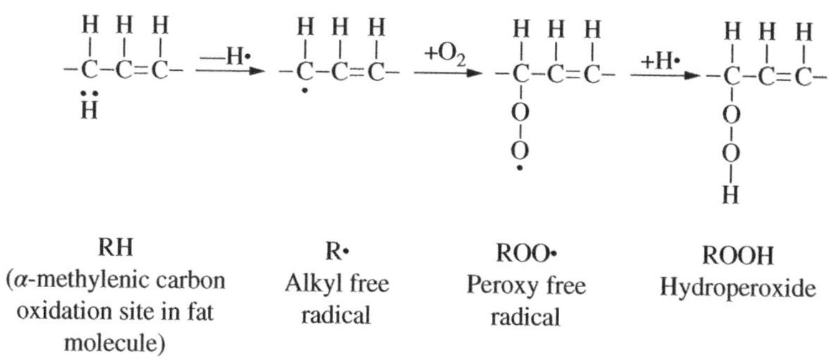

Figure 8.1 "Free radical" mechanism of oxidation of natural esters [9, 10].

8.1.2.2 Addition of Nanoparticles

The addition of a material to a base oil will change its inherent properties; however, this change can be beneficial or detrimental. Literature supports that when particles are added in appropriate size and quantity, it can improve the properties of the base fluid. The property enhancement of fluids by nanoparticle addition has resulted in varied domains of nanofluid application as shown in Figure 8.2 [12]. One such area of application of nanotechnology is the insulating oil in transformers often referred to as liquid nanodielectrics. This encouraged the development of natural ester nanofluids that exhibit improved dielectric and thermal characteristics when compared with their unfilled counterparts, thus increasing their suitability as alternate insulating oils for the conventional mineral oil. The preparation and characterization of natural ester nanofluids are detailed in Sections 8.2 and 8.3, respectively.

8.1.3 Commonly Used Nanoparticles

The choice of the nanofiller is based on the application and can be broadly classified as metallic oxides and hydroxides, silicon oxides, nanoclays (montmorillonite, hectorite), and carbon nanotubes [13]. Metal oxide nanoparticles, the commonly used fillers for property enhancement of dielectric materials, are of three types: the conductive particles are Fe_3O_4 and Fe_2O_3, semiconducting particles are TiO_2, CuO, CuO_2, and ZnO, whereas Al_2O_3, SiO_2, and BN are insulating nanoparticles. These particles differ in their permittivity and conductivity values. Table 8.2 lists

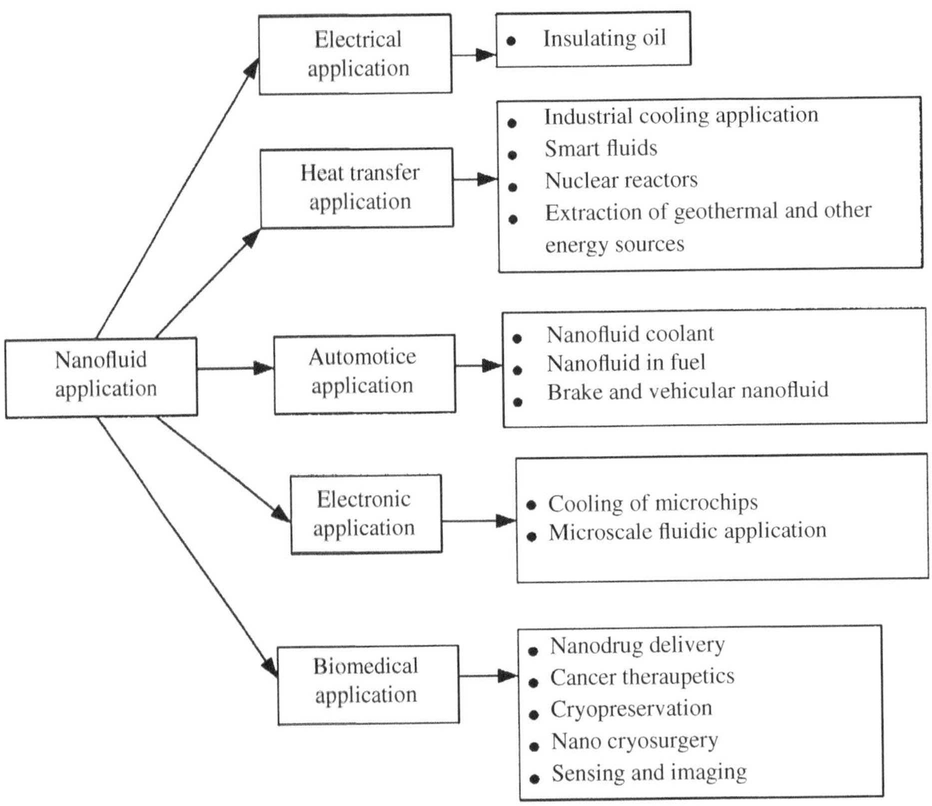

Figure 8.2 Applications of nanofluid. *Source:* Based on Wong and De Leon [12].

Table 8.2 Properties of commonly used nanoparticles.

Properties	Nanoparticle				
	Fe_3O_4	Al_2O_3	TiO_2	SiO_2	ZnO
Density (g/cm^3)	5.17	3.96	4.2–4.3	2.20	5.61
Relative permittivity	80	9.9	100	3.8	7.4–8.9
Relaxation time (s)	$7.47 * 10^{-14}$	12.2	77	$5.12 * 10^{-2}$	$1.05 * 10^{-11}$
Electrical conductivity (S/m)	10^4–10^5	10^{-12}	10^{-11}	$1.4 * 10^{-9}$	10–10^3
Thermal conductivity (W/mK)	4–8	30	—	1.4	23.4
Thermal expansion coefficient at 20 °C (μ/mmK)	9.2	—	—	30	2.9
Specific heat (J/kgK)	—	850	—	670	494

Sources: Based on Rafiq et al. [14] and Hwang et al. [15].

a few commonly used nanoparticles and their properties [14, 15]. Nanoparticle preparation methods are generally classified into two, the top-down and the bottom-up method. In the top-down approach, the nano-sized material is obtained from the bulk object by a series of processes while in the bottom-up approach, the nanoparticle is formed by interlocking of atoms and molecules. Some of the methods used in industry for nanoparticle preparation are mechanical methods involving abrasion and crushing, sol–gel technique, gas phase condensation, liquid solid-state reaction, etc. [16].

8.2 Preparation of Natural Ester Nanofluids and Stability Analysis

Among the many factors that affect nanofluids, stability is a major concern. A nanofluid is stable if the particles suspended in the base fluid remain uniformly dispersed irrespective of time. However, such an extent of stability is not practical and hence, stability can be defined by the extent of monodispersity of the suspended particles in nanofluids, without forming agglomerates. The stability of a nanofluid is dependent on many factors like the type of nanomaterials being used, the concentration of nanoparticles, the preparation methodology, etc. In this section, an overview of the different preparation methods and stability analysis of nanofluids is presented.

8.2.1 Preparation of Natural Ester Nanofluids

The preparation methods employed for mineral oil nanofluids can be extended to natural ester nanofluids with the required variations in control parameters. Hence, a discussion on the general methods of nanofluid preparation is presented here [17].

8.2.1.1 Different Methods of Nanofluid Preparation

Nanofluid preparation in general is classified into single-step preparation methodology and two-step preparation methodology.

8.2.1.1.1 Single-Step Preparation Methodology

As the name suggests, the nanofluid is prepared in a single step in which the nanoparticles are developed within the base fluid by different methods. In [18], VEROS (Vacuum Evaporation onto a Running Oil Stream) is explained, where metal is evaporated onto the surface of a liquid. This physical vapor deposition (PVD) or liquid chemical method is based on the fact that when metal is evaporated onto a solid surface, extremely fine solid particles are formed before a continuous film develops. This concept was applied to a liquid surface to develop ultrafine particles of ferromagnetic metals Fe, Co, and Ni. A modified VEROS using high-pressure magnetron sputtering for electrode evaporation was developed in [19] and was used for the preparation of Ag and Fe suspensions in different base fluids. Based on [18] and [19], a modified VEROS technique was developed at Argonne National Laboratory where nanocrystalline Cu particles were directly evaporated into flowing low-pressure ethylene glycol [20]. Another single-step preparation method was developed by Zhu et al. where Cu nanofluids were prepared by reducing $CuSO_4.5H_2O$ (Copper sulfate pentahydrate) with $NaH_2PO_2.H_2O$ (Sodium hypophosphate hydrate) in ethylene glycol under microwave irradiation [21]. Submerged Arc Nanoparticle Synthesis Systems (SANSS) is a popular single-step nanofluid preparation method. In [22], SANSS is employed to develop Copper oxide nanofluid where pure Cu electrode is submerged in a dielectric liquid (deionized water) in a vacuum chamber. A heating source produces an arc that melts and vaporizes the metal rod and also the deionized water at the region of contact with the arc. The high-pressure vapor quickly removes the vaporized metal which then undergoes nucleation, growth, and condensation, finally forming nanoparticles dispersed in deionized water. A nanofluid preparation method using laser heating where ablation of bulk copper placed in propanol is done using a laser to produce colloidal nanoparticles of copper oxide in propanol [23]. Reduced agglomeration of nanoparticles, avoidance of drying, storage and transportation of nanoparticles are some of the advantages of single-step preparation methods. However, the cost associated with the design setup and the difficulties in large-scale production shifts the focus of researchers toward the two-step preparation method.

8.2.1.1.2 Two-Step Preparation Methodology

In the two-step nanofluid preparation method, nanoparticles are synthesized separately and then mixed with the base oil by ultrasonication, mechanical or magnetic stirring methods. The type and period of mixing affect the proper dispersion of nanoparticles and hence the stability of nanofluid. Many works on nanofluid characterization employ the two-step preparation process due to the low cost of production. Eastman et al. followed the two-step preparation method to disperse Cu, CuO, and Alumina nanoparticles in water [20]. A similar procedure was followed by S. Lee et al. in [24], where Alumina and CuO nanofluids were prepared in water or ethylene glycol. In [25], the authors present three different ways (to reduce agglomeration) of two-step nanofluid preparation using Alumina and CuO in deionized water. In the first method, mechanical blending using a blending machine and ultrasonic bath was done, but it resulted in nanofluids containing nanoparticles both in the monodispersed and agglomerated form. The second method used a polymer coating (styrene-malic anhydride) to prevent agglomeration. In the third method, a filtration procedure was adopted to separate out the agglomerated particles (for diameter $> 1\,\mu m$). In [26], a two-step preparation procedure for Fe_3O_4 nanoparticle and rapeseed oil is carried out in the presence of ethanol as a dispersant and oleic acid as a surfactant. Characterization of nanofluid prepared using boron nanoparticle and kerosene oil in the presence of different types of surfactants like oleic acid, propylene glycol, sorbitan oleate, Tween 85, and CTAB is performed in [27]. The results show that the choice of the surfactant and dispersant based on its compatibility with the nanoparticle and base oil determines the agglomeration rate and in turn the stability of the prepared nanofluid. To improve the dispersion of nanoparticles in the base fluid and to reduce their agglomeration, the use of dispersants and surfactants is widely adopted in the two-step preparation procedure. Figure 8.3 shows the two-step preparation methodology.

The base oil is preheated at 115 °C to remove any moisture content. The required amount of nanoparticles is mixed with surfactant and dispersant by the sonication process. The mixture is then heated to remove all the excess dispersant leaving behind a viscous solution of nanoparticle and surfactant. To this mixture, the required volume of oil is added in steps and sonicated until a properly dispersed nanofluid is obtained. As natural esters are more viscous compared with mineral oil, sonication along with mechanical stirring would give better dispersion of nanoparticles in the base oil.

8.2.1.1.3 Other Methods of Nanofluid Preparation

Chemical Solution Method (CSM) is based on the manipulation of atoms or molecules in the liquid phase. The solution of two reactants A and B in the liquid phase yields a colloid containing a precursor C. Ultrasonication or microwave irradiation of the colloid in the presence of certain additives transforms it into a nanofluid [28]. In [29], nanofluids are prepared using a continuous flow microfluidic microreactor that has a very rapid mixing of reactant fluids resulting in a nanofluid. The phase transfer method, which is another novel nanofluid preparation technique, is discussed in [30–32].

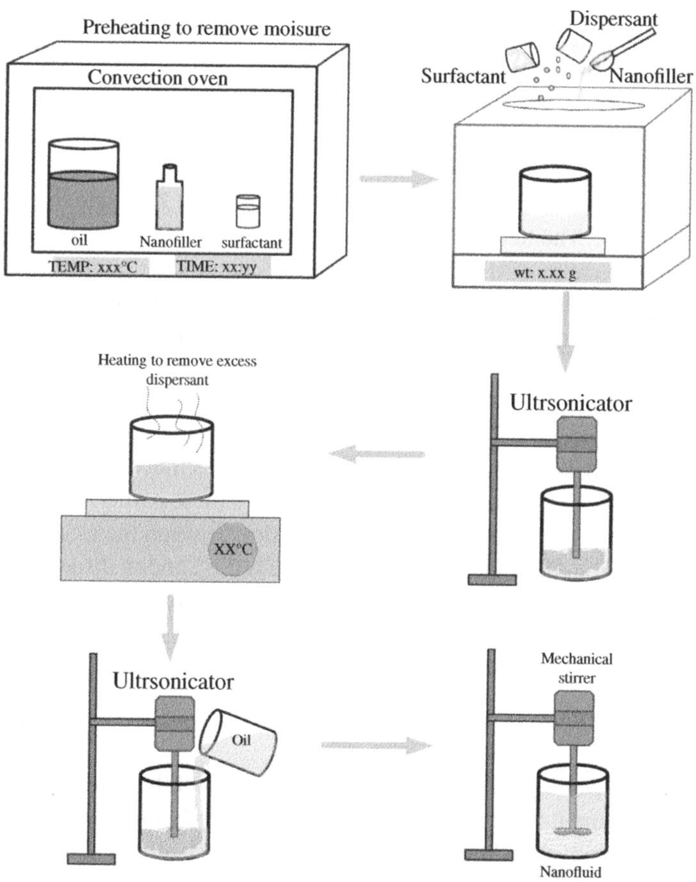

Figure 8.3 Two-step nanofluid preparation methodology.

8.2.2 Stability of Natural Ester Nanofluids

8.2.2.1 Stability of Nanofluids

One of the challenges faced by researchers involved in the development of nanofluids is the preparation of a stable nanofluid. Irrespective of its application, the expected nanofluid performance as an insulating oil or a coolant can be achieved only if the particles remain suspended in the oil. Stability can be attributed to the rate of agglomeration and sedimentation of nanoparticles in the base fluid. The particles dispersed in the base fluid exhibit different interactive forces which result in agglomeration or clustering. Agglomerates of large size thus developed do not show the required properties and may also sediment on surfaces resulting in additional damage to the system. For example, natural ester nanofluids that are proposed to be applicable in transformers as insulating oil provides better insulation and cooling than their unfilled counterparts. However, if the nanoparticles agglomerate, they would sediment on the surface of other materials inside a transformer. Such accumulation of particles on the surface of the solid insulation can drastically affect the performance of the cellulose material. Besides this, particle sedimentation in the cooling duct can cause undesirable clogging, disrupting the oil circulation inside the transformer. Hence, for a nanofluid, any property improvement unless accompanied by considerable stability is futile and conjoint research on nanofluid stability and characterization is preferred.

8.2.2.1.1 DLVO Theory

As the stability of nanofluids is crucial for its practical application, extensive research is being carried out to develop a nanofluid in which the particle remains suspended without forming agglomerates. The stability of nanofluids can be explained using the DLVO theory, named after the scientists Derjagin, Landau, Verwey, and Overbeek who postulated it in 1948. The theory explains that the stability of nanofluids is a balance between the Van der Waals attractive force and the electrostatic repulsive force [33]. Van der Waals attractive force is of three types: Keesome force, Debye force, and London dispersion force. The Keesome force is a dipole–dipole interaction occurring between two permanent dipoles because of the difference in their charge distribution. The Debye force or the dipole-induced dipole interaction occurs when a polar molecule having permanent dipoles approaches a nonpolar molecule and causes charge redistribution in it. The London dispersion force or induced dipole-induced dipole interaction or the electronic polarization force exists between nonpolar molecules. Natural esters have London dispersion forces as the predominant Van der Waals force in them.

When dispersed in a medium, due to Brownian motion, the neutral particles move and collide with one another. Transient polarization or electron transfer occurs between these particles resulting in electron-rich and electron-deficit particles. The attractive interaction between the electron-rich side of one particle and the electron-deficit side of another particle is the London dispersion force. Nonpolar molecules will have weak Van der Waals attractive force compared with polar molecules. The interaction energy is given by Eq. (8.6)

$$W(D) = \frac{-AR}{12D} \tag{8.6}$$

where A is the Hamaker constant that depends on the relative dielectric constant of the material and medium, D is the interparticle distance, and R is the radius of the particle. For a solution with a similar type of particles dispersed in it, A is positive and hence the Van der Waals force is always attractive.

Besides the attractive interaction, there exists a repulsive interaction due to the electron cloud overlap, which can be clearly explained based on the electrical double layer formation. When a nanoparticle is dispersed in a medium, an electrical double layer is formed. When the nanoparticle surface gets a net charge, oppositely charged ions concentrate in the vicinity of the particle. In the suspension, surrounding the particle is the stern layer of strongly bound oppositely charged ions. Beyond this is a layer of loosely bound ions called the diffuse layer. A boundary enclosing the particle, stern layer, and a portion of the diffuse layer (near the stern layer) called the slipping plane together form an entity in the medium as shown in Figure 8.4.

The ions beyond this boundary are a part of the bulk medium. The potential existing at this boundary between the entity and the bulk of the medium is called the zeta potential. The higher the zeta potential, the higher will be the repulsive force exerted by such entities on each other when they come in close contact due to Brownian motion. The repulsive forces prevent flocculation and hence agglomeration of the particles, thus improving the stability of the nanofluid. When particles come close, the diffuse layer overlaps and the ionic concentration in the overlap region increases, i.e. the ionic equilibrium or electroneutrality is lost. This results in an osmotic pressure buildup in this region forcing the liquid medium into this region. The liquid medium exerts a repulsive force between the particles, thus forcing them apart and preventing agglomeration. A simplified expression for the repulsive interaction is given by Eq. (8.7)

$$E = \frac{R}{2} Ze^{-kd} \tag{8.7}$$

where $Z = 9.22 * 10^{-11} \tanh^2 \frac{\psi_0}{103}$ at 25 °C, d is the distance between particles, R is the radius of the particle, k is the Debye length, and ψ_0 is the surface potential of the particle.

The stability of nanofluids based on DLVO theory can be explained using Figure 8.5.

The net force F_n on a particle is the result of the attractive and repulsive force as given by Eq. (8.8)

$$F_n = F_r - F_a \tag{8.8}$$

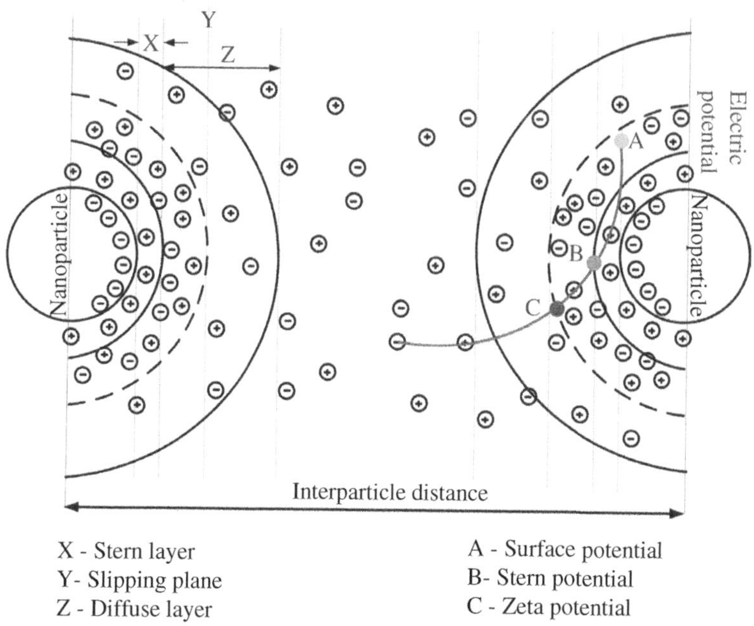

Figure 8.4 Electrical double-layer formation in a nanofluid. *Source:* Based on Lewis [7].

where F_r is the electrostatic repulsive force and F_a is the Van der Waals attractive force. When F_n is negative, the attractive force exceeds the repulsive force as in points "a" and "c" of Figure 8.5. Point "a" corresponds to the region of strong attractive force where permanent agglomerates are formed and "c" is the point where agglomerates formed can be dispersed again by stirring. When Fn is positive, the repulsive force exceeds the attractive force as in points "b" and "d" of Figure 8.5. At point "b," the stability is maximum and the rate of agglomeration is minimum. Also, at point "d," the fluid is relatively stable as the repulsive force exceeds the force of attraction but is of weaker magnitude than at point "b." Hence, it can be understood that for attaining a stable nanofluid, the nanoparticle concentration should neither be high nor should it be too low. It is the intermediate concentration that gives better stability compared with higher and lower nanoparticle concentrations [34].

8.2.2.2 Methods of Stability Improvement of Natural Ester Nanofluids

The most common methods for nanofluid stability enhancement are of two types: chemical methods and mechanical methods.

8.2.2.2.1 Chemical Method

The chemical method of stability enhancement includes the addition of any chemical agent that alters the surface properties of the nanofluid, thus affecting its dispersion and stability. Nanofluid stabilization can be done by adding surfactants (steric stabilization) or by varying the pH value (electrostatic stabilization). The two common chemical methods of stability improvement are briefly explained below:

Addition of Surfactant The contact improvement or wettability between the particle and the base fluid can be achieved by using suitable chemical additives called surfactants. These surfactants

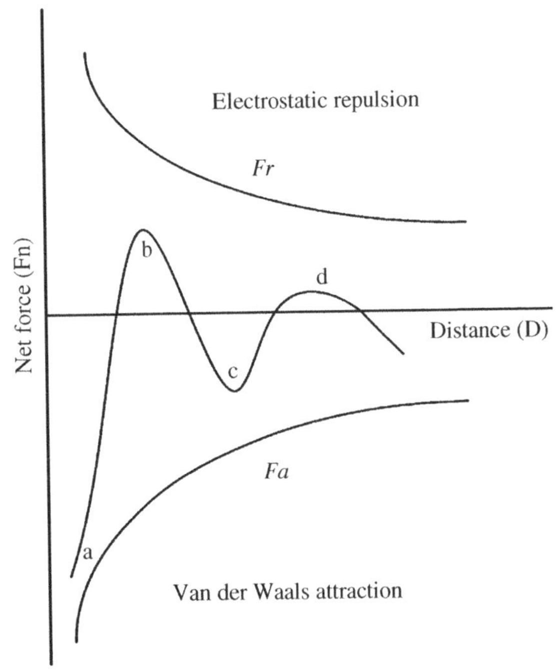

Figure 8.5 Forces existing in nanofluids based on DLVO theory [34].

reduce the interfacial tension at the solid–liquid interface, thus improving the particle suspension in the liquid. Surfactants have a hydrophilic head and a hydrophobic tail with which they provide continuity between the particle and the base fluid [14].

Surfactants can be nonionic, anionic, cationic, and amphoteric based on the composition of their head. The desired stability of a nanofluid is dependent on the choice of the surfactant. Polar fluid would require a water-soluble surfactant while natural esters and mineral oil would require an oil-soluble surfactant. The hydrophilic/lipophilic balance (HLB) value can be used to select the appropriate surfactant. A surfactant with high HLB indicating high water solubility can be used for polar fluids, while for oil-based nanofluids, surfactants with low HLB are preferred. The commonly used surfactants include oleic acid, stearic acid, propylene glycol, sorbitan oleate, Span and Tween, CTAB, SDBS, SDS, SOCT, DTAB, etc. For natural ester and mineral oil nanofluids, fatty acids like oleic acid, stearic acid, etc., are preferred as surfactants [17]. Adding a suitable surfactant during nanofluid preparation is the simplest and most cost-effective method of nanofluid stabilization. However, for applications involving elevated temperature, bond breakage between the particle and surfactant may occur accompanied by the formation of undesirable foam. Also, thermal resistance and hence thermal conductivity can be affected by the presence of surfactants [17, 35]. However, the choice of a suitable surfactant does show considerable improvement in the stability of the nanofluids.

pH Control The stability of a nanofluid can be enhanced by adjusting the pH value of the solution. This can be achieved by the addition of suitable acid or base solutions to the medium which does not interfere with their desired properties. A solution is stable when the zeta potential has a magnitude greater than 30 mV irrespective of whether it is positive or negative. A solution with a zeta potential value between −30 mV and +30 mV is an unstable solution with a high particle agglomeration rate as shown in Figure 8.6. The pH of the nanofluid can be adjusted by adding an alkali or acid medium to change the zeta potential of the nanofluid above +30 mV or below −30 mV. It is at the isoelectric point (IEP) that zeta potential is the minimum, hence for better stability of the solution, its pH should be much higher or lower than the pH corresponding to the IEP [36].

As shown in Figure 8.6, pH1 corresponds to the IEP (where zeta potential is zero), hence the addition of an acidic medium will reduce the pH of the solution toward pH2 and the addition of an alkaline solution would increase the pH toward pH3. For pH of the solution below pH2 (corresponding to zeta potential of −30 mV) and above pH3 (corresponding to zeta potential of +30 mV), the solution will be considered stable. In [37], a study is done on the change in the zeta potential of Al_2O_3–water and CuO–water nanofluid by varying its pH value using NaOH, NH_4OH,

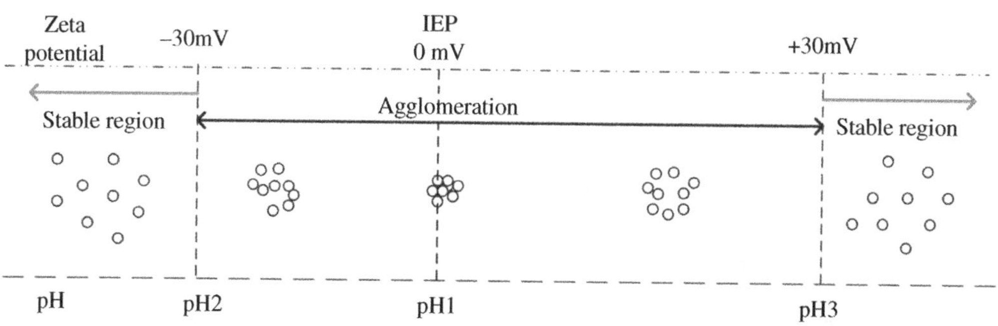

Figure 8.6 Effect of pH on the stability of nanofluids.

and acetic acid. It is observed that for Al_2O_3–water nanofluid, the IEP corresponds to a pH value around 7 and the zeta potential is maximum at pH of around 5 (acidic region). Similarly, for CuO–water nanofluid, the IEP was at a pH of around 6.5 while the maximum stability was attained around 5.5.

8.2.2.2.2 *Mechanical Method*

The mechanical method for improving stability is by stirring the nanofluid to break the clusters formed. Ultrasonication, magnetic stirring, and high shear stirring are different methods of homogenizing the nanofluid. However, stirring by ultrasonic probe or bath-type ultrasonication is preferred as it provides better uniformity in the particle dispersion. The duration for stirring depends on the type of base fluid, nanoparticle, and has to be optimized based on the sedimentation rate after stirring. Also, it is preferred by many researchers to apply a combination of the different chemical and mechanical methods for enhancing nanofluid stability.

8.2.2.3 Methods of Stability Analysis of Natural Ester Nanofluids

There are many methods to determine the stability of a nanofluid. Some of the commonly followed methods are described below:

8.2.2.3.1 *Sedimentation Method*

This is the simplest method of determining the stability of a colloidal solution wherein the solution is left undisturbed in glass tubes/bottles [34, 35, 38]. Depending upon its stability, the particles in the solution agglomerate and begin to sediment leaving a clear solution at the top and the sediments at the bottom. By measuring the time taken by the particles to sediment, a comparison of the stability of different solutions can be done. This method, however, does not provide a quantitative measure of the nanofluid stability. Also, for nanofluids with a low concentration of nanoparticles, the sedimentation is not visible and hence comparison becomes difficult. Figure 8.7a shows the sedimentation in a nanofluid over a certain period.

8.2.2.3.2 *Dynamic Light Scattering (DLS) Method*

This method uses the measurement of zeta potential and particle size distribution for determining the stability of nanofluids. It is reported in [35, 39, 40] that the zeta potential and hence stability of nanofluids increases with sonication time. The size of the nanocluster decreases with an increase in ultrasonication time. The reduction in the nanocluster size is higher during the initial period after which the rate decreases and finally becomes a constant. A representation of the size distribution and size range as shown in Figure 8.7b can be useful in studying the stability of nanofluids. It is seen that fluid B has more particles with a size in the lower dimension range compared with fluid A indicating reduced agglomeration and higher stability of fluid B. Another useful term is the polydispersity index (PDI) which when decreased below 0.3 indicates mono-dispersed particles, i.e. as ultrasonication time increases, the nanoclusters break apart and the cluster size decreases and thus the PDI value decreases below 0.3. This sonication time is the minimum time required for the mono dispersion of particles. A combination of the zeta potential, part size reduction rate, the intensity of light scattering, and PDI together can determine the ultrasonication time required for preparing a stable solution. The same parameters can be employed for comparing the stability of different nanofluid concentrations by keeping the ultrasonication time constant.

8.2.2.3.3 *UV-Vis Spectrum Method*

In this method, a UV-Vis spectrophotometer is used to measure the maximum absorbency of UV light by a colloidal solution [34, 35, 38, 39, 41]. A representation of the absorbance curve with

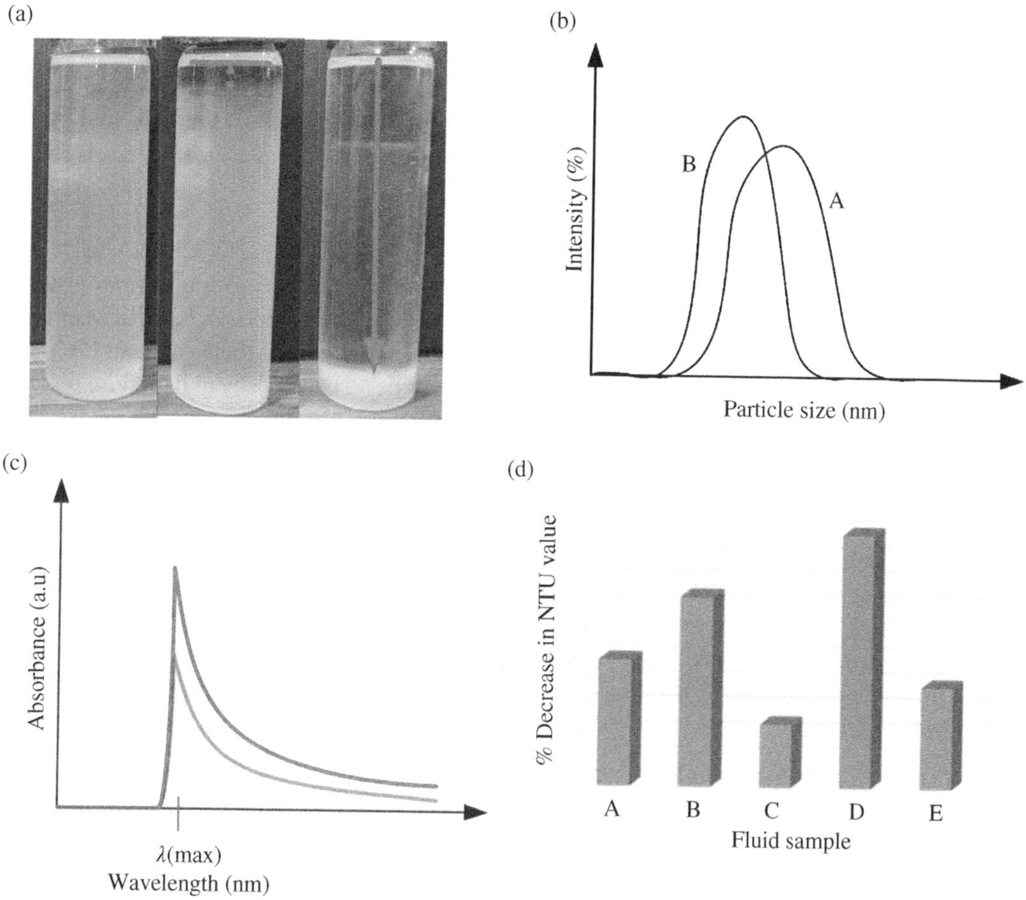

Figure 8.7 Stability analysis of nanofluids by (a) Sedimentation method, (b) DLS method, (c) UV-Vis spectrophotometer method, and (d) turbidity meter method.

respect to wavelength is shown in Figure 8.7c. The maximum absorbance (A_1) of the nanofluid immediately after preparation and the corresponding wavelength λ(max) is noted. As time progresses, the nanoparticles agglomerate and the size of the cluster increases. Light absorbency decreases with the increase of particle size. After a fixed time, the light absorbency (A_2) at λ(max) is again measured. The decrease in light absorbency from A_1 to A_2 is due to the formation of clusters. This in turn is a measure of the stability of the nanofluid. While comparing the stability of different concentrations of nanofluids, the nanofluid which has the lower percentage decrease in absorbency will be the most stable one.

8.2.2.3.4 Turbidity Measurement

Turbidity is the cloudiness of the suspended particles in a solution and is measured in terms of the Nephelometric Turbidity Unit (NTU). The NTU value of a colloid will be the highest just after the preparation as most of the particles will be in the suspended form. The NTU value of the solutions is measured after regular intervals. The NTU value decreases over time as particles begin to agglomerate. A representation of turbidity variation of five samples A, B, C, D, and E is shown in Figure 8.7d. The fluid sample C with the least percentage decrease in NTU value can be considered as the most stable sample [34, 42, 43].

8.3 Properties of Natural Esters and Natural Ester Nanofluids

A study of the properties of natural ester nanofluids and a comparison with mineral oil nanofluids is done in this section with a primary focus on the physical, electrical, thermal, and aging characteristics.

8.3.1 Physical Properties

The physical properties of nanofluids are studied to check their feasibility as transformer insulating oil. Density, viscosity, flash point, fire point, etc., are some of the properties that are commonly studied. The density of the nanofluid usually increases with the nanoparticle concentration but does not have any significant effect on the insulation properties. The viscosity of nanofluid is studied in detail as it affects the thermal properties of the nanofluid. Besides, the increase in viscosity on the addition of nanoparticles is alarming for a natural ester nanofluid as the unfilled ester oil itself is highly viscous when compared with the mineral oil. When viscosity increases, the free flow of the liquid is hindered which prevents effective heat dissipation further leading to localized hot spots and eventually thermal breakdown. However, an elevated temperature within the safe operating limits can provide a check on the viscosity increase to a great extent. The flash point and fire point of natural esters are almost twice that of mineral oil. Also, the addition of nanofillers does not create a noticeable increase or decrease in these values. The variations in the physical properties of mineral oil and esters by addition of different types and concentration of nanoparticles (NP) are shown in Table 8.3.

8.3.2 Electrical Properties

The study of the electrical properties of insulating oils is primarily focused on the permittivity, breakdown strength, and partial discharge. A detailed study of the variation of these parameters with the nanoparticle addition is presented in the section below.

8.3.2.1 Permittivity of Nanofluids

In the presence of an electric field, positive and negative ions in a nanoparticle separate and are uniformly distributed on the particle surface. This process of charge separation in a nanoparticle resembling a dipole is called polarization. Permittivity ($\mathcal{E}$) can be defined as a measure of the polarizability of material under the influence of an electric field. Polarization and hence permittivity is a complex function dependent on the frequency. The real part ($\mathcal{E}'$) is the charge storage capacity/polarizability of the material and the imaginary part ($\mathcal{E}''$) corresponds to the dielectric loss. To determine the polarizability of a mixed dielectric (such as nanofluid), the Maxwell-Garnett equation as shown in Eq. (8.9) was used.

$$\frac{\varepsilon_{rn}-\varepsilon_{r1}}{\varepsilon_{rn}+2\varepsilon_{r1}}=\phi\frac{\varepsilon_{r2}-\varepsilon_{r1}}{\varepsilon_{r2}+2\varepsilon_{r1}} \tag{8.9}$$

where ε_{rn}, ε_{r1}, ε_{r2} are the relative permittivity of nanofluid, oil, and particle, respectively, and ϕ is the volume fraction of the nanoparticle. However, this equation may not be suitable for nanofluids as they are usually prepared by dispersing a low volume fraction of nanoparticles. This is because, for lower volume fractions, the effect of ε_{r2}on ε_{rn} will be less. Hence, to consider the contribution of nanoparticle permittivity on the nanofluid permittivity, the Clausius–Mossotti equation is used.

Table 8.3 Comparison of physical properties of mineral oil and natural ester nanofluids.

Sl. no.	NP conc.	Oil type	NP type	Viscosity (cSt)		Flash point (°C)		Fire point (°C)		Ref.
				UF	NF	UF	NF	UF	NF	
1.	0.25 g/l	Mineral oil	Al_2O_3	15	18.9	175	179	190	195	[44]
			BN		19.7		183		194	
			Fe_3O_4		20.3		181		196	
		Honge oil	Al_2O_3	58	62.5	289	292	300	305	
			BN		61.3		2923		306	
			Fe_3O_4		62.7		294		306	
		Neem oil	Al_2O_3	60	62.9	288	303	297	310	
			BN		62.8		297		321	
			Fe_3O_4		62.5		306		314	
		Mustard oil	Al_2O_3	45	47.6	289	292	300	305	
			BN		47.5		293		306	
			Fe_3O_4		47.3		294		306	
		Punna oil	Al_2O_3	52	54.3	280	280	305	308	
			BN		54.5		281		311	
			Fe_3O_4		47.3		288		310	
2	0.05 g/l	Corn oil	SiO_2	210	180	210	320	350	350	[45]
		Coconut oil		160	170	270	280	310	310	
3	0.02 (wt.%)	Mineral oil	Al_2O_3	9.75	10	162	160	170	170	[34]
		Soybean oil		39.9	40.7	310	311	315	315	
4.	0.01 (wt.%)	Coconut oil	ZnO	86.45	83.15	235	290	270	315	[46]
			TiO_2	86.45	83.14	235	240	270	275	
		Corn oil	ZnO	117.37	111.86	225	235	310	315	
			TiO_2	117.37	115.17	225	230	310	315	
		Punga oil	ZnO	142.73	140.53	210	240	235	260	
			TiO_2	142.73	141.63	210	215	235	235	
		Sunflower oil	ZnO	109.21	88.68	220	275	310	310	
			TiO_2	109.21	101.48	220	220	310	315	
5.	0.25 (vol.%)	Mineral oil	Fe_3O_4	61% increase		13.87% increase		9.59% increase		[47]
			ZnO	69% increase		10.95% increase		7.53% increase		
			SiO_2	78% increase		8.76% increase		6.16%increase		
		Sunflower oil	Fe_3O_4	30% increase		—		—		
			ZnO	47% increase		—		—		
			SiO_2	52% increase		—		—		
		Rapeseed oil	Fe_3O_4	28% increase		—		—		
			ZnO	34% increase		—		—		
			SiO_2	39% increase		—		—		

In a nanofluid, the application of an electric field results in the polarization of the oil molecules, the inner polarization of the nanoparticle, and orientation polarization of the nanoparticle as depicted in Figure 8.8.

Clausius–Mossotti equation for nanofluids is given by Eq. (8.10)

$$\frac{\varepsilon_{rn}-1}{\varepsilon_{rn}+2}=\frac{1}{3\varepsilon_0}*\left(N_1\alpha_1+N_2\alpha_2+N_2\alpha_3\right) \tag{8.10}$$

where ε_0 is the permittivity of vacuum, α_1, α_2, α_3 are the polarizability of the transformer oil, inner nanoparticle polarization, and the orientation polarization of the nanoparticle, respectively, N_1 and N_2 are the number of oil molecules and nanoparticles, respectively, in a unit volume of the nanofluid. A study of the nanofluid polarization and permittivity variation is presented in [26]. At low frequency, the time period of half cycle will be high, which improves the polarizability and dielectric constant of a material. This results in increased electron trapping on the particle surface, thus decreasing the conductivity of the material. At high frequency, the polarizability will be low and hence the dielectric constant will be less, resulting in reduced electron trapping and improved conductivity.

To find the transformer oil polarizability $N_1\alpha_1$, the Clausius–Mossotti equation can be modified as in Eq. (8.11)

$$\frac{\varepsilon_{r1}-1}{\varepsilon_{r1}+2}=\frac{1}{3\varepsilon_0}*\left(N_1\alpha_1\right) \tag{8.11}$$

From the above equation, transformer oil polarizability $N_1\alpha_1$ can be obtained. Figure 8.9 represents the polarization of nanoparticles in nanofluids.

The internal electric field E_i is calculated using Eq. (8.12)

$$E_i=\frac{3\varepsilon_{r1}}{2\varepsilon_{r1}+\varepsilon_{r2}}E_0 \tag{8.12}$$

where ε_{r1}, ε_{r2} is the permittivity of the oil and nanoparticle, respectively, and E_0 is the applied electric field.

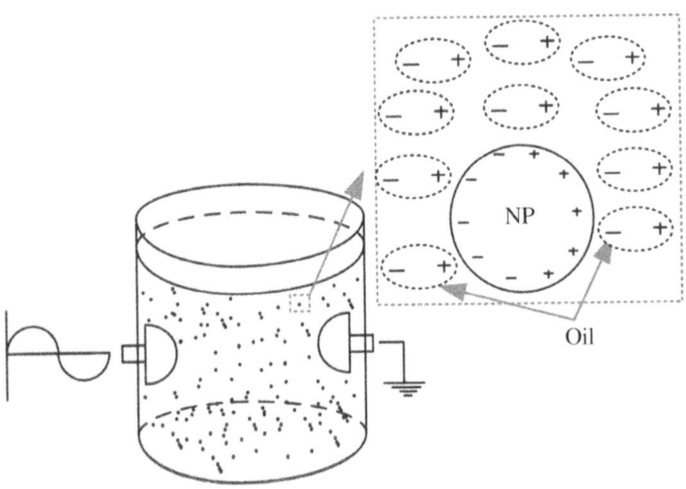

Figure 8.8 Polarization in nanofluids [48].

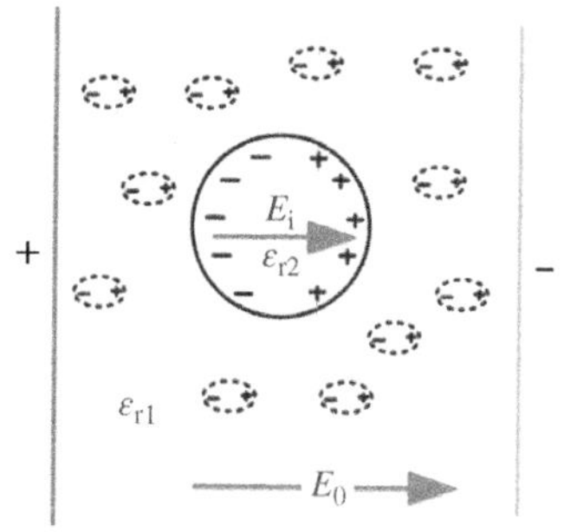

Figure 8.9 The internal electric field in a polarized nanoparticle [48].

Polarization vector P_i is calculated using Eq. (8.13)

$$P_i = \varepsilon_0 \left(\varepsilon_{r2} - 1\right) E_i \tag{8.13}$$

The vector sum of the dipole moment μ_i within a nanoparticle is calculated using Eq. (8.14)

$$\mu_i = P_i V \tag{8.14}$$

where V is the volume of the nanoparticle.

Nanoparticle polarizability, α_2, is calculated using Eq. (8.15)

$$\alpha_2 = \frac{\mu_i}{E_i} \tag{8.15}$$

The number of nanoparticles, N_2, is obtained from Eq. (8.16)

$$N_2 = \frac{\phi}{V} = \frac{3\phi}{4\pi a^3} \tag{8.16}$$

where a is the radius of the nanoparticle. Using Eqs. (8.15) and (8.16), we have $N_2\alpha_2$ of Eq. (8.10).

Orientation polarizability of the nanoparticle is calculated using Eq. (8.17)

$$\alpha_3 = \frac{\mu_c^2}{3KT} = \frac{4a^2Q^2}{3KT} \tag{8.17}$$

where μ_c is the electric dipole moment of the charged nanoparticle, K is the Boltzmann constant, and T is the thermodynamic temperature. Q is the magnitude of the total positive or negative charge on a nanoparticle surface given by $Q = \pi a^2 \varepsilon_0 E_0 \frac{\varepsilon_{r2} - \varepsilon_{r1}}{\varepsilon_{r2} + 2\varepsilon_{r1}}$.

The term $N_2\alpha_3$ of Eq. (8.10) can be obtained from Eqs. (8.16) and (8.17).

Incorporating Eqs. (8.11), (8.15)–(8.17) in Eq. (8.10), the Clausius–Mossotti equation for nanofluids can be written as in Eq. (8.18)

$$\frac{\varepsilon_{rn} - 1}{\varepsilon_{rn} + 2} = \frac{\varepsilon_{r1} - 1}{\varepsilon_{r1} + 2} + \frac{\phi}{3}\left(\varepsilon_2 - 1\right) + \frac{\phi}{3KT}\pi\varepsilon_0 a^3 E_0^2 \left(\frac{\varepsilon_{r2} - \varepsilon_{r1}}{\varepsilon_{r2} + 2\varepsilon_{r1}}\right)^2 \tag{8.18}$$

The validation of Eq. (8.9) is given in [48], where the permittivity of mineral oil nanofluid prepared using ZnO nanoparticles is determined experimentally and compared with the relative permittivity values obtained using Eqs. (8.9) and (8.18). The experimentally obtained permittivity values of the nanofluid are in adherence to those obtained using Eq. (8.18) and vary significantly from those obtained using Eq. (8.9). Also, from calculations, it is found that the value of the nanoparticle polarizability $N_2\alpha_2$ and orientation polarizability $N_2\alpha_3$ is much smaller than transformer oil polarizability $N_1\alpha_1$ as ϕ is very small in nanofluids. Hence, it is clear that the permittivity of the nanofluid is significantly influenced by the base oil permittivity when compared with that of the nanoparticle. However, $N_2\alpha_2$ is much higher than $N_2\alpha_3$ and hence the permittivity of nanofluid is comparatively higher than the unfilled counterpart due to the contribution of the inner polarization of nanoparticles compared with the orientation polarization.

Although the permittivity of nanofluids is higher than that of unfilled oil, it is not different at higher frequencies compared with lower ones. At high frequency, the time period and hence the polarizability of nanoparticles is less. So, the dielectric constant/ relative permittivity of nanofluids is similar to that of the unfilled oil. Similar will be the case of dissipation factor at high frequency. However, at low frequency, the polarizability of nanoparticles is high and hence the dielectric constant of nanofluids is higher when compared with the unfilled oil. The dissipation factor on the other hand will be higher for unfilled oil because of the predominance of conductance loss over polarization loss [26, 49].

8.3.2.2 Partial Discharge and Breakdown Voltage in Nanofluids

The presence of free electrons accompanied by a strong electric field can lead to localized discharges in the liquid medium referred to as partial discharge. Increased frequency and intensity of partial discharge in the liquid can subsequently lead to a continuous electron path or streamer between the electrodes. The ability of an insulating material to withstand the applied electric field and resist streamer development is measured in terms of the breakdown strength. Different theories have been postulated for the mechanism of streamer development and propagation in nanofluids, namely the electron scavenging theory, potential well model, shallow trap theory, and hydrophilicity theory [50].

8.3.2.2.1 Electron Scavenging Theory

When an electric field is applied to the nanofluid, the nanoparticles and the oil molecules are polarized. Positive and negative ions created by the nanoparticle polarization get uniformly distributed on the nanoparticle surface in the direction of the applied electric field. At the same time, the positive and negative ions created by the polarization of the oil molecules move toward the negative and positive electrodes, respectively. As the electrons are of low mass compared with the positive ions, they are fast moving with mobility of 1×10^{-4} m^2/V/s, while the heavier positive ions have considerably low mobility of 1×10^{-9} m^2/V/s. Hence, there will be fast-moving electrons toward the positive electrode while the positive ions remain more or less stationary. The free electrons on encountering a nanoparticle will be captured and trapped by the positive ions on the nanoparticle surface, thus neutralizing them. A point of saturation is reached where the nanoparticle cannot trap any more free electrons. Thus, the nanoparticles capture the free electrons and inhibit the formation of a continuous streamer between the electrodes. This in turn improves the breakdown strength of the nanofluid. The time required for the nanoparticles to polarize under the applied electric field is the relaxation time constant ($\boldsymbol{\tau}$) given by Eq. (8.19)

$$\tau = \frac{2\varepsilon_1 + \varepsilon_2}{\sigma_1 + \sigma_2} \tag{8.19}$$

where ε_1, σ_1 and ε_2, σ_2 are the relative permittivity and electrical conductivity of the nanofluid and the nanoparticle, respectively. The theory, however, can be applied only to nanoparticles whose relaxation time constant is lower than the time required for streamer development.

8.3.2.2.2 Potential Well Model

The relative permittivity difference between the nanoparticle and the oil can distort the electric field around the particle surface creating charge traps or potential wells. The addition of surfactant to the nanoparticle would increase the effective radius of the nanoparticle, which in turn would increase the potential gradient resulting in deep traps to capture free electrons. Hence, if in the nanofluid the trap depth on the particle surface is higher than that of the traps in the unfilled oil, then the breakdown strength of the nanofluid will be higher than the unfilled oil.

8.3.2.2.3 Shallow Trap Theory

In [51], Pulse Electro Acoustic (PEA) studies were conducted on unfilled and nanofilled ester oil to understand the charge accumulation and hence the electric field distribution. The ester oil samples were stressed at 16 kV/mm for half an hour. Results show that for pure ester, the charge density increases over time while for nanofluid it is decreasing. The internal electric field of pure ester increases and reaches 256% of the applied field while for the nanofluid, it remains almost constant and never exceeds 108% of the applied field. Similarly, the charge decay rate of unfilled and nanofilled ester oil is 47.3 and 64.8% in the first eight seconds. The higher charge decay rate in nanofluids prevents accumulation of charge resulting in a uniform electric field distribution. This rapid charge distribution in nanofluids could be due to high shallow trap density which results in electron trapping and de-trapping also called electron hopping. The electron hopping causes the electrons to slow down thus preventing charge accumulation and nonuniformity of the electric field, which delays the streamer development.

8.3.2.2.4 Hydrophilicity Theory

This is based on the affinity of the nanoparticle toward moisture. In [52], it is observed that the BDV of silica nanofluids is improved for higher levels of moisture content than the unfilled oil. The nanofluid prepared using silica nanoparticles have improved BDV as the moisture present in them is absorbed by the nanoparticles. To confirm the beneficial effect of the hydrophilicity of silica nanoparticles on BDV, silica nanoparticles were surface modified using surfactants. This coating of another material on the particle surface disturbs its hydrophilicity and hence the BDV of the nanofluid prepared using the surface-modified nanoparticles is not improved.

Literature reports extensive studies on the properties of natural ester nanofluids. Table 8.4 shows the AC breakdown voltage of unfilled (UF) and nanofilled (NF) natural esters. It is clear from the table that the addition of nanoparticles to the different ester oil increases the breakdown strength of the oil. However, there is an optimal concentration of nanoparticles beyond which the breakdown voltage might decrease due to the onset of agglomeration in the nanofluid. It is reported in [26] that the average positive and negative impulse breakdown voltage of natural ester nanofluids is 37 and 12% higher than the unfilled oil. Also, the time to breakdown by the application of positive and negative impulse breakdown voltage of the nanofluid is 21 and 14% higher than the unfilled oil. This is in accordance with the delay in streamer propagation observed by the addition of nanoparticles.

A study on the different factors that affect the breakdown strength of an insulating oil like moisture content, type and concentration of nanoparticles, etc., is presented in [14]. The presence of moisture in the oil can be in free form or bound form. It is seen that with an increase in moisture

Table 8.4 Comparison of AC breakdown voltage of mineral oil and natural ester nanofluids.

Sl. no.	NP	Base oil	Nanoparticle concentration	AC breakdown voltage (kV)		Standard	Ref.
				UF	NF		
1.	Fe_3O_4	Rapeseed oil	0.004 wt.%	49.9	59.8	IEC 60156	[26]
		FR3	0.008 w/w	65.4	77.7		[53]
		Mineral oil	0.25 vol.%	50% increase			[47]
		Sunflower oil		51% increase			
		Rapeseed oil		52.5% increase			
		Mineral oil	0.25 g/l	29	24		
		Honge oil		43	44		[44]
		Neem oil		43	45		
		Mustard oil		42	45		
		Punna oil		42	46		
2.	TiO_2	FR3	0.006 25 wt.%	62.71	82.23	ASTM D1816	[51]
3.	SiO_2	Sunflower oil	0.25 vol.%	42% increase		IEC 60156	[47]
		Rapeseed oil		42.5% increase			
		Corn oil	0.05 g/l	33.2	53.6		[45]
		Coconut oil	0.05 g/l	50	54.4		
4.	Al_2O_3	Mineral oil	0.25 g/l	29	22		[44]
		Honge oil		43	45		
		Neem oil		43	44		
		Mustard oil		42	45		
		Punna oil		42	46		
		Mineral oil	0.02 wt.%	41.45	43.30	ASTM D1816	[34]
		Soybean oil		52.81	57.47		
5.	BN	Mineral oil	0.25 g/l	29	24	IEC 60156	[44]
		Honge oil		43	43		
		Neem oil		43	44		
		Mustard oil		43	45		
		Punna oil		43	46		
		Cotton seed oil	0.1 wt.%	37.6	61.4	IEC 60156	[54]
6.	ZnO	Sunflower oil	0.25 vol.%	46% increase		IEC 60156	[47]
		Rapeseed oil	0.25 vol.%	47.5% increase			
		Punga oil	0.01 wt.%	26.4	45.2		[46]
		Corn oil	0.01 wt.%	20.8	57.6		
		Coconut oil	0.01 wt.%	36.8	59		
		Soybean oil	0.3 wt.%	23.6	28.1	ASTM D877	[55]
		Palm oil	0.5 wt.%	20.7	31.8		

content, the BDV of unfilled and Fe_3O_4 filled mineral oil decreases, but the presence of moisture has less effect on the ferrofluid. This might be due to the ability of the nanoparticle to bind water molecules on its surface, thus reducing the free water molecules in the oil. Similar results were obtained for TiO_2 nanofluids, where the higher breakdown strength compared with the unfilled oil was attributed to the semiconductive nature of TiO_2 which reduces the field distortion due to moisture content. SiO_2 nanofluids also showed high breakdown strength probably due to the hydrophilic nature of the nanoparticles. As the particle concentration increases, the BDV also increases. However, there is an optimum particle concentration beyond which the BDV begins to decrease. Agglomeration of particles at high concentrations could be the reason for this reduction in BDV. Surface-modified nanoparticles when dispersed in oils result in improved BDV of the oil, which could be due to the increased electron trapping by the nanoparticles, thus delaying the streamer formation.

Partial discharge tests on palm oil nanofluids prepared using Fe_3O_4 and TiO_2 nanoparticles are conducted in [56]. Weibull parameters of the obtained PD data show that the addition of nanoparticles has resulted in the localization of PD. Also, with the increase in particle concentration, the Partial Discharge Inception Voltage (PDIV) decreases for Fe_3O_4 nanofluid while for TiO_2 nanofluid, the PDIV increases. This could be due to the increased charge trapping capacity of Fe_3O_4 nanoparticles when compared with TiO_2 nanoparticles. For the same set of nanofluids, when PD studies were done at a fixed voltage of 26 kV, it is found that the number of PD events, total charge, mean charge, and maximum charge magnitude are lower for both Fe_3O_4 and TiO_2 nanofluids when compared with the unfilled oil. Also, it is observed that with an increase in nanoparticle concentration, for Fe_3O_4, the PD repetition and magnitude increases while for TiO_2 it decreases. As nanoparticle concentration increases, dipole–dipole interaction increases and results in increased conductivity in Fe_3O_4 nanofluids but due to the low conductivity of TiO_2 nanoparticles, the insulation mechanism predominates conduction.

8.3.3 Thermal Properties

Insulating oil is also intended to operate as a coolant and hence it must possess the required thermal properties to maintain the transformer operating temperature within safe limits. Solid metals have higher thermal conductivity than liquid as shown in Table 8.5. Hence the addition of solid particles to the cooling liquids was considered as one method of improving their thermal properties [8, 20, 24, 58]. This led to the suspension of micron-size particles in the coolants resulting in considerable improvement in the thermal properties. However, at higher concentrations, agglomeration of particles followed by sedimentation resulted in clogging of the cooling ducts. This problem was overcome by the development of nanofluids as a cooling agent which had particles of nanometric dimension that prevented clogging in the microchannels of the cooling ducts. Many factors are considered to affect the variation in thermal conductivity like Brownian motion of nanoparticles, formation of a thermally conducting layer at the fluid–particle interface, ballistic heat transfer within a nanoparticle and between nanoparticles, nanoparticle clustering, thermophoretic effect, and near field radiation [57, 58].

It was Maxwell who proposed that spherical particle suspensions in a liquid can cause enhancement in thermal conductivity. An increased surface area to volume ratio resulting from the reduced particle size can improve the effective thermal conductivity. Maxwell model for spherical particle suspension was extended by Hamilton and Crosser model for nonspherical particles. A few thermal conductivity models applicable for nanofluids are given in Eqs. (8.20)–(8.23) [25].

Table 8.5 Thermal conductivity of commonly used solids and liquids [57].

Type of material	Material	Thermal conductivity (W/mK)
Solid	Aluminum	237
	Copper	401
	Gold	317
	Silver	429
	Al_2O_3	20
	CuO	40
	SiC	120
	Carbon nanotubes	3000
Liquid	Water	0.55
	Engine oil	0.145
	Ethylene glycol	0.253
	Sodium at 644 K	72.3

1) Maxwell Model

$$\frac{k_e}{k_f} = 1 + \frac{3(\alpha - 1)\phi}{(\alpha + 2) - (\alpha - 1)\phi} \tag{8.20}$$

where k_e is the effective thermal conductivity, k_f is the thermal conductivity of the fluid, α is the ratio of thermal conductivity of particle to the thermal conductivity of fluid, and ϕ is the volume fraction of particles in the fluid. The model is accurate to the order ϕ^1 and is applicable for $\phi \ll 1$ or $|\alpha - 1| \ll 1$.

2) Jeffery Model

$$\frac{k_e}{k_f} = 1 + 3\beta\phi + \phi^2 \left[3\beta^2 + \frac{3\beta^2}{4} + \frac{9\beta^2(\alpha + 2)}{16(2\alpha + 3)} + \frac{3\beta^4}{2^6} + \cdots \right] \tag{8.21}$$

where $\beta = \dfrac{(\alpha - 1)}{(\alpha - 2)}$

The model is accurate to the order ϕ^2.

3) Davis Model

$$\frac{k_e}{k_f} = 1 + \frac{3(\alpha - 1)\left[\phi + f(\alpha)\phi^2 + 0\phi^3\right]}{(\alpha + 2) - (\alpha - 1)\phi} \tag{8.22}$$

The model is accurate to the order ϕ^2.

f(α) = 2.5 for α = 10; f(α) = 0.5 for α = infinity.

4) Lu and Lin Model

$$\frac{k_e}{k_f} = 1 + a\phi + b\phi^2 \tag{8.23}$$

For spherical particles, $a = 2.25$, $b = 2.27$ for $\alpha = 10$ and $a = 3$, $b = 4.51$ for $\alpha =$ infinity.

There are different methods of thermal conductivity measurement like the transient technique, thermal comparator method, and steady-state techniques as shown in Figure 8.10 [16]. In [34], mineral oil and soybean ester oil nanofluids of 0.1 wt.% are prepared using alumina nanoparticles. Thermal conductivity studies show that there is a 9.6% increase for natural ester nanofluids and an 8.3% increase for mineral oil nanofluids when compared with their unfilled counterparts. BN nanoparticle-filled vegetable oil nanofluid is prepared in [58]. It is observed that the temperature rise of nanofluid is slower than the unfilled oil. The maximum temperature of the unfilled and 0.1 wt.% nanofilled oil is 94.8 and 90.8 °C. Also, with the increase in particle concentration, the thermal conductivity increases, but it decreases with the increase in temperature. 0.1 wt.% nanofluid prepared using cottonseed oil and h-BN nanoparticle show a 33% increase in thermal conductivity than the unfilled oil. Also, it is obtained that the heating and cooling of nanofluids are rapid than the unfilled oil [54]. Fe_3O_4 nanofluid prepared using FR3 oil shows a 45% increase in thermal conductivity for 0.012 wt.% when compared with the unfilled oil [59].

8.3.4 Aging Study of Natural Ester Nanofluids

Transformers are intended to operate satisfactorily for over 30–40 years. However, about 75% of transformer collapse occurs due to insulation damage. Hence, the long-term performance of the insulation system must be studied. The aging studies of nanofluids are usually done using the ASTM D1934-95 standard test method for oxidative aging of electrical insulating petroleum oils by the open-beaker method. As only a few reports are available on the aging study of natural ester nanofluids, an extensive study of the different types of natural ester nanofluids must be carried out to assess their long-term safe operation. In [60, 61], TiO_2 and ZnO-filled vegetable oil nanofluids are aged for 312 hours at 150 °C. It is observed that as time progresses, the nanoparticles begin to aggregate which results in reduced stability of the nanofluid, particularly at elevated temperatures. At the end of the aging period, the breakdown voltage of aged ZnO and TiO_2 nanofluid is higher by 15 and 28%, respectively, as compared with the aged unfilled oil.

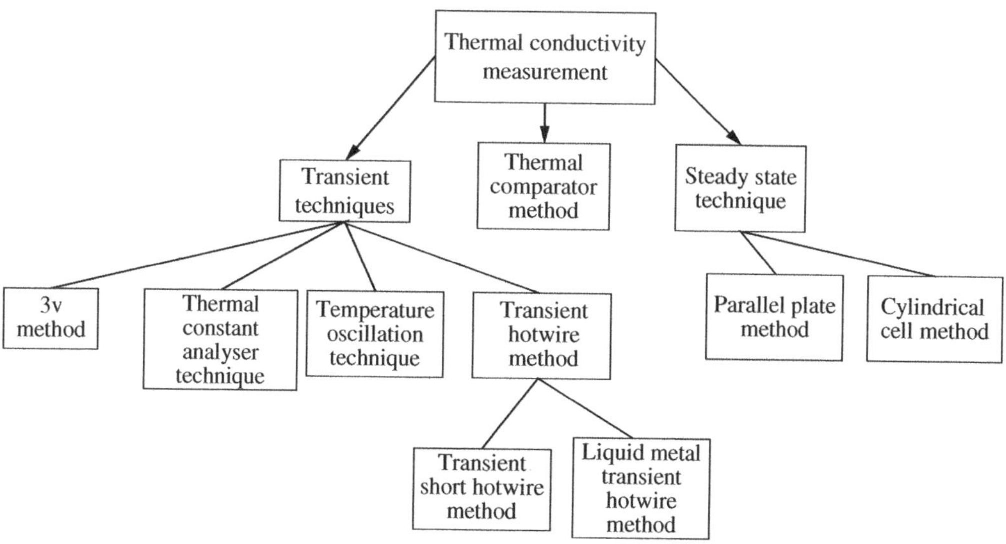

Figure 8.10 Methods of thermal conductivity measurement. *Source:* Based on Esfe and Afrand [16].

A representation of the variation of BDV with the moisture content of the oil follows the pattern shown in Figure 8.11. In the first 20 hours, the moisture content rises due to the degradation of oil and paper which in turn decreases the BDV value significantly. After that, the degradation rate and hence the moisture content in the oil decreases and correspondingly no further decrease of BDV value is recorded. The moisture absorption capability of vegetable oil nanofluids is higher than that of the unfilled oil. Hence, the nanofluid can absorb and hold more moisture from the paper, which reduces the degradation of paper insulation.

The moisture content in paper insulation, however, follows a different trend for unfilled and nanofilled vegetable oils. For both unfilled and nanofilled oil, the paper moisture decreases until around 100 hours after which the moisture content of the paper in nanofluid increases sharply until 200 hours and then reduces whereas that of the paper in unfilled oil remains almost constant as shown in Figure 8.12a. The degree of polymerization (DP) of paper in nanofilled oil is lesser than unfilled oil indicating the increased degradation rate in the presence of nanoparticles as shown in Figure 8.12b.

In [63], corona aging on five types of oil is studied. The five oil types are unfilled oil (A), FR3 oil with silica nanoparticle (B), FR3 oil with silica nanoparticle and surfactant CTAB (C), aged unfilled oil (D), and aged FR3 oil with silica nanoparticle and surfactant CTAB (E). The corona inception voltage (CIV) of C shows the maximum value followed by B, A, E, and D. This increased

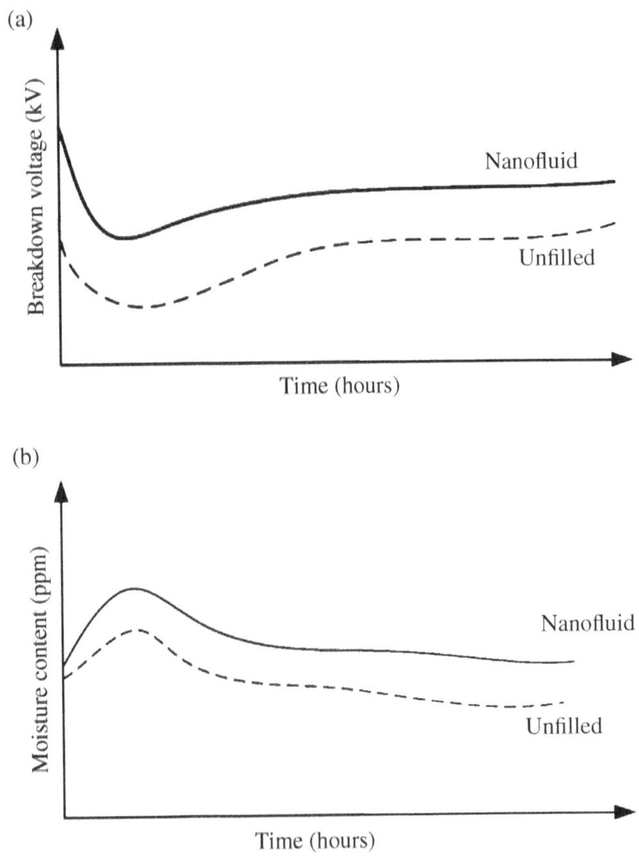

Figure 8.11 (a) BDV and (b) Moisture content of unfilled and nanofilled vegetable oil during ageing. *Sources:* Based on Fernandez et al. [60] and Aljure [61].

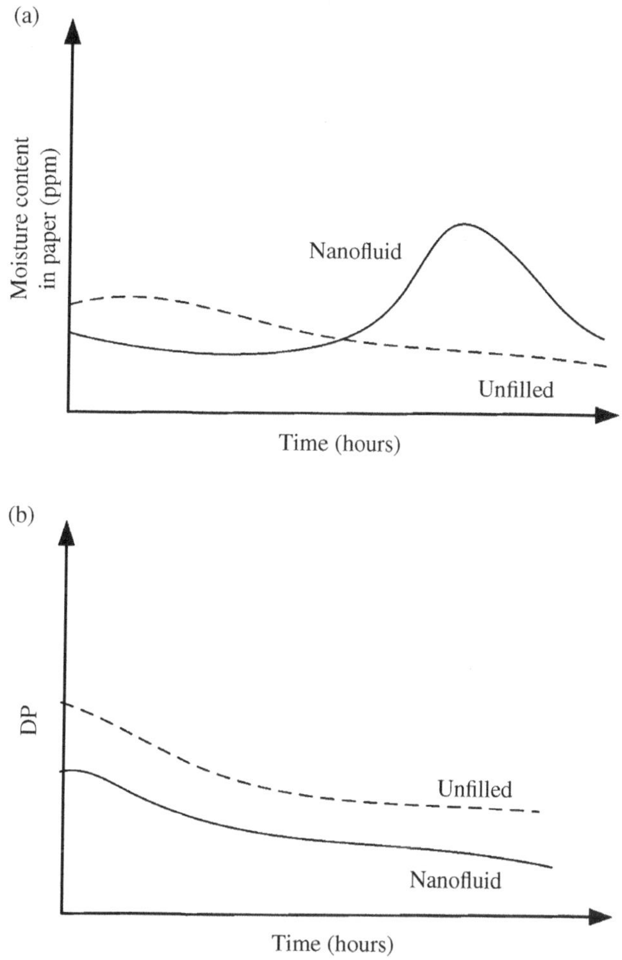

Figure 8.12 (a) Moisture content and (b) DP of paper aged in unfilled and nanofilled vegetable oil. *Sources:* Based on Fernandez et al. [60] and Aljure [61].

CIV of oil type C might be due to the addition of the insulating silica nanoparticle that enhances the dielectric properties of the oil by an electron trapping mechanism. Also, the agglomeration effect of the nanoparticles is reduced by the addition of surfactants. In [62, 64], a study of the properties of aged FR3 oil with 0.01 wt.% Eh-BN nanoparticles are done with copper catalyst, kraft paper, and pressboard. Weibull distribution analysis of unaged and aged oils at 63.5% failure probability shows that the AC BDV of unaged FR3 nanofluid is 2.86% higher than unaged FR3 oil and for aged FR3 nanofluid, it is 6.03% higher than the aged FR3 oil. The dielectric dissipation factor of unaged FR3 and FR3 nanofluid is 0.008 63 and 0.008 17 while that of aged FR3 and FR3 nanofluid is 0.0448 and 0.033, respectively. The dielectric constant of unaged FR3 and FR3 nanofluid is 2.83 and 2.85, respectively, whereas aged FR3 and FR3 nanofluid has higher values of 3.04 and 3.11, respectively. It is observed that with aging, the color of the oil deepens due to the products of degradation.

The work involves an extensive dissolved gas analysis of the oil samples. The analysis shows the formation of different types of gases like O_2, CO_2, N_2, H_2, CO, C_2H_4, C_2H_6, C_3H_6, C_3H_8, etc. The gases H_2, CO, C_2H_4, C_2H_6, C_3H_6, and C_3H_8 are produced by changes in the chemical bonds of the oil during aging. The other gases are formed by contamination from the environment. As the aging time progresses, the concentration of C_2H_6 gas in the oil increases but C_2H_6 is a stable gas unless present along with H_2. It is observed that after 500 hours of aging, H_2 gas is detected in the oil samples. However, the concentration of H_2 is low in the vegetable oil nanofluid than the unfilled mineral oil and vegetable oil. Hence, the vegetable oil nanofluids are less flammable than the unfilled oils, rendering them as suitable insulating oils for transformers in thickly populated areas. The heat-induced degradation of oil molecules results in the formation of CO_2. The degradation of paper results in the formation of CO gas which is the highest at 500 hours of aging. This is accompanied by the formation of sludges, flammable gases, and local hot spots in oil. However, the concentration of CO gas in vegetable oil nanofluid is the least indicating reduced degradation of paper in the nanofluid. Considering the chemical bonds in vegetable oil, it is found that chemical bonds are more stable in vegetable oil and hence the degradation rate is low. Mineral oil has CH_3, CH_2, and CH chemical groups linked by C-C bonds. Due to aging, free radicals are formed resulting in gases like H_2, CH_4, C_2H_6, C_2H_4, and C_2H_2. The DGA analysis shows that the gassing tendency of vegetable oil nanofluids is low compared with unfilled mineral oil and natural ester. Hence, the long-term performance of nanofluids can be considered superior to that of the unfilled oils.

8.3.5 Feasibility of Natural Ester Nanofluids as an Alternate Insulating Oil for Transformers

The application of natural ester nanofluids as insulating oil depends on their performance in comparison to conventional mineral oil. One of the major concerns regarding nanofluid is its stability, the time for which the nanoparticles will remain monodispersed in the oil. The stability of a nanofluid depends on many factors including the type of base oil, nanoparticle production method, nanofluid preparation method, the concentration of nanoparticle, type of surfactant, etc. The choice of the base oil and nanoparticle type is solely dependent on the application of the nanofluid. The different preparation methods detailed in Section 8.2.1 show that though the single-step preparation method provides stable nanofluids, for large-scale production, the two-step method would be cost-effective and simple. Besides, a considerable enhancement of stability can be achieved by the methods described in Section 8.2.2. Also, in transformers, the convection current produced by heating could assist the motion of the nanoparticles which otherwise would agglomerate and sediment in the transformer tank. Literature suggests that an optimum concentration of nanoparticles provides enhanced stability and property enhancement to the nanofluid.

The physical properties do not show much variation with the addition of nanoparticles. The viscosity of nanofluids is observed to increase with the increase in the concentration of nanoparticles, which can affect the thermal properties of the oil. However, it is expected that the operating temperature of the transformer will provide a check on the increase in viscosity beyond the threshold limit. Also, it can be seen from Table 8.3 that the flash point and fire point values of natural ester nanofluids are much higher than that of mineral insulating oil which facilitates their use in transformers located in thickly populated and risk zones.

The AC breakdown voltage of natural ester nanofluids is higher when compared with mineral insulating oils as shown in Table 8.4. It is reported that the addition of nanoparticles in some cases increases AC breakdown voltage while in some oils, it results in a reduction in the breakdown voltage. This can be attributed to the agglomeration of nanoparticles with the increase in nanoparticle concentration. Partial discharge characteristics of nanofluids show that the PDIV is higher than mineral oils which delay the streamer development and propagation in nanofluids. The thermal conductivity of natural ester nanofluids is higher than mineral insulating oil rendering them better cooling agents for transformers. The physiochemical, electrical, and thermal characterization of aged natural ester nanofluids, when compared with mineral oil, shows that the natural ester nanofluids are suitable for long-term use in transformers.

The present scenario of ultrahigh voltage transmission and distribution requires large-sized power transformers that are expensive. Improvement in the properties of the insulation system could eventually result in transformers of reduced size and cost. The biodegradable nature of natural esters combined with nanotechnology offers a feasible solution to this problem. Smaller and cost-effective transformers can be manufactured using stable natural ester nanofluids that have better dielectric properties, higher thermal conductivities, and that ensure the safe and prolonged operating life of the transformer.

8.4 Conclusion

The growing demand for a safe and reliable power supply requires the uninterrupted operation of the transformer. The operating efficiency of the transformer is dependent on the performance of the insulation system. Recent regulations regarding the safe disposal of mineral insulating oil have shifted the focus of researchers toward green insulating oils or natural esters. Further studies in this field with the introduction of nanotechnology have led to the development of nanofluids. Natural ester nanofluids are reported to exhibit enhanced dielectric and thermal performance which are the basic requirements of insulating oil. This can be attributed to the increased surface area and hence increased reactivity of the particles when dispersed in the oil.

There are many constraints to the practical application of natural ester nanofluids in a transformer, the major ones being the stability of the oil and its oxidation prone nature. Variations in the preparation methods with the addition of suitable dispersants and surfactants are expected to enhance the nanofluid stability. Also, the use of suitable antioxidants can improve the oxidation stability of the ester oil. In addition to this, it is required to assess the aging characteristics of ester nanofluids to ensure the safe operating life of the transformer. Besides, it is the oil paper composite that works as an insulation system in the transformer. Hence, the effect of ester oil on paper insulation is important. Continuous and effective research is being conducted to tackle the abovementioned concerns regarding natural ester nanofluids. Natural ester transformers are already replacing mineral oil-filled transformers worldwide and it is expected that natural ester nanofluids will emerge as the future insulating oil.

The promising concept of natural ester nanofluid has many challenges that can be dealt with as the future scope of work in this area [65].

8.4.1 Stability Enhancement of Natural Ester Nanofluids

Natural ester nanofluids are reported to have better stability than mineral oil nanofluids. The convection current within the transformer is expected to provide a motion to the dispersed nanoparticles. However, the higher viscosity of the ester oil can reduce this movement of ester oil, thus affecting the stability. Hence, it is required that the stability of the nanofluid be studied keeping in mind the effect of the physiochemical parameters of the oil and the effect on its dielectric and thermal properties. Also, all the studies reported in the literature refer to the stability of the nanofluid when in standstill conditions. Considering the convection currents in the transformer, stability studies involving a stirring motion in the nanofluid could be a better approach to the actual scenario inside a transformer.

8.4.2 Simulation Model for Nanofluids

Simulation studies on natural ester nanofluids are few and mostly restricted to a two-dimension model. Three-dimension models incorporating the particle tracing properties would provide an insight into agglomeration and its effects on the properties of the nanofluids. Also, very little is known about the interfacial characterization in nanofluids, knowledge of which would help understand the underlying mechanism of property variation in nanofluids.

8.4.3 Design of Transformers Using Natural Ester Nanofluids

The improved properties of natural ester nanofluids would result in transformers with reduced size and high voltage rating. Hence, it would be necessary to design the transformer and its insulation system accordingly. Also, the compatibility of the nanofluid with all the transformer parts has to be checked to prevent any unnecessary reaction between the oil and the transformer construction material.

8.4.4 Mixed Fluids and Multiparticle Nanofluids

The literature mostly reports studies on nanofluids with a single type of nanoparticle. Certain properties of one nanoparticle may excel others and hence a combination of nanoparticles could be tried so that there is a net positive effect of the particles on the properties of the oil. Investigation regarding the effect of more than one type of nanoparticle in the base oil is a research area that can be studied. Studies related to the use of a mixture of mineral oil and natural ester oil in the retro filling process of transformers are available in the literature. This can be extended to the combination of different types of natural ester oils as the base fluid for nanofluid preparation. It would enable the exploration of the properties of different types of particles in an oil or a combination of oils.

References

1 Rao, U.M., Fofana, I., Jaya, T. et al. (2019). Alternative dielectric fluids for transformer insulation system: progress, challenges, and future prospects. *IEEE Access* 7: 184552–184571.

2 Mehta, D.M., Kundu, P., Chowdhury, A. et al. (2016). A review of critical evaluation of natural ester vis-a-vis mineral oil insulating liquid for use in transformers: part II. *IEEE Transactions on Dielectrics and Electrical Insulation* 23 (3): 1705–1712.

3 Hernandez-Herrera, H., Silva-Ortega, J.I., Mejia-Taboada, M. et al. (2019). Natural ester fluids applications in transformers as a sustainable dielectric and coolant. *AIP Conference Proceedings* 2123 (1): 020049.

4 Rao, U.M., Sood, Y.R., and Jarial, R.K. (2016). Performance analysis of alternate liquid dielectrics for power transformers. *IEEE Transactions on Dielectrics and Electrical Insulation* 23 (4): 2475–2484.

5 Diestre Redondo, E.I., Pérez, F.S., VÁzquez Cantero, M.Á. et al. (2014). Soil ecotoxicity of natural ester transformer liquids. *IEEE 18th International Conference on Dielectric Liquids (ICDL)*, Bled, Slovenia (29 June–3 July 2014), pp. 1–4.

6 Lewis, T. (1994). Nanometric dielectrics. *IEEE Transactions on Dielectrics and Electrical Insulation* 1 (5): 812–825.

7 Lewis, T. (2004). Interfaces are the dominant feature of dielectrics at the nanometric level. *IEEE Transactions on Dielectrics and Electrical Insulation* 11 (5): 739–753.

8 Choi, S.U. and Eastman, J.A. (1995). Enhancing Thermal Conductivity of Fluids with Nanoparticles. *Tech. Rep*, ANL/MSD/CP-84938; CONF-951135-29, ON: DE96004174; TRN: 96:001707. Argonne National Lab., San Francisco, CA, United States (12–17 November 1995).

9 Azis, N. and Wang, Z.D. (2011). Acid generation study of natural ester. *XVII International Symposium on High Voltage Engineering*, Hannover, Germany (22 August 2011), pp. 22–26.

10 Sherwin, E.R. (1978). Oxidation and antioxidants in fat and oil processing. *Journal of the American Oil Chemists' Society* 55 (11): 809–814.

11 Kumar, S.S., Iruthayarajan, M.W., Bakrutheen, M., and Kannan, S.G. (2016). Effect of antioxidants on critical properties of natural esters for liquid insulations. *IEEE Transactions on Dielectrics and Electrical Insulation* 23 (4): 2068–2078.

12 Wong, K.V. and De Leon, O. (2010). Applications of nanofluids: current and future. *Advances in Mechanical Engineering* 2: 519659.

13 Preetha, P. (2016). Studies on epoxy nanocomposites as electrical insulation for high voltage power apparatus. Doctoral dissertation.

14 Rafiq, M., Lv, Y., and Li, C. (2016). A review on properties, opportunities, and challenges of transformer oil-based nanofluids. *Journal of Nanomaterials* 2016: 1–23.

15 Hwang, J.G., Zahn, M., O'Sullivan, F.M. et al. (2010). Effects of nanoparticle charging on streamer development in transformer oil-based nanofluids. *Journal of Applied Physics* 107 (1): 014310.

16 Esfe, M.H. and Afrand, M. (2019). An updated review on the nanofluids characteristics. *Journal of Thermal Analysis and Calorimetry* 138 (6): 4091–4101.

17 Yu, W. and Xie, H. (2012). A review on nanofluids: preparation, stability mechanisms, and applications. *Journal of Nanomaterials* 12: 1–17.

18 Akoh, H., Tsukasaki, Y., Yatsuya, S., and Tasaki, A. (1978). Magnetic properties of ferromagnetic ultrafine particles prepared by vacuum evaporation on running oil substrate. *Journal of Crystal Growth* 45: 495–500.

19 Wagener, M., Murty, B.S., and Günther, B. (1996). Preparation of metal nanosuspensions by high-pressure DC-sputtering on running liquids. *MRS Online Proceedings Library (OPL)* 457: 149–154.

20 Eastman, J.A., Choi, U., Li, S. et al. (1996). Enhanced Thermal Conductivity Through the Development of Nanofluids. *Tech. Rep.*, ANL/MSD/CP-90462; CONF-961202-94, ON: DE97003768; TRN: AHC29708%%43, vol. 457, Argonne National Lab., IL (United States), pp. 3–11.

21 Zhu, H.T., Lin, Y.S., and Yin, Y.S. (2004). A novel one-step chemical method for preparation of copper nanofluids. *Journal of Colloid and Interface Science* 277 (1): 100–103.

22 Lo, C.H., Tsung, T.T., Chen, L.C. et al. (2005). Fabrication of copper oxide nanofluid using submerged arc nanoparticle synthesis system (SANSS). *Journal of Nanoparticle Research* 7 (2–3): 313–320.

23 Campos, M.A., Gonzalez, E.C., Cervantes, A.G. et al. (2018). Synthesis of paramelaconite nanoparticles by laser ablation. *Journal of Laser Applications* 30 (1): 012012.

24 Lee, S., Choi, S.-S., Li, S., and Eastman, J. (1999). Measuring thermal conductivity of fluids containing oxide nanoparticles. *Journal of Heat Transfer* 121 (2): 280–289.

25 Wang, X., Xu, X., and Choi, S.U. (1999). Thermal conductivity of nanoparticle-fluid mixture. *Journal of Thermophysics and Heat Transfer* 13 (4): 474–480.

26 Li, J., Zhang, Z., Zou, P. et al. (2012). Preparation of a vegetable oil-based nanofluid and investigation of its breakdown and dielectric properties. *IEEE Electrical Insulation Magazine* 28 (5): 43–50.

27 Shariatmadar, F.S. and Pakdehi, S.G. (2016). Effect of various surfactants on the stability time of kerosene–boron nanofluids. *Micro & Nano Letters* 11 (9): 498–502.

28 Wang, L. and Wei, X. (2009). Nanofluids: synthesis, heat conduction, and extension. *Journal of Heat Transfer* 131 (3): 033102-1–033102-7.

29 Wei, X. and Wang, L. (2010). Synthesis and thermal conductivity of microfluidic copper nanofluids. *Particuology*, Elsevier 8 (3): 262–271.

30 Chen, Y. and Wang, X. (2008). Novel phase-transfer preparation of monodisperse silver and gold nanoparticles at room temperature. *Materials Letters* 62 (15): 2215–2218.

31 Feng, X., Ma, H., Huang, S. et al. (2006). Aqueous- organic phase-transfer of highly stable gold, silver, and platinum nanoparticles and new route for fabrication of gold nanofilms at the oil/water interface and on solid supports. *The Journal of Physical Chemistry B* 110 (25): 12311–12317.

32 Yu, W., Xie, H., Chen, L., and Li, Y. (2010). Enhancement of thermal conductivity of kerosene-based Fe_3O_4 nanofluids prepared via phase-transfer method. *Colloids and Surfaces A: Physicochemical and Engineering Aspects* 355 (1–3): 109–113.

33 Samuel, J.G. (2013). Colloids science-a toolbox for processing nanodielectrics. *IEEE Electrical Insulation Magazine* 29 (6): 16–20.

34 Jacob, J., Preetha, P., and Sindhu, T.K. (2020). Stability analysis and characterization of natural ester nanofluids for transformers. *IEEE Transactions on Dielectrics and Electrical Insulation* 27 (5): 1715–1723.

35 Ghadimi, A., Saidur, R., and Metselaar, H. (2011). A review of nanofluid stability properties and characterization in stationary conditions. *International Journal of Heat and Mass Transfer* 54 (17–18): 4051–4068.

36 Yu, F., Chen, Y., Liang, X. et al. (2017). Dispersion stability of thermal nanofluids. *Progress in Natural Science: Materials International* 27 (5): 531–542.

37 Umar, S., Sulaiman, F., Abdullah, N., and Mohamad, S.N. (2018). Investigation of the effect of pH adjustment on the stability of nanofluid. *AIP Conference Proceedings* 2031 (1): 020031.

38 Mukherjee, S. and Paria, S. (2013). Preparation and stability of nanofluids-a review. *IOSR Journal of Mechanical and Civil Engineering* 9 (2): 63–69.

39 Sadeghi, R., Etemad, S.G., Keshavarzi, E., and Haghshenasfard, M. (2015). Investigation of alumina nanofluid stability by UV–Vis spectrum. *Microfluidics and Nanofluidics* 18 (5–6): 1023–1030.

40 Missana, T. and Adell, A. On the applicability of DLVO theory to the prediction of clay colloids stability. *Journal of Colloid and Interface Science* 230 (1): 150–156.

41 Ghadimi, A., Metselaar, H., and Lotfizadehdehkordi, B. (2015). Nanofluid stability optimization based on UV-Vis spectrophotometer measurement. *Journal of Engineering Science and Technology* 10: 32–40.

42 Wee, S.C., Akhir, N.A.M., and Rozman, A.L. (2017). Investigation of nanoparticles dispersion in sodium hydroxide (NaOH) solvent. *MATEC Web of Conferences*, vol. 87), EDP Sciences, ENCON, Malaysia (24–25 October 2016), p. 03006

43 ASTM D7725 (2012). Standard test method for the continuous measurement of turbidity above 1 turbidity unit (TU).

44 Madavan, S.S.K. and Iruthyarajan, M.W. (2018). A comparative investigation on effects of nanoparticles on characteristics of natural esters-based nanofluids. *Colloids and Surfaces A: Physicochemical and Engineering Aspects* 556: 30–36.

45 Karthik, R. and Raymon, A. (2016). Effect of silicone oxide nano particles on dielectric characteristics of natural ester. *IEEE International Conference on High Voltage Engineering and Application (ICHVE)*, China (19–22 September 2016), pp. 1–3.

46 Srinivasan, M., Ragupathy, U., Sindhuja, K., and Raymon, A. (2016). Investigation and performance analysis of nanoparticles and antioxidants based natural ester. *International Journal of Advanced Engineering Technology* 2: 1007.

47 Madavan, R. and Balaraman, S. (2017). Investigation on effects of different types of nanoparticles on critical parameters of nano-liquid insulation systems. *Journal of Molecular Liquids* 230: 437–444.

48 Miao, J., Dong, M., Ren, M. et al. (2013). Effect of nanoparticle polarization on relative permittivity of transformer oil-based nanofluids. *Journal of Applied Physics* 113 (20): 204103.

49 Du, B., Li, J., Wang, F. et al. (2015). Influence of monodisperse Fe_3O_4 nanoparticle size on electrical properties of vegetable oil-based nanofluids. *Journal of Nanomaterials* 2016: 1–9.

50 Huang, Z., Li, J., Yao, W. et al. (2019). Electrical and thermal properties of insulating oil-based nanofluids: a comprehensive overview. *IET Nanodielectrics* 2 (1): 27–40.

51 Zhong, Y., Lv, Y., Li, C. et al. (2013). Insulating properties and charge characteristics of natural ester fluid modified by TiO_2 semiconductive nanoparticles. *IEEE Transactions on Dielectrics and Electrical Insulation* 20 (1): 135–140.

52 Aljure, M. (2019). Pre-breakdown phenomena in mineral oil based nanofluids. Ph.D. dissertation. KTH Royal Institute of Technology.

53 Charalampakos, V.P., Bakandritsos, A., Peppas, G.D. et al. (2017). A comparative study of natural ester based nanofluids with Fe_2O_3 and SiO_2 nanoparticles. *IEEE 19th International Conference on Dielectric Liquids (ICDL)*, Manchester, UK (25–29 June 2017), pp. 1–4.

54 Farade, R.A., Wahab, N.I.B.A., Mansour, D.-E.A. et al. (2020). Investigation of the dielectric and thermal properties of non-edible cottonseed oil by infusing h-BN nanoparticles. *IEEE Access* 8: 76204–76217.

55 Saenkhumwong, W. and Suksri, A. (2015). Investigation on voltage breakdown of natural ester oils based-on ZnO nanofluids. *Advanced Materials Research* 1119. Trans Tech Publ,: 175–178.

56 Makmud, M.Z.H., Illias, H.A., Chee, C.Y., and Dabbak, S.Z.A. (2019). Partial discharge in nanofluid insulation material with conductive and semiconductive nanoparticles. *Materials* 12 (5): 816.

57 Terekhov, V., Kalinina, S., and Lemanov, V. (2010). The mechanism of heat transfer in nanofluids:state of the art (review). part 1. synthesis and properties of nanofluids. *Thermophysics and Aeromechanics* 17 (1): 1–14.

58 Du, B., Li, X., Li, J., and Tao, X. (2015). Effects of BN nanoparticles on thermal conductivity and breakdown strength of vegetable oil. *IEEE International Conference on the Properties and Applications of Dielectric Materials (ICPADM)*, Australia (19–22 July 2015), pp. 476–479.

59 Zou, P., Li, J., Sun, C.X. et al. (2011). Dielectric properties and electrodynamic process of natural ester-based insulating nanofluid. *Modern Physics Letters B* 25 (25): 2021–2031.

60 Fernandez, I., Valiente, R., Ortiz, F. et al. (2020). Effect of TiO_2 and ZnO nanoparticles on the performance of dielectric nanofluids based on vegetable esters during their aging. *Nanomaterials* 10 (4): 692.

61 Diego, C., Santisteban, A., Fernandez, F. et al. (2017). Effect of TiO_2 nanoparticles on the performance of a natural ester dielectric fluid. *IEEE Electrical Insulation Conference (EIC)*, USA (11–14 June 2017), pp. 1–4.

62 Maharana, M., Baruah, N., Nayak, S.K. et al. (2019). Condition assessment of aged ester-based nanofluid through physicochemical and spectroscopic measurement. *IEEE Transactions on Instrumentation and Measurement* 68 (12): 4853–4863.

63 Thakur, S. and Sarathi, R. (2019). Investigating the impact of corona ageing on physicochemical properties of natural ester oil nanofluid. *2nd International Conference on High Voltage Engineering and Power Systems (ICHVEPS)*, IEEE, Indonesia (1–4 October 2019), pp. 282–286.

64 Baruah, N., Maharana, M., Dey, S.S., and Nayak, S.K. (2019). Behavioural assessment of aged natural ester based nanofluid using statistical technique. *IEEE Conference on Electrical Insulation and Dielectric Phenomena (CEIDP)*, USA (20–23 October 2019), pp. 653–656.

65 Jacob, J., Preetha, P., and Krishnan, S.T. (2020). Review on natural ester and nanofluids as an environmental friendly alternative to transformer mineral oil. *IET Nanodielectrics* 3 (2): 33–43.

9

Dielectric Properties of Silica-Based Synthetic Ester Nanofluid

G. D. P. Mahidhar[1], R. Sarathi[1], Nathaniel Taylor[2], and Hans Edin[2]

[1] *Department of Electrical Engineering, Indian Institute of Technology Madras, Chennai, India*
[2] *School of Electrical Engineering and Computer Science, KTH - Royal Institute of Technology, Stockholm, Sweden*

9.1 Introduction

Transformer is one of the important component in the power system network. Transformer's reliable operation is critical for the seamless power transfer in the modern day power system. Transformer insulation design and monitoring is one of the aspect that plays a crucial role in improving the reliability of the power system.

Liquid-filled transformers are the most commonly used transformers with wide range of applications from 10 kVA to 1500 MVA power rating and high voltages up to 1000 kV. The major insulating materials in it are the transformer liquid insulation and the solid cellulosic insulation. Mineral oil is commonly used as coolant and electrical insulant in transformer. The cellulosic material is used as solid insulation in the form of pressboard, which can provide mechanical support and electrical insulation. The transformer insulation experiences different type of electrical and thermal stresses over the years, which lead to degradation due to increased moisture content and oxidation by-products, causing failure of insulation in transformer [1, 2]. As per a survey by CIGRE WG 2.37, about 40% of transformer failures are due to degradation or aging of the dielectric insulating material [3]. It is thus important for the insulation engineers and scientists to develop superior materials that have improved life, which can improve the reliability of the transformer.

Conventionally, mineral oil is used due to its cheaper production cost, good insulating properties, and heat transfer properties, which make it being used widely in various type of liquid-filled transformers from over a century [4]. Based on the molecular composition of the type of hydrocarbons present in the mineral oil, they can be broadly classified as paraffinic oil, naphthenic oil, and aromatic oil. Paraffinic oil mainly contain long chains of alkanes and isoalkanes. Naphthenic oil contains cyclic alkane compounds. Aromatic compounds contain polycyclic aromatic compounds [5, 6]. One of the major disadvantage of the mineral oil is its low fire safety due to low flash point at around 150 °C. The presence of hotspots in the transformer can increase the risk of fire hazard leading to the failure of the entire transformer. To avoid such fatalities, researchers have started finding alternative insulating fluids with higher fire safety from 1950s.

As a replacement for the mineral oils, Askarels or polychlorinated biphenyls (PCBs) were introduced in the 1950s. PCB is a type of halogenated hydrocarbon. PCBs are obtained by chlorination

Alternative Liquid Dielectrics for High Voltage Transformer Insulation Systems: Performance Analysis and Applications, First Edition. Edited by U. Mohan Rao, I. Fofana, and R. Sarathi.

of biphenyls. They are considered nonflammable. They were primarily used in transformers and capacitors due to their fire retardant nature [6]. Even though, PCBs have high dielectric strength and nonflammable properties, they are nonbiodegradable and are highly toxic to the environment. Due to this reason, they are banned from usage from the 1970s [7, 8].

As an alternative to the PCBs, electrical industry started using silicone oil, which is nontoxic in nature and has higher flammability limit [8]. It has high chemical stability and very good dielectric properties. These are chemically known as polydimethyl siloxanes (PDMSs), which are obtained from the polymerization of dimethyl siloxane compound. PDMSs have higher fire safety, but they are non-biodegradable and can affect the environment [8–10].

Synthetic ester fluids (SEFs) are synthesized from organic acids and alcohols. Different type of ester groups can be synthesized with appropriate carboxylic acids and alcohols [5]. The commonly used synthetic ester is tetra-esters or pentaerythritol esters. SEFs have good dielectric properties and have high flammability limit. These have high oxidation stability and are also biodegradable leading to no accumulation of pollutants in the environment. These have high moisture saturation capability and can absorb more water content, but retain the same breakdown strength for higher concentrations of water in the liquid [11, 12]. Hence, these are becoming a replacement for the PCB, silicone oils, and mineral oil.

The cost of production of the SEF is high. As an alternative, researchers tried to produce insulating oils from naturally available seeds like canola/rapeseed, soybean, sunflower seeds, etc. These are naturally synthesized by esterification of the tri-alcohol glycerol with three fatty acids. These have high flash point and fire point, increasing the fire safety of the transformer. These are fully biodegradable and are also named as BIOTEMP fluids, but they have very poor oxidation stability. So, these are mainly used in sealed transformers or capacitor applications [11–13]. The concept of mixed insulating liquids with mixtures of mineral oil and ester fluids was also researched to reduce the costs and also improve the dielectric and thermal properties [14].

The researchers all over the world are making attempts at improving the physical and chemical characteristics of the insulating liquids by developing novel materials, which can reduce the cost and size requirements of the transformers and can improve the reliability of the power system.

9.1.1 Need for Nanofluids

With greater developments in the manufacturing technologies, generation of nanoparticles with sizes down to a few nanometers is possible with less efforts. The presence of the nanoparticles in a dielectric material influence the dielectric, thermal, and structural properties of the bulk material. This technology of nanodielectrics is finding potential applications in electronics, robotics, biomaterials, polymeric materials, fuels, etc. [15, 16]. Nanodielectrics are also finding application in power industry in the form of polymeric insulating nanocomposites that have superior dielectric and thermal properties [17]. With researchers finding good results with solid nanocomposite materials, focus was also put on the liquid nanodielectrics. In recent times, more research has been put up in dispersing nanoparticles in transformer insulating liquids to improve the thermal and dielectric properties of the transformer insulating liquids. The term nanofluids were coined for the two-phase systems where a small concentration of nanoparticles were dispersed in base liquid [18].

9.1.2 Methods of Property Enhancement of Nanofluids

Over the past few years, researchers have used different type of nanoparticles to improve the dielectric and thermal properties of the transformer oils. In general, researchers have dispersed conductive (Fe_3O_4, Fe_2O_3, Cu, C), semi conductive (TiO_2, ZnO, $BaTiO_3$), and insulating type (SiO_2, Al_2O_3) of nanoparticles in the transformer oils.

The experimental work conducted by Segal and Raj in 1998 was one of the earliest works that have carried out experiments to understand the thermal cooling performance of transformers using nanofluids [19]. Fe_3O_4 nanoparticles were dispersed in mineral oil to produce ferrofluid. There was a reduction in maximum temperature rise of up to 15% in the transformer model filled with ferrofluid. This study showed that usage of nanofluids can improve the heat transfer characteristics of the transformer oil. Choi et al. studied nanofluids with alumina (Al_2O_3) and aluminium nitride (AlN) nanoparticles dispersed in transformer oil [20]. The thermal conductivity of the nanofluid increases with nanoparticle concentration and due to the higher thermal conductivity of the nanoparticles. The study optimized the concentration of nanoparticle for optimum improvement in thermal properties. The AlN-based nanofluid with volume fraction of 0.5% can improve the thermal conductivity of the transformer oil by 8%. The overall heat transfer efficiency is improved by 20%.

Hwang et al. dispersed Fe_3O_4 nanoparticles in transformer oil and showed that there was an increment in the dielectric strength [21]. The study showed that nanoparticles acted like potential wells where electron get trapped leading to the reduction in the streamer propagation. Du et al. dispersed semi-conductive titania (TiO_2) nanoparticles in transformer nanofluid. There was an increment of 20% in impulse breakdown strength of transformer oil with addition of nanoparticles. Charge trap distribution characteristics of the nanofluid were studied using thermally stimulated current measurement, which showed that nanofluids have higher shallow trap density compared to that of the pure transformer oil. This increment in shallow trap density can slow down the fast electron during streamer propagation leading to the increment in the impulse breakdown strength [22]. Jin et al. have studied mineral-oil-based silica nanofluids with different concentrations. The AC breakdown voltage of the nanofluids was tested at different moisture content levels. Silica nanofluids exhibited improved breakdown strength, especially at high moisture content level. [23].

Even though ester fluids have good thermal properties, they have poor dielectric strength under impulse voltage due to the easier streamer initiation and higher streamer propagation velocity compared to mineral oil [24]. So, in recent times, the research on nanofluids based on the ester fluids are also of growing interest in order to improve the dielectric strength. Li et al. have studied Fe_3O_4-based natural ester nanofluid. They showed that there was increment in the AC breakdown voltage by about 20%. Nanofluid had higher positive and negative lightning impulse breakdown voltage [25]. Maharana et al. studied the impact of aging in the vegetable-based nanofluid. Surface-modified hexagonal boron nitride (h-BN) sheets were dispersed in the natural ester vegetable oil to improve the stability of the nanoparticles. Oxidative aging was performed and different optical spectroscopic measurements like Fourier transform infrared spectroscopy (FTIR), ultraviolet–visible spectroscopy, and fluorescence spectroscopy were performed to characterize the degradation in the aged oils. Studies showed that even after aging, the dielectric strength of the aged nanofluid was superior to mineral oil and vegetable oil [26].

It is evident that with addition of nanoparticles to the base liquid can improve the thermal and dielectric properties of the liquid. Another important aspect that needs to be taken care of is the stability of the nanoparticles in liquids. This factor is highly dependent on the physics of the particles at the nanometer level.

Darjaguin, Landan, Verwey, and Overbeek (DLVO) proposed a theory on the dispersion of solid particles in liquids [27, 28]. These studies showed a theoretical approach for understanding the stability of the solid particles in liquids. When the solid particles are dispersed, there will be electrical double-layer formation on the surface on the particles. This double layer has ions, which lead to electrostatic repulsive forces between the other particles. Apart from repulsive forces, there will be van der Waals attractive forces. These two type of forces play a major role in the stability of the nanofluids as well. As per the DLVO theory, the total interaction energy of the colloid is given by the sum of van der Waals attraction energy (E_{vdw}) and the electrostatic repulsive energy (E_R). The repulsion energy and the attraction energy both decrease with distance. When the interaction energy reaches maximum energy barrier of 15 kT, then it can be considered as a stable distance between the particles so that there is no agglomeration. The expressions for the van der Waals attraction energy and the electrostatic repulsion energy are given in equations (9.1) and (9.2) [29, 30].

$$E_{\mathrm{vdw}} = -\frac{A_{\mathrm{sl}}}{12\pi d^2} \tag{9.1}$$

$$E_R = 2\pi r \varepsilon_1 \varepsilon_0 \psi^2 ln\left(1 + e^{-\kappa d}\right) \tag{9.2}$$

In the equation (9.1), d is the distance between particles (surface to surface distance). The A_{sl} is Hamaker constant that is constant for each type of material. It was calculated using the dielectric constant of the material [31]. In equation (9.2), ψ is the surface potential at the nanoparticle. κ is the Debye-Hückel parameter and it is dependent on the ion concentration in the liquid [29, 32].

From these factors, it is understood that the stability is dependent on the surface condition on the nanoparticle. An optimum concentration of nanoparticles need to be added so that there is a balance in the attraction forces and the repulsion forces. If there are more nanoparticles than the optimum, there will be agglomerations that can be disadvantageous. The study by Choi et al. showed that with increase in agglomeration because of higher concentrations of nanoparticles can lead to degradation of the properties. To avoid such degradation, the study suggested the technique of surface modification of the nanoparticle by means of surfactant that can avoid agglomerations [20]. Saterlie et al. also conducted experiments with different type of surfactants like SDBS, cetyl trimethyl ammonium bromide (CTAB), and oleic acid. He concluded that optimum quantities of CTAB and Cu particles were able to improve the thermal conductivity [33]. Mansour et al. studied the effect of interfacial zone on the dielectric properties of the nanofluids. Al_2O_3 with cationic surfactant and TiO_2 with anionic surfactant were used to prepare two nanofluids. Different concentrations of nanoparticles and surfactant were studied for optimization. With optimized concentrations of Al_2O_3 and surfactant, there was 23% increment in breakdown voltage. For the case of TiO_2 nanofluid, it was 16.5% [34].

Based on the literature survey related to the transformer nanofluids and the type of property that was enhanced, a short summary of them is given in Table 9.1.

Many detailed studies have been performed using the conductive Fe_3O_4 and semiconductive TiO_2-based mineral oil nanofluids [35–37, 39, 42]. A study used Fe_3O_4 for the vegetable-oil-based nanofluid [25]. Studies on understanding fundamental properties of the nanofluids based on SEF are scarce in the literature. Hence, an attempt has been made by the author to understand the basic electrical properties of SEF and SEF dispersed with nanoparticles.

Past studies on nanocomposites have shown that silica nanocomposites have very good breakdown strength, lower dielectric dissipation losses, and good thermal conductivity properties [17]. Studies related to silica (SiO_2) nanofluids are scanty, and there was no work reported with silica-based ester nanofluid. Therefore, in the present study SiO_2 was chosen as nanoparticle to understand the performance of silica-based synthetic ester nanofluid.

In the present study, a methodical experimental study was performed to disperse the SiO_2 (silica) nanoparticles in SEF and to characterize its stability in ester fluid by physico-chemical studies. The nanoparticles dispersed in fluid can impact the dielectric properties and the conduction

Table 9.1 Literature survey on nanofluids.

Type of nanoparticle	Nanoparticle and base fluid	Remarks
Conductive	Fe_3O_4 in mineral oil [35]	Increment in impulse breakdown strength by 44.3%
	Fe_2O_3 in mineral oil [36]	Decrease in the partial discharge activity
	Fe_3O_4 in natural ester fluid [37]	Increment in AC breakdown strength by 23.2%
	C-60 in mineral oil [38]	Increment in electrical resistivity by 22% and increment in AC breakdown strength by 10%
Semi-conductive	TiO_2 in mineral oil [39]	Increment in AC breakdown strength by 28%
	ZnO in mineral oil [40]	Improvement in breakdown strength due to the improved I-V characteristics leading to reduced charge mobility
	$BaTiO_3$ in mineral oil [41]	Improvement in the heat transfer capability by 33%
	TiO_2 in natural ester fluid [42]	Improvement in the AC breakdown strength by 31% and also reduction in the partial discharge activity
Insulating type	Al_2O_3 in mineral oil [20]	Improvement in heat transfer efficiency by 20%
	SiO_2 in mineral oil [23]	Improvement in the AC breakdown strength by 20%
	BN in mineral oil [26]	Improvement in the AC breakdown strength and the heat transfer properties also
	BN in natural ester fluid [43]	Surface-modified BN sheets to improve the dielectric properties. It slows down the aging process as well

mechanisms. The frequency domain dielectric response is used to measure dielectric constant and tanδ. The time domain dielectric response characteristics were studied to measure the ionic mobility, conductivity, and other dielectric properties. The impact of high electric field on the conduction mechanisms is studied by means of I–U characteristics. Partial discharge activity can lead to the degradation of the insulating material. It is important to identify the partial discharges and understand the discharge activity in nanofluids. In order to understand the performance of the SEF and nanofluid under corona discharge, the corona inception voltage (CIV) and its discharge activity are studied by means of ultrahigh frequency (UHF) technique.

9.2 Nanofluid Preparation and Characterization

9.2.1 Nanoparticle Characterization

Silica nanoparticles are procured from Nanostructured and Amorphous Materials Inc. (USA) with mean particle size of 15 nm. X-ray diffraction (XRD) studies were carried out on silica nanoparticles by using Bruker D8 Discover powder XRD machine. Rietveld refinement was adopted to analyze the XRD data using PAN analytical X'pert High Score Plus software package. Figure 9.1a shows the XRD of silica nanoparticle. Broad peak suggests that silica is in amorphous phase and has small particle size. The peak (2θ) was observed at 21°, the characteristics of the silica. This also shows that the silica nanoparticles are pure. Transmission electron microscopy (TEM) of silica is

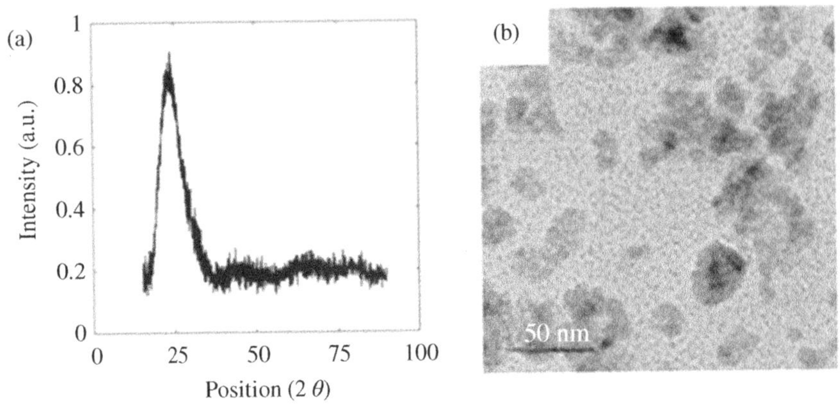

Figure 9.1 Silica nanoparticles, (a) XRD pattern and (b) TEM image.

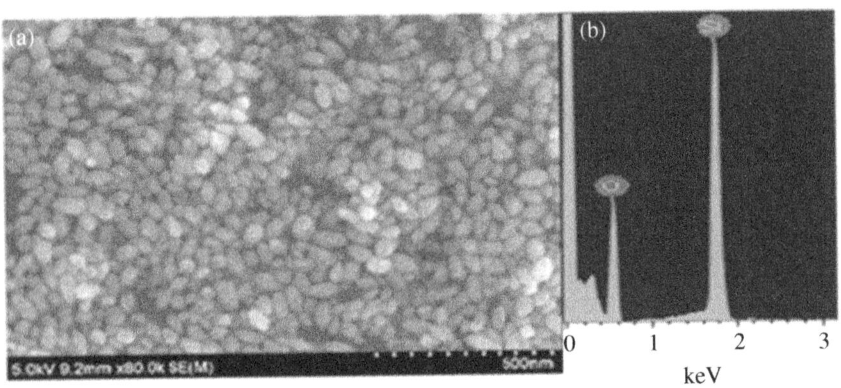

Figure 9.2 (a) SEM image and (b) EDAX of SiO_2 nanoparticles.

performed using the instrument JEM-2100. Figure 9.1b shows the TEM image of silica particles with a magnification of 10^6. Scanning electron microscopy (SEM) and energy dispersive spectroscopy (EDAX) analysis were carried out to determine the particle structure and the purity of the sample. Figure 9.2 shows the SEM image and EDAX spectrum of nanoparticles observed with a magnification of 10^5 using the instrument Hitachi S4800.

9.2.2 Nanofluid Preparation

Silica nanoparticles of known weight were added to Midel 7131 SEF to prepare the nanofluid. To have a uniform dispersion of nanoparticles, long-chained CTAB was used as surfactant. The moisture content in silica nanoparticles was removed by drying them in hot air oven at 150 °C for 8 hours. The nanoparticles and surfactant are added to SEF and mixed using magnetic stirring for 30 minutes and then the same solution was transferred for ultrasonication using Sonic Vibra Cell sonicator (500 W, 20 kHz). The temperature during this process was maintained at 40 °C. On completion of this process, the nanofluid was kept in vacuum desiccator to remove micro bubbles formed during the ultrasonication process and to avoid any moisture absorption/oxidation. No variation in color was observed at different concentrations of silica. Figure 9.3 shows the experimental procedure to prepare nanofluids with and without surfactant.

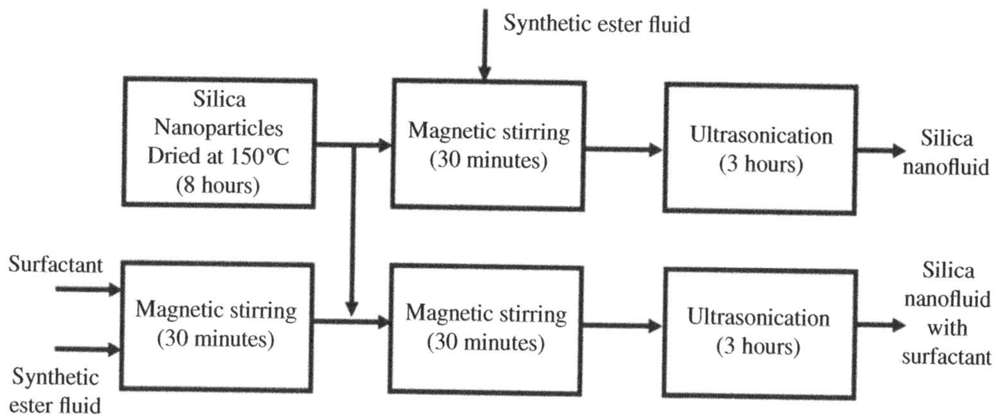

Figure 9.3 Schematic representation of sample preparation of nanofluid.

9.2.3 Nanofluid Stability

In order to determine the stability of the nanofluid, the particle size analysis and zeta potential measurements were carried out.

The size of the dispersed nanoparticles in ester fluid is measured using the instrument Horiba SZ-100 nanopartica, which works on the principle of dynamic light scattering (DLS) technique. The zeta potential was also measured using the same instrument. Zeta potential is the potential difference between the dispersion medium and the slipping plane. The higher the zeta potential is, the lower the tendency of nanoparticles to come together and agglomerate. Zeta potential greater than 30 mV is good for stability, 30–40 mV is stable, 40–60 mV has good stability, and greater than 60 mV has excellent stability [44].

Figure 9.4a shows the electrical double-layer formation of nanoparticle when dispersed in the liquid. Nanofluids with optimum concentration of silica was first determined and is named as Type-A sample. Then CTAB was added to the Type-A sample during preparation to improve the stability of the nanoparticles. Figure 9.4b shows the chemical structure of the CTAB. The optimum concentration of CTAB to be added is determined and is named as Type-B sample. Figure 9.4c shows the impact of CTAB on the electrical double-layer formation that can improve the stability of the nanofluid. Similar mechanism was elaborated in a study by Mansour et al. [34].

9.2.3.1 Particle Size Analysis

Figure 9.5 shows the particle size distribution of silica nanoparticles in the SEF with different concentrations. Particle size is increased as the concentration of the silica increases, which is due to agglomeration. Lower height of distributions indicates the particle may have agglomerated and resulting in the increase of size. With the addition of CTAB, the uniformity of the nanoparticles is increased, but higher concentration resulted in agglomeration. It is observed that the silica nanoparticles tend to form clusters of larger size than that of the silica nanoparticles when dispersed in the oil. These larger clusters get stabilized and dispersed in the SEF. Similar phenomenon was explained by Saterlie et al. by performing particle size analysis on copper nanofluid [33]. Jin et al. studied the influence of AC breakdown voltage of nano silica dispersed mineral oil and observed that the nano silica particle on dispersion in mineral oil exhibited improved breakdown strength, especially at high moisture content level [23]. Li et al. studied size of nanosilica particle on dispersion in vegetable oil and they have observed that on dispersion of nanofillers with oleic acid as

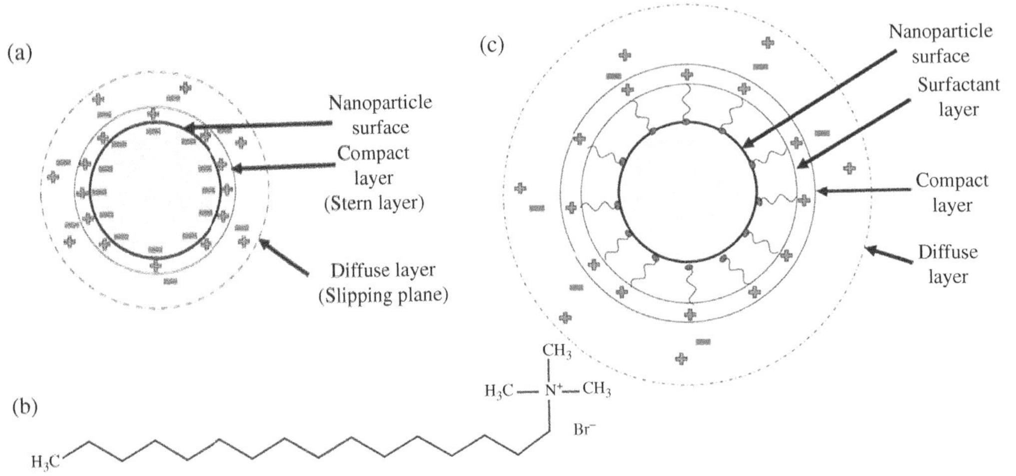

Figure 9.4 (a) Electrical double layer for nanoparticle in liquid, (b) chemical structure of CTAB, (c) electrical double layer for nanoparticle in the presence of CTAB.

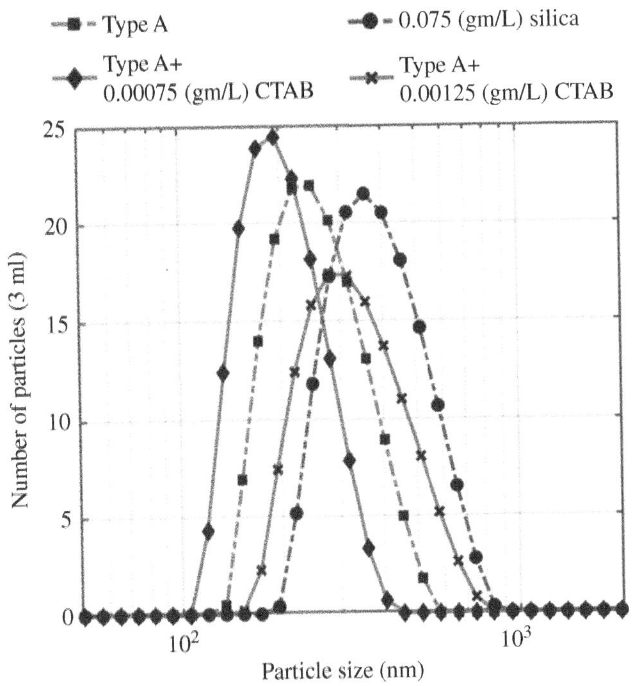

Figure 9.5 Particle size analysis of synthetic ester silica nanofluids in a 3 ml cuvette.

surfactant, the size of the particle has increased. They have explained that the cause is due to surfactant. Thus, the nanoparticles gets agglomerated to form larger size [25]. In general, it is observed that Type B nanofluid with optimum concentrations of both the nanoparticles and the surfactant is found to have lowest mean size of particles dispersed in SEF.

9.2.3.2 Zeta Potential Analysis

Micro electrophoresis cell was used and five stable measurements were carried out to arrive at a mean value of zeta potential. Zeta potential explains about the nanoparticle stability using their electrophoretic behavior. It is measured using the principle of laser Doppler electrophoresis [45]. The presence of surface charges on the dispersed particles leads to development of ionic characteristics. The opposite charges get attached to the surface of the dispersed particles forming a layer of charged ions defined as the Stern layer. Around this rigid layer, there is a layer called slipping plane. Zeta potential is the potential difference between the dispersion medium and the slipping plane. In the present study, it can be seen that silica nanoparticles are very stable. Figure 9.6 shows the variation of the zeta potential in Case 1 and Case 2. Case 1 shows the variation of CTAB in 0.05 g/l of silica-based SEF: (i) 0.0005 g/l, (ii) 0.00075 g/l, (iii) 0.00125 g/l, (iv) 0.00175 g/l. Case 2 shows the variation with respect to the concentration of silica: (i) 0.025 g/l, (ii) 0.05 g/l, (iii) 0.075 g/l, (iv) 0.1 g/l. As the concentration increased, the stability also increased, but at higher concentration, a slight decrease in zeta potential is seen which could be due to agglomeration of silica particles. A marginal increase in the stability can be observed when CTAB is added as surfactant.

Based on the particle size analysis and zeta potential measurements, the sample with 0.05 g/l of silica was optimum and is named Type A sample. The sample with 0.05 g/L and 0.00075 g/l of CTAB was the sample with both optimum concentrations of silica and surfactant and is named as Type B sample.

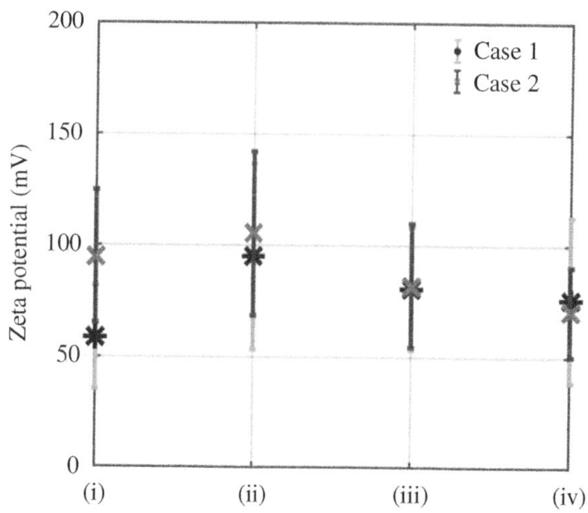

Figure 9.6 Comparison of zeta potential of synthetic ester silica nanofluid with and without surfactant.

9.2.3.3 Viscosity Measurement

Viscosity of SEF is one of the important parameters to be examined as it helps in understanding one of the fundamental property of SEF to be a coolant. Viscosity measurement has been carried out using Physica MCR-301 Rheometer. Variation in viscosity of SEF and nanofluid at different temperatures is shown in Figure 9.7. Jin et al. also showed that silica nanofluid has viscosity near the base oil's viscosity [23]. They also observed that addition of silica nanoparticles in mineral oil has minimal effect on the viscosity. Similar observation is noticed when silica nanoparticles are added to SEF. Thus, there was not much variation in viscosity observed on addition of nanoparticles to ester fluid.

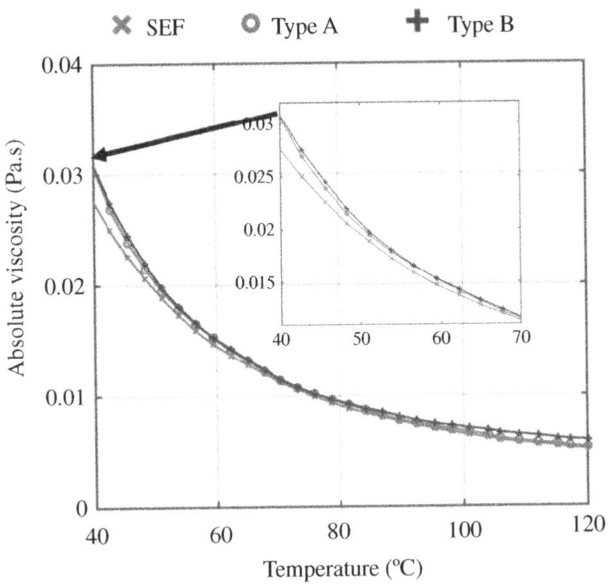

Figure 9.7 Absolute viscosity variation with temperature of the SEF and nanofluids.

9.3 Frequency Domain Dielectric Response

9.3.1 Experimental Setup

The frequency domain and the time domain dielectric response measurements for the liquids were performed using the IEC 60247-3 terminal oil test cell [46]. The capacitance of the empty test cell is 55 pF. The frequency domain dielectric spectroscopy was performed using the Omicron's Dirana in the frequency range of 0.1 Hz–1 kHz at 200 V. The gap distance between the electrodes is 2 mm. The guard electrode is used for reduction in the surface leakage currents of the test cell. The dielectric constant and the tanδ measurements were carried out at 30 °C, 50 °C, 70 °C, and 90 °C. Figure 9.8 shows the experimental setup used for the frequency domain dielectric response.

Based on the fundamentals discussed in IEC 60247 and IEC 61620 [46, 47], important parameters like dielectric constant and the dissipation factor can be measured. Dielectric constant (ε_r) and dielectric loss factor (tanδ) for each of the oil samples were measured. Dielectric loss factor is the ratio of leakage current or conduction current density to the dielectric displacement current density. These current densities in liquids are influenced by many factors like (i) movement of ions or charged particles in the liquid, (ii) rotation of permanent dipoles, (iii) rotation of induced dipoles. ε_r and tanδ are important parameters that are highly affected by the physical and chemical properties of the liquid [5].

For a general dielectric material, which is considered between two parallel plate electrodes, the ratio of the capacitances of the parallel plates with dielectric (C_x) and without dielectric (C_0) is called as dielectric constant and is given by the following equation

$$\varepsilon_r = \frac{C_x}{C_0} \tag{9.3}$$

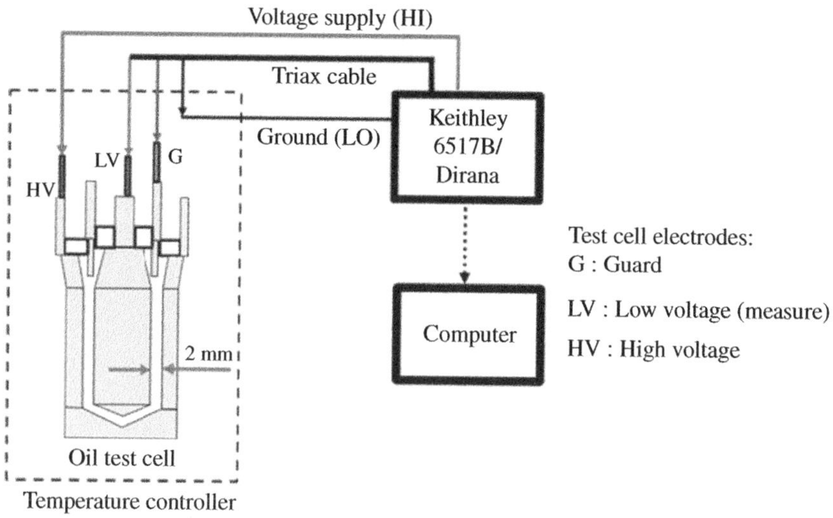

Figure 9.8 Experimental setup for frequency domain dielectric spectroscopy.

The dielectric dissipation factor or tanδ is the measure of the electrical conductivity and the polarization losses of the material. It is given by the ratio of the conduction current and the dielectric displacement current. The expression for it can be given by the following equation.

$$tan\delta = \frac{\sigma}{\varepsilon_0 \varepsilon_r \omega} \tag{9.4}$$

9.3.2 Dielectric Constant

Figure 9.9 shows the variation of dielectric constant of SEF, Type A, and Type B at 30 °C. Due to the presence of insulating nanoparticles, there is a slight increment in the dielectric constant. The presence of nanoparticles increases the capacitance of the nanofluid because of the charging of nanoparticles. There is more increment at the lower frequency in nanofluids because of the interfacial polarization between the nanoparticles and the interaction between the liquid and the nanoparticles in the liquid. A similar observation has been made in a study where ZnO nanoparticles were dispersed in mineral oil [48] where the dielectric constant increased with increment in the concentration of the nanoparticles.

9.3.3 Dissipation Factor

Figure 9.10 shows the variation of the tanδ of the oil samples at different temperatures in the frequency range of 0.1–1000 Hz. tanδ reduces with increase in frequency. It tends to increase with increase in temperature for all the samples. This is because of the thermal activation of the oil, which increases the conduction current of the oil samples. tanδ for Type A sample is least because of the presence of the insulating nanoparticles that block the movement of electrons in turn reducing the conductivity. tanδ for the Type B sample, which differs by addition of surfactant, is higher. This can be because of the CTAB that is formed as a layer on the nanoparticle surface. CTAB consists of long-chained molecules and its interaction with the nanoparticles creates an interfacial

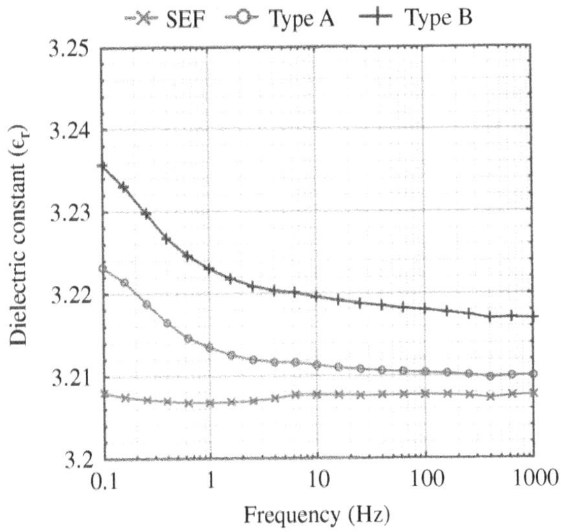

Figure 9.9 Variation in dielectric constant of SEF and nanofluids with frequency at 30 °C.

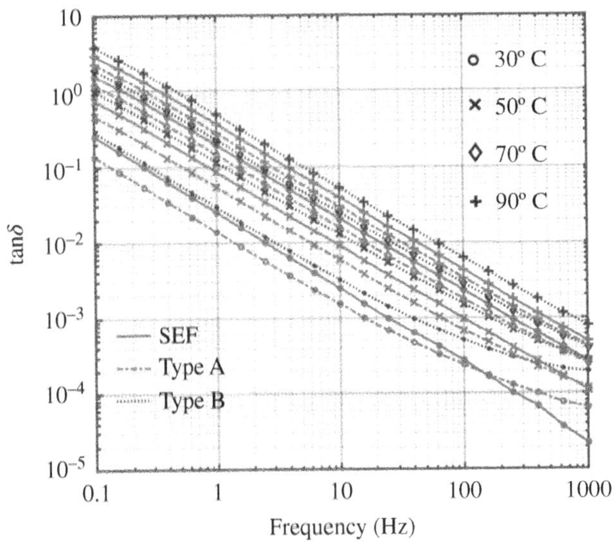

Figure 9.10 tanδ variation with frequency of SEF, Type A, Type B measured at (a) 30 °C, (b) 50 °C, (c) 70 °C, and (d) 90 °C.

layer with a dipole moment in certain direction [34]. In order for the nanoparticle and the layer of CTAB to orient itself to the direction of the electric field, more energy is required, which leads to the increment in the tanδ [5]. In general, because of the charge accumulation at the metal electrode which leads to the interfacial polarization between the metal electrode and the oil, the dielectric constant increases, but in the present case, there is a distinct variation in the onset of this interfacial polarization because of the nanoparticles and surfactant. Using the information of dielectric constant and tanδ, conductivity can be calculated using Equation (9.4) and Equation (9.5), where ε is the permittivity of the oil and ω is the angular frequency.

$$\sigma = \varepsilon\omega\tan\delta \tag{9.5}$$

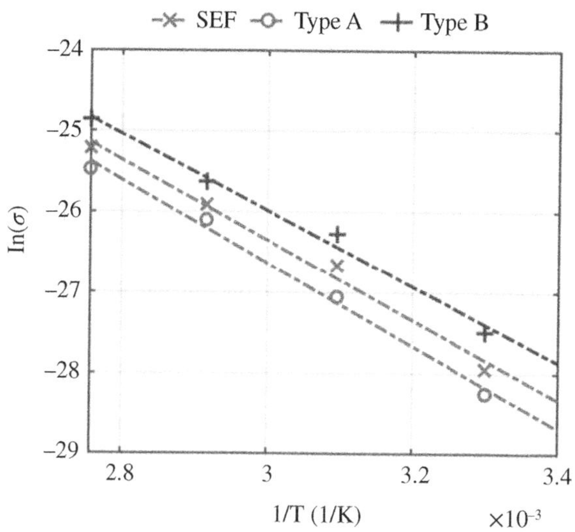

Figure 9.11 Variation of conductivity with respect to temperature of synthetic ester nanofluids.

Table 9.2 Arrhenius parameters for synthetic ester nanofluids.

Sample	E_a (eV)	σ_0 (μS/m)
SEF	0.4245	9.426
Type A	0.4402	12.237
Type B	0.4043	6.736

This conductivity is varying with temperature, so this relationship can be expressed by means of Arrhenius equation given by equation (9.6), where σ is the conductivity, σ_0 is conductivity at T_0, E_a is activation energy, k is the Boltzmann constant, and T is temperature.

$$\sigma = \sigma_0 \exp\left(\frac{-E_a}{k}\left(\frac{1}{T} - \frac{1}{T_0}\right)\right) \tag{9.6}$$

Figure 9.11 shows the variation of conductivity at 50 Hz and the inverse of temperature. The straight-line fit shows that the conductivity is temperature activated. The activation energy (E_a) and conductivity constant (σ_0) is shown in Table 9.2.

9.4 Time Domain Dielectric Response

9.4.1 Experimental Setup

The same IEC 60247 test cell used in previous study shown in Figure 9.8 is used for conduction current measurements. The conduction current is measured using the Keithley 6517B electrometer under polarity reversal DC voltages at 25, 50, 100, and 200 V. Studies have shown that DC polarity reversal method has more advantages like stable results and less rest times due to less accumulation of charges at electrodes between two consecutive tests [48, 49]. Voltages were restricted to 200 V or electric fields lower than 0.1 kV/mm, because at higher voltages, due to the electrohydrodynamic (EHD) motion, the measured current is distorted at higher voltages. This can also be due to charge accumulation at the electrode–oil interface and also charge depletion in the bulk part of the liquid. The DC polarity reversal cycle was for 1000 s and 500 s for each polarity.

9.4.2 Ion Mobility

Insulating liquids are generally nonpolar in nature, but the presence of water molecules or other impurities that are polar in nature leads to the formation of free ions under the influence of electric field. These ions are a result of the molecular dissociation where the first step is ion pair generation and then the dissociation [5, 49, 50]. The mobility of these ions is influenced by the molecular structure of the chemicals present in the oils. Upon application of external electric field, as per the Stokes law, the force due to the electric field is balanced by the viscous force. Viscosity of the bulk of the liquid is an important parameter that affects this viscous force.

Time-dependent conduction current or polarization current measurement with polarity reversal can help in measuring the ionic mobility. The applied voltage alternates between steady positive and negative values of the same magnitude, and the conduction current in the oil is measured. Ionic mobility is defined in terms of the motion of ions from one electrode to the other electrode. This motion is captured in the conduction current signal in the form of transit time where the current reaches the peak, reduces and settles to a constant value. The alternating polarity reversal method is a preferred method due to its repeatability and due to the low impact of the residual charges. The ionic mobility (μ) is given by equation (9.7), in which U is the applied voltage, d is the gap distance, and T_t is the transit time or time of flight [49–51]. Figure 9.12a shows a typical current signal that is used to calculate the ionic mobility.

$$\mu = \frac{d^2}{U \bullet (T_t)} \tag{9.7}$$

In the present study, the experiments were performed by application of four cycles of 25, 50, 100, and 200 V alternating polarity voltages. The parameters measured from the current signal were peak current (I_P), settled current (I_S), and transit time (T_t), which are shown in Figure 9.12a. The above parameters were measured from the second, third, and fourth cycle, and their average values were used for calculating the ionic mobility. The first cycle is omitted from the measurement because, in the uncharged oil, the free ions in it are dispersed randomly in the oil. In the first cycle, upon application of voltage, they are attracted toward the electrodes, but in the subsequent cycles, the ions only move from one electrode to another. The condition of the oil in the first cycle is different from the subsequent cycles leading to deviation in the characteristics of current. The applied voltages in the measurement of ionic mobility were restricted to 200 V only because at higher voltages, due to the EHD motion, the measured current is distorted. Figure 9.12b shows the typical current signals obtained at 500 V. The current signals are distorted, and also there is a variation between the current measured in all the cycles but tends to settle in the third and fourth cycle. The transit time is also very small, which makes the measurement slightly inaccurate at the peak as the sampling is performed at 1 sample/s for higher accuracy of current measurement.

Figures 9.13 and 9.14 show the variation of I_P and I_S under positive polarity and negative polarity voltages, respectively. The current increases fairly linear with increase in voltage in both cases under low electrical stress of 0.1 kV/mm. This is because under low electrical stress, only ionic and ohmic conduction mechanisms are prevalent, and the impact of space charge is not present. I_P and I_S are least in Type A sample and SEF and Type B sample tends to have nearly same value at each voltage because of higher conduction current.

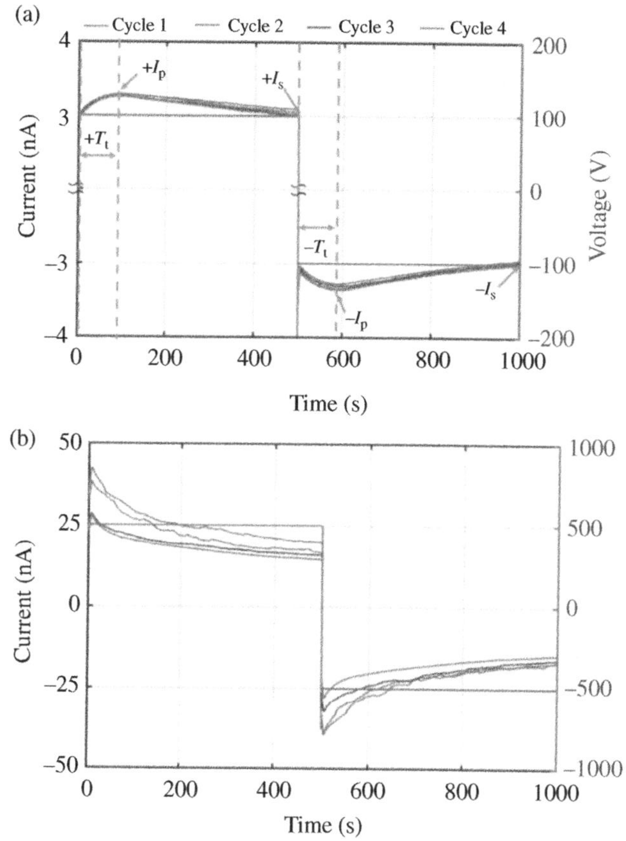

Figure 9.12 Typical current signals recorded in SEF at an alternating polarity voltage of (a) 100V and (b) 500V.

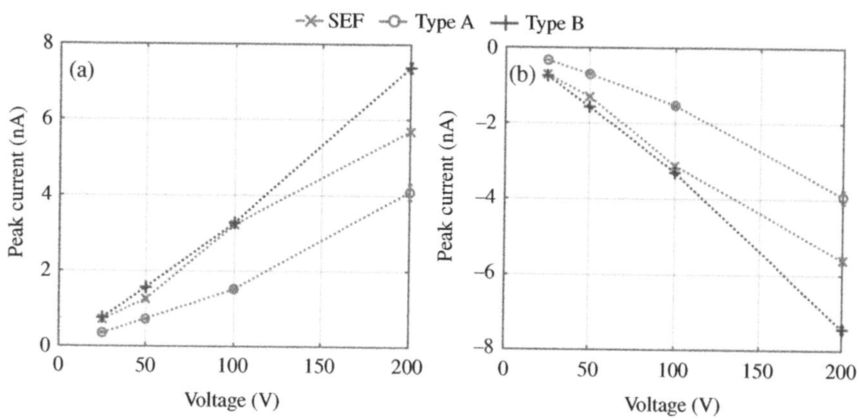

Figure 9.13 Variation of peak current (I_P) with voltage in SEF, Type A, and Type B samples under (a) positive polarity and (b) negative polarity.

Figure 9.15 shows the variation of transit time with voltage in oil samples. It reduces with increase in voltage as the peak of the current appears sooner in the case of higher voltages. Figure 9.16 shows the variation of ionic mobility with voltage in oil samples. There is a marginal variation in the ionic mobility in the voltage range of 25–100V, but there is an increment at 200V.

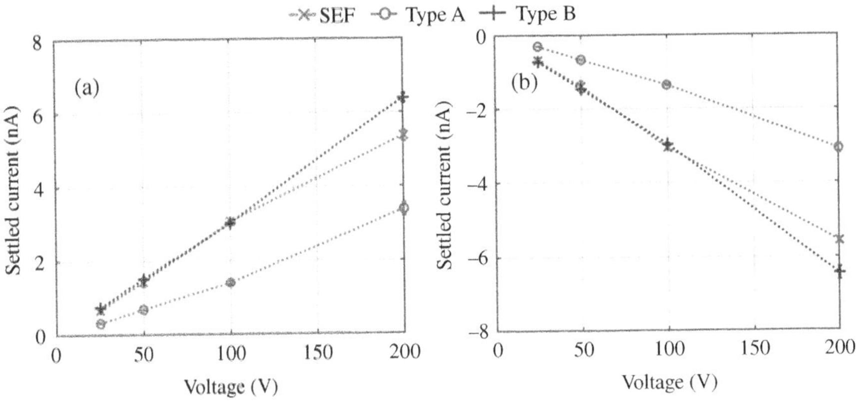

Figure 9.14 Variation of settled current (I_S) with voltage in SEF, Type A, and Type B samples under (a) positive polarity and (b) negative polarity.

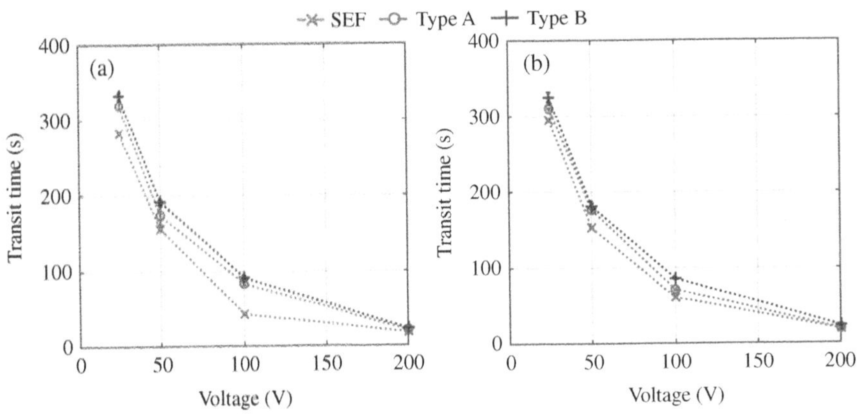

Figure 9.15 Variation of transit time (T_t) with voltage in SEF, Type A, and Type B samples under (a) positive polarity and (b) negative polarity.

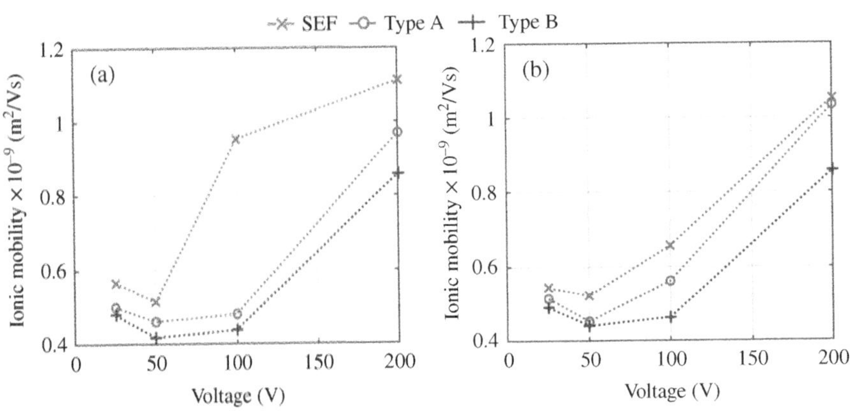

Figure 9.16 Variation of Ionic mobility (μ) with voltage in SEF, Type A, and Type B samples under (a) positive polarity and (b) negative polarity.

A similar observation was made in [49]. The ionic mobility is least in Type B sample and highest in SEF. The motion of ions is slowed down in the oil because of the presence of the nanoparticles leading to higher transit time and lower ionic mobility. A similar study has been conducted with TiO_2 nanoparticle in mineral oil, finding that the ionic mobility in nanofluid was less compared to the mineral oil at higher electrical stress [52]. Here, the charging of the nanoparticles is also leading to the slowdown of the ion movement in the oil.

9.4.3 Conductivity and Other Dielectric Properties

The conduction current measured is time dependent, and it becomes nearly constant after a certain duration, and conductivity also follows a similar trend. The conductivity (σ) in the present case is calculated at $t = 500\,\text{s}$. Equation (9.8) is used for calculating the conductivity, where C_0 is the capacitance of the empty test cell.

$$\sigma = \frac{I(t)\varepsilon_0}{C_0 U} \tag{9.8}$$

Figure 9.17 shows the variation of conductivity in different oil samples at different voltages. Conductivities of the oil samples are nearly constant with respect to voltage. Conductivity is highest in Type B and least in Type A. This increment in the conductivity in the Type B sample compared to Type A is because of the presence of CTAB, which is increasing the dielectric loss in the oil as more energy is required for the polarization of the stable nanoparticles in the oil gap.

Under low electric field stress, dissociation and recombination processes are dominant [5, 49–51]. They depend on the conductivity and the ionic mobilities of the oil samples. The ionic concentration n_{eq} can be calculated by Equation (9.9), where q is the charge of an electron, and μ^+ and μ^- are ionic mobilities at positive polarity and negative polarity, respectively.

$$n_{eq} = \frac{\sigma}{q(\mu^+ + \mu^-)} \tag{9.9}$$

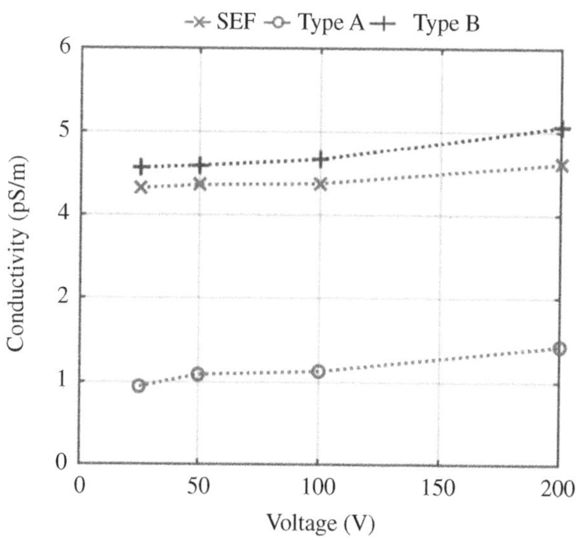

Figure 9.17 Variation of conductivity with voltage in SEF, Type A, and Type B samples under (a) positive polarity and (b) negative polarity.

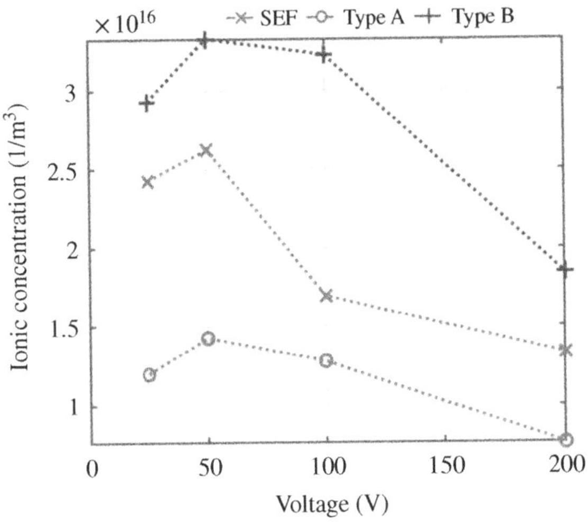

Figure 9.18 Variation of ionic concentration with voltage in SEF, Type A, and Type B samples.

Figure 9.18 shows the variation in ionic concentration in oil samples under different voltages. Ionic concentration is least in Type A and highest in Type B for all the voltages. It tends to be constant up to 100 V but reduces at 200 V, in all the oil samples. This can be because of the higher ionic mobility, which is caused by the EHD motion, leading to the recombination of ions.

Ionic mobility is influenced by the viscosity of the oil samples. The dynamic viscosity for the SEF, Type A, and Type B samples at 30 °C are 0.0377, 0.0408, and 0.0412 Pa.s, respectively. The relation with ionic mobility and viscosity is given by Stokes' Law and Walden's rule, which is given by the following Equations (9.10) and (9.11),

$$\mu = \frac{e}{6\pi\eta a} \tag{9.10}$$

$$\mu\eta = \text{constant} \tag{9.11}$$

Where a is the ionic radius, μ is the ionic mobility, and η is the dynamic viscosity. There is an inverse relation between the ionic mobility and the viscosity. Using the above relations, the ionic radius can be calculated. Figure 9.19 shows the variation of ionic radius of oil samples in positive polarity and negative polarity. As viscosity is constant, the inverse relationship with the ionic mobility gives values with an opposite trend to that of the ionic mobility. A study conducted on ionic mobility of insulating oils showed that ionic mobility and the derived quantities follow the Arrhenius law [49].

9.5 Conduction at High Electric Field

9.5.1 Experimental Setup

The main conduction mechanisms before the occurrence of partial discharge and breakdown are (i) ohmic conduction, (ii) tunneling, (iii) space charge saturation [5], which are electronic processes. In the low electric field condition, there will be no charge generation from the electrode. Ohmic conduction is more prevalent. During the high electric field condition, there will be charge injection from the electrodes due to field emission or field ionization, which causes space charge formation in the electrode gap. These phenomenon can be better understood using the I–U characteristics [53].

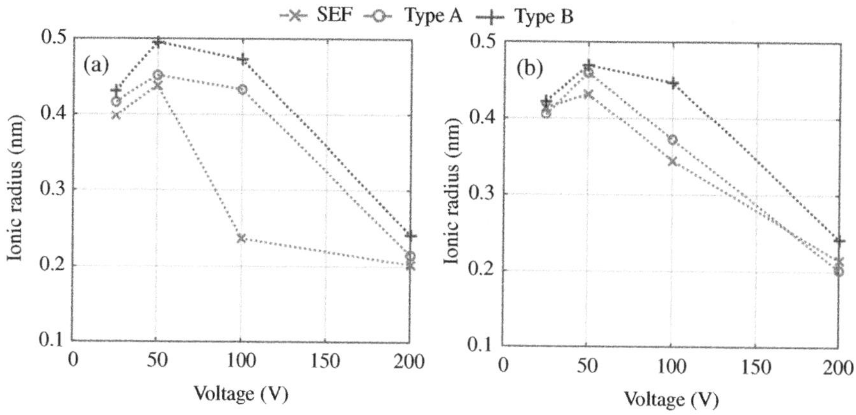

Figure 9.19 Variation of ionic radius with voltage in SEF, Type A, and Type B samples under (a) positive polarity and (b) negative polarity.

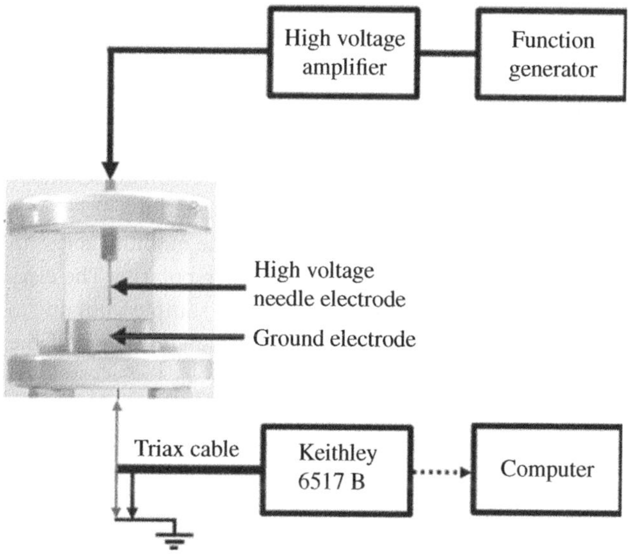

Figure 9.20 Experimental setup for measurement of I–U characteristics.

Figure 9.20 shows the experimental setup used for the measurement of I–U characteristics. The needle plane electrode configuration used for studying the I–U characteristics under high electric field conditions. The radius of curvature for the high-voltage stainless steel needle electrode was 50 μm. The plane electrode, or ground electrode, has a diameter of 5 cm and is made up of aluminum. The approximate amount of oil required for testing was 75 ml. The conduction current was measured using a Keithley 6517b electrometer under positive and negative DC voltages. The conduction current was measured for a duration of 1 minute at different voltages from 1 to 15 kV in steps of 1 kV.

9.5.2 I–U Characteristics

In the previous sections, the conduction mechanisms in the oil are only ohmic in nature, and the I–U characteristics are linear. There is no influence of space charge accumulation in the bulk liquid on the conduction current. The main underlying mechanisms in the oil gap under high electric field are

(i) ohmic conduction, (ii) tunneling, and (iii) space charge saturation. I–U characteristics help in understanding the type of mechanism that is present at each voltage [53]. Studies in the past have performed similar analysis for understanding the conduction mechanism under needle plane configuration [40, 54]. These studies have indicated that apparent mobility of the ions generated at the needle tip can be a factor, which determines the corona discharge and breakdown voltage of the oil gap. The I–U characteristics are initially plotted with x-axis as voltage and y-axis as the square root of current. Then the plot fitted to a straight line tends to follow a straight-line equation given in Equation (9.12). "y" is $\sqrt{(I/[nA])}$, x is U/[kV]. The slope (m) of the equation is directly proportional to the mobility and its relation is given in Equations (9.13) and (9.14). Equations (9.13) and (9.14) are derived based on the calculations of the apparent ion mobility in a needle plane configuration [40, 53–55].

$$I = mU + c \tag{9.12}$$

$$m = \sqrt{\frac{2\mu\varepsilon_r\varepsilon_0}{d}} \tag{9.13}$$

$$\mu = \frac{m^2 d}{2\varepsilon_r\varepsilon_0} \tag{9.14}$$

Figures 9.21 and 9.22 show the I–U characteristics of the oil samples under positive and negative polarity, respectively. In the present study, the voltage is restricted to 15 kV under both polarities up to 15 kV and mainly to understand the ion mobility before the occurrence of corona discharges. The straight lines are fitted to the data, and the mobility is calculated for each case. Table 9.3 shows the slope of the fitted straight line and the mobility. It can be observed that the mobility is less in the Type B sample compared to Type A and SEF in both positive and negative polarity. The electrostrictive force that is exerted from the needle tip is nearly same in all the oil samples, but due to the presence of nanoparticles the EHD motion is slowed down, leading to slowing down of the ions, in turn to lower the apparent mobility in nanofluids.

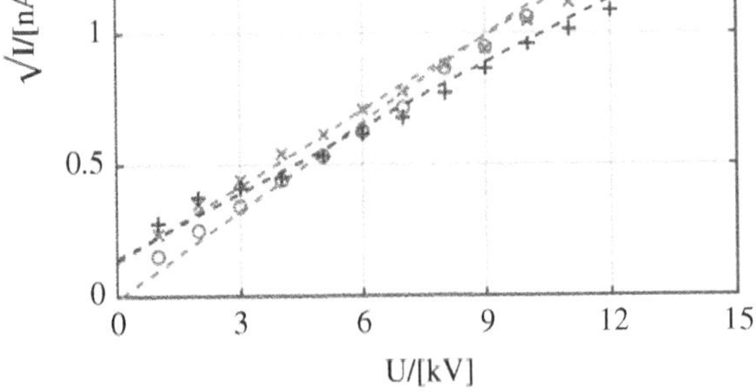

Figure 9.21 I–U characteristics of SEF, Type A, and Type B samples under positive DC voltage.

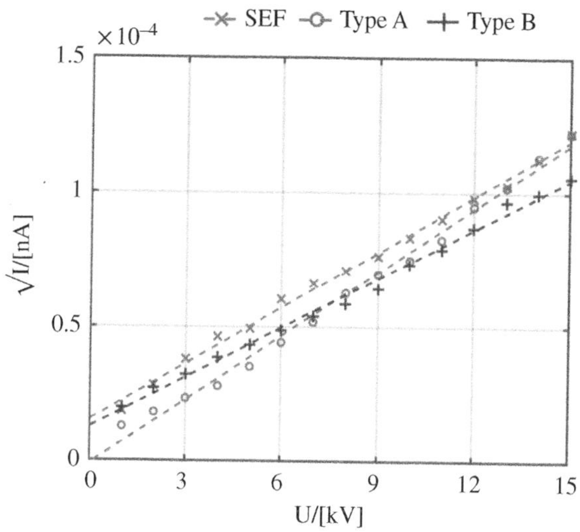

Figure 9.22 I–U characteristics of SEF, Type A, and Type B samples under negative DC voltage.

Table 9.3 Slope "m" and mobility under positive and negative polarity.

	Positive polarity		Negative polarity	
Sample	**$m \times 10^{-9}$**	**$\mu+$ (n.m2/Vs)**	**$m \times 10^{-9}$**	**$\mu-$ (n.m2/Vs)**
SEF	9.489	7.945	6.916	4.220
Type A	11.108	10.887	7.910	5.521
Type B	8.207	5.943	6.127	3.312

In a needle plane configuration, under positive polarity, the negative ions get attracted to the needle tip leading to formation of space charge that adds up to the applied electric field, which reduces the effective gap distance leading to higher ion mobility and lower discharge inception voltage. Under negative polarity, the negative ions injected from the needle tip form space charge at the ground electrode. This reduces the applied electric field, leading to increased apparent gap distance, and the accumulated space charge impedes the injecting negative ions leading to reduced ion mobility, which can affect the CIV under both the polarities.

9.6 Corona Inception Voltage

9.6.1 Experimental Setup

The same needle plane electrode test cell used for studying the I–U characteristics is used for studying the corona discharge activity. Figure 9.23 shows the experimental setup for CIV measurement. In order to generate high voltages for measurement of CIV, the desired voltage waveform was generated through a function generator (Tektronix AFG3051C 5GSa/s, 50 MHz). Output of function generator is given to a High Voltage Trek Amplifier (Trek model 30/20A), which amplifies the output voltage 6000 times.

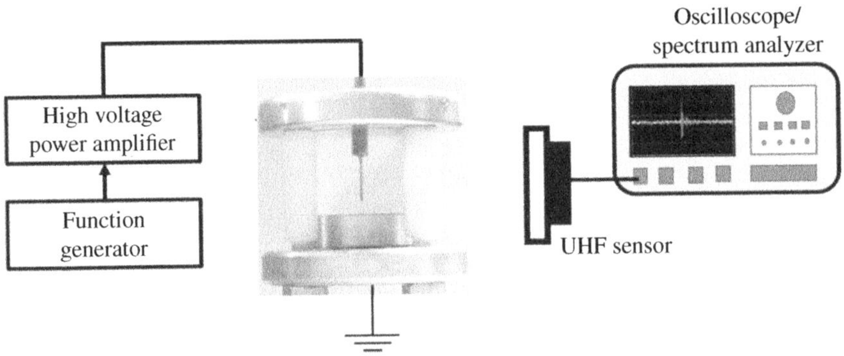

Figure 9.23 Experimental setup for corona discharge detection using UHF sensor.

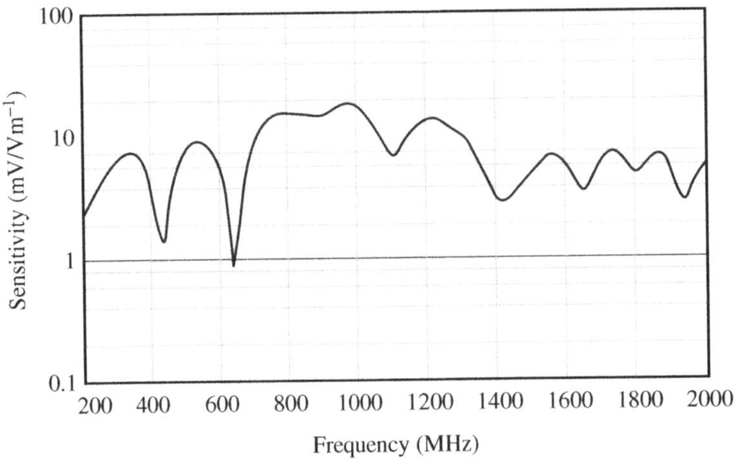

Figure 9.24 Sensitivity of UHF sensor [57].

A broad band UHF sensor was used in the following study. Figure 9.24 shows the sensitivity of the UHF sensor used for the partial discharge. The design aspects of UHF sensor and their sensitivity response of the broadband sensor is detailed in the study by Judd et al. [56]. Similar sensor has been used in the study by Sarathi et al. [57]. The detection sensitivity for the onsite tests of HV test objects needs to be carefully chosen to provide a signal only if the UHF signal from the coupler surpasses a signal level range of −75 to −70 dBm (0 dBm = 224 mV). A high-frequency DSO (Lecroy Wavepro 7 z series 4 channel digital real-time oscilloscope) of 3.5 GHz bandwidth operated at 40 GSa/s (10 GSa/s per each channel) and spectrum analyzer (Agilent CX9000) with an input impedance of 50 Ω were used for the studies.

9.6.2 CIV Results and Discussion

Figure 9.25a shows the typical UHF signal recorded during corona discharge. Figure 9.25b shows the FFT of the UHF signal recorded. The frequency content of the signal can be observed to be between 0.8 and 2.2 GHz. The dominant frequency can be seen at 0.9 GHz.

Figure 9.26a shows the variation in CIV of SEF under AC and DC voltages with variation in concentration of silica. Figure 9.26b shows the variation in CIV of type-A nanofluid with AC and

DC voltages with variation in CTAB concentration. Corona discharge activity is an intermittent process, which occurs due to the localized high electric field stress. In the present study, the CIV was arrived at based on the first UHF signal measured in oscilloscope on application of voltage to the needle plane electrode gap filled with transformer oil/nanofluids. Twenty such voltage measurements were made, and the average of it is represented as CIV. It is observed that negative DC voltage have high CIV followed with positive DC and AC voltage. It is also observed that CIV is highest at 0.05 g/l, under AC and DC voltages. Increase in the concentration of silica above 0.05 g/l leads to reduction in CIV, much lower than that of SEF. The cause for it could be due to agglomeration of nanoparticles or the agglomerated nanoparticle acquiring certain charge, thereby forming nonuniformity in the electrode gap, incepting discharge at much lower voltages. From this result also it can be seen that Type A with 0.05 g/l of silica is optimum for improving the CIV. With addition of 0.00075 g/l of CTAB, the CIV increases further.

From the section 9.5.2, which shows the I–U characteristics under high electric field conditions, it can be seen that there is a reduction in the mobility of the ions with nanoparticles in the liquid. Because of the presence of nanoparticles, the ions get attached on the nanoparticle surface leading to the slowdown of the ions under high electric field. The same reasoning can be applicable at the

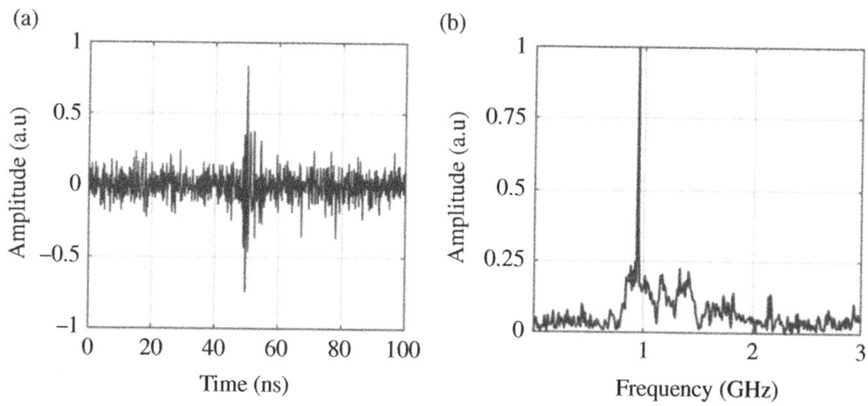

Figure 9.25 Typical (a) UHF signal formed due to corona activity and (b) corresponding FFT analysis.

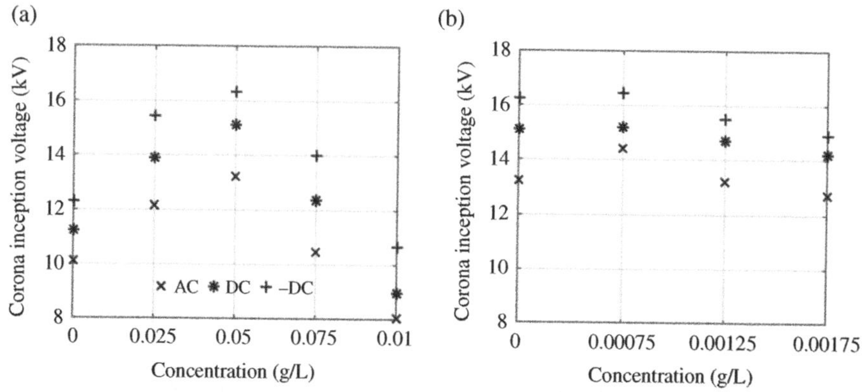

Figure 9.26 Variation in corona inception of SEF (a) with variation in concentration of silica nanoparticles and (b) with variation of surfactant with 0.05 g/l of silica.

condition of CIV where the reduction in mobility of the ions is leading to the increment in the CIV of the stable nanofluid. In any fluids, the local electric field variations due to nanoparticles alter the ionization field of the liquid enhancing the discharge inception voltage, irrespective of polarity of the needle electrode. In the present study, it is observed that the discharge inception voltage is less under positive DC voltage compared with negative DC voltage. The electrons generated in the electrode gap get trapped by the nanoparticle, thereby producing counter electric field near the needle electrode and reducing the electric field for discharge inception, especially predominant under negative DC voltage.

Jin et al. observed 20% higher inception voltage with silica nanofluids compared with mineral oil [23]. Under negative DC voltage, the CIV with silica nanofluids and the transformer oil is nearly same. They also have observed that the magnitude of discharges is less with silica nanofluids due to its inherent capability of absorbing moisture, acids, and other additives, thereby reducing the influence of contaminants on breakdown strength of oil. Beroual et al., observed that the discharges in liquids under positive polarity have multiple branches and travel faster. When the pointed electrode is negative, thicker root discharge originates and rate of discharge propagation is highly limited due to lower diffusion /drift of generated charges [58]. This can be because of the space charge accumulation on the ground plates that inhibit the injection of the ions from needle electrode under negative polarity that leads to apparent increase in gap distance leading to increased CIV in negative polarity compared to that of the positive polarity [59].

9.6.3 Incipient Discharge Activity

9.6.3.1 Corona Discharge Activity Under Harmonic AC Voltages

In order to understand the incipient discharge activity, phase-resolved partial discharge (PRPD) analysis was carried out. Figure 9.27 shows the comparison between the PRPD of SEF and Type B nanofluid. CIV for SEF under the harmonic AC with third and fifth harmonic voltages with 2.5 and 40% THD are 7.8 and 6.1 kV, respectively. CIV for Type B nanofluid under the harmonic AC with third and fifth harmonic voltages under 2.5 and 40% THD are 10.8 and 8.6 kV, respectively. CIV for SEF under the harmonic AC with third, fifth, and seventh harmonic voltages with 2.5 and 40% THD are 6.7 and 5.3 kV, respectively. CIV for Type B nanofluid under the harmonic AC with third, fifth, and seventh harmonic voltages under 2.5 and 40% THD are 10 and 6.7 kV, respectively.

The output voltage of The UHF sensor output is connected to the spectrum analyzer and operated in zero span mode to acquire PRPD analysis. The spectrum analyzer and the supply voltage were synchronized to acquire discharge location in the phase of the applied AC voltage. In the present study, the spectrum analyzer operated in zero span mode by having center frequency as 1 GHz. It can be observed that PD activity is significantly reduced in the case of nanofluid. The higher PD activity under harmonic voltages can be due to voltage waveform factors like crest factor, slew rate, and shape factor [60, 61]. Steepness of the voltage waveform also effects the PD [62]. The extreme THD condition of each case (2.5 and 40%) are shown in Figure 9.27. It is observed from Figure 9.27 that discharges occur at the peak zone of the applied voltage in the case of lower THDs, whereas in the case of higher THDs, dV/dt is also influencing the higher corona discharge near the zero crossing, where the dV/dt magnitude is high. It is observed that in SEF, the corona discharge activity is predominant where rate of rise of voltage is high. Also, it is noticed that the number of discharges in negative half cycle is high in SEF, under AC and harmonic AC voltages, irrespective of THDs. In nanofluid, the number of discharges has considerably reduced and also a characteristic reduction is observed in the negative half cycle of the applied AC voltages with different THD. The other details related to number of discharges are included in Table 9.4. It is clear that increase in harmonic content shows increase in number of pulses, and number of discharges

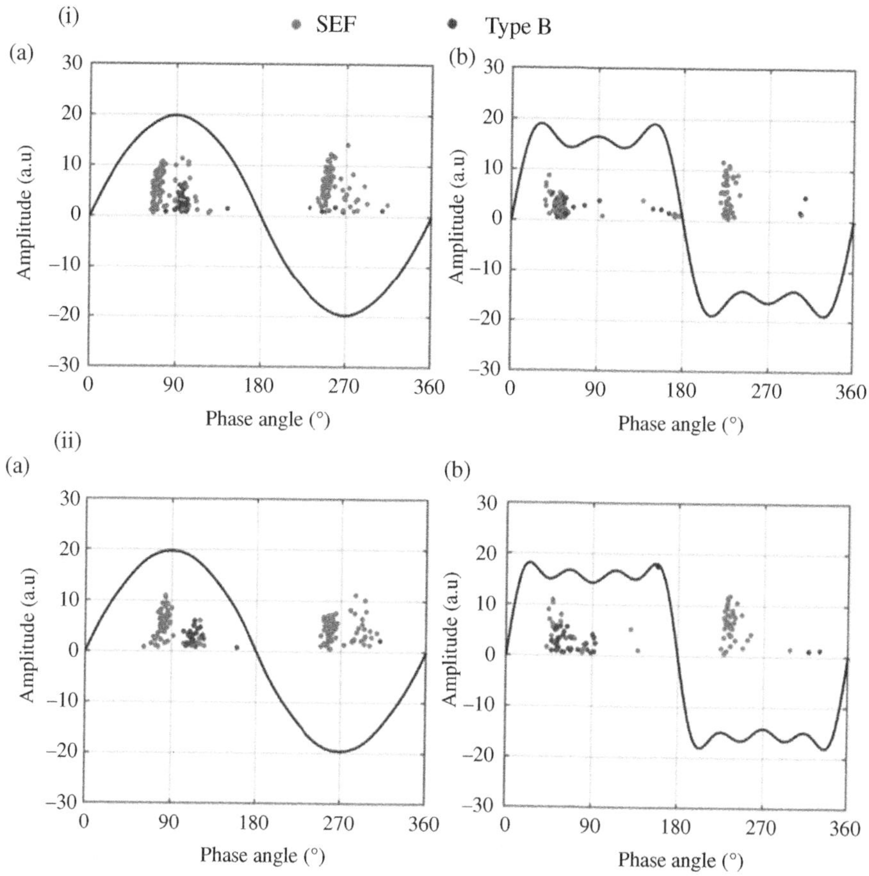

Figure 9.27 PRPD analysis (a) 2.5% THD and (b) 40% THD, (i) third and fifth harmonic combined voltage, (ii) third, fifth, and seventh harmonic combined voltage.

Table 9.4 Number of corona discharges activity per second with synthetic ester fluids and nanofluid under harmonic AC voltages with different THDs.

Voltage profile	Polarity	SEF (2.5%)	Type B (2.5%)	SEF (40%)	Type B (40%)
AC with third and fifth harmonic	Positive	61	30	48	36
	Negative	75	3	54	2
AC with third, fifth, and seventh harmonic	Positive	72	33	31	25
	Negative	78	4	45	2

occurring in the positive half cycle is high compared with negative half cycle. Also the number of pulses formed is less with nanofluid, in both half cycle of the AC voltage.

9.6.3.2 UHF Signal Energy Analysis

In order to understand the type of defect, the energy of the discharge can be one of the important parameters to detect the type of defect. Reid et al. was able to classify different type of defects based on the UHF signal energy content [63]. PD activity involves multiple current pulses, each contributing some charge to the cumulative charge q. The energy of a particular UHF signal (E_i) is

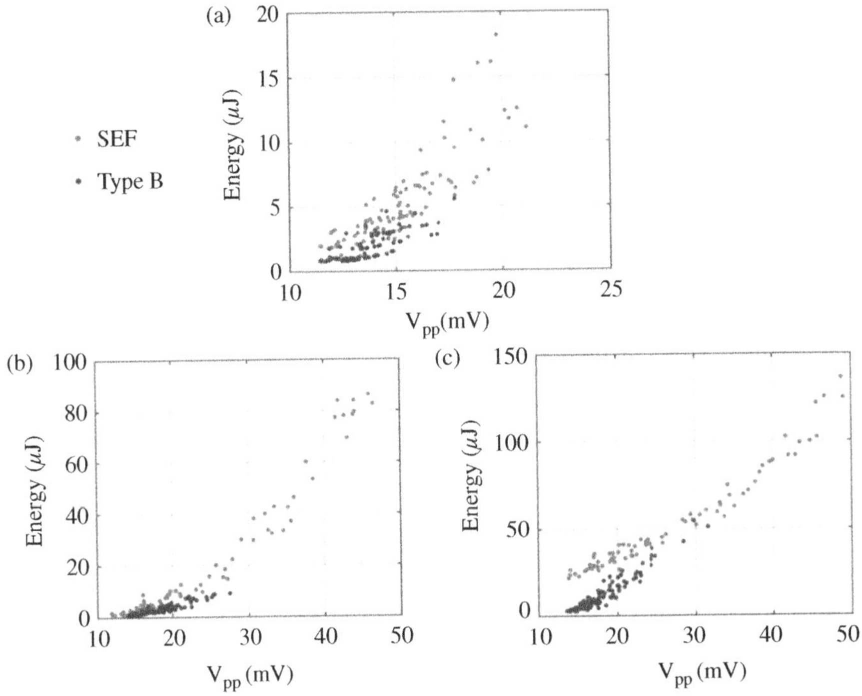

Figure 9.28 Variation in energy content of UHF signal formed due to corona activity in SEF and Type B nanofluid under AC and DC voltages: (a) AC, (b) +DC, and (c) −DC.

proportional to q^2. The UHF signal magnitude (V) is directly proportional to the q [63, 64]. The expression for the UHF signal energy of a particular signal is defined by:

$$E_j = E_{j-1} + \Delta T \sum_{i=1}^{n} \frac{V_i^2}{R_L} \tag{9.15}$$

where E_j is the energy of the UHF signal at PD, E_{j-1} is the total energy at the previous data point, $\Delta T = 0.1$ ns is the time interval between samples, V_i is the voltage of the UHF signal, and R_L is the measurement impedance (50 Ω), 100 signals were separately analyzed.

Figure 9.28 shows variation in UHF signal voltage peak to peak generated during corona discharge activity and its energy content. It is observed that energy of the UHF signal generated due to corona activity SEF and Type B oil follows the sequence $AC < DC < -DC$. It is also observed that the energy of UHF signal in Type B sample is less compared to SEF. Due to the presence of nanoparticles, there is inhibition of the discharge occurrence as the ions get charged on the nanoparticle surface and lowering the mobility leading to the reduction in the discharge magnitude as well. Due to reduction in the magnitude, the energy of the discharges with Type B nanofluid are much lower compared to that of the discharges in SEF.

9.7 Conclusions and Future Scope

A stable synthetic ester silica nanofluid was produced, which has better dielectric strength compared to SEF. Usage of CTAB surfactant improved the stability. The results of dielectric properties of nanofluid helped to understand the underlying mechanism of the nanofluid under different electrical stress. Increase in dielectric constant value with reduced tanδ observed with nano ester fluid. The presence of nanoparticles inhibit the motion of the ions leading to lowering of ionic mobility. Due to lower mobility of ions, for the occurrence of partial discharges, higher voltages were needed leading to increased CIV. These properties could be observed by means of I–U characteristics of the liquids insulants. Usage of silica nanoparticle ester fluid in transformer can reduce corona discharge activity and can improve the reliability of the transformer. In summary, the following are some of the key results obtained with the present study.

- SEF with 0.05 g/l of silica nanoparticles with surfactant 0.00075 g/l of CTAB makes a stable nanofluid. The optimum concentration of nanoparticles and surfactant was to improve the stability of the particles in the liquid and was determined with the help of particle size analysis and zeta potential measurements.
- Viscosity is not affected very much on the addition of nanoparticles and surfactant. Zeta potential increases upon the addition of the surfactant, indicating the increased stability of nanoparticle dispersed ester fluid.
- Dielectric constant in nanofluid is increased slightly because of the interfacial polarization between the nanoparticles and the liquid medium.
- tanδ in the nanofluid without surfactant was less than the plain SEF because of Type A's insulating nature, but upon addition of the long-chained surfactant CTAB, more energy is needed for polarization leading to increased tanδ of the nanofluid Type B. Conductivity of the oils are thermally activated and it increases with increase in temperature.
- In the low electric stress region ($E < 0.1$ kV/mm), only ohmic conduction is prevalent and the currents under positive and negative polarity tend to have a linear variation with voltage; conductivity tends to remain constant in this region.
- The transit time (T_t) tends to decrease with increase in the applied voltage, leading to slight increment in the mobility.
- There is a reduction of ionic mobility upon addition of nanoparticles and surfactant.
- In the high electric field region, because of the space charge, the I–U characteristics show that $\sqrt{I}$ is directly proportional to applied voltage U.
- The apparent mobility caused because of the EHD motion is lower in the case of the nanofluid with surfactant because the stable nanoparticles dispersed in the liquid slow down the motion of ions. This can be a reason behind the increment in the CIV.
- There was about 30% increment in the CIV from that of the pure SEF. The UHF signal generated due to corona discharges with silica nanoparticle-based SEF has the dominant frequency at 0.9 GHz.
- Corona discharge activity is much lower in the case of nanofluid compared to SEF. CIV reduces with increase in harmonic content of supply voltage. Corona discharge activity increases with the increase in the harmonic content and THD. It is observed that the dV/dt of the supply voltage have high impact on corona discharge activity.
- The energy content of the UHF signal formed due to corona activity under AC and DC voltages shows the trend AC < DC < -DC. Also the magnitude of UHF signal formed due to corona activity in SEF is high compared with that in nanofluid.

As a future work, it is important to understand the impact of different type of nanoparticles on the variation on the dielectric, thermal, physical, and chemical properties of the SEF, and also to understand the impact of different type of surfactant on the improvement of stability of the nanoparticles in the liquid. To understand the long-term behavior of the nanofluids, the accelerated ageing studies will be carried out. The temperature distribution in transformers by use of nanofluid in a transformer needs to be understood. Also, it is important to study the inception and breakdown characteristics under various practically occurring transient voltages.

References

1 Saha, T.K., Darveniza, M., Yao, Z.T. et al. (1999). Investigating the effects of oxidation and thermal degradation on electrical and chemical properties of power transformers insulation. *IEEE Transactions on Power Delivery* 14 (4): 1359–1367.

2 Lundgaard, L.E., Hansen, W., Linhjell, D., and Painter, T.J. (2004). Aging of oil-impregnated paper in power transformers. *IEEE Transactions on Power Delivery* 19 (1): 230–239.

3 Cigre Technical Brochure 642 (2015) Transformer reliability survey, working group A2.37.

4 Rouse, T.O. (1998). Mineral oil in transformers. *IEEE Electrical Insulation Magazine* 14 (3): 6–13.

5 Bartnikas, R. (1994). *Electrical Insulating Liquids, Engineering Dielectrics*, vol. 3. American Society for Testing and Materials.

6 Berger, N., Randoux, M., Ottmann, G., and Vuarchex, P. (1997). Review of insulating fluids. Cigré WG 15.02, *Electra*, 171.

7 Carballeira, M. and Crine, J. (1989). PCBs replacement liquids-how to determine their toxicity? *IEEE Electrical Insulation Magazine* 5 (3): 27–29.

8 Claiborne, C.C. and Pearce, H.A. (1989). Transformer fluids. *IEEE Electrical Insulation Magazine* 5 (4): 16–19.

9 Miller, R.E. (1981). Silicone transformer liquid: use, maintenance, and safety. *IEEE Transactions on Industry Applications* 17 (5): 463–468.

10 Crine, J.P. (1986). Silicone oil as replacement fluid for PCBs in transformers. *Canadian Electrical Engineering Journal* 11 (3): 110–113.

11 McShane, C.P. (2000). New safety dielectric coolants for distribution and power transformers. *IEEE Industrial Applications Magazine*: 24–32.

12 Cigre Technical Report 436 (2010) Experiences in service with new insulating liquids.

13 Oommen, T.V. (2002). Vegetable oils for liquid-filled transformers. *IEEE Electrical Insulation Magazine* 18 (1): 6–11.

14 Perrier, C., Beroual, A., and Bessede, J.L. (2006). Improvement of power transformers by using mixtures of mineral oil with synthetic esters. *IEEE Transactions on Dielectric and Electrical Insulation* 13 (3): 556–564.

15 Lewis, T.J. (1994). Nanometric dielectrics. *IEEE Transactions on Dielectrics and Electrical Insulation* 1 (5): 812–825.

16 Frechette, M.F., Trudeau, M.L., Alamdar, H.D., and Boily, S. (2004). Introductory remarks on nanodielectrics. *IEEE Transactions on Dielectrics and Electrical Insulation* 11 (5): 808–818.

17 Tanaka, T. (2005). Dielectric nanocomposites with insulating properties. *IEEE Transactions on Dielectrics and Electrical Insulation* 12 (5): 914–928.

18 Yu, W., France, D.M., Choi, S.U.S., and Routbort, J.L. (2007). *Review and Assessment of Nanofluid Technology for Transportation and Other Applications*. Oak Ridge, TN: Argonne National Laboratory is a U.S. Department of Energy, Office Science and Technology Info.

19 Segal, V. and Raj, K. (1998). An investigation of power transformer cooling with magnetic fluids. *Indian Journal of Engineering and Materials Sciences* 5 (6): 416–422.

20 Choi, C., Yoo, H.S., and Oh, J.M. (2008). Preparation and heat transfer properties of nanoparticle-in-transformer oil dispersions as advanced energy-efficient coolants. *Current Applied Physics* 8 (6): 710–712.

21 Hwang, J.G., O'Sullivan, F., Zahn, M. et al. (2008) Modeling of streamer propagation in transformer oil-based nanofluids. *Conference on Electrical Insulation and Dielectric Phenomena, CEIDP*, Quebec, QC, Canada (26–29 October 2008), 361–366.

22 Du, Y., Lv, Y., Li, C. et al. (2011). Effect of electron shallow trap on breakdown performance of transformer oil-based nanofluids. *Journal of Applied Physics* 110 (10): 104104 (1-4). https://doi.org/10.1063/1.3660783.

23 Jin, H., Andritsch, T., Tsekmes, I.A. et al. (2014). Properties of mineral oil based silica nanofluids. *IEEE Transactions on Dielectrics and Electrical Insulation* 21 (3): 1100–1108.

24 Dang, V.H., Beroual, A., and Perrier, C. (2012). Investigations on streamers phenomena in mineral, synthetic and natural ester fluids under lightning impulse voltage. *IEEE Transactions on Dielectrics and Electrical Insulation* 19 (5): 1521–1527.

25 Li, J., Zhang, Z., Zou, P. et al. (2012). Preparation of a vegetable oil-based nanofluid and investigation of its breakdown and dielectric properties. *IEEE Electrical Insulation Magazine* 28 (5): 43–50.

26 Maharana, M., Baruah, N., Nayak, S.K. et al. (2019). Condition assessment of aged ester-based nanofluid through physicochemical and spectroscopic measurement. *IEEE Transactions on Instrumentation and Measurement* 68 (12): 4853–4863.

27 Derjaguin, B.V. and Landau, L.D. (1941). A theory of the stability of strongly charged lyophobic sols and the coalescence of strongly charged particles in electrolytic solution. *Acta Physics* 14: 216–235.

28 Verwey, E.J.W. and Overbeek, T.G. (1948). *Theory of the Stability of Lyophobic Colloids*. Amsterdam, The Netherlands: Elsevier.

29 Hogg, R., Healy, T.W., and Fuerstenau, D.W. (1966). Mutual coagulation of colloidal dispersions. *Transactions of the Faraday Society* 62: 1638–1651.

30 Missana, T. and Andres, A. (2000). On the applicability of DLVO theory to the prediction of clay colloids stability. *Journal of Colloid and Interface Science* 230: 150–156.

31 Hamaker, H.C. (1937). The London-Van Der Waals attraction between special particles. *Physica IV* 10: 1058–1072.

32 Dai, J., Dong, M., Wen, F. et al. (2015). The molecular dynamic simulation investigation of the dispersion stability of nano-modified transformer oil. *IEEE Conference on Electrical Insulation and Dielectric Phenomena (CEIDP)*, Ann Arbor, MI (18–21 October 2015), 475–478.

33 Saterlie, M., Sahin, H., Kavlicoglu, B. et al. (2011). Particle size effects in the thermal conductivity enhancement of copper-based nanofluids. *Nano scale Research Letters* 6 (1): 1–7.

34 Mansour, D.A., Elsaeed, A.M., and Izzularab, M.A. (2016). The role of interfacial zone in dielectric properties of transformer oil-based nanofluids. *IEEE Transactions on Dielectrics and Electrical Insulation* 23 (6): 3364–3372.

35 Hwang, J.G., Zahn, M., O'Sullivan, F.M. et al. (2010). Effects of nanoparticle charging on streamer development in transformer oil-based nanofluids. *Journal of Applied Physics* 107: 014310 (1–17). https://doi.org/10.1063/1.3267474.

36 Cavallini, A., Karthik, R., and Negri, F. (2015). The effect of magnetite, grapheme oxide and silicone oxide nanoparticles on dielectric withstand characteristics of mineral oil. *IEEE Transactions on Dielectrics and Electrical Insulation* 22 (5): 2592–2600.

37 Peppas, G.D., Charalampakos, V.P., Pyrgioti, E.C. et al. (2016). Statistical investigation of AC breakdown voltage of nanofluids compared with mineral and natural ester fluid. *IET Science, Measurement & Technology* 10 (6): 644–652.

38 Huang, Z., Wang, F., Wang, Q. et al. (2019). Significantly enhanced electrical performances of eco-friendly dielectric liquids for harsh conditions with fullerene. *Nanomaterials* 9: 989.

39 Atiya, E.G., Mansour, D.A., Khattab, R.M., and Azmy, A.M. (2015). Dispersion behavior and breakdown strength of transformer oil filled with TiO2 nanoparticles. *IEEE Transactions on Dielectrics and Electrical Insulation* 22 (5): 2463–2472.

40 Li, Y., Dong, M., Zhang, C. et al. (2018). Investigation on charge-carrier transport characteristics of transformer oil-based nanofluids. *IEEE Transactions on Dielectrics and Electrical Insulation* 25 (6): 2443–2451.

41 Mansour, D.A., Shaalan, E.M., Ward, S.A. et al. (2019). Multiple nanoparticles for improvement of thermal and dielectric properties of oil nanofluids. *IET Science, Measurement & Technology* 13 (7): 968–974.

42 Zhong, Y., Lv, Y., Li, C. et al. (2013). Insulating properties and charge characteristics of natural ester fluid modified by TiO2 semi conductive nanoparticles. *IEEE Transactions on Dielectrics and Electrical Insulation* 20 (1): 135–140.

43 Du, B. (2017). Investigating the influence of different types of nanoparticles on thermal and dielectric properties of insulation in converter transformer. *Chapter from Properties and Applications of Polymer Dielectrics*. Intechopen.

44 Dey, D., Kumar, P., and Samantaray, S. (2017). A review of nanofluid preparation, stability, and thermo-physical properties. *Heat Transfer–Asian Research* 46: 1413–1442.

45 Horiba nanoparticle Analyzer, Horiba nanopartica SZ-100 brochure.

46 IEC 60247 (2004). Insulating liquids - Measurement of relative permittivity, dielectric dissipation factor (tan δ) and d.c. resistivity.

47 IEC 61620 (1998). Insulating liquids- determination of the dielectric dissipation factor by measurement of the conductance and capacitance- test method.

48 Miao, J., Dong, M., Ren, M. et al. (2013). Effect of nanoparticle polarization on relative permittivity of transformer oil-based nanofluids. *Journal of Applied Physics* 113: 2013.

49 Yang, L., Gubanski, S., Serdyuk, Y.V., and Schiessling, J. (2012a). Dielectric properties of transformer oils for HVDC applications. *IEEE Transactions on Dielectrics and Electrical Insulation* 19 (6): 1926–1933.

50 CIGRE Technical Report 646 (2016). HVDC transformer insulation: oil conductivity, working group A2/D1.41.

51 Alj, A., Denat, J., Gosse, P., and Gosse, B. (1985). Creation of charge carriers in nonpolar liquids. *IEEE Transactions on Electrical Insulation* 20 (2): 221–231.

52 Lv, Y., Du, Q., Wang, L. et al. (2017). Effect of TiO2 nanoparticles on the ion mobilities in transformer oil-based nanofluid. *AIP Advances* 7: 105022.

53 Butcher, M., Neuber, A.A., Cevallos, M.D., and Dickens, J.C. (2006). Conduction and breakdown mechanisms in transformer oil. *IEEE Transactions on Plasma Science* 34 (2): 467–475.

54 Jing, Y., Timoshkin, I.V., Wilson, M.P. et al. (2014). Dielectric properties of natural ester, synthetic ester Midel 7131 and mineral oil Diala D. *IEEE Transactions on Dielectrics and Electrical Insulation* 21 (2): 644–652.

55 McCluskey, F.M.J. and Denat, A. (1996). The behavior of small bubbles generated by electrical current impulses over a wide range of applied pressures. *Journal of Applied Physics* 80: 2049.

56 Judd, M.D. and Farish, O. (1998). A pulsed GTEM system for UHF sensor calibration. *IEEE Transactions on Instrumentation and Measurement* 47 (4): 875–880.

57 Sarathi, R., Reid, A.J., and Judd, M.D. (2008). Partial discharge study in transformer oil due to particle movement under DC voltage using the UHF technique. *Electrical Power System Research* 78 (11): 1819–1825.

58 Beroual, A., Zahn, M., Badent, A. et al. (1998). Propagation and structure of streamers in liquid dielectrics. *IEEE Electrical Insulation Magazine* 14 (2): 6–17.

59 Kreuger, F.H. (1989). *Partial Discharge Detection in High Voltage Equipment*. Butterworths.

60 Florkowski, M. and Florkowska, B. (2006). Distortion of partial-discharge images caused by high-voltage harmonics. *IEE Proceedings - Generation, Transmission and Distribution* 153 (2): 171–180.

61 Bodega, R., Cavallini, A., Morshuis, P.H.F., and Wester, F.J. (2002). The effect of voltage frequency on partial discharge activity. *Annual Report Conference of Electrical Insulation and Dielectric Phenomenon*, Cancun, Mexico (20–24 October 2002), 685–689.

62 Khanali, M. and Jayaram, S.H. (2014). Partial discharge activities in transformer insulation under steep front voltages. *Annual Report Conference of Electrical Insulation and Dielectric Phenomenon*, Des Moines, IA, USA (19–22 October 2014) 204–207.

63 Reid, A.J., Judd, M.D., Fouracre, R.A. et al. (2011). Simultaneous measurement of partial discharges using IEC60270 and radio-frequency techniques. *IEEE Transactions on Dielectrics and Electrical Insulation* 18 (2): 444–455.

64 Cleary, G.P. and Judd, M.D. (2006). UHF and current pulse measurements of partial discharge activity in mineral oil. *IEE Proceedings on Science Measurement Technical* 153 (2): 47–54.

10

Behavior of Ester Liquids Under Various Operating Fault Conditions

U. Mohan Rao, I. Fofana, and L. Loiselle

Research Chair on the Aging of Power Network Infrastructure (ViAHT), University of Quebec at Chicoutimi, Chicoutimi, QC, Canada

10.1 Introduction

From a statistical point of view, the fleet of transformers in service worldwide is getting older with operating times. Some of them are getting replaced or refurbished when problems are suspected or detected. Age statistics [1, 2] shows that there are lots of in-service transformers of over 50 years old. It is important to maintain the fleet in a good working service as it is highly expensive to replace. So, appropriate diagnostic strategies are mandatory to plan the optimum time for the replacement or repair of the different units. It is not only an economic gain, but it is an environmental gain as fewer resources are going to be used for the replacement of the transformers. Power transformers are the most expensive equipment on the grid and can represent up to 60% of the investment [3]. The investment on transformers used in the grid worldwide are calculated in hundreds of billions of dollars.

The cost of changing the transformers is expensive, and it may take up to 2 years to manufacture a transformer and put it into service [2]. The goal is to predict as accurately as possible to the moment of the failure using predictive tools while attaining the designed life. The strategy is to improve the time the transformers are in service and predict the replacement of the units before the breakdown happens. Further, with proper condition monitoring activities and maintenance strategies, a long service life may be achieved and reduce failure rates.

The cost of a failure can be really high in terms of replacing the outage, loss of productions, and the impact on the life of people as the modern society relies on electricity. To quote an experience, on 6 July 6 2010, at the Laterrière plant in Rio Tinto Alcan, two transformers exploded. These two explosions deprived half of the electrolysis pot rooms of the plant. The factory technical teams were able to restart the electrolysis rooms, two months after the event, at a cost of more than 50 million. The transformers were recovered, reused, and restarted. Fortunately, the cost of these 2 transformers would have been staggering in the event of a replacement. Also, on 11 December 2010 at the Chute-des-Passes power plant in Rio Tinto Alcan, Canada, a transformer was ignited due to moisture saturation limit decrease at low temperatures [4].

This resulting oil leakage has a salty cost. Fines of US$ 180 000 and US$ 270 000 were imposed for a total of US$ 450 000, excluding multi-million-dollar decontamination costs, given the remote location and lost production [5].

Alternative Liquid Dielectrics for High Voltage Transformer Insulation Systems: Performance Analysis and Applications, First Edition. Edited by U. Mohan Rao, I. Fofana, and R. Sarathi.

There is no method that demonstrates the influence of liquid insulation degradation on the gassing tendency of power transformers. Even though aging of oil can be measured by interfacial tension (IFT) or acidity. So, as the acidity and IFT are increasing and decreasing with aging, respectively, and the gassing tendency of liquid is affected by these parameters. Thus, the objective is to demonstrate that the quantity and type of gas generated as the oil has degraded with time as the transformers are in service over the years. This method could improve the diagnosis to get a better picture of the state of the in-service transformers. Actually, there is no adaptable index for the quantity of gas, according to the aging markers (acidity and IFT) [6].

In recent years, many researchers have been developing improved tools to perform the diagnosis of in-service transformers. However, few researchers dedicated their works on the root cause of gassing. This development is done even if some tools are not responding to the actual defaults on in-service transformers and especially for gassing.

The production of gases is affected by different factors. With these factors, the actual age of the oil (that can be measured by acidity and interfacial tension) is influencing the quantity of gases produced. Unfortunately, there are no tools that take into account the aging of the oil and the gassing tendency of in-service transformers. In addition, this gassing tendency under various conditions will also influence the decay particles generated in the bulk liquids with aging and thus demanding for studies on understanding the gassing tendency and behavior of the decay particles in various insulating liquids.

The actual age of transformers in service is coupled with the ever-increasing demand for energy, increases the potential risks of unanticipated breakdowns, accompanied by extremely high monetary losses. Replacing these equipment with new ones–only because of its age–is clearly out of reach for any company. Some transformers in service may thus fail, leading to significant cost implications for the operator and, in extreme cases, an explosion with the consequent threat of serious damage or death of workers, and significant impact on the environment. The goal is to diagnose the behavior of the transformer to prevent the breakdown with the diagnosis and prognosis testing of liquid. The most common test performed for transformers in service is dissolved gas analysis (DGA). DGA interpretation techniques are used for a long period of time as they provide information on diagnosis of the situation prevailing with the unit. Unfortunately, these techniques do not take into consideration the aging of the fluid, and it is believed that it can lead to a misinterpretation of the results. So, the hypothesis is if a fluid is fresh (out of the barrel) and another one is aged, they are not going to develop the same gassing trends or tendency of gassing. The tests performed in this research are aimed to understand the influence of the liquid degradation (aging by-products) on gassing tendencies. In addition, the operational faults occurred due to various reasons also influence the degradation and the gassing tendency as well. Thus, to be more precise, the first hypothesis of this research is directed in a way to enumerate gassing tendency of insulation liquids under different faults at different states (aging) of the liquid.

The intent of this work is to understand the influence aging by-products on the tendency of gassing and liquid degradation. Emphasis has been laid on electrical and thermal faults on the liquid degradation and the gassing tendency at different aging conditions (based on acidity values). In this chapter, the influence of low-energy discharge faults, high-energy discharge faults, and thermal faults in mineral oil, synthetic esters, and natural esters onto the degradation and the gassing tendency at different aging factors have been reported.

10.2 Dissolved Gas Analysis and Transformer Faults

The analysis of dissolved gases (dissolved gas analysis) in oil, often abbreviated as DGA, is a diagnostic method for power transformers. It consists of looking for gases dissolved in the oil of transformers to detect partial discharges, aging of the insulation, or the formation of electric arcs in a transformer. The critical aspect of fault gases analysis is the accurate diagnosis of the faults that generate the detected fault gases. There are many methods that have been developed over the years for the interpretation of the fault type from the dissolved gases. However, a few are developed for the diagnosis of the fault gas in ester-filled transformers. Below are the details of some of the diagnostic methods that are applicable for the dissolved gas analysis of ester liquids.

10.2.1 Duval's Triangle

The Duval triangle uses a triangular representation of three hydrocarbons: CH_4, C2H4, and C_2H_2 in terms of relative proportions from 0 to 100% of the three gases. The triangle method is presented in Figure 10.1. There are six faults and a mixed fault (electrical and thermal (DT)) represented in Table 10.1 (PD, D1, D2, T1, T2, T3, or DT). For high-energy faults (like arcs), C_2H_2 and C_2H_4 are used in all analysis methods and high-temperature faults.

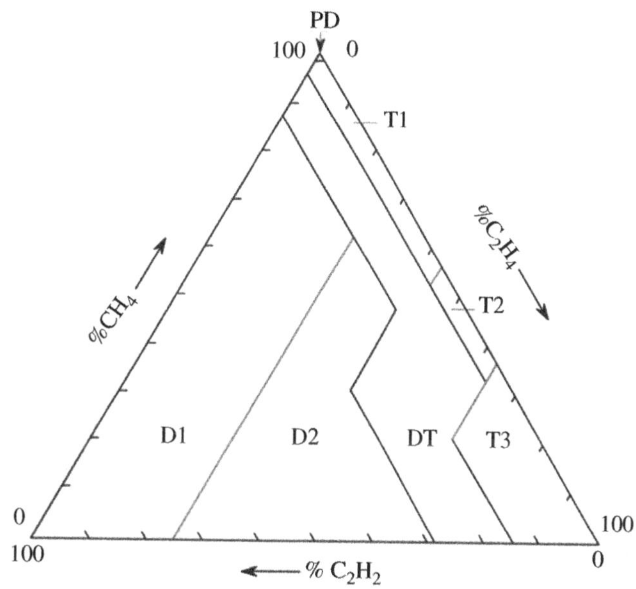

Figure 10.1 Duval's triangle 3 for natural esters. *Source:* Xiang et al. [7].

Table 10.1 Example of fault detected by Duval's triangle.

Symbol	Fault
PD	Partial discharges
D1	Low-energy discharges
D2	High-energy discharges
T1	Thermal fault T < 300 °C
T2	Thermal fault 700 °C<T<700 °C
T3	Thermal fault T> 700 °C
DT	Mixtures of faults

10.2.2 Duval's Pentagon

The Duval Pentagon is similar to the Duval triangle. The Pentagon uses five (5) gases instead of the three (3) ones used by the triangle [8]. The five main hydrocarbons used are H_2, CH_4, C_2H_6, C_2H_4, and C_2H_2. The Pentagon method is presented in Figure 10.2. There are seven faults represented in Table 10.2 (PD, D1, D2, T1, T2, T3, or S). Duval Pentagon may be used alone as the Duval triangle, and it is not intended to replace the Duval triangle for oils; rather, it is to reveal complementary information.

10.2.3 Research Progress on Various Faulty Conditions

It is to be recalled that the type of gasses evolved are dependent on the fault conditions and the state of the liquid. Gassing tendency based on the state of the liquid has been least emphasized in the literature. However, tendencies of liquids based on the fault conditions have been focused and are summarized in this section.

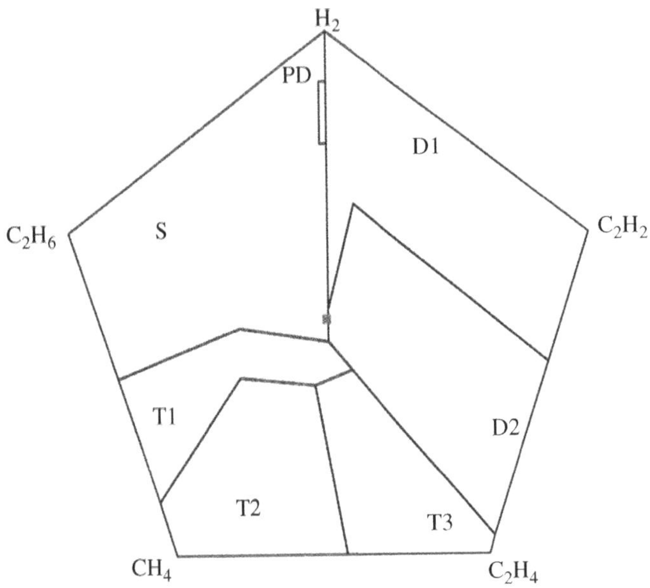

Figure 10.2 Duval's Pentagon 3 method for ester fluids. *Source:* Duval and Lamarre [8] / with permission of IEEE.

Table 10.2 Example of fault detected by Duval's Pentagon

Symbol	Fault
PD	Partial discharges
D1	Low-energy discharges
D2	High-energy discharges
T1	Thermal fault T < 300 °C
T2	Thermal fault 700 °C < T < 700 °C
T3	Thermal fault T > 700 °C
S	Stray gassing

(a) Low-Energy Discharge

Wang et al. [9] investigated the gas generated in mineral oil and natural esters for partial discharge and sparking faults. They found that the key gases generated under PD faults are hydrogen and acetylene in both mineral oil and natural ester. This means that they suggested methods for diagnosing PD and sparking faults in mineral oil are applicable in natural esters. They mention that the fault gasses under PD was 50 times higher for natural esters than mineral oil, the total volume of fault gases per unit fault energy is similar for the two liquids. Korraya Jongvilaikasem et al. [10] tested gas generation for various partial discharges like corona discharge, surface discharge, and internal discharge in natural esters. All gases follow an increasing trend as the voltage is increasing and the radius is decreasing. For ethane, the increase in gas is highly dependent on the voltage applied. For carbon dioxide, the gas tendency is much more dependent on the radii of the electrode than the voltage applied. For hydrogen, methane, ethylene, acetylene, and carbon monoxide, the gassing tendency is increasing exponentially while the voltage is increasing and the radius of the electrode is decreasing. In general, there is an increasing in gassing tendency when the ester in non-impregnated compared to an impregnated ester. Only carbon dioxide that does not show major increases/changes over time.

Imad et al. [11] examined the impact of alternative fluid on diagnostic gas analysis (DGA). Mineral oil, synthetic esters, and natural esters were taken for analysis. Different aged oil was aged at 90, 150 °C for 3 days and 14 days, and 200 °C for one hour. Acetylene is the key gas for low-energy electrical faults and hydrogen is the key gas for partial discharges. Even in electrical faults, esters generate less gas compared to mineral oil. For both types of fault, the small quantity in the total gas generation will make the identification of the problem more challenging at an initial phase in practical applications. The Duval triangle is providing an appropriate diagnostic in the thermal faults in all three oils: mineral oil (MO), synthetic ester (SE), and natural ester (NE). For the esters, the temperature range of the fault is overestimated.

Low-energy discharges for electrical faults were correctly diagnosed under all methods. Significant quantity of hydrogen is generated in all cases for corona discharges. The fingerprints on DGA for esters and mineral oils are the same, but the criteria of faults gas ratio need to be redefined to provide correct diagnosis. In order to compare the gassing tendency for mineral oils and ester fluids, they were put under partial discharge or low-energy discharge in [12–14]. The key indicator remains C_2H_2 even if the quantity in natural esters is increased for this last one and for H_2 as well.

(b) High-Energy Discharge

High-energy discharges are mainly due to arcing caused by short-circuiting faults and winding deformations. These discharges involve the excessive generation of hydrogen and acetylene with traces of methane and ethylene. If this fault occurs in cellulose insulants, carbon monoxide and carbon dioxide will also be traced. High-energy discharges are detrimental to the insulation system [15]. High-energy discharges are also associated with high temperatures that may cause premature thermal degradation of oil/paper insulation. Also, influence of high-energy discharges on the gassing tendency of dielectric fluids has been reported in [7, 12, 16]. Similar to low-energy discharge faults, H_2 and C_2H_2 are noticed due to arcing faults. However, noticeable traces of C_2H_2 are generated as compared to low-energy discharging activity.

C. Xiang et al. [7] tested gas generation in mineral oil and vegetable oil under typical electrical and thermal fault. It was found that a large amount of C_2H_2 is generated under high-energy electrical breakdown both in vegetable insulating oil and mineral insulating oil, which shows that the main dissolved gas under electrical breakdown in vegetable insulating oil and mineral insulating oil is the same.

(c) Thermal Faults (Hotspot)

The gassing tendency of ester fluids under simulated thermal fault condition was also explored and is reported in numerous publications such as [7, 12, 13, 17, 18]. It is noticed that the natural ester liquid generates high dissolved gases at high temperatures (>300 °C) and as the temperature increases, traces of C_2H_6 increases. For thermal faults of ester fluids, it is found that key gasses are CH_4, C_2H_4, CO, and C_2H_6. As various methods exist for the simulation of a hotspot thermal faults simulation at different temperatures, the gassing tendency will act accordingly. The details of which are analyzed and reviewed in [19]. Xiang et al. [7] investigated on thermal faults on mineral oil and vegetable insulating oils and found that:

- CH_4 is the main gas obtained for overheating in mineral oil, C_2H_6 in FR3 is the main gas, and H_2 and C_2H_6 are the main gases composition for camellia insulating oil. The existence of insulating paper significantly increases the amount of CO and CO_2 dissolved in oil.
- As the temperature increases, the percentage of main dissolved gases increases. The percentage of CH_4 and C_2H_4 increases as the temperature increases in mineral oil; the percentage of C_2H_6 increases as the temperature increases in FR3 oil; the percentage of H_2 increases while the percentage of C_2H_6 fluctuates with the increase of temperature in camellia oil.

10.3 Simulation of Various Faults in Laboratory Environment

10.3.1 Low-Energy Discharges (Surface Discharges)

The low-energy discharging often called as partial discharges could be an internal discharge, surface discharge, and a corona discharge. In this work, a local setup as per the ASTM D6180 is used for a surface discharge of 9 kV. This surface discharges are simulated for five hours above the surface of insulating liquid. The discharge is generated by using a copper electrode (cylindrical). The copper electrode is situated in the center of the liquid surface at a distance of approximately one inch above the surface. The details of the experimental setup are shown in Figure 10.3. The discharge is initiated under vacuum in test cell, and the pressure changes are recorded to report the gassing tendency of an insulating liquid. This method of testing is allowed in understanding the potential of the stability of an insulating liquid under electrical discharges.

10.3.2 Thermal Faults (Hotspot)

Thermal faults or hotspots in mineral oil (MO), natural ester (NE), and synthetic ester (SE) are simulated by a local laboratory test setup shown in Figure 10.8. Constantan is used as the heating element due to its flat resistivity over a wide range of temperature. The variac is excited and by recording the current and voltage, average temperature simulated by the coil (hotspot temperature) may be estimated. To cross validate this temperature estimation, a thermocouple (high-temperature resistant). Also, the aluminum funnel structure is placed above the heating coil. Thus, the gas bubbles generated due to hotspot enters into the pipette connected to the end of the funnel on the other side. For each test, 800 ml of liquid is introduced and is allowed to heat for periods of 30 minutes and the volume of gas generated is recorded. The temperature reached to values close to 250 °C, which for the mineral is risky as the flashpoints are around 220 °C. For security purposes, the current is reduced to zero until the liquid temperature reaches the ambient temperature (~24 °C). This test cycle of 30 minutes of heating and cooling down to room temperature has been repeated for 12 times. For every 30 minutes of the heating cycle, the volume of gas is recorded. To

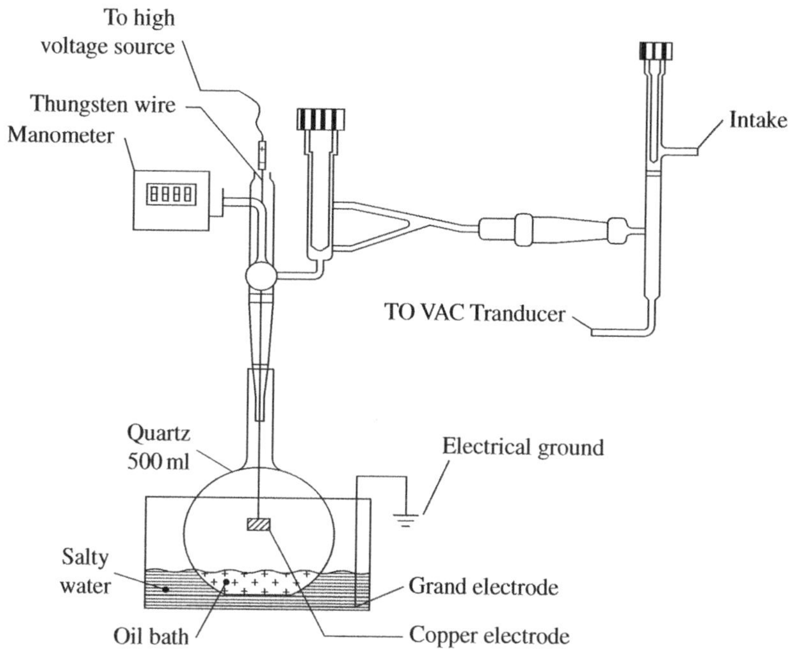

Figure 10.3 Experimental setup for the surface discharge activity [20].

analyze the liquid degradation influence, UV/Vis and turbidity testings are performed before and after 12 cycles. The view of the hotspot simulation setup is shown in Figure 10.4.

10.3.3 High-Energy Discharges (Arcing)

High-energy discharge, i.e. arcing faults are stimulated by creating repeated breakdowns in the bulk volume of the oil. The breakdown testing of the insulation liquid is carried according to ASTM D 877. Continually, 100 breakdowns are simulated with 2 minutes break in between each breakdown. To understand the impact of high-energy discharge electrical faults, dissolved gas analysis is performed after 100 continuous breakdowns. To analyze the liquid degradation influence, UV/Vis and turbidity testings are performed both before and after 100 breakdowns. The details of the breakdown test set up are presented in Figure 10.5.

10.4 Influence of Different Faults on the State of Liquid and Gassing Tendency

To understand the gassing behavior as a function of the state of the liquid, insulating liquids at different aging factors have been prepared. Initially, fresh mineral oil (MON), fresh natural ester (NEN), and fresh synthetic ester (SEN) are subjected to accelerated thermal aging at 115 °C. Thermal aging is performed in presence of copper (3 g/l) and cellulose (1:20) under open beaker condition. After thermal stressing (2000 hours), degraded oil/paper insulation is subjected for fine mixing using Qsonica Q1375 Sonicator for 30 seconds. A less sonication period is adopted to allow protecting the chemical properties of the liquids [21]. This sonication allowed to obtain fine solution of

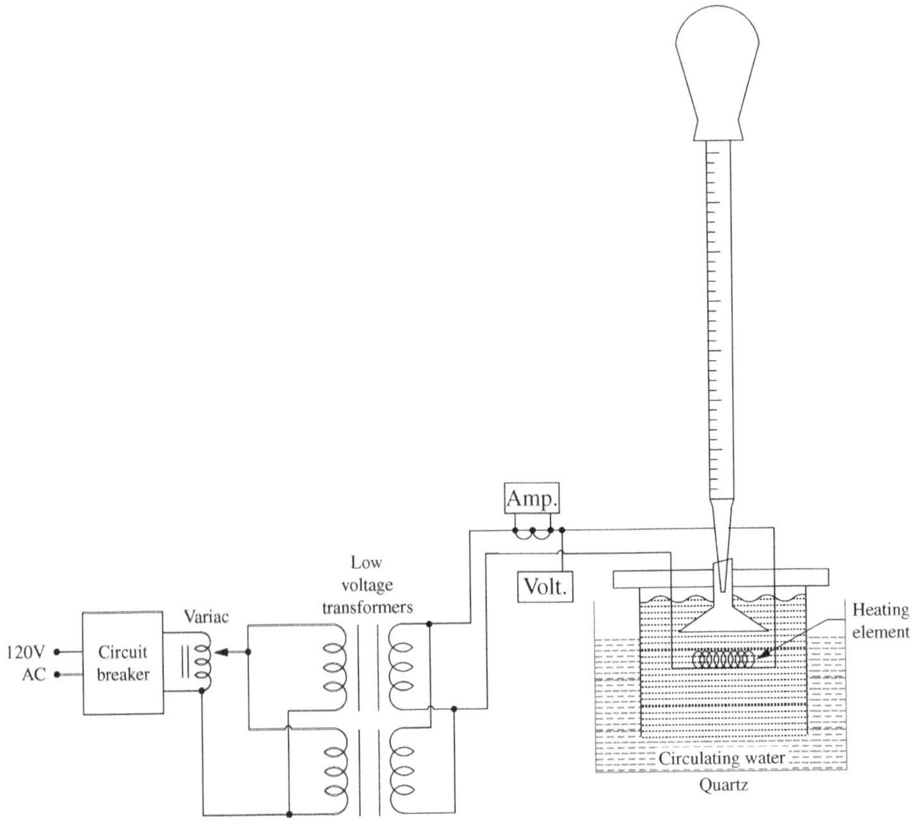

Figure 10.4 View of the hotspot fault setup.

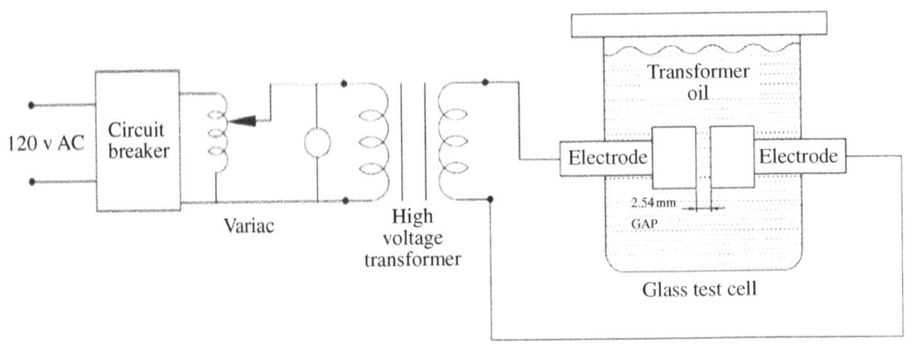

Figure 10.5 View of the arcing fault setup.

degraded oil and paper, normally treated as sludge (decay content). Later, the decay content is introduced to fresh liquids at appropriate proportions to attain different levels of neutralization number. Finally, the solution is subjected to sonication (30 seconds) followed magnetic stirring (5 minutes) to ensure gentle mixing. The process of test sample preparation is presented in Figure 10.6.

Fresh and unused oils are used to prepare different class liquids based on the neutralization number. Later, mineral oil and synthetic esters of class new/fresh (N), proposition A (Prop. A), and bad class (B) are used for the present experimental studies. Table 10.3 and Figure 10.7 highlights

Figure 10.6 Process for the preparation of the samples.

Table 10.3 Liquids classification and acidity values (measured).

Sample	Acidity test 1mg KOH/g	Acidity test 2mg KOH/g	Average	Std. deviation	Class
MOPA	0.0818	0.082	0.0819	0.0001	Prop A
MOB	0.2757	0.2675	0.2716	0.0058	Bad oil
SEPA	0.1034	0.092	0.0977	0.081	Prop A
SEB	0.2314	0.2573	0.24435	0.0183	Bad oil
NEB	0.3016	0.2801	0.29085	0.0152	Bad oil

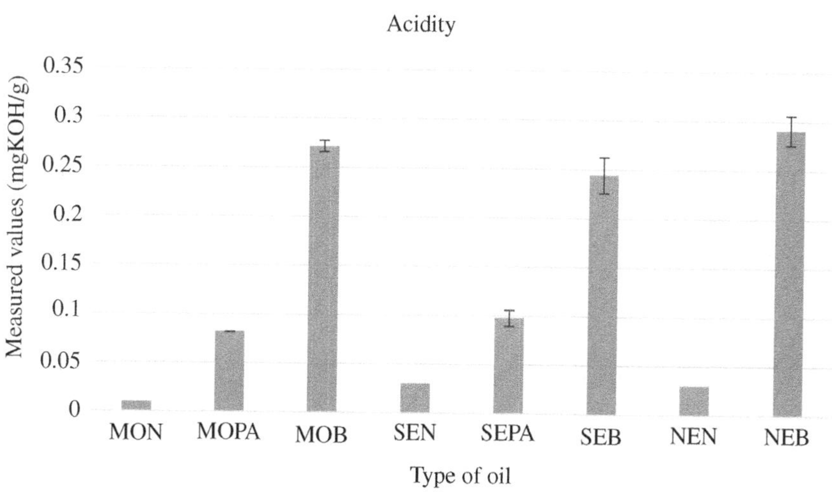

Figure 10.7 Acidity of the oil samples at different oil classes.

the details of the acidity of mineral oil fresh (MON), natural ester fresh (NEN), and synthetic ester fresh (SEN). Measurements are performed to ensure bad oil classes for mineral oil (MOB), bad class for synthetic ester (SEB), and bad class for natural ester (NEB).

Later the liquids at different aging states are subjected to various fault conditions outlined in the previous section. The liquids are tested for dissolved gas analysis after the fault simulation to understand the gassing tendency with the state of the liquid. The influence of the liquid degradation is also investigated by suitable diagnostic characterizations.

10.4.1 Effect on Gassing Tendency

DGA results are presented in Figure 10.8. The concentration of most of the gasses is noticed to be higher in case of aged oil in case of thermal and electrical faults. For arcing faults, H_2, CH_4, and C_2H_2 are found to increase in all the dielectric fluids. However, O_2 is reduced and significant

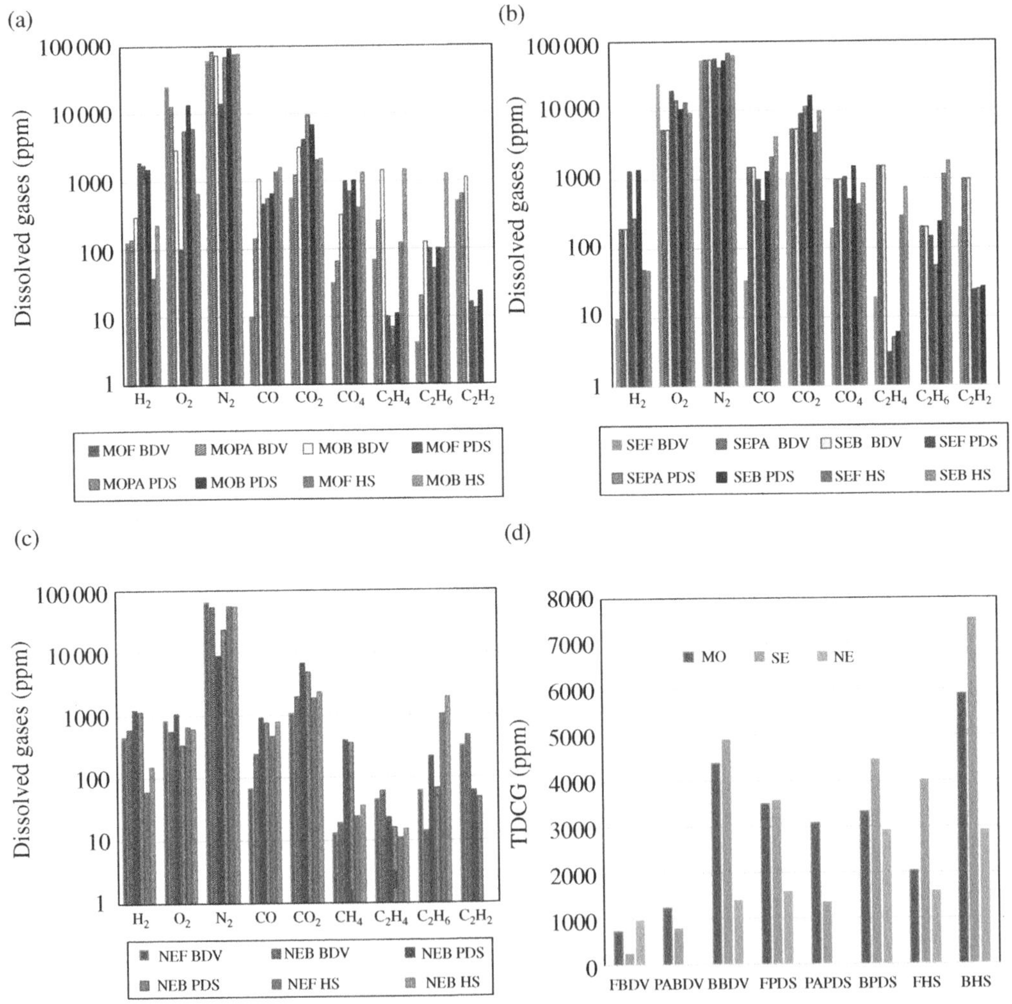

Figure 10.8 Illustration of dissolved gasses under fault conditions: (a) Mineral oil; (b) synthetic ester; (c) natural ester; (d) TDCG.

change is not noticed in the traces of N_2 for fresh and aged oils. Thus, H_2, CH_4, and C_2H_2 gas concentrations vary with the state condition of the oil under arcing faults for mineral oils and non-mineral oils. For partial discharge faults, a much less change in H_2 and C_2H_2 is noticed for fresh and aged oils. Due to the reason that decay particles act as a filtering medium under the influence of low-energy discharge [21]. Yet, there is no change in the concentrations of CH_4, C_2H_4, and C_2H_6 for fresh and aged mineral oil. For the ester fluids, a noticeable change in the traces of CH_4, C_2H_4, and C_2H_6 is noticed. Therefore, under partial discharging activity, CH_4, C_2H_4, and C_2H_6 vary with the condition of the ester fluids and H_2 will be generated in common for both the fluids according to the state of the oil. For thermal faults, there is a significant change in H_2, CH_4, C_2H_4, and C_2H_6 in mineral oil, while in ester fluids, there is a small change in CH_4, C_2H_4, and C_2H_6. Under thermal faults with lower temperatures, H_2, CH_4, C_2H_4, and C_2H_6 gases will be generated in common for both the fluids according to the state of the oil. Also, combustible gasses generation increased with increasing degradation of liquid. However, generation of combustible gasses are higher for partial discharge and thermal hotspot activities. This is because the degradation of dielectric fluids is accelerated by cellulose decay particles and other conductive particles. Therefore, excessive amounts of CO and H_2 increase easily in any event of the fault, thus aiding in the increase in the total dissolved combustible gases (TDCG). TDCG is the summation of H_2, CH_4, C_2H_2, C_2H_4, C_2H_6, and CO and is represented in Figure 10.8d. The illustration of Duval's Pentagon and triangle analysis is shown in Figure 10.9 for mineral oil, natural ester, and synthetic ester.

10.4.2 Effect on Degradation

The degradation of fresh and aged oils while subjected to electrical and thermal faults is of concern. Fresh oils are technically clean and are free from decay particles and cellulose particles, and other aging by-products. Whereas aged oils contain cellulose particles, dissolved decay products, acids and other aging by-products. Therefore, when a fresh oil and aged oil are subjected to different fault conditions, the undergoing chemical perspectives are expected to be different. Subsequently, the influence of these faults on the dielectric fluids will be different. Hence absorbance, dissolved decay products and turbidity measurements are performed before and after simulating fault conditions. The details have been presented in Figure 10.10.

In fresh oils, the changes that occur due to the application of thermal and electrical faults depend majorly on the thermal stability of the chemical bonds and the ionization potential, respectively. In case of aged oils, along with the said factors, the concentration of colloidal and soluble decay particles plays a significant role in further degrading a dielectric fluid. Therefore, in the present research, the concentration of the dissolved decay products and turbidity measurements have been reported. From Figure 10.10, it is seen that aged oils have a high turbidity, decay content concentration, which is obvious. It is to be noticed that the application of fault conditions (both electrical and thermal) had a different impact for fresh and aged oils. In Figure 10.10a, from fresh mineral oil, the influence of partial discharges is high as compared to arcing and hotspot. Similarly, in the case of aged oils, the hotspot influence is superior to that of the partial discharges and arcing. However, it is seen that the absorbance of aged oil is reduced with arcing. This may be due to the splitting or de-agglomeration of the cellulose particles present in the oil, which reduces the absorption of light. To further understand this, the chemical function group analysis is to be understood. The absorbance and turbidity of the aged oil are increasing with partial discharges and thermal faults. From Figure 10.10b, it is noticed that the influence of hotspot is higher than that of the partial discharges and arcing for fresh and aged synthetic esters. Also, the influence of partial discharges and arcing is too low in the case of fresh and aged fluid samples. One may understand that the

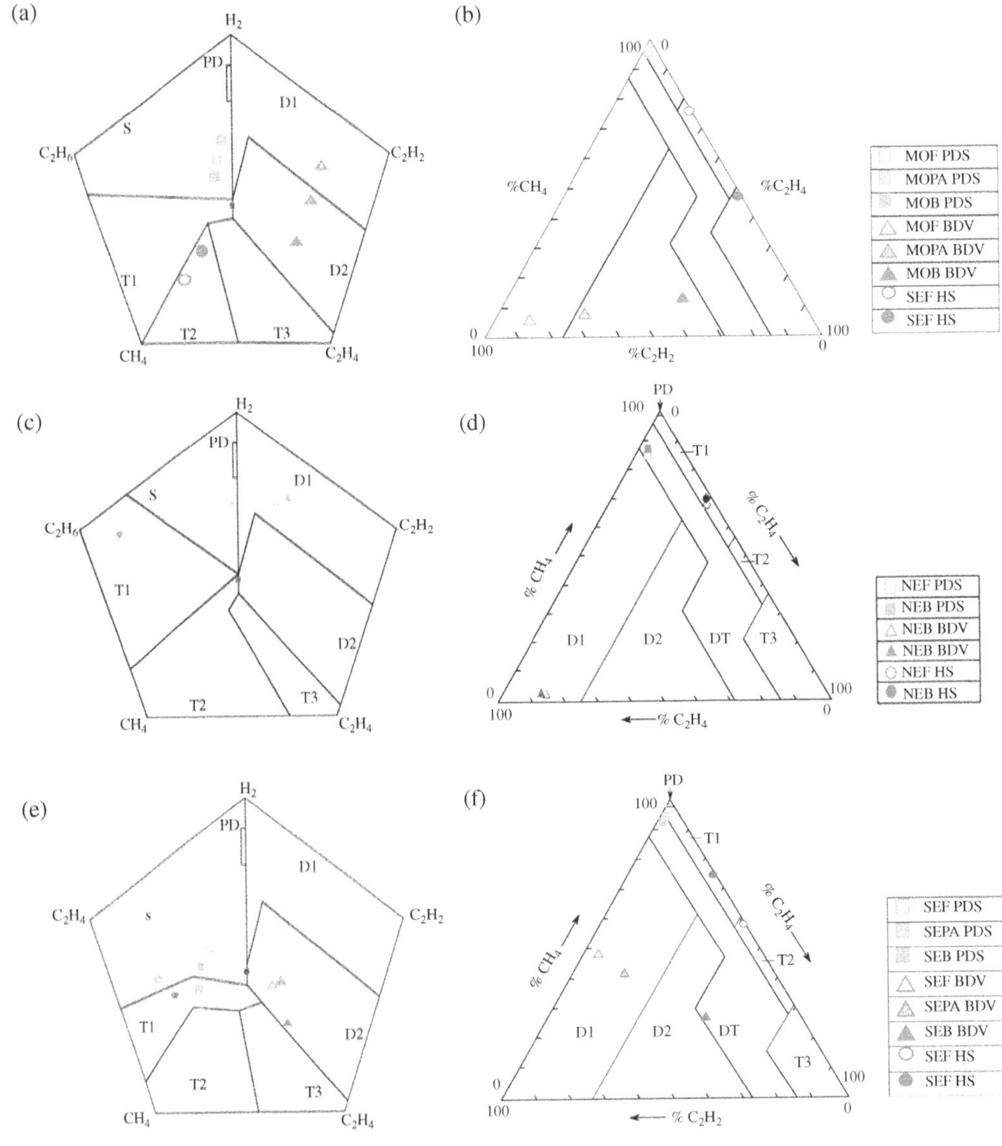

Figure 10.9 Representation of fault gasses in Duval's triangle and Duval's Pentagon for mineral oil, natural esters, and synthetic esters at different aging factors: (a) Duval's Pentagon for mineral oil; (b) Duval's triangle for mineral oil; (c) Duval's Pentagon for natural esters; (d) Duval's triangle for natural esters; (e) Duval's Pentagon for synthetic esters; (f) Duval's triangle for synthetic esters.

influence of electric faults is low when compared to thermal faults in the case of synthetic esters. However, it is to be assured that the intensity of the fault nature is uniform in all three cases. Also, in Figure 10.10c, the influence of the hotspot is higher when compared to the other faults in the case of fresh natural esters. To understand the influence, faults of much higher intensities are to be evaluated. It is to be recalled that thermal faults involve an increase in local temperature and promoting oxidation of the oil/paper insulation. These generated acids trigger a self-sustained process called hydrolysis, which in turn degrades the solid insulation. But, in case of electrical faults, the

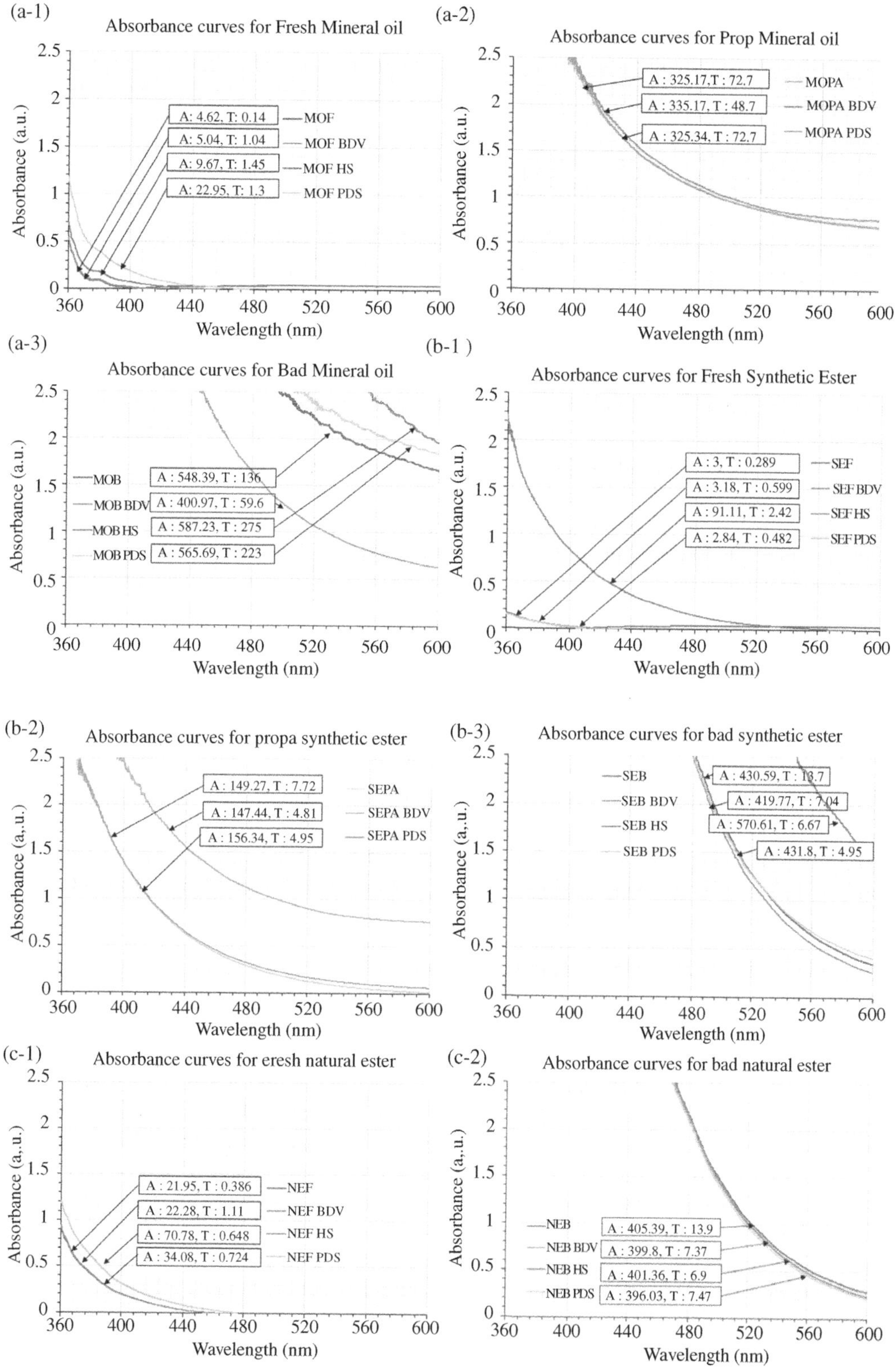

Figure 10.10 Illustration of the degradation of the investigated dielectric fluids under different fault conditions: (a) Mineral oil; (b) synthetic ester; (c) natural ester. "**A**" stands for area of absorbance and "**T**" for turbidity.

influence is more associated to a generation of acids and free radicals, which leads the degradation. However, both electrical and thermal faults are involved with the generation of dissolved gasses.

10.5 Conclusions and Future Scope

It was found that as the fluid is increasing in acidity (which is the condition criterion to determine the aging of the fluids), under the different stress: thermal, low electrical discharge, and high electrical discharge, the gassing was affected. The detailed conclusion of the above experimental are reported in [22–24] and are outlined briefly below.

a) *On a more specific base for low-energy (partial discharge) faults, it was found that*:
From the fresh class of oil to a bad class of oil, there is an O_2 increase of 134 times. There is also an increase in N_2 of almost seven times. The changes in other gases go from 0.8 to less than two times, which is not significant as compared to O_2 and N_2. For synthetic ester, there is an increase of some gases such as CO_2, CH_4, C_2H_4, C_2H_6, and C_2H_2. For natural ester, all gases are decreasing and particularly C_2H_6 of more than three times. All these changes in the quantity of gas will affect the results using the traditional diagnostic methods such as Duval triangle and Duval Pentagon. It is to mention that the representation of natural ester is the less affected by the displacement in the Duval tools.

b) *On a more specific base for high-energy faults, it was found that*:
From the fresh class of oil to bad class of oil, for synthetic ester, all gases were increased except O_2 and N_2. CO_2 increased 4 times, CO increased 44 times, C_2H_6 increased 198 times, C_2H_4 increased by 81 times, C_2H_2 increased 5 times, and H_2 increased 20 times.

Similarly, for mineral oil, all the gases were increasing other than O_2 and N_2. CO increased 108 times from new oil to bad oil. While C_2H_6 increased 31 times, C_2H_4 increased 21 times. C_2H_2 increased 2 times, and H_2 increased 2 times. It is to be mentioned that reference for per unit is based on the gas values for fresh oil. For natural ester, all gases increase except O_2, N_2, and C_2H_6. These results are affecting the traditional diagnostic methods such as Duval triangle and Duval Pentagon. It is to mention that the representation of natural ester is less affected by the displacement in the Duval tools.

c) *On a more specific base for thermal faults, it was found that*:
From the fresh class of oil to bad class of oil, all gases increased except the unchanged of C_2H_2 and the increase of O_2. Some gases increase substantially such as C_2H_4 by 12 times, C_2H_6 13 times, and H_2 6 times. The synthetic ester is less affected by the thermal faults, but there are definitely some changes. H_2 and C_2H_2 are not changed. O_2 and N_2 are decreased and the quantity of the other gases is increased. For NE, C_2H_2 is not changed, O_2 and N_2 are decreased, while the quantity of all the other gases is increased. These results are affecting the traditional diagnostics methods such as Duval triangle and Duval Pentagon. It is to mention that the representation of natural ester is the less affected by displacement in the Duval's tools. The proof of concept that gassing tendency changes with state of the liquid has been reported in this chapter. In order to develop the tools required for maintenance, it is necessary to push these various points. The aging of fluid that was under investigation was limited in term of acidity values. To get much more information on the variations of gassing, it will be interesting to reach a higher level of acidity, especially for natural and synthetic ester. Also, the quantity of electrical high-energy discharge can be extended to different repetition values. The length of low-energy discharges can be adjusted to various time and various voltage. The geometry of the electrode can be changed as well. The temperature on the thermal

stress (hot spot apparatus) can be changed to reach different values. A lot of data is required in order to be able to process into machine learning technique's algorithms. Further experimentations are required to tune the different tools that need to be developed.

References

1 Tee, S., Liu, Q., Wang, Z. et al. (2016). An early degradation phenomenon identified through transformer oil database analysis. *IEEE Transactions on Dielectrics and Electrical Insulation* 23: 1435–1443.

2 Martin, D., Marks, J., Saha, T.K. et al. (2018). Investigation into modeling Australian power transformer failure and retirement statistics. *IEEE Transactions on Power Delivery* 33: 2011–2019.

3 Fofana, I., Hemmatjou, H., and Farzaneh, M. (2010). Low temperature and moisture effects on polarization and depolarization currents of oil-paper insulation. *Electric Power Systems Research* 80: 91–97.

4 Radio-Canada. (2010). Incendie à la centrale de Chute-des-Passes. https://ici.radio-canada.ca/nouvelle/496737/explosion-chute-des-passes-incendie (accessed 11 December 2010).

5 M. St-Hilaire. (2010). Centrale Chute-des-Passes de RTA: Un transformateur s'enflamme. http://www.lapresse.ca/le-quotidien/201012/10/01-4351381-centrale-chute-des-passes-de-rta-un-transformateur-senflamme.php (accessed 11 December 2010).

6 Betie, A., Rao, U.M., Fofana, I. et al. (2019). Influence of cellulose paper on gassing tendency of transformer oil under electrical discharge. *IEEE Transactions on Dielectrics and Electrical Insulation* 26 (6): 1729–1737.

7 Xiang, C., Zhou, Q., Li, J. et al. (2016). Comparison of dissolved gases in mineral and vegetable insulating oils under typical electrical and thermal faults. *Energies* 9: 312.

8 Duval, M. and Lamarre, L. (2014). The duval pentagon-a new complementary tool for the interpretation of dissolved gas analysis in transformers. *IEEE Electrical Insulation Magazine* 30: 9–12.

9 Wang, Z., Wang, X., Yi, X. et al. (2013). Gas generation in natural ester and mineral oil under partial discharge and sparking faults. *IEEE Electrical Insulation Magazine* 29: 62–70.

10 Jongvilaikasem, K., Maneerot, S., Jariyanurat, K., and Pattanadech, N. (2019). Comparison of dissolved gases in natural ester under partial discharges. *2019 IEEE 20th International Conference on Dielectric Liquids (ICDL)*, Roma, Italy (23–27 June 2019), pp. 1–4.

11 Khan, I.U., Wang, Z., Cotton, I., and Northcote, S. (2007). Dissolved gas analysis of alternative fluids for power transformers. *IEEE Electrical Insulation Magazine* 23: 5–14.

12 Meira, M., Álvarez, R.E., Catalano, L.J. et al. (2018). Comparación de aceites dieléctricos minerales y vegetales en relación a la producción de gases. *Congreso Bienal de IEEE Argentina ARGENCON 2018*, Tucumán (6 al 8 de junio de 2018), Bangalore, India, 10–12 December 2015.

13 Meira, M., Ruschetti, C., Álvarez, R. et al. (2019). Dissolved gas analysis differences between natural esters and mineral oils used in power transformers: a review. *IET Generation, Transmission & Distribution* 13: 5441–5448.

14 Perrier, C., Marugan, M., and Beroual, A. (2012). DGA comparison between ester and mineral oils. *IEEE Transactions on Dielectrics and Electrical Insulation* 19: 1609–1614.

15 Mohan Rao, U., Sood, Y.R., and Jarial, R.K. (2015). Review on ester based dielectric liquids for transformer insulation technology *International Conference on Condition Assessment Techniques in Electrical Systems (CATCON)*, Bangalore, India (10–12 December 2015).

16 Hamid, M., Ishak, M., Ariffin, M. et al. (2017). Dissolved gas analysis (DGA) of vegetable oils under electrical stress. *2017 International Conference on High Voltage Engineering and Power Systems (ICHVEPS)*, Denpasar, Indonesia (2–5 October 2017), pp. 29–34

17 Dongjin, K. and Kyosun, K. (2012). The criteria of dissolved gas analysis in natural Ester fluid. *2012 IEEE International Conference on Condition Monitoring and Diagnosis*, Bali, Indonesia (23–27 September 2012), pp. 901–904.

18 Liu, Y., Li, J., and Zhang, Z. (2013). Fault gases dissolved in vegetable insulating oil under electrical faults. *2013 Annual Report Conference on Electrical Insulation and Dielectric Phenomena*, Chenzhen, China (20–23 October 2013), pp. 198–201.

19 Przybylek, P. and Gielniak, J. (2019). Analysis of gas generated in mineral oil, synthetic ester, and natural ester as a consequence of thermal faults. *IEEE Access* 7: 65040–65051.

20 ASTM D 6180 - 05 standard test method for stability of insulating oils of petroleum origin under electrical discharge, ed, 2005, p. 6.

21 Electrical transformers containing electrical insulation fluids comprising high oleic acid oil compositions. U.S. patent nos. 5 949 017, 1999, 6 274 067, August 14, 2001, and 6 312 623, November 6, 2001.

22 Loiselle, L., Mohan Rao, U., and Fofana, I. (2020). Influence of aging on oil degradation and gassing tendency under electrical faults for mineral oil and synthetic ester under low energy discharges. *MDPI Energies* 13 (3): 595.

23 Loiselle, L., Mohan Rao, U., and Fofana, I. (2020). Gassing tendency of fresh and aged mineral oil and ester fluids under electrical and thermal fault conditions. *MDPI Energies* 13: 3472.

24 Loiselle, L., Mohan Rao, U., and Fofana, I. (2020). Influence of aging on oil degradation and gassing tendency under high energy electrical discharge faults for mineral oil and synthetic ester. *IET High Voltage* 5 (6): 731–738.

11

In-Service Performance of Natural Esters

D. Martin[1] *and L. McPherson*[2]

[1] *ETEL Transformers, Auckland, New Zealand*
[2] *Essential Energy, Port Macquarie, New South Wales, Australia*

11.1 Introduction

In Australia, two utilities have commissioned power transformers filled with natural ester, one utility since 2005 and the other 2008. According to CIGRE, natural esters have only been used in power transformers since 2002 [1], and so the units presented in this chapter are among the longest operating worldwide. Data from these transformers provided a unique insight in how natural esters will perform over long periods compared to laboratory-based accelerated aging. The units have operated without incident and are enabling learning about the long-term in-service performance of natural esters.

During installation, there was not comprehensive coverage of natural esters in standards such as the IEEE guide for diagnostic field testing of fluid-filled power transformers, which was only updated to also include values for natural esters in 2013 [2]; IEEE Guide for acceptance and maintenance of natural ester insulation liquid in transformers revised in 2018 [3]; and IEC 62975 (still in draft at time of writing) in 2020 [4]. Therefore, the utilities commissioned university research to evaluate any changes in fluid properties and understand why these were occurring and any impacts on the transformer's reliability [5, 6]. Vegetable oils were among some of the first types of dielectric liquid used in the power industry, for instance George Westinghouse introduced caster and linseed oils in the 1880s [7]. However, these fluids were later considered unsuitable because they readily oxidized, and so mineral oils were adopted. In the 1990s, there was renewed interest in using vegetable-oil-based dielectrics by blending in antioxidants [8]. Thus, the problem of unwanted oxidation was resolved. First, these fluids were only used in small distribution transformers. As the industry became more confident, these fluids began to be used in ever larger power transformers.

In this chapter, a review is done on in-service performance of these transformers that, at the time of writing, have operated up to fifteen years. First, an introduction is given as to why these two utilities chose to use natural esters as transformer dielectrics. Second, the transformer ratings are presented. Third, a summary is provided of the research, which was performed to support the operation of these transformers by providing the utilities with new understanding. Fourth, a case study is done on the installation and early operation of a 50 MVA transformer which had an online

Alternative Liquid Dielectrics for High Voltage Transformer Insulation Systems: Performance Analysis and Applications, First Edition. Edited by U. Mohan Rao, I. Fofana, and R. Sarathi.

DGA monitor attached. Finally, an analysis of a fleet of natural-ester-filled power transformers in operation since 2005 is given.

11.2 Reasons Why These Utilities Chose a Natural Ester

One utility began the risk assessment of two mobile multi-ratio power transformers that were specifically designed to assist with fast outage restoration. The engineers noted that due to the units being frequently transported fully assembled, with radiators and accessories, there was a higher risk that fluid could leak during transport. The utility would then be liable for subsequent clean-up costs. Natural ester was therefore chosen for its biodegradability and nontoxicity to aquatic lifeforms. Through the experience of owning this transformer, the utility's engineers became familiar with the use of the natural ester and thus became comfortable with using it on other parts of their network. This utility operates networks passing through national parks. Even though a spillage of mineral oil was unlikely, the utility strove to lower the risk of ecological damage from oil discharge by choosing to use a natural ester.

Another safety benefit from using natural esters is that their fire point is 360 °C, much higher than that of mineral oil (180 °C). For this reason, they are referred to as K-class high fire point fluids. Another Australian utility noted that using K-class fluids can be more cost-effective than using mineral oil in the long run due to different clearances within a substation [9]. In one case, the utility decided to increase the capacity of a substation supplying a town, which required the installation of two larger transformers. To remain compliant with the Australian fire prevention standard, AS 2067 [10], two options were to move the transformers further from the nearby substation building or to build a fire wall. The utility opted to build the fire wall at a cost of US$82 000 per transformer. Set against the cost of filling a 20 000 liter transformer with natural ester of approximately US$40 000 more than the cost of filling it with mineral oil, it would have probably been cheaper specifying a natural ester from the start than building the wall.

The K-class fire point has also been found favorable for using natural esters in substations in dense cities under commercial buildings [5].

11.3 Transformers Under Study

Details on the transformers analyzed in this study are given in Table 11.1. Notable is the transformer that started operating in 2005 because according to CIGRE [1] the first power transformer was filled with natural ester in 2002, and thus, it is one of the units worldwide with the longest operating experience. The "K" in the cooling type denotes the K-class high fire point fluid. Two units are identified as S1 and S2, which were spares. Given that these have not seen as significant usage it is useful to plot their data to see how the properties of the fluid changed with time even when the transformer has not been operational.

There were no differences in the functional specification of these power transformers from filling them with natural ester instead of mineral oil. The utility left any required changes to the design to the manufacturer. The transformers were all sealed since it was not recommended to use natural esters in a free breathing environment due to oxidation. Clear labelling of the fluid used was ensured to prevent inadvertent topping up with mineral oil. Mixing compromises the fire point, which begins to fall from about 8% mixture of mineral oil [11]. One utility chose to paint their natural-ester-filled transformer green for this identification [12].

Table 11.1 Power transformers analyzed in this study.

Site	Rated voltage (kV)	Rated power (MVA)	Year of manufacture	Cooling type	Site	Rated voltage (kV)	Rated power (MVA)	Year of manufacture	Cooling type
1	33/11	5/8	2008	KNAF	13	33/11	10/16	2005	KNAF
2	33/11	5/8	2007	KNAF	14	33/11	10/16	2007	KNAF
3	33/11	5/8	2007	KNAF	15	33/11	5/8	2008	KNAF
4	33/11	5/8	2009	KNAF	16	33/11	5/8	2011	KNAF
5	66/11	20	2014	KNAN	17	33/11	7.5/10	2013	KNAF
6	66/11	20	2014	KNAN	18	132/33	45/90	2006	KNAF
7	33/11	5/8	2007	KNAF	19	132/11	50	2008	KDWF
8	33/11	5/8	2007	KNAF	20	132/11	50	2008	KDWF
9	33/11	5/8	2006	KNAF	21	132/11	50	2012	KDWF
10	33/11	5/8	2010	KNAF	22	132/11	50	2012	KDWF
11	33/11	5/8	2010	KNAF	S1	66/11	20/28	2004	KNAF
12	132/11	30	2013	KNAN	S2	66/11	20/28	2004	KNAF

11.4 Summary of Research Applied to Manage These Transformers

Various research projects were funded and supported by these utilities to support the operation and condition monitoring of these transformers.

The thermal properties of a natural ester are important to understand heating, and subsequent effects. An experiment was set to investigate how these fluids would behave during transformer overload conditions when bubbling of water from the paper may occur [13]. Investigations were conducted to determine how fluid properties changed during normal operation, as it was felt at the time that most researchers were focusing on behavior during fault conditions.

Interactions between water and effect on breakdown voltage were explored to help set limits. Both AC and lightning impulse dielectric strength was considered. Given that the oxidative stability of a natural ester is not as high as mineral oil, conditions representative for refurbishment and maintenance activities were explored to determine if these were appropriate for use with a natural-ester-filled transformer.

Esters are more hygroscopic than mineral oils (e.g. a natural ester dissolves around 1000 ppm of water at room temperature versus 50 ppm in mineral oil). Consequently, the rate at which ester-impregnated insulation paper adsorbs water was investigated. This helped set limits on how long oily transformer windings could be exposed to air during maintenance.

While laboratory investigations were being performed to understand how different transformer faults would affect dissolved gas generation, there was still an absence of data on the levels of dissolved gases expected for normal fault-free operation. A study was therefore performed investigating dissolved gas generation in the transformer fleet. These aspects are discussed in the next sections.

11.5 Fluid Temperature at Rated Load

The cooling ability of a fluid is related to its thermal properties such as viscosity and coefficient of thermal expansion, with lower viscosity associated with better heat conduction [14]. This is noteworthy because the viscosity of the Envirotemp FR3 fluid is higher than that of standard mineral oil. Figure 11.1 is a plot showing the kinematic viscosity of the two fluids as a function of temperature. As can be seen the rate of change of kinematic viscosity with temperature falls as the temperature rises. Consequently, it follows that any change in transformer thermal performance due to oil viscosity will be less pronounced at higher temperatures. This is important to understand as the utilities use thermal protection on transformers, so if an ester-filled transformer operated at a higher temperature than expected, there could be unintended trips. While increasing the maximum operational temperature might be considered for a natural-ester-filled transformer, it is then necessary to understand the general effects of higher temperatures on insulation aging and reliability.

An investigation was performed to understand any likely increase in fluid temperature caused by differences in fluid properties. This is important because the utilities often have thermal alarms and trips on their transformers, which could be set considering the type of fluid used.

Two nominally identical 45/90 MVA ONAN/ONAF transformers were manufactured for a British utility, one filled with Envirotemp FR3 and the other with mineral oil [15]. During factory test, they were operated at rated load, providing the opportunity to study the different temperature rises. The transformers were individually run at both 45 and 90 MVA loads in ONAN and ONAF cooling modes until a steady-state temperature rise had been reached. In accordance with IEC-recommended procedure, this steady state was reached once the rate of increase of top oil temperature rise had fallen below 1 K per hour and had remained below that rate for three hours [16].

As can be seen in Table 11.2, there are two loadings given, one 45 MVA representing ONAN cooling mode and the other ONAF. The top oil temperature of the more viscous Envirotemp FR3 unit was only four degrees warmer than that of the mineral oil unit in both 45 and 90 MVA loading configurations. While the difference in the 90 MVA loading was very slightly more (0.2 °C) than the 45 MVA ones, the percentage difference was less, which is consistent with the observation that the change in thermal performance is less at higher temperature due to the lesser change in viscosity. Two general effects of operating transformer insulation at higher temperatures are as follows.

One is there is a general relationship that the aging rate of Kraft paper insulation (immersed in mineral oil) doubles for approximately every temperature increase in 6 °C (within temperature

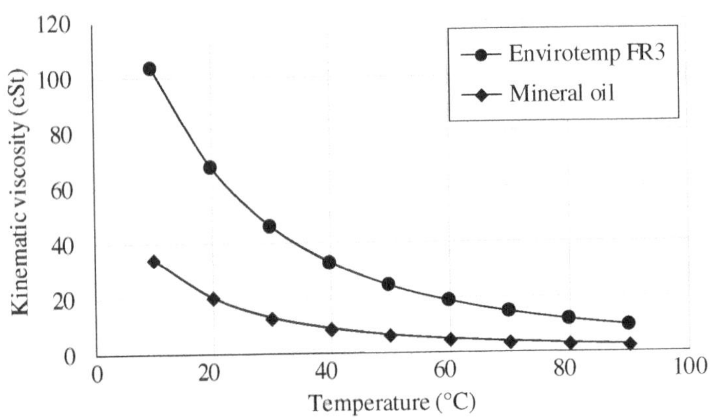

Figure 11.1 Comparison of fluid viscosity as function of temperature.

Table 11.2 Comparison of heat run temperatures.

	Top oil rise (K)	Mean oil rise (K)
90 MVA loading		
Mineral oil	52.7	33.8
FR3	57.1	34.2
Difference	4.4 (7.7%)	0.4
45 MVA loading		
Mineral oil	34.5	26.2
FR3	38.7	28.0
Difference	4.2 (12%)	1.8

Source: Modified from Martin et al. [15].

range 80–140 °C). However, as noted in IEC standard 60076/14 and other research [17, 18], the rate of aging of Kraft paper in natural ester is 5–8 times slower than that in mineral oil. Consequently, despite there being a difference in temperature, the rate of aging has still been reduced.

Two is for gas bubbles to be ejected from the paper at elevated temperatures during transformer overload [19, 20]. The thermal properties of a fluid, such as its density, affect the ability for bubbles to form and grow, and so a laboratory-based experiment was performed to investigate the behavior of insulation paper immersed in natural ester [13, 21, 22]. A coil was manufactured representing a section of transformer winding, where paper-wrapped copper conductor was used and then the system was immersed in natural ester (Figure 11.2). Thermocouples were embedded into the turns to measure temperature. The water content of the paper was controlled by conditioning sections of pressboard to have a desired water content and then immersing this pressboard into the system. The adsorbed water would then migrate and equilibrate with the paper. A variac supplied voltage to the current transformer, which provided the current needed to heat up the coil. When bubbles were observed, the temperature reached was noted. This investigation found that the bubbling temperature of paper immersed in Envirotemp FR3 is 5–10 °C higher than when immersed in mineral oil, as shown in Figure 11.2c. The conclusion of this work was that while the bubbling temperature of this natural ester was higher than that of mineral oil, possible increases in hotspots within a natural-ester-filled transformer should be checked to ascertain whether their temperature could be great enough to cause bubbling.

11.6 Breakdown Voltage and Water Content

It is well known that the breakdown voltage of oil is impacted by water and particles. Given that if transformer insulation becomes too wet it can be dried out, it is useful to understand the relationship between water and breakdown voltage to determine when this operation should be performed.

In one experiment, the AC and lightning impulse breakdown voltage of Envirotemp FR3 was investigated as a function of water [23, 24]. These tests were performed in accordance with ASTM D1816 [25] using a 1 mm gap and ASTM D3300 [26] using a 3.8 mm gap with a sphere-sphere arrangement.

As can be seen in Figure 11.3a, water only affected the AC breakdown voltage at around 400 ppm, while this level did not affect the lightning impulse breakdown voltage (Figure 11.3b). Within the context of setting limits for these transformers, 400 ppm was therefore considered as a value when

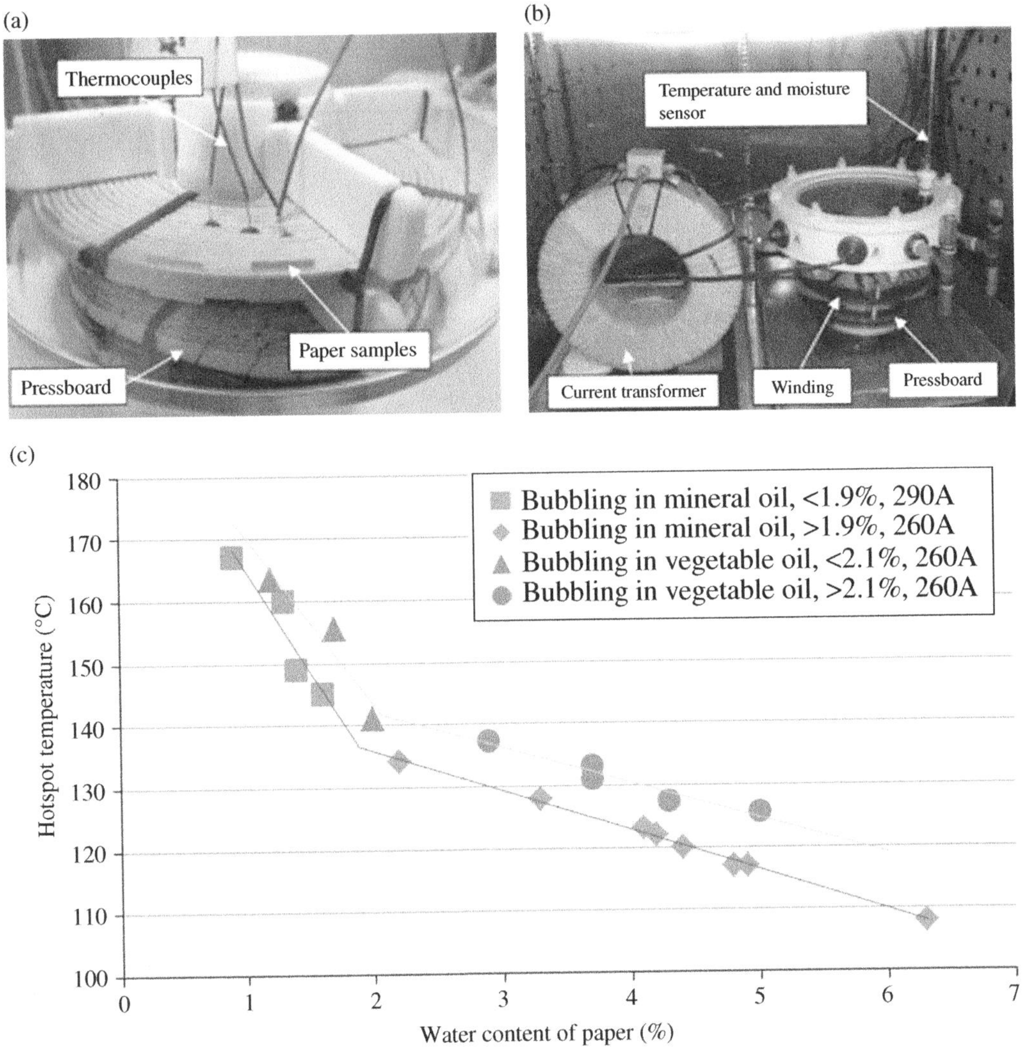

Figure 11.2 Experiment to investigate bubbling at high load, (a) showing the coil, (b) system was heated by applying current from transformer, (c) comparison of bubbling temperature in mineral oil and Envirotemp FR3 natural ester. *Source:* Perkasa et al. [13].

the utilities should consider further investigation to check the sealing. This was corroborated by the manufacturer's product information guide [11].

11.7 Investigations into Oxidation and Handling Fluid-Impregnated Paper

During manufacturing, repairing, or refurbishing an ester-containing transformer, the fluid-impregnated cellulosic insulation is likely to be exposed to the air for short periods. Some investigations were performed to understand any differences in handling for natural esters. These were to measure the rate of adsorption of water from the atmosphere [9] and to observe the effect of heat

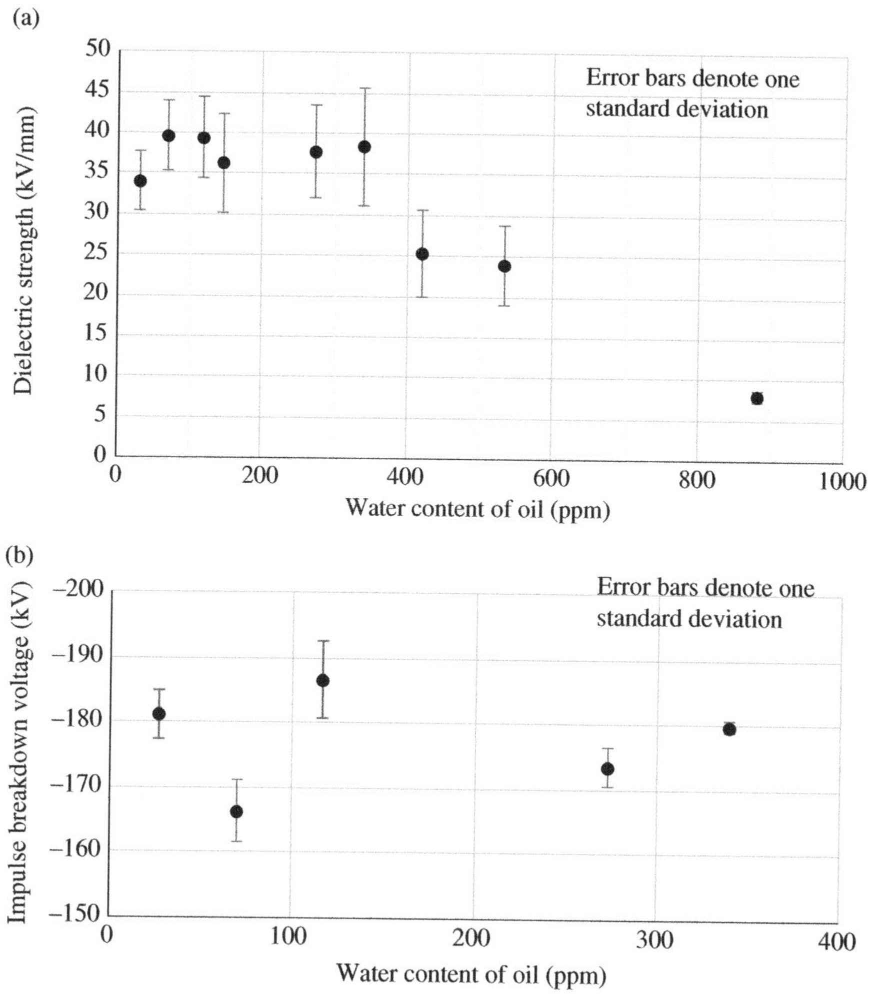

Figure 11.3 (a) Effect of water on AC dielectric strength, measured according to standard ASTM D1816-2004. (b) Effect of water on lightning impulse dielectric strength, measured according to standard ASTM D3300-2000.

from air-circulating ovens [27]. This was performed since esters dissolve more water than mineral oil, and so ester-impregnated cellulose might adsorb atmospheric moisture quicker than when mineral oil is used. However, while natural ester dissolves more water than mineral oil (1000 ppm versus 50 ppm at room temperature), these differences are comparatively small when considering that cellulosic paper adsorbs several percent by mass of water, where 1% = 10 000 ppm.

Samples of paper were dried by heating in an oven with a helium atmosphere. Helium was used to prevent oxidation by excluding atmospheric oxygen. The paper was then transferred to a high-precision balance and left to adsorb moisture from the air. The increase in mass due to the water was recorded for unimpregnated paper, then the paper was impregnated with oil, and then the drying and weighing repeated.

Figure 11.4 shows the uptake in water content. Over a two-hour period, as can be seen, the type of fluid impregnating the paper has had an insignificant impact on the rate of water adsorption by the paper. The experiment was left to run overnight and then weighing was continued the following

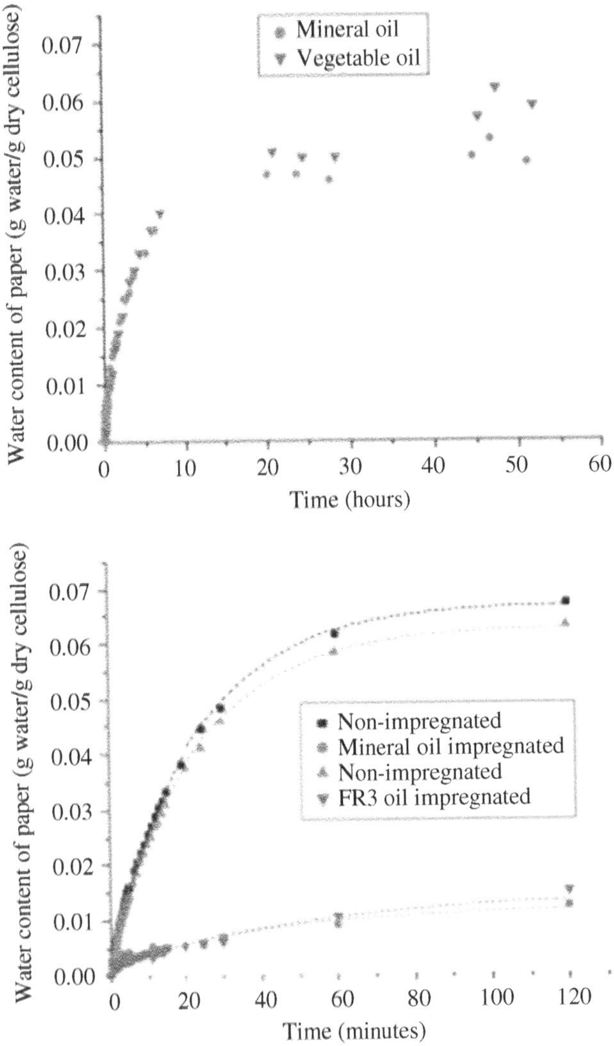

Figure 11.4 Increase in water content of fluid-impregnated insulation paper. *Source:* Modified from Martin et al. [9].

morning. Eventually, the vegetable-oil-immersed paper absorbs more water. A conclusion from this work was that if a utility or manufacturer returns impregnated paper insulation to the transformer tank within seven hours, the difference in the water content of the paper will be negligible.

Natural esters are not recommended for use in free breathing transformers because of their lower oxidation stability compared to that of mineral oil. Given that at some point in the life of a transformer the windings are likely to be exposed to atmospheric oxygen, tests were performed to determine any difference in treatment which might be required. Factors investigated were as follows:

- The tendency for oil puddles at the bottom of a drained tank to solidify.
- Effects of leaving pumps, or other ancillary equipment, unused for long periods of time (e.g. months).
- Propensity for natural ester to gel on paper exposed to air at both ambient and high temperatures (e.g. an air circulating oven).

To simulate puddles of fluid at the bottom of a drained tank, a 15-cm diameter petri dish was filled with natural ester to a 5 mm depth and left in the laboratory [27]. After one month, no

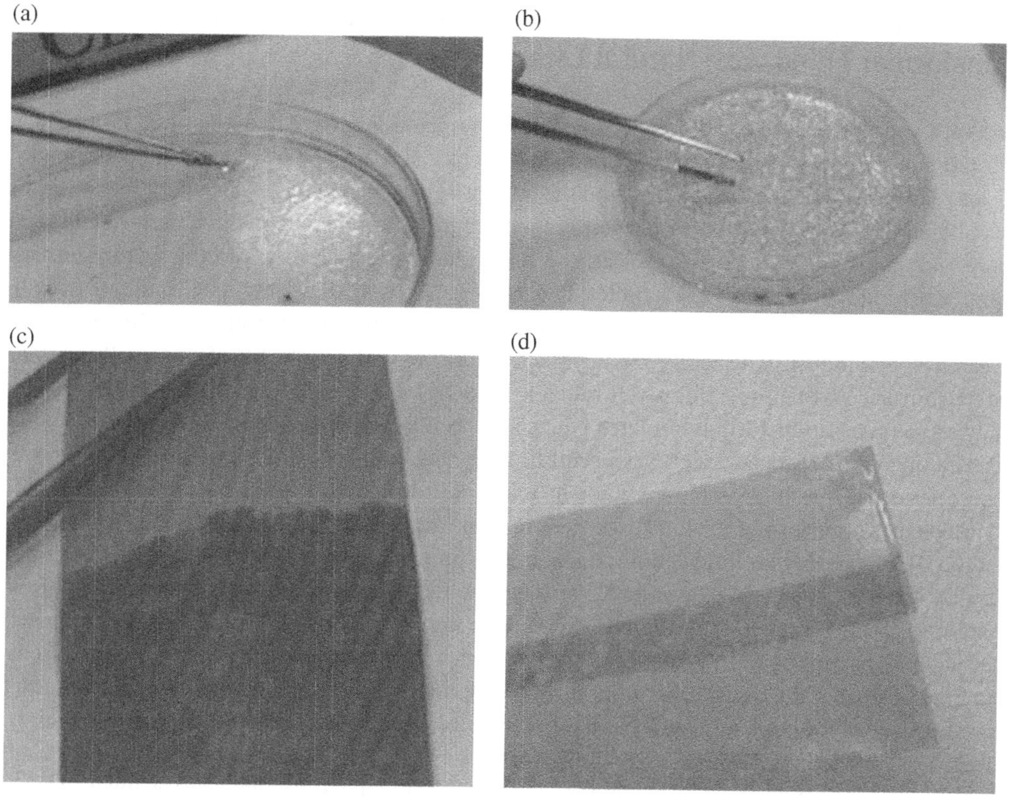

Figure 11.5 Aging investigations into long-term aging of a natural ester (a) after 2 years and (b) 2 years 5 months. *Source:* van Hest et al. [27]. (c) and (d) shows the results of drying impregnated paper in an air-circulating oven to simulate drying.

changes were observed; however, it was noted that insects seemed to be attracted to the ester, falling into the fluid and dying. This could be a problem for transformer maintenance if insects were attracted into an open tank and led to dielectric failure of an oil gap. Two years passed before a skin of gel was observed on the surface of the oil, as shown in Figure 11.5a and b.

Consequently, it was concluded as follows:

- Gelling is unlikely to form in puddles at the bottom of an empty tank, while exposed to air during maintenance, because the tank will not be left open for this duration.
- Ancillary equipment occasionally used to service a transformer, such as an oil pump, are unlikely to become clogged with gel unless unused for several months. Flushing the equipment with an ester is recommended.

Drying transformer insulation by heating in an air-circulating oven is another process used by industry. To show the effect of using hot air to dry natural-ester-impregnated paper, samples were heated in an air-circulating oven. A strip of paper was dipped into a natural ester and then clamped vertically in the oven, allowing the fluid to gather at the bottom of the paper. After heating at 110 °C for five hours, the fluid was noted to have dried on the surface of the paper, similar to a varnish (Figure 11.5c and d). These results confirmed to the utilities that hot air should not be used to dry insulation impregnated with natural ester.

11.8 Study on Installation and Early Operation of a Power Transformer Filled with Natural Ester

In 2008, two vegetable-oil-filled transformers, of the same design, were manufactured for use in a substation in Sydney [5, 6]. These were rated 132/11/11 kV 50 MVA KWDF (K-class fluid and water cooled). Only one was installed in 2008 and the other was placed in storage. The operating unit was exceptional in that it was fitted with a comprehensive monitoring system which performed DGA every four hours. As part of this project, the researchers also had access to the factory during manufacture and test to collect oil samples. The transformer was extensively tested and satisfied all acceptance criteria. Consequently, it was considered to be operating normally. This gave the researchers and industry the ability to study the gases formed early in the life of a normally operating transformer. At that time, this was invaluable because most research had been concerned with studying gas generation during simulated faults, such as [28–31], and not gas levels which are typical. The unenergized transformer was useful to study gas levels of a transformer which was not being operated. Thus, interpretations could be made of fluid changes which happen naturally.

Various measurements of the fluid quality were made over a two-year period. Figure 11.6a shows that the DDF, a marker for fluid condition, did not vary significantly compared to the IEEE recommended maximum value of 0.03 at 25 °C. Measurements were also made at higher temperatures because in the IEEE guide [32], there is no recommendation although some laboratories report DDF at 90 °C. An online DGA system was present, which measured gas levels every four hours. Oil was circulated from the transformer tank through the monitor for analysis, then back to the tank. The advantage of the online measurement is that possible inaccuracy due to fluid handling is minimized. The appropriate Ostwald gas-solubility coefficients for this fluid were used in the dissolved gas measurements. If coefficients for mineral oil are used, the measurements will be incorrect.

Online DGA measurements are shown in Figure 11.6b, where time indicates days after installation. The dashed lines show when the monitor was offline. Of significance was the increase in ethane content, which is not expected in mineral-oil-filled transformers. However, the chemical processes leading to ethane production appear to have ceased fairly soon after installation. This was further investigated.

Three independent Australian laboratories were used to measure the DGA content offline. Techniques used to sample mineral oil are also suitable for natural ester. However, because a natural ester is more viscous than mineral oil, extra care must be taken to ensure that any air bubbles are ejected from the collecting container, which in this instance were 50 ml glass syringes. If this is not performed, then the air gases could cause an imprecise measurement.

Table 11.3 shows the dissolved gas content of the transformer through various stages of testing to operation at the substation. As the transformer passed all HV, temperature rise, and overload tests, the manufacturer was confident that the transformer was operating in accordance with its specification.

While most of the hydrocarbon gases remained low, of interest were the increases in ethane and hydrogen which this study then focused on explaining. As shown in Table 11.3, hydrogen and hydrocarbon gases were consistently very low during testing and only appeared to have begun rising when at the substation. The increase in ethane and hydrogen continued for approximately one month before levelling off (Figure 11.6b). This was not completely unexpected because a comparatively high rate of generation of hydrogen and ethane had been observed by Lewand, Khan, Duval, and noted in the Envirotemp FR3 DGA guide [28–30, 33].

References [34, 35] point to the generation of ethane being caused by a chemical reaction with oxygen. In particular, some of the components of soybean vegetable oil, such as linolenic acid, generate ethane by reacting with oxygen [34]. Another suggestion is that ethane is generated by the lipid peroxidation mechanism, which can occur in all omega-3 unsaturated fatty acids [35]. Since

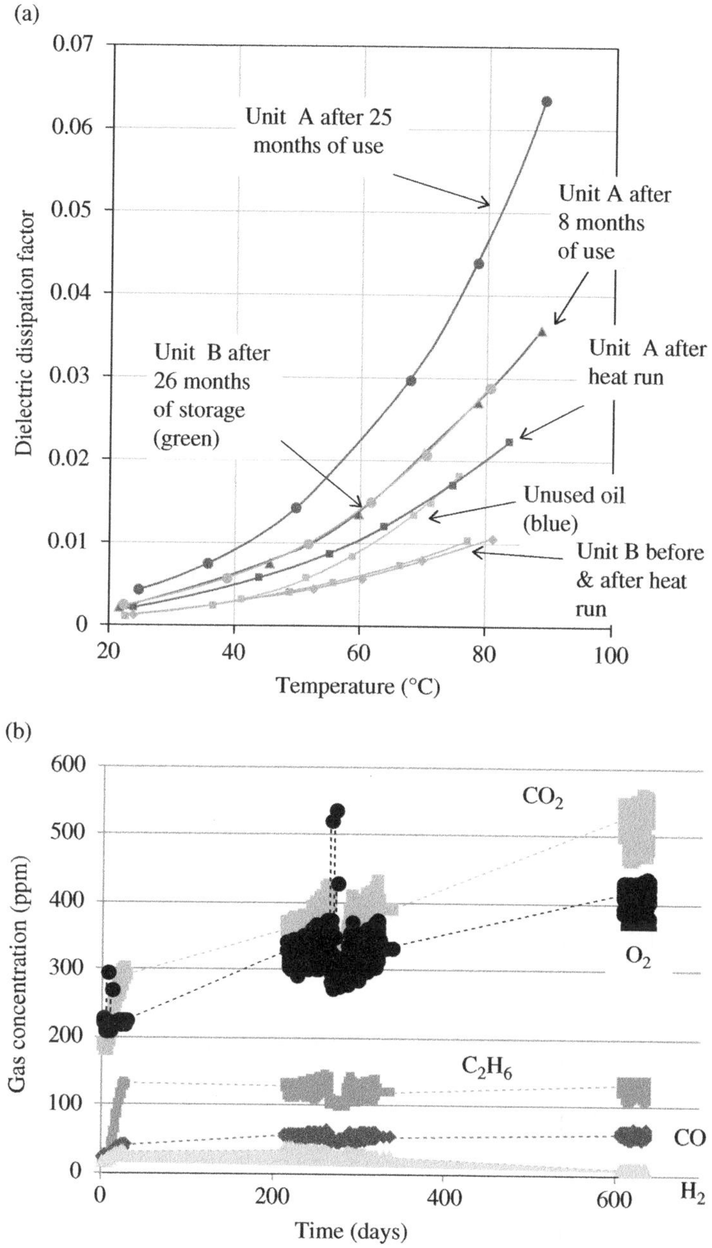

Figure 11.6 (a) Changes in fluid DDF. Unit A was in service, unit B was stored. These values are far below the IEEE-recommended maximum limit of 0.03 at 25 °C. (b) Change in dissolved gas content early in operation of transformer. *Source:* Martin et al. [6].

different types and blends of vegetable oil can be used in the feedstock of a natural-ester dielectric, this could imply that the rate of ethane generation differs across products and their batches.

An ingress of air and rise in ethane during installation is consistent with the transformer being delivered to site in parts, which were connected, and so some air entered the system. An increasing level of ethane can therefore be an indication of air entering a transformer, e.g. if unexpected could indicate that the sealing has failed. This is advantageous because it provides an early warning if a

Table 11.3 Dissolved gas content of oil in parts per million (ppm), taken new from the drum and from the transformer before and after factory tests, and at the substation.

1.	2.	3.	4.	5.	6.	7.	8.	9.	10.	11.	
Sample	**New from drum**	**Before all tests**	**After HV tests**	**Before tempera ture rise test**	**After tempera ture rise test**	**After overload test**	**After all tests**	**1 Sep 08 at substation**	**10 Oct 08 at substation**	**11 Mar 09 at substation**	
Lab	**Lab 1**	**Lab 2**	**Lab 2**	**Lab 2**	**Lab 2**	**Lab 2**	**Lab 2**	**Lab 1**	**Lab 2**	**Lab 1**	**Lab 3**
Sample temp	**Ambient**	**Ambient**	**Ambient**	**Ambient**	**Ambient**	**Ambient**	**Ambient**	**Ambient**	**34 °C**	**34.5 °C**	**33 °C**
Unit status	**–**	**At factory**	**At factory**	**At factory**	**At factory**	**At factory**	**At factory**	**Before installation**	**After installation, before energized**	**In operation**	**In operation**
H_2	29	<1	1	2	3	1	2	10	71	47	52
C_2H_6	0.8	<1	<1	<1	<1	<1	<1	8	26	113	61
CH_4	1	<1	<1	1	2	1	1	0.7	1	3	2.2
C_2H_4	0.9	<1	<1	<1	<1	1	<1	0	<1	2	1.5
C_2H_2	0.4	<1	<1	<1	<1	1	<1	0	<1	0	0
CO	15	<1	<1	<1	<1	<1	<1	19	17	77	60
CO_2	501	9	15	194	93	37	34	177	154	313	200
O_2	1798	1169	1179	11 270	6951	1469	1,760	1929	5244	321	448
N_2	60 959	4492	4472	33 890	21 807	5874	6874	43 423	29 927	40 510	32 100
TDCG[a]	47.1	0	1	3	5	4	3	36	115	242	177

[a] TDCG: total dissolved combustible gases, which is the combined content of H_2, C_2H_6, CH_4, C_2H_4, C_2H_2, and CO.
Source: Modified from Martin et al. [5].

transformer is no longer sealed. Although an elevating concentration of dissolved oxygen will probably also be noted, this could be attributed to contamination of a sample by air. Ethane as a marker is preferable because this gas is not present in the air to a significant degree.

Elevated levels of ethane will also be shown in the later section analyzing dissolved gases in a fleet of transformers over many years.

Table 11.4 shows the fluid conditions, confirming that the transformer are operating as expected. All these values show normal operation according to the fluid test guide [11]. The acidity shows no difference between the in-service and stored transformer, indicating that little chemical change has occurred. The interfacial tension of the in-service transformer shows no change and is very similar to the unused transformer. The oil–water content of the stored transformer is lower than that of the in-service transformer. However, this may be explicable by the temperature of the in-service transformer being higher. When cellulosic materials are warmed, such as by the heat generated by transformer operation, water desorbs entering the oil where it is then measured. In contrast, the cooler cellulose in the unused transformer will have adsorbed water from the oil. In any case, since a natural ester dissolves 1000 ppm of water at room temperature, these differences are small.

11.9 Fleet Measurements

Data on this power transformer fleet was analyzed in 2015 [36]. This data set has been revised and extended for use in this chapter. Changes in dielectric dissipation factor (DDF), interfacial tension (IFT), and acid number are indicative of chemical processes occurring in the fluid. Since there are differences in molecular structure between triglycerides (in natural ester) and the naphthenic, paraffinic, and aromatic components in mineral oil, the DDF, IFT, and acid number values should not be directly compared.

Table 11.4 Fluid properties.

	In-service transformer			Stored transformer
Test	**On energization of transformer**	**After 5 months of operation**	**After 25 months of operation**	**After 26 months**
Breakdown voltage (kV) IEC 60156; VDE 0370 electrodes	75	68.2	67.2	—
Moisture (ppm) Karl Fisher titration	30; ASTM D1533	22; IEC 60814	22; IEC 60814	7; IEC 60814
Acidity (mg of KOH/g of oil)	0.03; ASTM D974	0.03; AS 1767.2.1	0.03; AS 1767.1	0.03; ASTM D974
Interfacial tension (mN/m) ASTM D971	21.6	21.6	21.6	24.5
Dielectric dissipation factor 50Hz;IEC 60247	0.00273 at 25 °C	0.0042 at 30 °C; 0.0358 at 90 °C	0.0503 at 90 °C	—

Source: Modified from Martin et al. [5].

11.9.1 Dielectric Dissipation Factor, Interfacial Tension, and Acid Number

For DDF, the fluid manufacturer recommends 1% as the value for continued service [33]. As can be seen in Figure 11.7a, all of the fluid samples have remained below 1%, and furthermore, there is no apparent upward trend over time which would show long-term oil aging.

Values for IFT are shown in Figure 11.7b. One transformer (#5) showed a slightly lower IFT just after commissioning than the limit proposed for continued service. However, at the time of writing, this unit has operated successfully for six years. Its DGA did not indicate improper operation and so it may have been a measurement anomaly.

The acid number of an oil is the mass of the alkali potassium hydroxide (KOH) required to fully neutralize the acids in one gram of oil. However, this measure should not necessarily be compared across different types of oil because it does not indicate the aggressiveness of these acids, which may differ. Some of the chemical compounds in natural ester react with water or oxygen to form acids, which may be aggressive or benign to paper insulation depending on their molecular structure and the reactions involved. In general, oxidation reactions generate short-chain acids, which degrade the paper insulation, whereas, hydrolysis rations with water create long-chain acids, which do not degrade the paper insulation [18].

The manufacturer of one natural ester proposes that hydrolysis may be beneficial because it consumes the water, which might have otherwise degraded the paper.

The acid number of the fleet is shown in Figure 11.7c. A key point is that there has not been an upward trend, which would indicate chemical changes (either oxidation or hydrolysis).

11.9.2 Water Content of Oil

The water content of a transformer oil varies significantly due to temperature and time. As cellulose is heated water migrates into the oil, and then during cooling water is adsorbed. The diffusion time constants for cellulose will vary between days and months depending on temperature. The load profile of many Australian power transformers is cyclic, on a 24 hour period, with peak load often around noon coinciding with air-conditioning load. Consequently, the water content of oil is usually lagging temperature, and thermodynamic equilibrium with the cellulosic insulation is not usually reached. This means that occasionally high values of water content are measured, which then fall later. Many of the water content measurements are below 50 ppm (Figure 11.8a). Another measure of the wetness of an oil is its percentage saturation, which numerically is

$$\mathrm{PS} = \frac{\mathrm{WCO}(\mathrm{ppm})}{\mathrm{WCO}(\mathrm{saturation\,ppm})} \tag{11.1}$$

This is the ratio of mass of water to the mass of water at saturation. The solubility of oil in ppm at a known temperature is calculated using Equation (11.2), where A and B are two coefficients, and T is temperature in °C.

$$\mathrm{Solubility}(T) = 10^{\left(A - \frac{B}{T+273}\right)} \tag{11.2}$$

For this work, samples of the natural ester were collected from the transformer manufacturer, and the saturation coefficients were determined by heating up the oil to various temperatures and then measuring the vapor pressure and WCO by Karl Fischer titration [9]. Coefficients were determined as $A = 6.000$ and $B = 881.39$. These coefficients differ slightly from those used by the manufacturer, $A = 5.3318$ and $B = 684$ [28]. It follows that the oil sample analyzed would dissolve less water than indicated by the manufacturer at high temperatures.

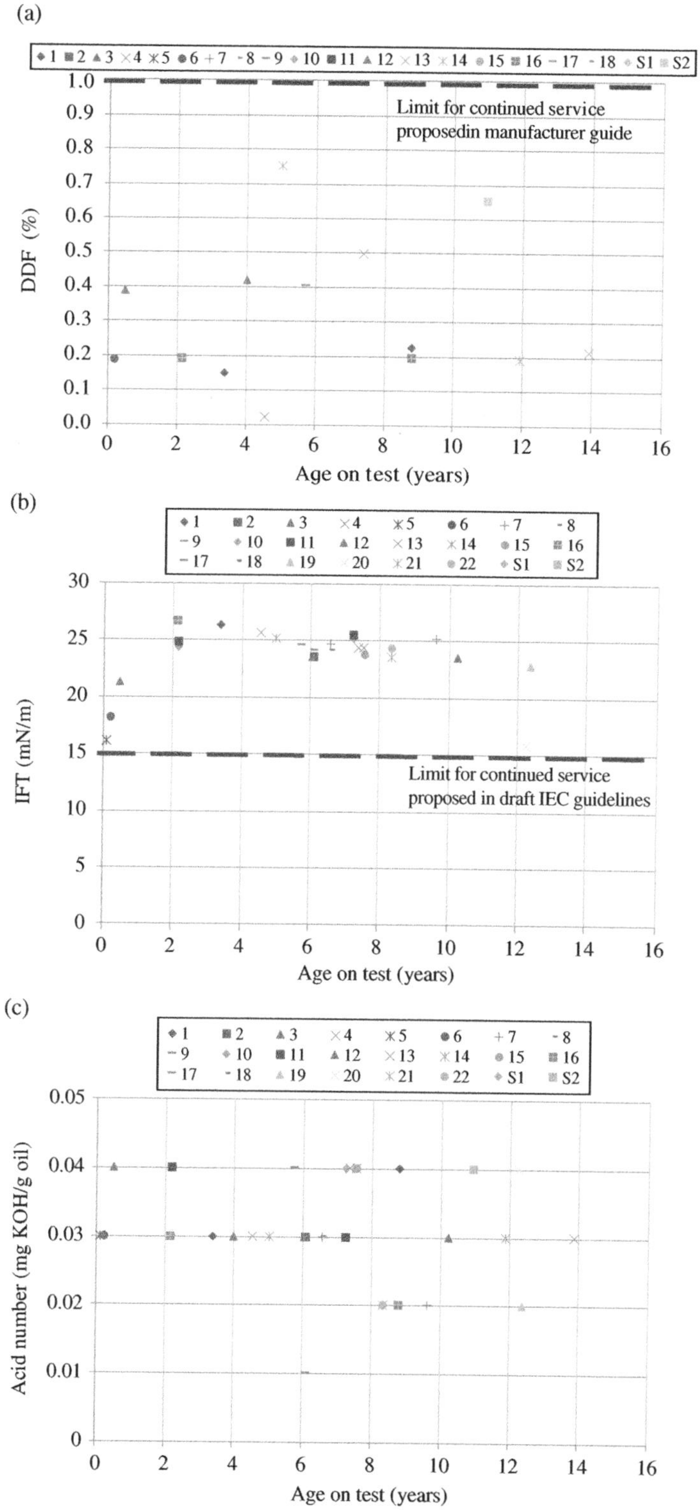

Figure 11.7 Oil quality measures as a function of transformer age (a) dielectric dissipation factor (b) interfacial tension and (c) acid number.

The temperature of the oil on sampling was recorded by the utility, Figure 11.8b, which was then used to calculate the percentage saturation in Figure 11.8c. Transformer #9 has had several high readings of water which fell. The utility was unaware of the cause, so it may have been due to sampling or handling technique.

11.9.3 Breakdown Voltage of Oil

The breakdown voltages of the natural ester were measured in accordance with IEC 60156, which uses a 2.5-mm electrode gap. These values, Figure 11.8d, were compared to the limits for mineral oil recommended in IEC 60422. Since a high breakdown voltage is a necessary property of insulation, meeting the same limits is necessary regardless of the type of fluid. The minimum breakdown voltage recommended by the IEC depends on the rated voltage of the transformer [37]. This is either 40 kV for a transformer rated below 72 kV of 50 kV for a transformer rated between 72 and 170 kV. As can be seen in Figure 11.8d, all of the measurements are above 50 kV, even for the 33 kV transformers, and so these units are functioning according to standard. There is no downward trend over time, which if were observed could indicate deterioration.

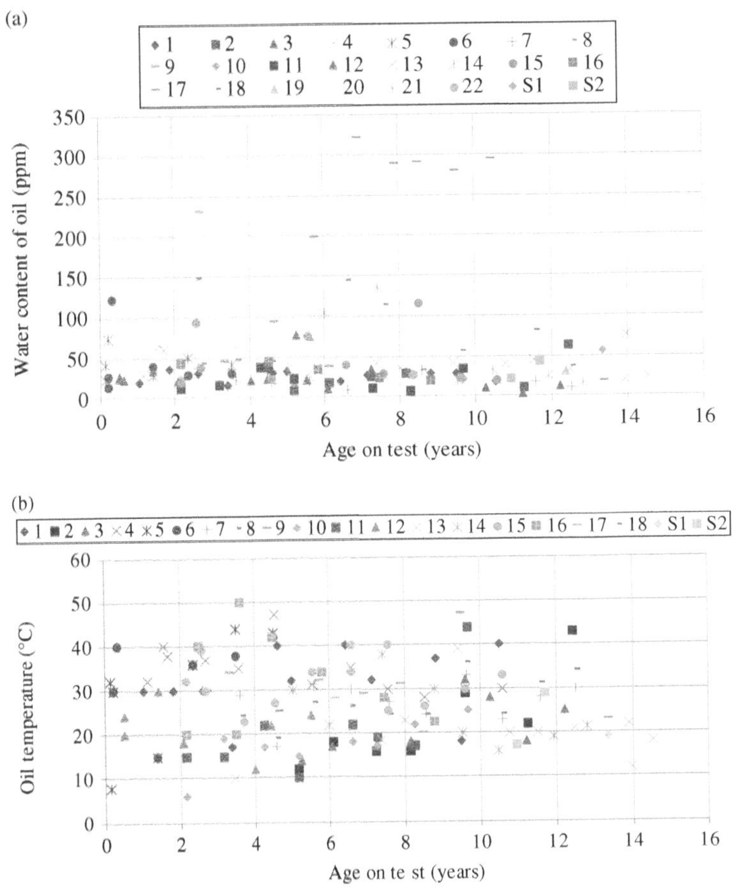

Figure 11.8 (a) Water content of oil, (b) oil temperature, (c) percentage saturation, and (d) breakdown voltage.

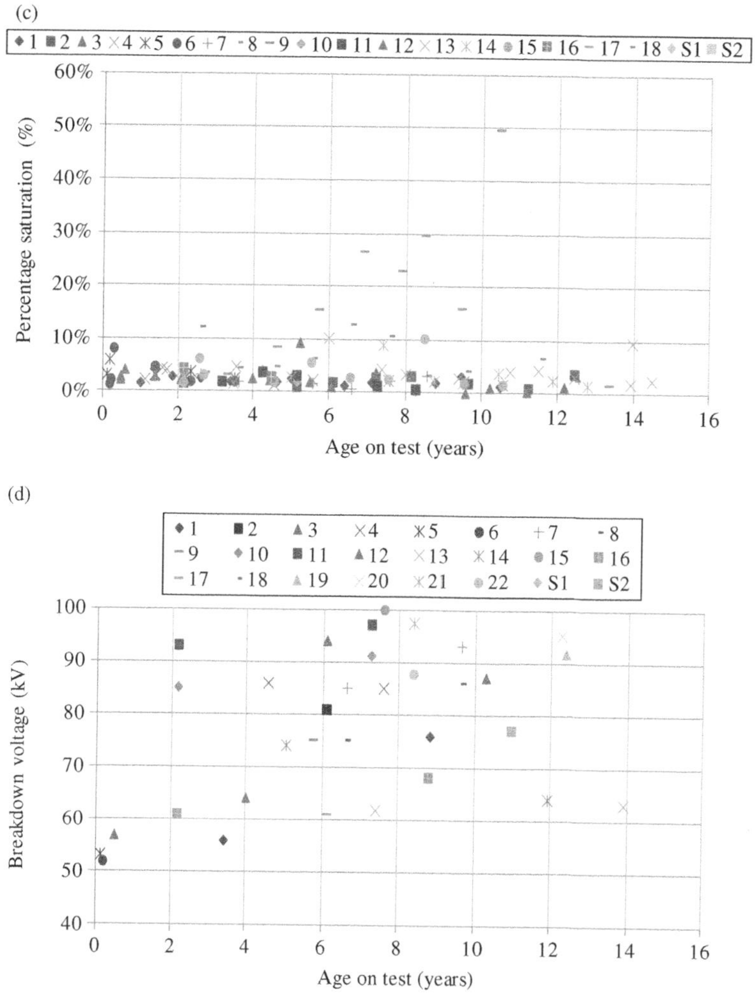

Figure 11.8 (Continued)

In summary, these measurements indicate that the natural ester used in these transformers has continued to operate effectively over the years. In the next section, the dissolved gas analysis (DGA) of these transformers is presented.

11.9.4 Dissolved Gas Analysis

The ratios of gases generated during faults have been noted to slightly differ depending on the type of fluid [28, 29]. Duval [30] has updated triangles for use with natural ester fluids, where zone boundaries have been moved based on observations. Key findings for this type of natural ester (Envirotemp FR3) have been that it generates the following:

- The same gases in mineral oil in the presence of faults, but with different concentration ratios.
- These gases (except ethane) in lower concentrations than mineral oil.
- Ethane in larger concentrations than mineral oil.

In total, 158 samples of DGA have been performed on these transformers over the years. When analyzing data for mineral oil, 90th percentile typical values are given by both the IEC and IEEE, which is applied by industry to detect abnormal operation. The 90th percentiles for the studied

Table 11.5 90th percentile values determined from data.

	Hydrogen	Methane	Ethane	Ethylene	Acetylene	Carbon monoxide
Concentration (ppm)	380	4	215	4	0.5	72

data set are given in Table 11.5. It is stressed that a sample having a dissolved gas concentration in excess of the 90th percentile limit does not necessarily indicate a fault but only highlights the top 10% of samples for further analysis. It is used to show for a population of transformers where the gas levels are placed within the population. Compared to mineral oil, the hydrogen and ethane 90th percentile values are significantly higher.

Laboratory DGA on the Envirotemp FR3-filled transformers are given in Figures 11.9 and 11.10. In general, no long-term upward trend is observed, and thus, the 90th percentile values for dissolved gases is not expected to change over this 15-year period.

As can be seen in the figures below the methane, ethylene and acetylene concentrations are very low and so no further comment is given. The carbon monoxide level was elevated in one transformer, #9, however dropped back down without intervention by the utility, so the cause is unknown.

As discussed earlier, the ethane content of vegetable oil was observed to be higher than that of mineral oil. This is confirmed in Figure 11.10a over the complete fleet. Notable features of the figure are as follows: firstly, the increase in ethane appears to have peaked between the first and second year of operation; however, this may be due to sampling period. Secondly, the two spare transformers S1 and S2 also have a scattered ethane content significant because this shows that the transformer does not have to be operational.

In Figure 11.10b, the hydrogen concentration for transformer #12 is seen to increase steeply up to nearly 10 years of age and then fall off sharply to below 500 ppm. The utility noted that around the 10-year mark, a high-impedance fault was found in the termination of the connecting cable. The cable was operating hot, where this heat would have been conducted into the transformer and raised the temperature of the local fluid volume, possibly generating hydrogen. Once the termination was repaired, the fluid was not degassed; i.e. the fall in hydrogen was not caused by oil processing but fell once it was no longer being produced. Unit #5 had one hydrogen point at 1200 ppm after three years. On inspection, it was found that there was a loose connection in the OLTC which was generating heat; the surface temperature of the OLTC tank was measured to be 140 °C by an infrared imager. Even though the OLTC was in a separate compartment, it is likely that the oil on the other side of the barrier in the main tank was also heated, and then generated hydrogen. Transformer #16 also generated some hydrogen early on, however, without intervention this had returned to typical levels by the fourth year of operation.

11.9.5 Electrical Testing of Transformers

The insulation resistance of a transformer is occasionally measured to check condition. IEEE specifies typical resistance of >500 MΩ for new equipment, falling to between 10 and 100, which is indicative of insulation deterioration. As can be seen in Figures 11.11. and 11.12, all measurements are above 500 MΩ, and thus the insulation has performed as expected.

The winding dielectric dissipation factor (DDF), another measure of the condition of the insulation, was also periodically measured and is shown in Figure 11.13. The IEEE standard recommends a limit of 1%, slightly higher than the corresponding value for mineral-oil-filled transformers because of the higher DDF of the ester fluid [2].

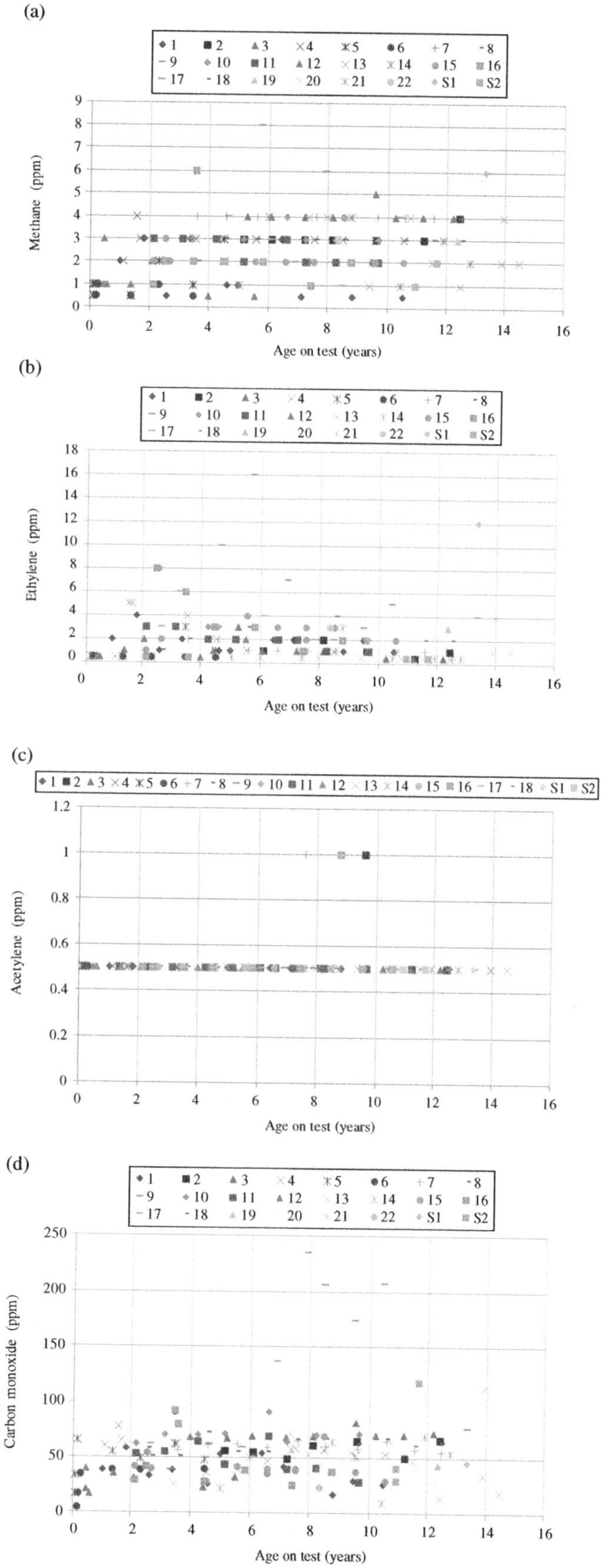

Figure 11.9 Dissolved gas concentrations for (a) methane, (b) ethylene, (c) acetylene, and (d) carbon monoxide.

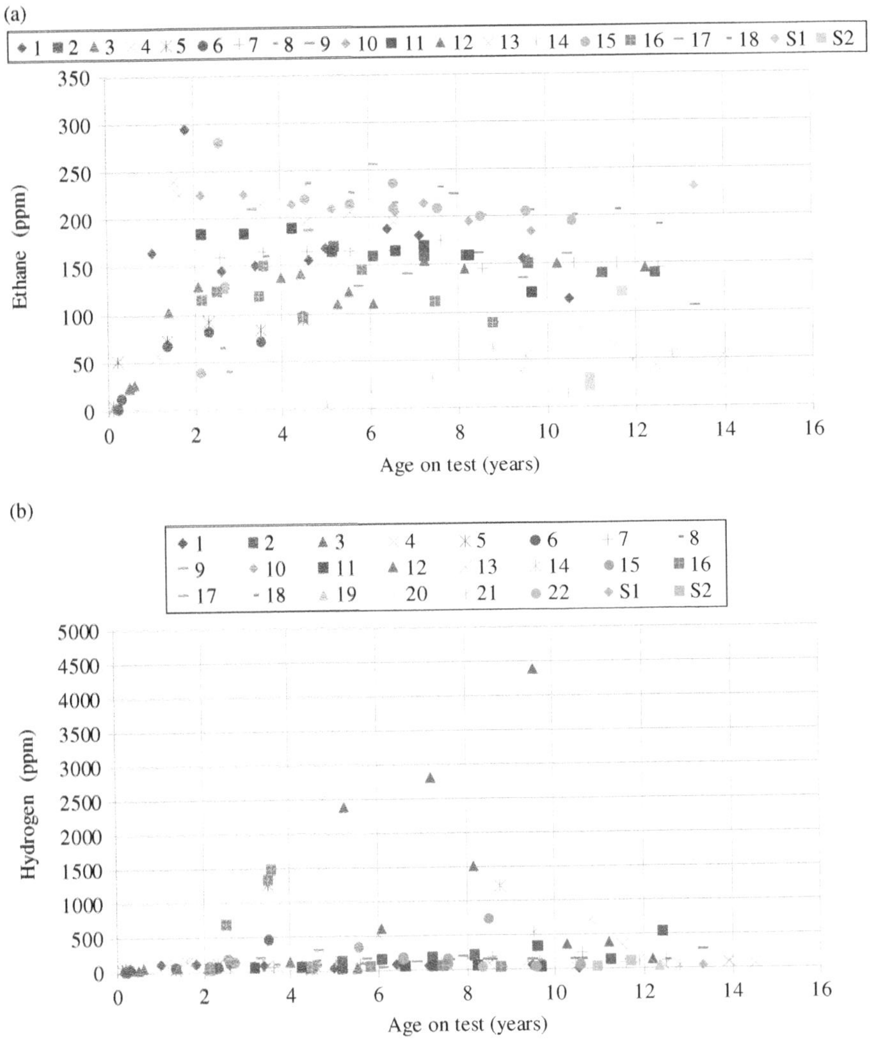

Figure 11.10 Dissolved gas concentrations for (a) ethane and (b) hydrogen.

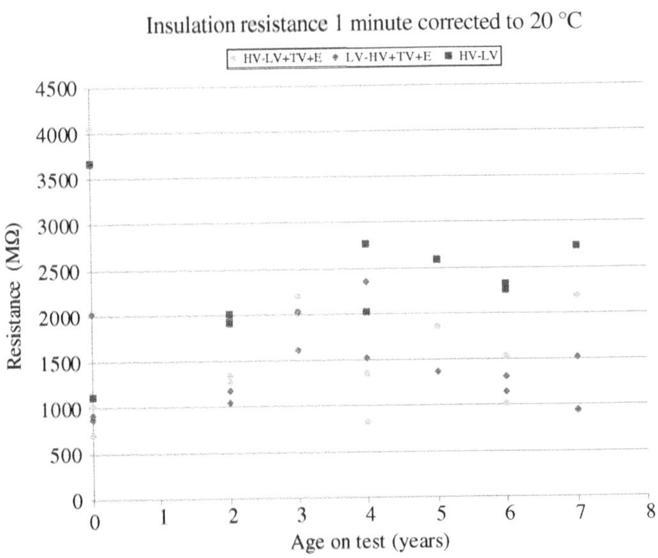

Figure 11.11 One-minute insulation resistance of transformer fleet.

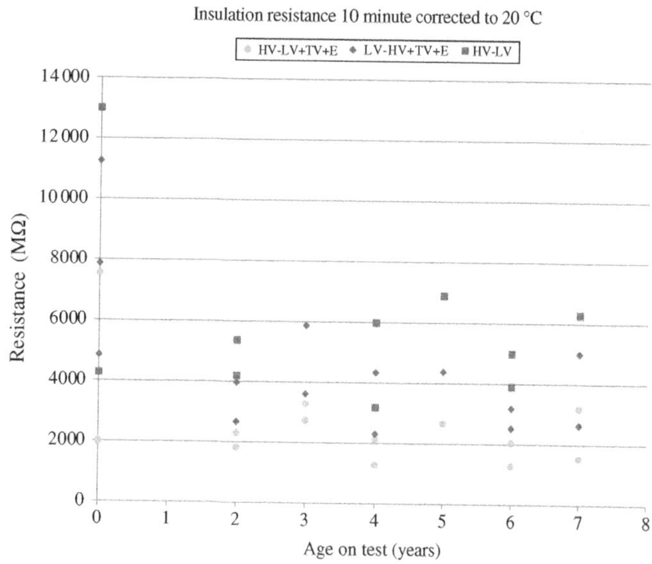

Figure 11.12 Ten-minute insulation resistance of transformer fleet.

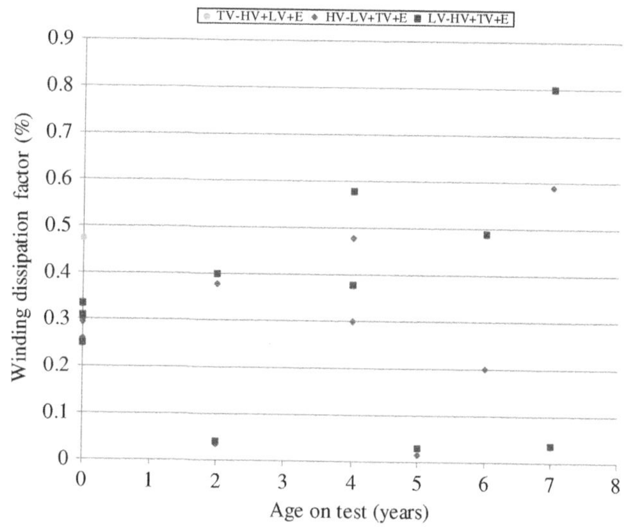

Figure 11.13 Winding dielectric dissipation factor.

11.10 Summary

Asset management of natural-ester-filled transformers can be based on the same tests as used with mineral oil. However, in many cases values indicating normal operation will differ. For instance, the DDF, IFT, water content, and acid number are different, reflective of the different molecular structure of a natural ester.

The ethane concentration of natural ester has a tendency to rise if the fluid dissolves oxygen. While this has appeared harmless in the operation of these transformers, an increase may indicate that the sealing has failed. Rising ethane could therefore be used as an indicator of air ingress.

While natural esters are known to have a lower oxidative stability than mineral oil, the data analyzed in this chapter strongly indicates that this fluid is effective as insulation over the long term, at least 15 years, as long as the tank is sealed. Even over this time, the tests have not indicated a long-term degradation of the natural ester, and so it is likely to last for a while into the future. (The breakdown voltage of the natural esters has not fallen.)

References

1 CIGRE A2.35 working group (2010). *Experiences in Service with New Insulating Liquids*. Paris, France: CIGRE.

2 IEEE Std C57.152-2013 (2013). IEEE Guide for diagnostic field testing of fluid-filled power transformers, regulators, and reactors, IEEE Power and Energy Society.

3 IEEE Std C57.147-2018 (2018). IEEE guide for acceptance and maintenance of natural ester insulating liquid in transformers. IEEE Power and Energy Society.

4 Draft IEC 62975 (2019). Natural esters – guidelines for maintenance and use in electrical equipment.

5 Martin, D., Lelekakis, N., Davydov, V., and Odarenko, Y. (October 2010). Preliminary results for dissolved gas levels in a vegetable oil–filled power transformer. *IEEE Electrical Insulation Magazine* 26 (5): 41–48.

6 Martin, D., Lelekakis, N., Guo, W., and Odarenko, Y. (October 2011). Further studies of a vegetable-oil-filled power transformer. *IEEE Electrical Insulation Magazine* 27 (5): 6–13.

7 Lewand, L. (Fall 2004). Natural ester dielectric liquids. *NETA WORLD*: 1–4.

8 McShane, C.P., Gauger, G.A., and Luksich, J. (1999). Fire resistant natural ester dielectric fluid and novel insulation system for its use. *IEEE/PES Transmission Distribution Conference* 2: 890–894.

9 Martin, D., Lelekakis, N., Wijaya, J., and Williams, K. (September 2013). Water uptake rates of transformer paper insulation impregnated with FR3 fluid. *IEEE Electrical Insulation Magazine* 29 (5): 56–61.

10 Australian Standard AS 2067 (2008).Substations and high voltage installations exceeding 1 kV a.c.

11 Cooper Power Systems (2004). *Envirotemp FR3 Oil Testing Guide*. Waukesha, WI: Cooper Industries Inc.

12 Martin, D., Guo, W., Lelekakis, N., and Heyward, N. Using a remote system to study the thermal properties of a vegetable oil filled power transformer: how does operation differ from mineral oil. *IEEE PES Innovative Smart Grid Technologies Asia (ISGT)*, Perth, 2011.

13 Perkasa, C., Lelekakis, N., Czaszejko, T. et al. (2014). A comparison of the formation of bubbles and water droplets in vegetable and mineral oil impregnated transformer paper. *IEEE Transactions on Dielectrics and Electrical Insulation* 21 (5): 2111–2118.

14 Grubb, R.L., Hudis, M., and Traut, A.R. (1981). A transformer thermal duct study of various insulating oils. *IEEE Trans. Power App. Syst.* PAS-100: 466–473.

15 Martin, D., Wijaya, J., Lelekakis, N. et al. (January 2014). Thermal analysis of two transformers filled with different oils. *IEEE Electrical Insulation Magazine* 30 (1): 39–45.

16 IEC Standard 60076-2 (1997). Power transformers—Part 2: temperature rise.

17 IEC Standard 60076-14 (2013). Power transformers—Part 14: liquid-immersed power transformers using high-temperature insulation materials.

18 McShane, C.P., Rapp, K., Corkran, J. et al. (2001). Aging of paper insulation in natural ester dielectric fluid. *IEEE Transmission Distribution Conference and Exposition* 2: 675–679.

19 Heinrichs, F.W. (1979). Bubble formation in power transformer windings at overload temperatures. *Transmission Power Apps Systtem* 98: 1576–1582.

20 Oommen, T.V. and Lindgren, S.R. (2001). Bubble evolution from transformer overload. *IEEE/PES Transmission and Distribution Conference Exposition* 1: 137–142.

21 Perkasa, C.Y., Lelekakis, N., Wijaya, J., and Martin, D. (2012). Investigating bubble formation in vegetable and mineral oil impregnated transformer paper insulation systems. *22nd Australasian Universities Power Eng. Conf. (AUPEC)*, Bali, Indonesia (26–29 September 2012), pp. 1–5.

22 Perkasa, C.Y., Wijaya, J., Lelekakis, N., and Martin, D. (2013). Preliminary study of bubble formation in vegetable oil filled power transformer. *IEEE Conference Electrical Insulation and Dielectrrical Phenomena*, Shenzhen, China (20–23 October 2013), pp. 551–554.

23 Martin, D. (2008). Evaluation of the dielectric capability of ester based oils for power transformers. Doctoral thesis. University of Manchester, UK.

24 Martin, D. and Wang, Z.D. (2006). A comparative study of the impact of moisture on the dielectric capability of esters for large power transformers. *IEEE Annual Report Conference on Electrical Insulation and Dielectric Phenomena*, Kansas City, MO, USA (15–18 October 2006), pp. 409–412, USA.

25 ASTM Standard D1816 (2004). *Test Method for Dielectric Breakdown Voltage of Insulating Oils of Petroleum Origin Using VDE Electrodes*. USA: ASTM International.

26 ASTM Standard D3300 (2000). *Test Method for Dielectric Breakdown Voltage of Insulating Oils of Petroleum Origin Under Impulse Conditions*. USA: ASTM International.

27 van Hest, J., Lelekakis, N., Martin, D., and Williams, K. (2012). Exposure of vegetable oil impregnated transformer windings to air, *Australasian Universities Power Engineering Conference*, Indonesia (26–29 September 2012).

28 Lewand, L. (2001). Laboratory evaluation of several synthetic and agricultural based dielectric liquids. *Proceedings of Sixty-Eighth Annual International Conference of Doble Clients*.

29 Khan, I., Wang, Z.D., Cotton, I., and Northcote, S. (2007). Dissolved gas analysis of alternative fluids for power transformers. *IEEE Electrical Insulation Magazine* 23 (5): 5–14.

30 Duval, M. (2008). The Duval triangle for load tap changers, non-mineral oils and low temperature faults in transformers. *IEEE Electrical Insulation Magazine* 24 (6): 22–29.

31 IEEE C57.104-2019 (2019). IEEE guide for the interpretation of gases generated in mineral oil-immersed transformers.

32 IEEE Standard C57.147-2008 (2008). IEEE guide for acceptance and maintenance of natural ester fluids in transformers.

33 Cooper Power Systems (2007). *Envirotemp FR3 Fluid Guide*. Cooper Power Systems.

34 Schaich, K.M. (2005). Lipid oxidation: theoretical aspects. In: *Bailey's Industrial Oil and Fat Products*, 6e (ed. F. Shahidi), 269–355. New York, NY: John Wiley and Sons Inc.

35 Atanasova-Hoehlein, I., Hammer, Th., and Schaeffer, M. (2010). Diagnostic markers for oxidation condition of mineral oil and ester insulating fluids, Cigre Session, paper D1_231, Paris, France.

36 Martin, D., Saha, T., and McPherson, L. (2017). Condition monitoring of vegetable oil insulation in in-service power transformers: some data spanning 10 years. *IEEE Electrical Insulation Magazine* 33 (2): 43–50.

37 IEC standard 60422:2013 (2013). Mineral insulating oils in electrical equipment—supervision and maintenance guidance.

Index

Alternative Liquid Dielectrics for High Voltage Transformer Insulation Systems: Performance Analysis and Applications, First Edition. Edited by U. Mohan Rao, I. Fofana, and R. Sarathi.

Books in the IEEE Press Series on Power Engineering

Series Editor: ME El-Hawary, Dalhousie University, Halifax, Nova Scotia, Canada.

The mission of the IEEE Press Series on Power and Energy Systems is to publish leading-edge books that cover a broad spectrum of current and forward-looking technologies in the fast-moving area of power and energy systems including smart grid, renewable energy systems, electric vehicles and related areas. Our target audience includes power and energy systems professionals from academia, industry and government who are interested in enhancing their knowledge and perspectives in their areas of interest.

1. *Electric Power Systems: Design and Analysis, Revised Printing*
   Mohamed E. El-Hawary
2. *Power System Stability*
   Edward W. Kimbark
3. *Analysis of Faulted Power Systems*
   Paul M. Anderson
4. *Inspection of Large Synchronous Machines: Checklists, Failure Identification, and Troubleshooting*
   Isidor Kerszenbaum
5. *Electric Power Applications of Fuzzy Systems*
   Mohamed E. El-Hawary
6. *Power System Protection*
   Paul M. Anderson
7. *Subsynchronous Resonance in Power Systems*
   Paul M. Anderson, B.L. Agrawal, J.E. Van Ness
8. *Understanding Power Quality Problems: Voltage Sags and Interruptions*
   Math H. Bollen
9. *Analysis of Electric Machinery*
   Paul C. Krause, Oleg Wasynczuk, and S.D. Sudhoff
10. *Power System Control and Stability, Revised Printing*
    Paul M. Anderson, A.A. Fouad
11. *Principles of Electric Machines with Power Electronic Applications*, Second Edition
    Mohamed E. El-Hawary
12. *Pulse Width Modulation for Power Converters: Principles and Practice*
    D. Grahame Holmes and Thomas Lipo
13. *Analysis of Electric Machinery and Drive Systems*, Second Edition
    Paul C. Krause, Oleg Wasynczuk, and S.D. Sudhoff

Alternative Liquid Dielectrics for High Voltage Transformer Insulation Systems: Performance Analysis and Applications,
First Edition. Edited by U. Mohan Rao, I. Fofana, and R. Sarathi.

14. *Risk Assessment for Power Systems: Models, Methods, and Applications*
Wenyuan Li

15. *Optimization Principles: Practical Applications to the Operations of Markets of the Electric Power Industry*
Narayan S. Rau

16. *Electric Economics: Regulation and Deregulation*
Geoffrey Rothwell and Tomas Gomez

17. *Electric Power Systems: Analysis and Control*
Fabio Saccomanno

18. *Electrical Insulation for Rotating Machines: Design, Evaluation, Aging, Testing, and Repair*
Greg C. Stone, Edward A. Boulter, Ian Culbert, and Hussein Dhirani

19. *Signal Processing of Power Quality Disturbances*
Math H. J. Bollen and Irene Y. H. Gu

20. *Instantaneous Power Theory and Applications to Power Conditioning*
Hirofumi Akagi, Edson H. Watanabe and Mauricio Aredes

21. *Maintaining Mission Critical Systems in a 24/7 Environment*
Peter M. Curtis

22. *Elements of Tidal-Electric Engineering*
Robert H. Clark

23. *Handbook of Large Turbo-Generator Operation and Maintenance,* Second Edition
Geoff Klempner and Isidor Kerszenbaum

24. *Introduction to Electrical Power Systems*
Mohamed E. El-Hawary

25. *Modeling and Control of Fuel Cells: Distributed Generation Applications*
M. Hashem Nehrir and Caisheng Wang

26. *Power Distribution System Reliability: Practical Methods and Applications*
Ali A. Chowdhury and Don O. Koval

27. *Introduction to FACTS Controllers: Theory, Modeling, and Applications*
Kalyan K. Sen, Mey Ling Sen

28. *Economic Market Design and Planning for Electric Power Systems*
James Momoh and Lamine Mili

29. *Operation and Control of Electric Energy Processing Systems*
James Momoh and Lamine Mili

30. *Restructured Electric Power Systems: Analysis of Electricity Markets with Equilibrium Models*
Xiao-Ping Zhang

31. *An Introduction to Wavelet Modulated Inverters*
S.A. Saleh and M.A. Rahman

32. *Control of Electric Machine Drive Systems*
Seung-Ki Sul

33. *Probabilistic Transmission System Planning*
    Wenyuan Li
34. *Electricity Power Generation: The Changing Dimensions*
    Digambar M. Tagare
35. *Electric Distribution Systems*
    Abdelhay A. Sallam and Om P. Malik
36. *Practical Lighting Design with LEDs*
    Ron Lenk, Carol Lenk
37. *High Voltage and Electrical Insulation Engineering*
    Ravindra Arora and Wolfgang Mosch
38. *Maintaining Mission Critical Systems in a 24/7 Environment, Second Edition*
    Peter Curtis
39. *Power Conversion and Control of Wind Energy Systems*
    Bin Wu, Yongqiang Lang, Navid Zargari, Samir Kouro
40. *Integration of Distributed Generation in the Power System*
    Math H. Bollen, Fainan Hassan
41. *Doubly Fed Induction Machine: Modeling and Control for Wind Energy Generation Applications*
    Gonzalo Abad, Jesús López, Miguel Rodrigues, Luis Marroyo, and Grzegorz Iwanski
42. *High Voltage Protection for Telecommunications*
    Steven W. Blume
43. *Smart Grid: Fundamentals of Design and Analysis*
    James Momoh
44. *Electromechanical Motion Devices, Second Edition*
    Paul Krause, Oleg Wasynczuk, Steven Pekarek
45. *Electrical Energy Conversion and Transport: An Interactive Computer-Based Approach, Second Edition*
    George G. Karady, Keith E. Holbert
46. *ARC Flash Hazard and Analysis and Mitigation*
    J.C. Das
47. *Handbook of Electrical Power System Dynamics: Modeling, Stability, and Control*
    Mircea Eremia, Mohammad Shahidehpour
48. *Analysis of Electric Machinery and Drive Systems, Third Edition*
    Paul C. Krause, Oleg Wasynczuk, S.D. Sudhoff, Steven D. Pekarek
49. *Extruded Cables for High-Voltage Direct-Current Transmission: Advances in Research and Development*
    Giovanni Mazzanti, Massimo Marzinotto
50. *Power Magnetic Devices: A Multi-Objective Design Approach*
    S.D. Sudhoff

51. *Risk Assessment of Power Systems: Models, Methods, and Applications, Second Edition*
    Wenyuan Li

52. *Practical Power System Operation*
    Ebrahim Vaahedi

53. *The Selection Process of Biomass Materials for the Production of Bio-Fuels and Co-Firing*
    Najib Altawell

54. *Electrical Insulation for Rotating Machines: Design, Evaluation, Aging, Testing, and Repair, Second Edition*
    Greg C. Stone, Ian Culbert, Edward A. Boulter, and Hussein Dhirani

55. *Principles of Electrical Safety*
    Peter E. Sutherland

56. *Advanced Power Electronics Converters: PWM Converters Processing AC Voltages*
    Euzeli Cipriano dos Santos Jr., Edison Roberto Cabral da Silva

57. *Optimization of Power System Operation, Second Edition*
    Jizhong Zhu

58. *Power System Harmonics and Passive Filter Designs*
    J.C. Das

59. *Digital Control of High-Frequency Switched-Mode Power Converters*
    Luca Corradini, Dragan Maksimoviæ, Paolo Mattavelli, Regan Zane

60. *Industrial Power Distribution, Second Edition*
    Ralph E. Fehr, III

61. *HVDC Grids: For Offshore and Supergrid of the Future*
    Dirk Van Hertem, Oriol Gomis-Bellmunt, Jun Liang

62. *Advanced Solutions in Power Systems: HVDC, FACTS, and Artificial Intelligence*
    Mircea Eremia, Chen-Ching Liu, Abdel-Aty Edris

63. *Operation and Maintenance of Large Turbo-Generators*
    Geoff Klempner, Isidor Kerszenbaum

64. *Electrical Energy Conversion and Transport: An Interactive Computer-Based Approach*
    George G. Karady, Keith E. Holbert

65. *Modeling and High-Performance Control of Electric Machines*
    John Chiasson

66. *Rating of Electric Power Cables in Unfavorable Thermal Environment*
    George J. Anders

67. *Electric Power System Basics for the Nonelectrical Professional*
    Steven W. Blume

68. *Modern Heuristic Optimization Techniques: Theory and Applications to Power Systems*
    Kwang Y. Lee, Mohamed A. El-Sharkawi

69. *Real-Time Stability Assessment in Modern Power System Control Centers*
    Savu C. Savulescu

70. *Optimization of Power System Operation*
Jizhong Zhu

71. *Insulators for Icing and Polluted Environments*
Masoud Farzaneh, William A. Chisholm

72. *PID and Predictive Control of Electric Devices and Power Converters Using MATLAB®/ Simulink®*
Liuping Wang, Shan Chai, Dae Yoo, Lu Gan, Ki Ng

73. *Power Grid Operation in a Market Environment: Economic Efficiency and Risk Mitigation*
Hong Chen

74. *Electric Power System Basics for Nonelectrical Professional, Second Edition*
Steven W. Blume

75. *Energy Production Systems Engineering*
Thomas Howard Blair

76. *Model Predictive Control of Wind Energy Conversion Systems*
Venkata Yaramasu, Bin Wu

77. *Understanding Symmetrical Components for Power System Modeling*
J.C. Das

78. *High-Power Converters and AC Drives, Second Edition*
Bin Wu, Mehdi Narimani

79. *Current Signature Analysis for Condition Monitoring of Cage Induction Motors: Industrial Application and Case Histories*
William T. Thomson, Ian Culbert

80. *Introduction to Electric Power and Drive Systems*
Paul Krause, Oleg Wasynczuk, Timothy O'Connell, Maher Hasan

81. *Instantaneous Power Theory and Applications to Power Conditioning, Second Edition*
Hirofumi, Edson Hirokazu Watanabe, Mauricio Aredes

82. *Practical Lighting Design with LEDs, Second Edition*
Ron Lenk, Carol Lenk

83. *Introduction to AC Machine Design*
Thomas A. Lipo

84. *Advances in Electric Power and Energy Systems: Load and Price Forecasting*
Mohamed E. El-Hawary

85. *Electricity Markets: Theories and Applications*
Jeremy Lin, Jernando H. Magnago

86. *Multiphysics Simulation by Design for Electrical Machines, Power Electronics and Drives*
Marius Rosu, Ping Zhou, Dingsheng Lin, Dan M. Ionel, Mircea Popescu, Frede Blaabjerg, Vandana Rallabandi, David Staton

87. *Modular Multilevel Converters: Analysis, Control, and Applications*
Sixing Du, Apparao Dekka, Bin Wu, Navid Zargari

88. *Electrical Railway Transportation Systems*
Morris Brenna, Federica Foiadelli, Dario Zaninelli

89. *Energy Processing and Smart Grid*
James A. Momoh

90. *Handbook of Large Turbo-Generator Operation and Maintenance, 3rd Edition*
Geoff Klempner, Isidor Kerszenbaum

91. *Advanced Control of Doubly Fed Induction Generator for Wind Power Systems*
Dehong Xu, Dr. Frede Blaabjerg, Wenjie Chen, Nan Zhu

92. *Electric Distribution Systems, 2nd Edition*
Abdelhay A. Sallam, Om P. Malik

93. *Power Electronics in Renewable Energy Systems and Smart Grid: Technology and Applications*
Bimal K. Bose

94. *Distributed Fiber Optic Sensing and Dynamic Rating of Power Cables*
Sudhakar Cherukupalli, and George J Anders

95. *Power System and Control and Stability, Third Edition*
Vijay Vittal, James D. McCalley, Paul M. Anderson, and A. A. Fouad

96. *Electromechanical Motion Devices: Rotating Magnetic Field-Based Analysis and Online Animations, Third Edition*
Paul Krause, Oleg Wasynczuk, Steven D. Pekarek, and Timothy O'Connell

97. *Applications of Modern Heuristic Optimization Methods in Power and Energy Systems*
Kwang Y. Lee and Zita A. Vale

98. *Handbook of Large Hydro Generators: Operation and Maintenance*
Glenn Mottershead, Stefano Bomben, Isidor Kerszenbaum, and Geoff Klempner

99. *Advances in Electric Power and Energy: Static State Estimation*
Mohamed E. El-hawary

100. *Arc Flash Hazard Analysis and Mitigation, Second Edition*
J.C. Das

101. *Maintaining Mission Critical Systems in a 24/7 Environment, Third Edition*
Peter M. Curtis

102. *Real-Time Electromagnetic Transient Simulation of AC-DC Networks*
Venkata Dinavahi and Ning Lin

103. *Probabilistic Power System Expansion Planning with Renewable Energy Resources and Energy Storage Systems*
Jaeseok Choi and Kwang Y. Lee

104. *Power Magnetic Devices: A Multi-Objective Design Approach, Second Edition.*
Scott D. Sudhoff

105. *Optimal Coordination of Power Protective Devices with Illustrative Examples*
Ali R. Al-Roomi

106. *Resilient Control Architectures and Power Systems*
Craig Rieger, Ronald Boring, Brian Johnson, and Timothy McJunkin

107. *Alternative Liquid Dielectrics for High Voltage Transformer Insulation Systems: Performance Analysis and Applications*
Edited by U. Mohan Rao, I. Fofana, and R. Sarathi

Printed and bound by CPI Group (UK) Ltd, Croydon, CR0 4YY
21/09/2026
14965841-0001